POPE PIUS XII LIBRARY, ST. JOSEPH COL.

3 2528 10407 4978

S0-AMK-892

WITHDRAWN
USJ Library

Pesticide Protocols

METHODS IN BIOTECHNOLOGY™

John M. Walker, SERIES EDITOR

METHODS IN BIOTECHNOLOGY™

Pesticide Protocols

Edited by

José L. Martínez Vidal

Antonia Garrido Frenich

Department of Analytical Chemistry, Faculty of Sciences
University of Almería, Almería, Spain

HUMANA PRESS ✳ TOTOWA, NEW JERSEY

© 2006 Humana Press Inc.
999 Riverview Drive, Suite 208
Totowa, New Jersey 07512

www.humanapress.com

All rights reserved. No part of this book may be reproduced, stored in a retrieval system, or transmitted in any form or by any means, electronic, mechanical, photocopying, microfilming, recording, or otherwise without written permission from the Publisher. Methods in Biotechnology™ is a trademark of The Humana Press Inc.

All papers, comments, opinions, conclusions, or recommendations are those of the author(s), and do not necessarily reflect the views of the publisher.

This publication is printed on acid-free paper. ∞
ANSI Z39.48-1984 (American Standards Institute)

Permanence of Paper for Printed Library Materials.

Cover design by Patricia F. Cleary
Cover illustration provided by José L. Martínez Vidal and Antonia Garrido Frenich.

For additional copies, pricing for bulk purchases, and/or information about other Humana titles, contact Humana at the above address or at any of the following numbers: Tel.: 973-256-1699; Fax: 973-256-8341; E-mail: orders@humanapr.com; or visit our Website: www.humanapress.com

Photocopy Authorization Policy:
Authorization to photocopy items for internal or personal use, or the internal or personal use of specific clients, is granted by Humana Press Inc., provided that the base fee of US $30.00 per copy is paid directly to the Copyright Clearance Center at 222 Rosewood Drive, Danvers, MA 01923. For those organizations that have been granted a photocopy license from the CCC, a separate system of payment has been arranged and is acceptable to Humana Press Inc. The fee code for users of the Transactional Reporting Service is: [1-58829-410-2/06 $30.00].

Printed in the United States of America. 10 9 8 7 6 5 4 3 2 1

eISBN 1-59259-929-X

Library of Congress Cataloging-in-Publication Data

Pesticide protocols / edited by José L. Martínez Vidal, Antonia Garrido Frenich.
       p. cm. -- (Methods in biotechnology ; 19)
    Includes bibliographical references and index.
    ISBN 1-58829-410-2 (alk. paper) -- ISBN 1-59259-929-X (eISBN)
       1. Pesticides--Analysis--Laboratory manuals. I. Vidal, José L. Martínez.
   II. Frenich, Antonia Garrido. III. Series.

   RA1270.P4P4685 2005
   363.17'92--dc22

                                          2005046201

Preface

Pesticides are a broad class of bioactive compounds used in crop protection, food preservation, and human health. They differ from other chemical substances because they are spread deliberately into the environment. Presently, about 1000 active ingredients have been registered that can be grouped into more than 40 classes of chemical families. Exposure to pesticides through the most important routes of uptake (oral, dermal, and inhalation) depends on the physicochemical characteristics of the pesticide and the nature of the contact, varying with the edge, lifestyle, and working conditions. The level of pesticides in different environmental compartments—such as water, agricultural foods, and products of animal origin—has became a relevant issue. Moreover, analytical measurements of dermal exposure and exposure by inhalation have become as important as analytical measurements of internal dose.

Unlike other contaminants, pesticides may affect both workers and the general population as a result of the consumption of contaminated food and water, domestic use, and proximity to agricultural settings. Information about actual human exposure to pesticides has important uses, including informing risk assessments, helping predict the potential consequences of exposures, and developing exposure criteria for regulations and other public policy guidance.

Pesticide exposure can be measured through the biomonitoring of the parent compounds and/or metabolites in such body fluids as urine, blood, serum, and saliva, among others. Indoor exposure may take place through treated furniture, or such home structures as fitted carpets or wood-treated walls. Regarding outdoor exposure, the main sources are represented by spray drifts of pesticides from agricultural and industrial areas and by the atmospheric dispersal of pesticides evaporated from treated surfaces. Very little information is available on dermal and inhalation exposure to pesticides. Contamination of food represents one of the most pervasive sources of pesticide exposure for the general population.

Pesticide analysis has been affected by the recent detection of parent or metabolite compounds, thus driving the demand for techniques that can measure lower and lower levels of concentration. In recent years, criteria to support in a solid way the steps corresponding to the identification, confirmation, and quantification of the analyte have became more frequently used.

During the last decade, noticeable changes in multiresidue methods have taken place. Chromatography remains the workhorse technique for pesticides. The development of different types of injection techniques, columns, stationary phases, and detectors has allowed for the improvement in the sensitivity and selectivity of the analytical determinations. The availability in analytical laboratories of mass spectrometry detectors coupled to gas chromatography, as well as to liquid chromatography, has increased the degree of confidence in the identification of organic compounds. Other techniques, such as capillary electrophoresis, are promising candidates for a relevant role in this area. The current use of powerful analytical tools coupled with the application of quality control/quality assurance criteria has resulted in an increase in the reliability of an analysis. However, special emphasis is needed on the development of multiresidue methods for the analysis of as many pesticides as possible in one analytical run.

Pesticide Protocols contains methods for the detection of specific compounds or their metabolites useful in biological monitoring and in studies of exposure via food, water, air, and skin. Liquid and gas chromatography coupled to mass spectrometry detection, and other classic detectors, are the most widely used techniques, although such others as capillary electrophoresis and immunochemical or radioimmunoassay methods are also proposed. Chapters cover the varied array of analytical techniques applied to the analysis of several families of pesticides. The extractions and cleanup procedures have been focused in order to use more automated and miniaturized methods, including solid-phase extraction, solid-phase micro-extraction, microwave-assisted extraction, or on-line tandem liquid chromatography (LC/LC) trace enrichment, among others.

All methods have been written by scientists experienced in pesticide analysis in different matrixes. Each chapter describes a specific method, giving the analytical information in sufficient detail that a competent scientist can apply it without having to consult additional sources. Our book will prove valuable as a general reference and guide for students and postgraduates, as well for researchers and laboratories alike.

We would like to express our personal gratitude to all the authors for the quality of their contributions. Thanks are also owed to Professor John Walker and to Humana Press for allowing us to edit this volume.

José L. Martínez Vidal
Antonia Garrido Frenich

Contents

Contributors

CRISTINA APREA • *Unità Funzionale di Igiene Industriale e Tossicologia Occupazionale (Department of Industrial Hygiene and Occupational Toxicology) Laboratorio di Sanità Pubblica, Azienda USL 7 (Public Health Laboratory, National Health Service Local Unit 7), Siena, Italy*

PATRICIA ARAQUE • *Laboratory of Medical Investigations, Hospital Clínico, University of Granada, Granada, Spain*

FRANCISCO J. ARREBOLA LIÉBANAS • *Department of Analytical Chemistry, Faculty of Sciences, University of Almería, Almería, Spain*

ANTONELLA AUSILI • *Department of Environmental Quality Monitoring, Istituto Centrale per La Ricerca Scientifica e Tecnologica Applicata al Mare (Institute for Scientific and Applied Marine Research), Rome, Italy*

CLAUDIO BAGGIANI • *Dipartimento di Chimica Analitica, Università di Torino, Torino, Italy*

DAMIA BARCELÓ • *Department of Environmental Chemistry, IIQAB-CSIC, Barcelona, Spain*

STEVEN A. BARKER • *Analytical Systems Laboratories, School of Veterinary Medicine, Louisiana State University, Baton Rouge, LA*

DANA B. BARR • *National Center for Environmental Health, Centers for Disease Control and Prevention, Atlanta, GA*

JOHN R. BARR • *National Center for Environmental Health, Centers for Disease Control and Prevention, Atlanta, GA*

EDITH BERGER-PREISS • *Fraunhofer Institute of Toxicology and Experimental Medicine, Hannover, Germany*

SARAH BIRINDELLI • *International Centre for Pesticides and Health Risk Prevention, Ospedale Universitario Luigi Sacco–Busto Garolfo, Milan, Italy*

NANDA BOZZI • *Unità Funzionale di Igiene Industriale e Tossicologia Occupazionale (Department of Industrial Hygiene and Occupational Toxicology) Laboratorio di Sanità Pubblica, Azienda USL 7 (Public Health Laboratory, National Health Service Local Unit 7), Siena, Italy*

ROBERTO BRAVO • *National Center for Environmental Health, Centers for Disease Control and Prevention, Atlanta, GA*

DERK H. BROUWER • *TNO Chemistry, Food and Chemical Risk Analysis, Zeist, The Netherlands*

LAURA CAMPO • *Department of Occupational and Environmental Health, University of Milan and Ospedale Policlinico, Mangiagallie Regina Elena, Milan, Italy*

SHAOGANG CHU • *Great Lakes Institute for Environmental Research, University of Windsor, Windsor, Ontario, Canada*

JANE C. CHUANG • *Battelle, Columbus, OH*

ANNA M. CICERO • *Department of Environmental Quality Monitoring, Istituto Centrale per La Ricerca Scientifica e Tecnologica Applicata Al Mare (Institute for Scientific and Applied Marine Research), Rome, Italy*

CLAUDIO COLOSIO • *International Centre for Pesticides and Health Risk Prevention, Ospedale Universitario Luigi Sacco–Busto Garolfo, Milan, Italy*

ADRIAN COVACI • *Toxicological Center, University of Antwerp, Universiteits-Plein, Wilrijk, Belgium*

ALFONSO DI MUCCIO • *Formerly at Laboratory of Applied Toxicology, Istituto Superiore di Sanità (National Institute of Health), Rome, Italy*

STEFANO DI MUCCIO • *Department of Environmental Quality Monitoring, Istituto Centrale per La Ricerca Scientifica e Tecnologica Applicata al Mare (Institute for Scientific and Applied Marine Research), Rome, Italy*

ELLEN DIJKMAN • *Laboratory for Analytical Chemistry, National Institute for Public Health and The Environment, Bilthoven, The Netherlands*

FRANCISCO J. EGEA GONZÁLEZ • *Department of Analytical Chemistry, Faculty of Sciences, University of Almería, Almería, Spain*

LUTZ ELFLEIN • *Fraunhofer Institute of Toxicology and Experimental Medicine, Hannover, Germany*

MARC P. FERNANDEZ • *Regional Contaminants Laboratory, Institute of Ocean Sciences, Fisheries and Oceans Canada, Sidney, British Columbia, Canada*

SILVIA FUSTINONI • *Department of Occupational and Environmental Health, University of Milan and Ospedale Policlinico, Mangiagallie Regina Elena, Milan, Italy*

ANTONIA GARRIDO FRENICH • *Department of Analytical Chemistry, Faculty of Sciences, University of Almería, Almería, Spain*

CRISTINA GIOVANNOLI • *Dipartimento di Chimica Analitica, Università di Torino, Torino, Italy*

C. RICHARD GLASS • *Environmental Biology Group, Central Science Laboratory, York, UK*

MANUEL J. GONZÁLEZ RODRÍGUEZ • *Department of Analytical Chemistry, Faculty of Sciences, University of Almería, Almería, Spain*

WOLFGANG GRIES • *Department SUA–GHA–GSS, Institute of Biomonitoring, Bayer Industry Services GmbH and CoOHG, Leverkusen, Germany*

ELBERT HOGENDOORN • *Laboratory for Analytical Chemistry, National Instiute for Public Health and The Environment, Bilthoven, The Netherlands*

CHIA-SWEE HONG • *Wadsworth Center, New York State Department of Health, and School of Public Health, State University of New York at Albany, Albany, NY*

MICHAEL G. IKONOMOU • *Regional Contaminants Laboratory, Institute of Ocean Sciences, Fisheries and Oceans Canada, Sidney, British Columbia, Canada*

ISABEL C. S. F. JARDIM • *Departamento de Química Analítica, Instituto de Química, Universidade Estadual de Campinas, Campinas, SP, Brazil*

ROGER JEANNOT • *Service Analyse et Caractérisation Minérale, BRGM, Orleans, France*

HIROYUKI KATAOKA • *Laboratory of Applied Analytical Chemistry, Department of Biological Pharmacy, School of Pharmacy, Shujitsu University, Okayama, Japan*

DIETMAR KNOPP • *Institute of Hydrochemistry and Chemical Balneology, Technical University Munich, München, Germany*

ANNA KOUKOURIKOU • *Pesticide Science Laboratory, Aristotle University of Thessaloniki, Thessaloniki, Greece*

STEVEN J. LEHOTAY • *Agricultural Research Service, US Department of Agriculture, Eastern Regional Research Center, Wyndmoor, PA*

GABRIELE LENG • *Department SUA–GHA–GSS, Institute of Biomonitoring, Bayer Industry Services GmbH and CoOHG, Leverkusen, Germany*

MARIA J. LÓPEZ DE ALDA • *Department of Environmental Chemistry, IIQAB-CSIC, Barcelona, Spain*

LIANA LUNGHINI • *Unità Funzionale di Igiene Industriale e Tossicologia Occupazionale (Department of Industrial Hygiene and Occupational Toxicology) Laboratorio di Sanità Pubblica, Azienda USL 7 (Public Health Laboratory, National Health Service Local Unit 7), Siena, Italy*

M.-PILAR MARCO • *Department of Biological Organic Chemistry, IIQAB-CSIC, Barcelona, Spain*

A. MARÍN • *Department of Analytical Chemistry, Faculty of Sciences, University of Almería, Almería, Spain*

JOSÉ L. MARTÍNEZ VIDAL • *Department of Analytical Chemistry, Faculty of Sciences, University of Almería, Almería, Spain*

FRANCISCO MERINO • *Department of Analytical Chemistry, Faculty of Sciences, University of Córdoba, Córdoba, Spain*

KURIE MITANI • *Laboratory of Applied Analytical Chemistry, Department of Biological Pharmacy, School of Pharmacy, Shujitsu University, Okayama, Japan*

LARRY L. NEEDHAM • *Centers for Disease Control and Prevention, National Center for Environmental Health, Atlanta, GA*

MIKAELA NICHKOVA • *Department of Biological Organic Chemistry, IIQAB-CSIC, Barcelona, Spain*

KEVIN N. T. NORMAN • *Central Science Laboratory, York, UK*

M. FÁTIMA OLEA-SERRANO • *Department of Nutritional and Food Sciences, University of Granada, Granada, Spain*

NICOLAS OLEA • *Lab of Medical Investigations, Hospital Clínico, University of Granada, Granada, Spain*

ANDERS O. OLSSON • *National Center for Environmental Health, Centers for Disease Control and Prevention, Atlanta, GA*

SEAN H. W. PANTON • *Central Science Laboratory, York, UK*

EMMANUIL NIKOLAOS PAPADAKIS • *Pesticide Science Laboratory, Aristotle University of Thessaloniki, Thessaloniki, Greece*

EUPHEMIA PAPADOPOULOU-MOURKIDOU • *Pesticide Science Laboratory, Aristotle University of Thessaloniki, Thessaloniki, Greece*

JOHN PATSIAS • *Aristotle Pesticide Science Laboratory, University of Thessaloniki, Thessaloniki, Greece*

JANUSZ PAWLISZYN • *Department of Chemistry, University of Waterloo, Waterloo, Canada*

DOLORES PÉREZ-BENDITO • *Department of Analytical Chemistry, Faculty of Sciences, University of Córdoba, Córdoba, Spain*

YOLANDA PICÓ • *Laboratory of Bromatology and Toxicology, Faculty of Pharmacy, University of Valencia, Valencia, Spain*

JOSEANE M. POZZEBON • *Departamento de Química Analítica, Instituto de Química, Universidade Estadual de Campinas, Campinas, SP, Brazil*

SONIA C. N. QUEIROZ • *Laboratório de Dinâmica de Agroquímicos, Embrapa Meio Ambiente, Jaguariúna, SP, Brazil*

SARA RODRÍGUEZ-MOZAZ • *Department of Analytical Chemistry, Faculty of Sciences, University of Córdoba, Córdoba, Spain*

SOLEDAD RUBIO • *Department of Analytical Chemistry, Facultad De Ciencias, Edificio Anexo Marie Curie, Córdoba, Spain*

HASSAN SABIK • *Food Research and Development Center, Agriculture and Agri-Food Canada, St-Hyacinthe, Quebec, Canada*

MITSUSHI SAKAMOTO • *Tokushima Prefectural Institute of Public Health and Environmental Sciences, Tokushima, Japan*

GIANFRANCO SCIARRA • *Unità Funzionale di Igiene Industriale e Tossicologia Occupazionale (Department of Industrial Hygiene and Occupational Toxicology) Laboratorio di Sanità Pubblica, Azienda USL 7 (Public Health Laboratory, National Health Service Local Unit 7), Siena, Italy*

CARLOS SONNENSCHEIN • *Department of Anatomy and Cellular Biology, Tufts University School of Medicine, Boston, MA*

ANA M. SOTO • *Department of Anatomy and Cellular Biology, Tufts University School of Medicine, Boston, MA*

MASAHIKO TAKINO • *Yokogawa Analytical Systems Inc., Tokyo, Japan*

TAIZOU TSUTSUMI • *Tokushima Prefectural Institute of Public Health and Environmental Sciences, Tokushima, Japan*

KATINKA E. VAN DER JAGT • *TNO Chemistry, Food and Chemical Risk Analysis, Zeist, The Netherlands; currently, European Medicines Agency, London, UK*

JEANETTE M. VAN EMON • *Methods Development and Research Branch, National Exposure Research Laboratory, US Environmental Protection Agency, Las Vegas, NV*

JOOP J. VAN HEMMEN • *TNO Chemistry, Food and Chemical Risk Analysis, Zeist, The Netherlands*

ZISIS VRYZAS • *Pesticide Science Laboratory, Aristotle University of Thessaloniki, Thessaloniki, Greece*

JI Y. ZHANG • *GlaxoSmithKline, King of Prussia, PA*

I

ANALYTICAL METHODOLOGIES TO DETERMINE PESTICIDES AND METABOLITES IN HUMAN FAT TISSUES AND BODY FLUIDS

1

Analysis of Endocrine Disruptor Pesticides in Adipose Tissue Using Gas Chromatography–Tandem Mass Spectrometry

Assessment of the Uncertainty of the Method

José L. Martínez Vidal, Antonia Garrido Frenich, Francisco J. Egea González, and Francisco J. Arrebola Liébanas

Summary

A multiresidue method based on extraction with organic solvents, cleanup by preparative liquid chromatography, and detection by gas chromatography (GC) using tandem mass spectrometry (MS/MS) mode is described for the determination of α- and β-endosulfan and three main metabolites (sulfate, ether, and lactone) in human adipose tissue samples. The analytical methodology is verified, and the values of some performance characteristics, such as linearity, limit of detection (LOD), limit of quantification (LOQ) limits, precision (intraday and interday), and accuracy (recovery) are calculated. The high efficiency of the cleanup step for the elimination of interference allows reaching detection limits at low micrograms per kilogram (parts per billion, ppb) concentration levels. In addition, an estimation of measurement uncertainty, using validation data, is presented for each target compound. The results show that the sources of largest uncertainty are those relative to the balance calibration, from the gravimetric step, and both the relative uncertainty associated with the recovery and the intermediate precision of the method.

Key Words: Cleanup; endocrine disruptor; endosulfan; gas chromatography; metabolites; human adipose tissue; measurement uncertainty; organochlorine compounds; pesticides; preparative liquid chromatography; tandem mass spectrometry.

1. Introduction

Organochlorine pesticides, such as endosulfan, are persistent environmental contaminants and tend to accumulate in humans and other animals *(1,2)*. Endosulfan is still widely used in agricultural activities in developed countries because of its low relative persistence in comparison with other organochlorinated insecticides and its

From: *Methods in Biotechnology, Vol. 19, Pesticide Protocols*
Edited by: J. L. Martínez Vidal and A. Garrido Frenich © Humana Press Inc., Totowa, NJ

excellent insecticidal action. However, endosulfan accumulates in adipose tissue along the food chain because of its high stability and liposolubility. It is frequently found in both environmental and biological samples *(3–6)*. In addition, endosulfan has estrogenic effects on humans and is considered an endocrine-disrupting chemical *(7–9)*.

Humans are exposed to endosulfan residues mainly in the workplace and through diet. The best way to measure human exposure is the direct determination of its residues in adipose tissue, although levels of organochlorine compounds in serum are frequently used as indicators of total body burden. Obviously, serum is a more accessible matrix for ascertaining residue levels of organochlorine compounds. However, a direct relationship between residues in serum and adipose tissue is not always found *(10,11)*.

The technical product of endosulfan is a mixture of two isomers, α- and β-endosulfan, that is metabolized by oxidation routes within the organisms, yielding metabolic compounds such as endosulfan sulfate, alcohol, ether, or lactone. As a consequence, reliable analytical methodologies are necessary for determination of endosulfan and its metabolites in human adipose tissue to ascertain exposure levels and avoid effects on public health. Generally, effective solvent extraction methods followed by cleanup steps and gas chromatographic (GC) determination are applied to the determination of nonpolar pesticide residues in complex biological samples *(4–6,12–20)*. Mass spectrometry (MS), especially the tandem MS/MS operation mode, is the preferred detection technique because it allows the identification, quantitation, and confirmation of the detected residues. In addition, the use of MS/MS improves the sensitivity and selectivity of the technique with a drastic reduction of the background and without losing identification capability. Most matrix interferences are avoided, and the target compounds are identified by their secondary spectra by comparison with MS/MS libraries.

It is now recognized that analytical results cannot be acceptable without calculating the measurement uncertainty *(21)*, which is the confidence that can be placed in the result. Formally, uncertainty is defined as a value associated with the result of a measurement that characterizes the dispersion of the values that could reasonably be attributed to the measurand *(22)*. Uncertainty can be expressed in two different forms, standard and expanded. The standard uncertainty $u(x_i)$ corresponds with the uncertainty of the result x_i of a measurement expressed as a standard deviation. When the standard uncertainty of the result y of a measurement derives from different sources of uncertainty, it is referred to as combined standard uncertainty $u_c(y)$. It is equal to the positive square root of a sum of terms. The expanded uncertainty U represents an interval around the measurement result, which contains the unknown true value with a defined probability. U is obtained by multiplying $u(y)$ by a coverage factor k *(23)*.

Three approaches are proposed for the estimation of the uncertainty: bottom-up *(23,24)*, top-down *(21,25)*, and in-house validation methods *(26,27)*. The bottom-up method estimates each individual uncertainty for every step of the measurement process and obtains the combined standard uncertainty from the sum of each contribution *(28–30)*. The top-down method is based in the use of interlaboratory information *(30,31)*, and the third approach considers the information obtained from in-house validation of analytical methods *(32)*.

This chapter describes the simultaneous determination of α- and β-endosulfan and their metabolites sulfate, ether, and lactone in human adipose tissue samples by gas GC coupled to MS/MS, including the evaluation of the uncertainty of the method to determine the critical steps of the analytical process. For this aim, a combination of the bottom-up approach with in-house validation data for estimating the uncertainty of each stage of the analytical method is used *(5,17,33–36)*.

2. Materials

1. Ultrapure water is prepared by distillation and then by Milli-Q SP treatment.
2. Pesticide quality solvents *n*-hexane, methanol, diethyl ether, and 2-propanol.
3. Standards of the pesticides with purity higher than 99% (*see* **Note 1**).
4. Heptachlor (purity 99%) used as internal standard (ISTD) (*see* **Notes 1** and **2**).
5. Ultrahigh purity helium (minimum purity 99.999%).
6. Alumina (Al_2O_3) (Merck, Darmstadt, Germany) 90 (70–230 mesh) no. 1097 (*see* **Note 3**).
7. Liquid chromatograph (e.g., Waters 990, Milford, MA) with a constant-flow pump (e.g., Waters 600 E) and a Rheodyne six-port injection valve with a 1-mL sample loop.
   a. Ultraviolet-visible photodiode array detector (e.g., Waters 990).
   b. Software for data acquisition and data analysis (e.g., Waters 991).
   c. Liquid chromatographic (LC) column: 250 mm long × 4 mm id, 5 μm particle size (e.g., Lichrospher Si column from Merck).
8. Gas chromatograph (e.g., Varian 3800, Sunnyvale, CA) with a split/splitless programmed temperature injector and an autosampler (e.g., Varian Model 8200).
   a. Ion trap mass spectrometer (e.g., Saturn 2000 from Varian).
   b. Software for data acquisition and data analysis (e.g., Saturn 2000 from Variant), including an MS/MS library especially created for the target analytes in our experimental conditions (*see* **Note 4**).
   c. GC capillary column: 30 m long × 0.25 mm id × 0.25 μm film thickness (e.g., DB5-MS, J&W Scientific, Folsom, CA).

3. Methods
3.1. Preparation of Stock Solutions

1. Primary solutions: Weigh 50 mg of each pesticide standard and of the ISTD into a 100-mL volumetric flask and fill the flask with *n*-hexane to the level (*see* **Notes 5** and **6**).
2. Secondary solutions: Make a 1:100 dilution with *n*-hexane to obtain a work solution containing all the target pesticides (*see* **Notes 5** and **7**). Make also a 1:100 dilution with *n*-hexane of the primary ISTD solution (*see* **Note 7**).
3. Dilute the secondary solution with *n*-hexane to obtain the GC calibration solutions (the standards) in the 1.0- to 250-μg/L range with each containing 100 μg/L of the secondary solution of the ISTD (*see* **Note 8**).

3.2. Extraction

1. Weigh 500 mg of adipose tissue sample and extract five times with 4 mL of *n*-hexane and shake in a vortex mixer for 1 min.
2. Pass the extract through the alumina column (*see* **Note 3**). Next, preconcentrate the eluate at reduced pressure and then under a stream of nitrogen at 40°C and adjust the final residue to 1 mL with *n*-hexane.

Table 1
MS/MS Parameters

Pesticide	Activation Time (min)	m/z Range	Excitation Amplitude (V)	Excitation storage Level (m/z)
Endosulfan-ether	13.5–15.0	85–280	77	80
ISTD	15.0–16.5	85–290	80	100
Endosulfan-lactone	16.5–19.0	85–330	92	141
α-Endosulfan	19.0–20.5	85–250	81	80
β-Endosulfan	20.5–24.5	85–290	79	80
Endosulfan-sulfate	20.5–24.5	85–290	62	80

3.3. Instrumental Conditions

3.3.1. High-Performance Liquid Chromatographic System

The mobile phase, under gradient conditions, is as follows: initially 2-min isocratic gradient with 100% phase A (*n*-hexane); 15-min linear gradient to 60% phase A, 40% phase B (*n*-hexane:methanol:2-propanol, 40:45:15 v/v); 20-min linear gradient to 100% phase B; and 30-min linear gradient to 100% phase A. An additional time of 5 min with this composition of mobile phase is enough to return the system to the initial conditions for subsequent analysis. The mobile phase is set at a flow rate of 1 mL/min, and the diode array detector is used at 280 nm for monitoring lipid elution on-line. The fraction corresponding to the first 11 min (*see* **Note 9**) eluting from the high-performance liquid chromatographic (HPLC) system is collected, dried under a nitrogen stream, and eluted with 1 mL of *n*-hexane (*see* **Note 10**). Of this extract, 2 µL are injected into the GC–MS/MS instrument.

3.3.2. GC System

The GC has a septum-equipped, temperature-programmable injector that is initially held at 90°C for 0.1 min before ramping to 280°C at a rate of 200°C/min. The GC oven is initially held at 80°C for 2.5 min, then ramped at 50°C/min to 140°C, and finally from 140°C is increased at 5°C/min to 260°C and held for 3 min. The ion trap mass spectrometer is operated in the electron ionization mode, and the MS/MS option is used. The GC–MS conditions are as follows: 11-min solvent delay; 70-eV electron impact energy; 0.6-scans/s scan rate; 85–450 scanned range *m/z*. The transfer line is kept at 260°C and the ion trap manifold at 200°C. The automatic gain control is switched on with a target fixed at 5000 counts. Helium at a flow rate of 1 mL/min is used as the carrier and collision gas. The MS/MS parameters are shown in **Table 1**.

3.4. Cleanup

Inject 1 mL of the final residue in *n*-hexane into the HPLC system. Collect the fraction corresponding to the first 11 min (*see* **Note 9**) eluting from the HPLC system, dry under a nitrogen stream, and elute with 1 mL of *n*-hexane (*see* **Note 10**). Inject 2 µL of this extract into the GC–MS/MS instrument.

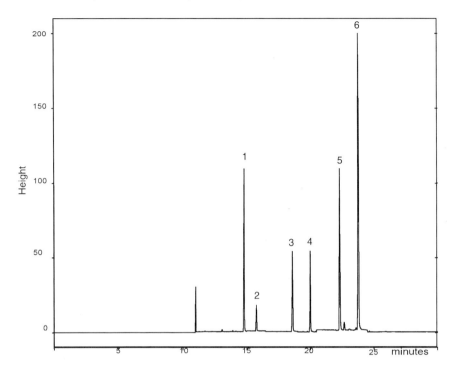

Fig. 1. GC–MS/MS chromatogram of a standard solution of the target pesticides in *n*-hexane at 100 µg L^{-1}: 1, endosulfan-ether; 2, ISTD; 3, endosulfan-lactone; 4, α-endosulfan; 5, β-endosulfan; and 6, endosulfan-sulfate.

3.5. GC–MS/MS Analysis

3.5.1. Verification of the Analytical Method (see **Note 11**)

1. Inject 2-µL aliquots of the GC calibration solutions, each containing 100 µg/L of the ISTD. **Figure 1** shows a chromatogram of a mixture containing the target pesticides and the ISTD.
2. Check the linearity of the detector response over the concentration range 1.0–250 µg/L (*see* **Note 12**) using relative areas of the target compounds to the internal standard. The correlation coefficients must have a minimum value of 0.99.
3. Inject, 10 times, 2-µL aliquots of a GC calibration solution containing 100 µg/L of the ISTD to calculate the retention time windows (RTWs) (*see* **Note 13**).
4. Check the selectivity, or the existence of potential interference in the chromatograms from the biological samples, by running blank samples in each calibration (*see* **Note 14**).
5. Obtain the reference spectrum, used for the confirmation of positive results in the analysis of real samples, by analyzing 10 blank spiked samples (*see* **Note 15**) at the 200-µg/kg concentration level (*see* **Note 16**).
6. Check the limit of detection (LOD) and limit of quantification (LOQ), calculated as 3 and 10 times the respective standard deviation (10 injections) of the baseline signal corresponding to a blank matrix extract chromatogram at the analyte retention times divided by the respective slopes of the calibration curves of the analytes (*see* **Note 17**).

7. Check the recovery of the target pesticides and the intraday and interday precision of spiked samples (*see* **Notes 15** and **18**). The extraction recovery of target pesticides is determined by comparing the peak area ratios (for the analytes relative to the ISTD) in samples spiked with the analytes prior to extraction with those for samples to which the analyte is added postextraction (*see* **Note 19**). The intraday precision, or repeatability, and the interday precision, or intermediate precision, are estimated by the analysis of different aliquots ($n = 5$) of the same spiked sample within day or between days (five different days), respectively (*see* **Note 20**).

3.5.2. Sample Analysis (see **Note 21**)

1. Inject 2 µL of the last extract obtained, redissolving with 1 mL of *n*-hexane the fraction eluted from the HPLC system, after drying, and run the chromatogram.
2. Integrate the chromatogram and report the compounds detected and its peak area.
3. Identify the pesticides detected using the RTW values, which means that the retention time of the compound must be in the previously established RTW when the method is verified.
4. Quantify the positive results by the ITSD method. The value *V* of the target compound in the analyzed sample is calculated according to the following expression:

$$V_{(\mu g / Kg)} = \frac{C}{m} V_{\mathrm{f}}$$

   where *C* is the analytical concentration obtained from the analytical curve, *m* is the mass of adipose tissue sample, and V_{f} is the sample dilution volume for analysis.
5. The confirmation of the previously identified compound can be made by comparing the MS/MS spectrum obtained in the sample with another stored as a reference spectrum in the same experimental conditions (*see* **Note 22**). If the fit value is higher than the threshold fit value previously established (*see* **Note 16**), the compound is positively confirmed.
6. Express the result with the uncertainty (U), that is, as R ± U.

3.5.3. Assessment of the Uncertainty of the Measurement

The estimation of the uncertainty, a validation parameter, it is explained in more depth because it is less known. In estimating the overall uncertainty, the main sources of uncertainty have to be identified and separately studied to obtain its contribution. In the present method, the components are (1) the gravimetric step $ur(g)$ and (2) the chromatographic quantification step, which comprises two components: statistical evaluation of the relative uncertainty associated with recovery $ur(c_{recovery})$ and that associated with the repeatability of the method $ur(ip)$.

1. Gravimetric step. This step results from the combination of the following three components:
   a. The relative reference standard uncertainty $u_r(s)$:

$$u_r(s) = \frac{\left(\dfrac{Tol_{ref}}{2}\right)}{m}$$

   where *Tolref* represents the tolerance of the reference standard given by the supplier, and *m* represents the weight of the reference standard.

b. The relative balance calibration standard uncertainty $u_r(b)$:

$$u_r(b) = \frac{\left(\dfrac{Tol_{bal}}{2}\right)}{m}$$

where Tol_{bal} represents the reported tolerance of the balance for the range used. This source of uncertainty is counted twice because the weighing process involves a difference.

c. The gravimetric sample relative uncertainty $ur(ms)$:

$$u_r(m_s) = \frac{\left(\dfrac{Tol_s}{\sqrt{3}}\right)}{m_s}$$

where Tol_s represents the reported tolerance of the balance for the range used, and m_s represents the weight of the adipose tissue sample (500 mg).

The components are combined into the following equation *(23)*:

$$u_r(g) = \sqrt{[u_r(s)]^2 + [u_r(b)]^2 \cdot 2 + [u_r(m_s)]^2}$$

2. Chromatographic quantification step, which results from the combination of the following two sources:
   a. Statistical evaluation of relative uncertainty associated with recovery from 10 equal quantities of the sample matrix spiked at a single concentration level (200 mg/kg) and analyzed intraday $ur(c_{recovery})$. It results from the combination of two components: The first is the relative uncertainty associated with the calibration curve $ur(cal)$

$$u_r(cal) = \frac{\left[\dfrac{\left(\dfrac{S_{res}}{b}\right)^2}{n} + d_c^2 \cdot \left(\dfrac{S_b}{b}\right)^2\right]^{1/2}}{c_c}$$

where s_{res} is the residual standard deviation of the calibration curve, s_b is the standard deviation of the calibration curve, b is the calibration curve slope, n is the number of the calibration standards, and d_c is the difference between the average concentration of the calibration standards and the representative concentration (100 µg/L) of the sample c_c. The second is the relative uncertainty associated with the repeatability of the method $u_r(rep)$

$$u_r(rep) = \frac{\left(\dfrac{SD_{rep}}{\sqrt{n}_{rep}}\right)}{\bar{R}_{rep}}$$

where SD_{rep} represents the standard deviation from the recoveries obtained of the replicate analyses, n_{rep} is the number of replicates analyzed, and $\overline{R}_{rep}$ is the mean recovery obtained.

Both sources are combined into the equation:

$$u_r\left(c_{recovery}\right)=\sqrt{\left[u_r(cal)\right]^2+\left[u_r(rep)\right]^2}$$

b. Statistical evaluation of the relative uncertainty associated with the intermediate precision of the method and calculated from 10 equal quantities of the sample matrix spiked at the same single concentration level (200 µg/kg) and analyzed interday $u_r(ip)$.

$$u_r(ip)=\frac{\left(\dfrac{SD_{ip}}{\sqrt{n}_{ip}}\right)}{\overline{R}_{ip}}$$

where SD_{ip} represents the standard deviation from recoveries obtained of replicate analyses, n_{ip} is the number of replicates analyzed, and $\overline{R}_{ip}$ is the mean of the recovery obtained.

3. Calculation of combined and expanded uncertainty. Once the parameters and their associated uncertainties that contribute to the uncertainty for the method as a whole are listed, the individual uncertainties are combined in the uncertainty budget to give $u_c(y)$:

$$u_c(y)=\sqrt{\left[u_r(g)\right]^2+\left[u_r\left(c_{recovery}\right)\right]^2+\left[u_r(ip)\right]^2}$$

The expanded uncertainty U is obtained by multiplying $u_c(y)$ by a coverage factor k, assuming a normal distribution of the measurand. The choice of this factor is based on the level of confidence desired. Usually, a value of $k = 2$ is used, which provides an approximate level of confidence of 95%.

$$U=u_c(y)\cdot k$$

4. The quantification of these sources (*see* **Note 23**) showed that, in the selected experimental conditions, the largest sources of uncertainty were the ur(b) from the gravimetric step and the ur($c_{recovery}$) and $ur(ip)$ from the chromatographic quantification step (*see* **Note 24**). **Figure 2** shows the contributions to the measurement uncertainty for endosulfan-β. U values lower than 24% are obtained for all the pesticides.

4. Notes

1. Standards of the pesticides and the ISTD must be kept in the refrigerator (approx 2–4°C). Owing to their toxicity, some precautions in relation to contact with skin and eyes and inhalation must be observed.
2. Heptachlor is recommended for ISTD because it is not encountered in biological samples and does not coelute with target pesticides during GC separation.

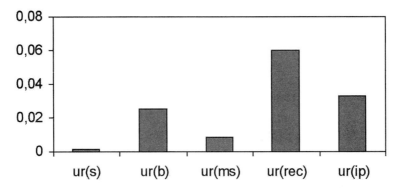

Fig. 2. Contributions to the measurement uncertainty for the determination of β-endosulfan in an adipose tissue sample.

3. Alumina activated at 600°C can be stored at room temperature for up to 6 mo. For gravity flow elution of the chlorinated pesticides, a deactivation of approx 5% water has been found to be satisfactory for alumina. For that, pipet 100 µL of distilled water into an Erlenmeyer flask. Rotate the flask gently to distribute water over its surface. Add 2 g of activated alumina and shake the flask containing the mixture for 10 min on a mechanical shaker. Prepare cleanup columns by plugging the glass column (15 cm long × 0.5 cm id) with a small wad of glass wool. Add 1 g of granular anhydrous Na_2SO_4 and then the 2 g of deactivated alumina.
4. A parent ion is chosen for each analyte, from its full scan spectra, by taking into consideration its *m/z* and its relative abundance (both as high as possible) to improve sensitivity. The parent ions selected of the target analytes are the following typical values: 239 for endosulfan-ether, 272 for the ISTD, 321 for endosulfan-lactone, 239 for endosulfan-α, 239 for endosulfan-β, and 272 for endosulfan-sulfate. Next, the selected ions are subjected to collision-induced dissociation to obtain secondary mass spectra. The object is to generate spectra with the parent ion as their molecular peaks (between 10 and 20% of relative abundance). The MS/MS spectra of the pesticides in our experimental conditions are stored in our own electron ionization MS/MS library.
5. All standards should be prepared in clean, solvent-rinsed volumetric glassware (A class) and stored in a freezer when not in use.
6. If kept in the refrigerator, the primary stock solutions of the target pesticides and the ISTD (approx 2–4°C) can be used for at least 6 mo.
7. If kept in the refrigerator at 4°C, the secondary solutions may be used for at least 1 mo.
8. The calibration solutions are stable at ambient temperature up to at least 24 h.
9. Three pooled fractions (α, x, and β) can be separated by HPLC. The α-fraction is collected in the first 11 min, the x fraction is collected between minutes 11 and 13, and the β-fraction is collected between minutes 13 and 25. Xenoestrogens, such as the target organochlorinated compounds, elute in fraction α; natural estrogens elute in fraction β. Neither xenoestrogens nor natural estrogens are detected in fraction x.
10. Do not forget to add the ISTD to this 1 mL. The use of the ISTD increases the repeatability of the analytical signal measured into the GC–MS/MS.

11. Internal method validation is the first step before the application of an analytical method. It consists of the validation steps carried out within one laboratory to verify that the measurement chemical process is under statistical control and to ensure that the method is "fit to purpose." For this aim, performance characteristics such as accuracy, precision, detection limit, quantification limit, linear range, or selectivity must be checked. The values of these validation parameters will depend on the instrument, column, environmental conditions, and so on for the analytical laboratory in which they are obtained. In this chapter, we present the values obtained in our laboratory.

12. Five calibration standards are prepared with the following concentrations: 1, 25, 50, 100, and 250 µg/L. Calibration standards must be injected in triplicate if higher precision in the calibration step is desired. In this case, the calibration curve is obtained using all the values at each concentration level.

13. The RTWs are calculated as the average of the retention times plus or minus three standard deviations of the retention times for 10 measurements. RTW values of 20.64–21.42 min for endosulfan-ether, 23.39–24.20 min for the ISTD, 27.96–28.89 min for endosulfan-lactone, 34.23–35.24 min for α-endosulfan, 39.95–41.06 min for β-endosulfan, and 45.37–46.51 min for endosulfan–sulfate are typical values obtained.

14. The absence of any chromatographic component at the same retention times as target pesticides indicates that no chemical interference is occurring. It must be mentioned that the MS/MS detection mode can determine up to six compounds that coelute.

15. In the fortification step, it is necessary to use a volume of standard as low as possible; to be sure that the spiked sample is homogenized; and to let the spiked sample dry at least for 30 min before extraction.

16. From the 10 spectra obtained for each compound under the same analysis conditions, select 1 as a reference spectrum and compare the other 9 spectra with it. The product of the comparison is 9 fit values (from 0 to 1000 for best match) and an average fit value. A threshold fit value is obtained by subtracting three times the value of the standard deviation (estimated from the 9 fit values) to the average fit. This subtraction is done to compensate for the spectral variation caused by the routine analysis of samples, which dirty the instrument and require maintenance operations that would slightly affect the detector response and therefore the spectra.

17. LOD (LOQ) limits in the matrix of about 0.4 (2) µg kg⁻¹ for endosulfan-ether, 1.2 (4) µg kg⁻¹ for endosulfan-lactone, 2.4 (8) µg kg⁻¹ for α-endosulfan, 5 (16) µg kg⁻¹ for β-endosulfan, and 1.6 (6) µg kg⁻¹ for endosulfan-sulfate are easily obtained. These limits are sufficiently low for the trace analysis of the target pesticide residues in human adipose samples.

18. The use of samples spiked with target compounds is necessary to carry out reliable studies about the recovery achieved by the procedure because of the lack of certified reference samples.

19. Recoveries must be between 65 and 120% for adipose tissue samples spiked with 40–200 µg kg⁻¹.

20. Intraday precision and interday precision, as measured by relative standard deviation, must be lower than 10 and 20%, respectively.

21. Laboratory reagents blank, laboratory spiked blank, and a calibration curve must be analyzed with each set of real samples. The laboratory reagents blank checks any interference contamination caused by reagents during processing samples. Analyses of samples are carried out if recoveries of laboratory-spiked blanks are between 60 and 130%.

22. The reference spectrum is obtained during the verification process (*see* **Note 17**). The results of this comparison (fit) allow checking that the spectra obtained have not changed since the verification process. The fit value obtained in this comparison must be higher than the threshold fit value established (*see* **Note 17**).
23. We have not taken into account the contribution to uncertainty corresponding to the dilution of the primary standard. Our experience in this field allows us to conclude that the contribution of this step is not significant.
24. Obviously, to decrease the uncertainty of the method, it would be adequate to act on the components with a high contribution [ur(b), ur($c_{recovery}$), and ur(ip)]. On one hand, increasing the amount of sample weighed for the analysis or the amount of solid standard weighed for the preparation of primary standard solution. On the other hand, ur($c_{recovery}$) and ur(ip) by trying to improve the precision of the method or to increase the concentration levels or the number of calibration points.

References

1. Barr, D. B. and Needham, L. L. (2002) Analytical methods for biological monitoring of exposure to pesticides: a review. *J. Chromatogr. A* **778**, 5–29.
2. Strandberg, B., Strandberg, L., Bergqvist, P. A., Falandysz, J., and Rappe, C. (1998) Concentrations and biomagnification of 17 chlordane compounds and other organochlorines in harbour porpoise (*Phocoena phocoena*) and herring from the southern Baltic Sea. *Chemosphere* **7**, 2513–2523.
3. Garrido Frenich, A., Pablos Espada, M. C., Martínez Vidal, J. L., and Molina, L. (2001) Broad spectrum analysis of pesticides in groundwater samples by gas chromatography with ECD, NPD and MS/MS detectors. *J. AOAC*, **84**, 1–12.
4. Martínez Vidal, J. L., Moreno Frías, M., Garrido Frenich, A., Olea-Serrano, F., and Olea, N. (2000) Trace determination of α and β endosulfan and three metabolites in human serum by GC–ECD and GC–MS–MS. *Rapid Commun. Mass Spectrom.* **14**, 939–946.
5. Martínez Vidal, J. L., Moreno Frías, M., Garrido Frenich, A., Olea-Serrano, F., and Olea, N. (2002) Determination of endocrine-disrupting pesticides and polychlorinated biphenyls in human serum by GC–ECD and GC–MS/MS and evaluation of the contributions to the uncertainty of the results. *Anal. Bioanal. Chem.* **372**, 766–775.
6. Hernández, F., Pitarch, E., Serrano, R., Gaspar, J. V., and Olea, N. (2002) Multiresidue determination of endosulfan and metabolic derivatives in human adipose tissue using automated liquid chromatographic cleanup and gas chromatographic analysis. *J. Anal. Toxicol.* **26**, 94–103.
7. Soto, A. M., Chung, K. L., Sonnenschein, C. (1994) The pesticides endosulfan, toxaphene, and dieldrin have estrogenic effects on human estrogen-sensitive cells. *Environ. Health Perspect.* **102**, 380–383.
8. Olea, N., Pazos, P., and Exposito, J. (1998) Inadvertent exposure to xenoestrogenes. *Eur. J. Cancer Prev.* **7**(Suppl. 1), S17–S23.
9. Gascón, J., Oubiña, A., and Barceló, D. (1997) Detection of endocrine-disrupting pesticides by enzyme-linked immunosorbent assay (ELISA): application to atrazine. *Trends Anal. Chem.* **16**, 554–562.
10. Aronson, K. J., Miller, A. B., Woolcott, C. G., et al. (2000) Breast adipose tissue concentrations of polychlorinated biphenyls and other organochlorines and breast cancer risk. *Cancer Epidemiol. Biomarkers Prev.* **9**, 55–63.

11. Kohlmeier, L. and Kohlmeier, M. (1995) Adipose tissue as a medium for epidemiology exposure assessment. *Environ. Health Perspect.* **103**, 99–106.

12. Ludwicki, J. L. and Goralczyk, K. (1994) Organochlorine pesticides and PCBs in human adipose tissues in Poland. *Bull. Environ. Contam. Toxicol.* **52**, 400–403.

13. Asakawa, A., Jitsunari, F., Shiraishi, H., Suna, S., Takeda, N., and Kitamado, T. (1996) Accumulation of chlordanes in adipose tissues of mice caused by long exposure of low level technical chlordane. *Bull. Environ. Contam. Toxicol.* **57**, 909–916.

14. Bucholski, K. A., Begerow, J., Winneke, G., and Duneman, L. (1996) Determination of polychlorinated biphenyls and chlorinated pesticides in human body fluids and tissues. *J. Chromatogr. A* **754**, 479–485.

15. Garrido Frenich, A., Martínez Vidal, J. L., Moreno Frías, M., Olea-Serrano, F., and Olea, N. (2000) Quantitative determination of endocrine-disrupting polychlorinated biphenyls and organochlorinated pesticides in human serum using GC/ECD and tandem mass spectrometry. *J. Mass Spectrom.* **35**, 967–975.

16. Moreno Frías, M., Garrido Frenich, A., Martínez Vidal, J. L., Olea, F., Olea, N., and Mateu, M. (2001) Analysis of endocrine-disrupting compounds lindane, vinclozolin, aldrin, p-p′ DDE, p-p′ DDT in human serum using GC–ECD and tandem mass spectrometry. *J. Chromatogr. B* **760**, 1–15.

17. Moreno Frías, M., Garrido Frenich, A., Martínez Vidal, J. L., et al. (2003) Determination of endocrine disrupting pesticides in serum by GC–ECD and GC–MS/MS techniques including an evaluation of the uncertainty associated with the results. *Chromatographia* **57**, 213–220.

18. Moreno Frías, M., Jiménez Torres, M., Garrido Frenich, A., Martínez Vidal, J. L., Olea-Serrano, F., and Olea, N. (2004) Determination of organochlorine compounds in human biological samples by GC–MS/MS. *Biomed. Chromatogr.* **18**, 102–111.

19. Pauwels, A., Wells, D. A., Covaci, A., and Schepens, P. J. C. (1999) Improved sample preparation method for selected persistent organochlorine pollutants in human serum using solid-phase disk extraction with gas chromatographic analysis. *J. Chromatogr. B* **723**, 117–125.

20. Röhrig, L. and Meisch, H.-U. (2000) Application of solid phase micro extraction for the rapid analysis of chlorinated organics in breast milk. *Fresenius J. Anal. Chem.* **366**, 106–111.

21. Analytical Methods Committee. (1995) Uncertainty of measurement—implications for its use in analytical science. *Analyst* **120**, 2303–2308.

22. International Organization for Standardization. (1993) *International Vocabulary of Basic and General Terms in Metrology*, International Organization for Standardization, Geneva.

23. International Organization for Standardization. (1993) *Guide for the Expression of Uncertainty in Measurements*, International Organization for Standardization, Geneva.

24. EURACHEM Guide. (2000) *Quantifying uncertainty in analytical measurement*, 2nd ed., http//www.vtt.fi/ket/eurachem/quam2000-p1.pdf.

25. Wernimont, G. T. (1985) *Use of Statistics to Develop and Evaluate Analytical Methods*, AOAC, Arlington, V.A.

26. Thompson, M., Ellison, S. L. R., Wood, R. (2002) Harmonized guidelines for single-laboratory validation of methods. *Pure Appl. Chem.* **74**, 835–855.

27. Hill, A. R. and Reynolds, S. L. (1999) Guidelines for in-house validation of analytical methods for pesticide residues in food and animal feeds. *Analyst* **124**, 953–958.

28. Quintana, J., Martí, I., and Ventura, F. (2001) Monitoring of pesticides in drinking and related waters in EN Spain with a multiresidue SPE–GC–MS method including an estimation of the uncertainty of the analytical results. *J. Chromatogr. A* **938**, 3–13.

29. Bettencourt da Silva, R. J. N., Santos, J. R., and Camões, M. F. G. F. C. (2003) Evaluation of the analytical method performance for incurred samples. *Anal. Chim. Acta* **485,** 241–252.
30. Hund, E., Massart, D. L., and Smeyers-Verbeke, J. S. (2003). Comparison of different approaches to estimate the uncertainty of a liquid chromatographic assay. *Anal. Chim. Acta* **480,** 39–52.
31. Dehouck, P., Vander Heyden, Y., Smeyers-Verbeke, J., et al. (2003) Determination of uncertainty in analytical measurements from collaborative study results on the analysis of a phenoxymethylpenicillin sample. *Anal. Chim. Acta* **481,** 261–272.
32. Maroto, A., Boqué, R., Riu, J., and Rius, F. X. (1999) Evaluating uncertainty in routine analysis. *Trends Anal. Chem.* **18,** 577–584.
33. Maroto, A., Boqué, R., Riu, J., and Rius, F. X. (2000) Critical discussion on the procedures to estimate uncertainties in chemical measurements. *Quim. Anal.* **19,** 85–94.
34. Bettencourt da Silva, R. J. N., Joäo Lino, M., Sanots, J. R., and Camões, M. F. G. F. C. (2000) Estimation of precision and efficiency mass transfer steps for the determination of pesticides in vegetables aiming at the expression of results with reliable uncertainty. *Analyst* **125,** 1459–1464.
35. Lisinger, T. P. J., Führer, M., Kandler, W., and Schuhmacher, R. (2001) Determination of measurement uncertainty for thedetermination of triazines in groundwater from validation data. *Analyst* **126,** 211–216.
36. Cuadros-Rodriguez, L., Hernández Torres, M. E., Almansa López, E., et al. (2002) Assessment of uncertainty in pesticide multiresidue analytical methods: main sources and estimation. *Anal. Chim. Acta* **454,** 297–314.

2

Determination of Pyrethroids in Blood Plasma and Pyrethroid/ Pyrethrin Metabolites in Urine by Gas Chromatography– Mass Spectrometry and High-Resolution GC–MS

Gabriele Leng and Wolfgang Gries

Summary

In this chapter, two analytical methods are presented suitable for the determination of pyrethroids in blood plasma and pyrethroid/pyrethrin metabolites in urine. As pyrethroids such as cyfluthrin, cypermethrin, deltamethrin, permethrin, and bioallethrin are metabolized very fast, they can only be detected within about 24 h after exposure; that is, the method shown should only be applied in case of intoxication. After solid-phase extraction, the sample is analyzed by high-resolution gas chromatography–negative chemical ionization mass spectrometry (HRGC–NCIMS) with a detection limit of 5 ng/L blood plasma. In all other cases of exposure (occupational surveillance, environmental, biological monitoring programs, etc.), the determination of metabolites in urine by gas chromatography–mass spectrometry (GC–MS) or HRGC–MS should be preferred. The urine method is adequate for the simultaneous determination of the pyrethroid metabolites *cis*- and *trans*-3-(2,2-dichlorovinyl)-2,2-dimethylcyclopropane carboxylic acid, *cis*-3-(2,2-dibromovinyl)-2,2-dimethylcyclopropane carboxylic acid, 3-phenoxybenzoic, and 4-fluoro-3-phenoxybenzoic acid as well as of the pyrethrin/bioallethrin-specific metabolite *trans*-chrysanthemumdicarboxylic acid (-CDCA). After acid hydrolysis and sample extraction with tert-butyl-methylether, the residue is derivatized with 1,1,1,3,3,3-hexafluoroisopropanol and analyzed by HRGC–MS (detection limit 0.1 µg/L urine).

Key Words: Bioallethrin; biomonitoring; blood plasma; cyfluthrin; cypermethrin; deltamethrin; derivatization; insecticide; GC–MS; hexafluoroisopropanol; HRGC–NCIMS; metabolites; permethrin; pyrethroids; pyrethrum; solid-phase extraction; -chrysanthemum-dicarboxylic acid; urine.

1. Introduction

Synthetic pyrethroids such as cyfluthrin, cypermethrin, deltamethrin, and permethrin originate from the botanical insecticide pyrethrum, an extract obtained from the flowers of *Chrysanthemum cinerariaefolium*. Pyrethrins as one of the natural

From: *Methods in Biotechnology, Vol. 19, Pesticide Protocols*
Edited by: J. L. Martínez Vidal and A. Garrido Frenich © Humana Press Inc., Totowa, NJ

esters of pyrethrum, and the synthetic pyrethroids are among the insecticides most often used worldwide.

In mammals, pyrethroid esters are rapidly detoxified by ester hydrolysis and hydroxylation, partially conjugated, and finally eliminated, mainly in the urine (**Fig. 1**). The main metabolites are *cis*- and *trans*-3-(2,2-dichlorovinyl)-2,2-dimethylcyclopropane carboxylic acid (*cis*-DCCA and *trans*-DCCA), *cis*-3-(2,2-dibromovinyl)-2,2-dimethylcyclopropane carboxylic acid (-DBCA), 3-phenoxybenzoic acid (3-PBA), and 4-fluoro-3-phenoxybenzoic acid (FPBA). The biological half-lives of the different pyrethroids vary between 2.5 and 12 h in blood plasma *(1–3)*. Half-lives of 6.44 h were found for the urinary excretion of the metabolites *cis*-DCCA, *trans*-DCCA, and FPBA after oral or inhalation exposure to cyfluthrin in volunteers. Of the metabolites, 94% were excreted renally during the first 48 h after exposure *(4)*.

Chrysanthemate insecticides like natural pyrethrins or (S)-bioallethrin are also metabolized by hydrolysis, oxidation and finally conjugation with the major metabolite eliminated in the urine *(5,6)*. **Figure 2** shows that the major metabolite is -(E)-chrysanthemumdicarboxylic acid (*trans*-CDCA). Interestingly, *cis*-CDCA as well as *trans*-chrysanthemic acid are not found in humans. Following (S)-bioallethrin exposure, maximum peak excretion of trans-CDCA was within the first 24 h after exposure, and 72 h later the concentration of trans-CDCA was below the limit of detection *(6)*.

In humans, a variety of reversible symptoms, such as paraesthesia, irritations of the skin and mucosa, headache, dizziness, and nausea, are reported following pyrethroid/pyrethrin exposure *(1,7,8)*. For these adverse health effects, the original pyrethroid/pyrethrin and not the detoxified metabolites is responsible. Therefore, from the medical point of view, it is useful to determine the pyrethroid/pyrethrin in plasma. Exposure to high pyrethroid doses, as seen in cases of acute intoxication, leads to detectable pyrethroid concentrations in blood plasma during the first hour after exposure, rapidly decreasing within 24 h *(9)*. In persons occupationally exposed to pyrethroids as well as in persons exposed in their private surroundings, pyrethroid plasma levels are always below the detection limit, although detectable amounts of metabolites can be found in urine *(10–12)*. Therefore, for routine biological monitoring of persons exposed to pyrethrins or pyrethroids, the determination of the corresponding metabolites in urine is most often described in literature *(4,9–16)*.

1.1. Determination of Pyrethroids in Blood Plasma

With the method described here, all relevant pyrethroids (i.e., cyfluthrin, cypermethrin, deltamethrin, permethrin, and bioallethrin) can be determined in 1 mL blood plasma (*see* **Note 1**). After cleanup, sample enrichment with solid-phase extraction and elution with hexane/dichloromethane, the sample is analyzed by high-resolution gas chromatography–negative chemical ionization mass spectrometry (HRGC–NCIMS) (5 ng/L blood plasma detection limit) (*see* **Note 2**). The analysis in negative chemical ionization (NCI or CI⁻) mode is more sensitive than the most-often-used positive electron impact (EI⁺) mode. This is based on the weaker ionization process in NCI and results in lower mass fragmentation, which enables lower detection limits.

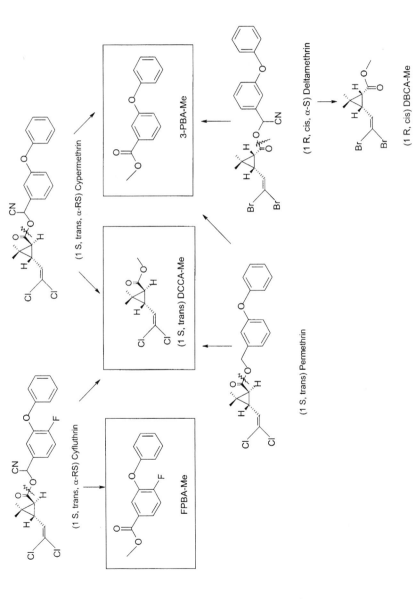

Fig. 1. Metabolism of the pyrethroids cyfluthrin, cypermethrin, deltamethrin, and permethrin in humans. The corresponding metabolites found in urine are shown in brackets. cis-DCCA and trans-DCCA: *cis*- and *trans*-3-(2,2-dichlorovinyl)-2,2-dimethylcyclopropane carboxylic acid; *cis*-DBCA: *cis*-3-(2,2-dibromovinyl)-2,2-dimethylcyclopropane carboxylic acid; 3-PBA: 3-phenoxybenzoic acid; FPBA: 4-fluoro-3-phenoxybenzoic acid.

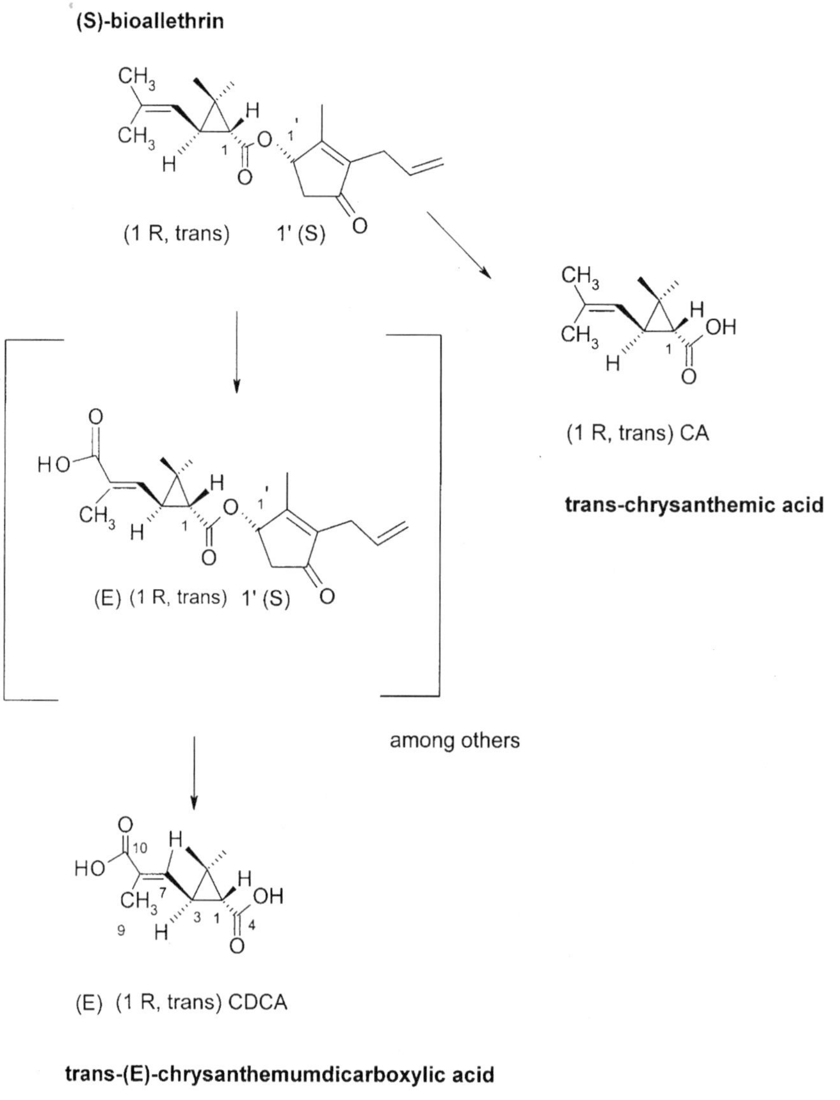

(S)-bioallethrin

(1 R, trans)   1' (S)

(1 R, trans) CA

trans-chrysanthemic acid

(E) (1 R, trans) 1' (S)

among others

(E) (1 R, trans) CDCA

trans-(E)-chrysanthemumdicarboxylic acid

Fig. 2. Metabolism of (S)-bioallethrin in humans.

1.2. Determination of Pyrethrin/Pyrethroid Metabolites (cis/trans CDCA, cis/trans-DCCA, cis-DBCA, FPBA, 3-PBA) in Urine

This method is developed for the simultaneous determination of the metabolites of synthetic pyrethroids (*cis*-DCCA, *trans*-DCCA, *cis*-DBCA, 3-PBA, and FPBA) together with the metabolite of pyrethrin chrysanthemumdicarboxylic acid (trans-CDCA) (*see* **Note 3**). After acid hydrolysis, the sample is derivatized with 1,1,1,3,3,3-

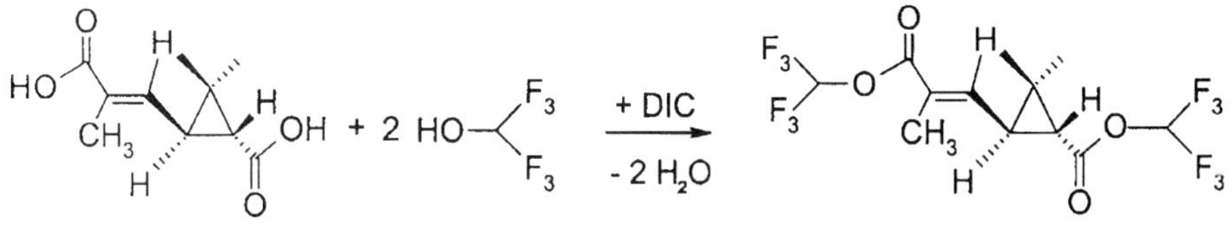

Fig. 3. Esterification of CDCA with 1,1,1,3,3,3-hexafluoroisopropanol.

hexafluoroisopropanol (HFIP) in the presence of N,N'-diisopropylcarbodiimide (DIC). Detection is done by HRGC–MS after separation on a Rtx 65 fused silica capillary column (0.1 µg/L urine detection limit) (*see* **Note 4**).

The reaction scheme of CDCA esterification with hexafluoroisopropanol is shown in **Fig. 3**.

2. Materials
2.1. Determination of Pyrethroids in Blood Plasma

1. Microliter pipets, adjustable between 1 and 1000 µL (e.g., Eppendorf, Hamburg, Germany).
2. 10-mL tubes with Teflon-sealed screw caps.
3. Nitrogen evaporator.
4. Microvials (e.g., Agilent, Palo Alto, CA).
5. Microevaporator unit.
6. Solid-phase extraction (SPE) column station with column drying option (e.g., Supelco, Bellefonte, PA).
7. Oasis HLB [hydrophilic–lipophilic balance] cartridges, 6 mL/200 mg (Waters, Milford, MA).
8. GC–MS system with NCI equipment (e.g., AutoSpec Ultima, Micromass/Waters, Milford, MA).
9. Helium 5.0.
10. Capillary column, 30 m × 0.25 mm × 0.1 µm DB5 (Durabond 5; Agilent).
11. Cyfluthrin (e.g., Dr. Ehrenstorfer GmbH, Augsburg, Germany).
12. Deltamethrin (e.g., Dr. Ehrenstorfer).
13. Cypermethrin (e.g., Dr. Ehrenstorfer).
14. Permethrin (e.g., Dr. Ehrenstorfer).
15. Bioallethrin (e.g., Dr. Ehrenstorfer).
16. Fenvalerat (e.g., Dr. Ehrenstorfer), used as internal standard (ISTD).
17. Dichloromethane (Supra-Solv).
18. Hexane (Supra-Solv).
19. Methanol (Supra-Solv).
20. For conditioning of Oasis HLB cartridges, First wash each column with 4 mL methanol at atmospheric pressure. After methanol is rinsed through the column, repeat the same procedure with 6 mL water.
21. To prepare the standard solutions, about 10 mg of each compound (or proportionally more if purity < 100%) is weighed into separate 10-mL flasks. Each flask is diluted to volume with acetonitrile. The concentration of these standard starting solutions is 1000 mg/L. The following dilutions were performed with these starting solutions.

Table 1
Necessary Fortification Levels for Pyrethroids in Blood Plasma

Concentration (µg/L)	Stock solution	Spike volume (µL) in 1 mL plasma	Spike volume ISTD dilution 0.1 mg/L (µL)
Blank value	—	—	10
0.01	4	10	10
0.02	4	20	10
0.05	4	50	10
0.10	3	10	10
0.20	3	20	10
0.50	3	50	10
1.00	2	10	10

a. For stock solution 1 of 1.0 mg/L, 100 µL of each standard starting solution is added to a 100-mL flask, which is filled to volume with acetonitrile (1:1000 dilution).

b. For stock solution 2 of 0.1 mg/L, 1000 µL of stock solution 1 is added to a 10-mL flask, which is filled to volume with acetonitrile (1:10,000 dilution).

c. For stock solution 3 of 0.01 mg/L, 100 µL of stock solution 1 is added to a 10-mL flask, which is filled to volume with acetonitrile (1:100,000 dilution).

d. For stock solution 4 of 0.001 mg/L, 100 µL of dilution 2 is added to a 10-mL flask, which is filled to volume with acetonitrile (1:1,000,000 dilution).

e. For ISTD solution of 0.1 mg/L, preparation is done with separate dilutions in comparison to stock solution 2 described in **a** above.

For the calibration experiment, defined volumes of dilution 1, 2, 3, or 4 are added to 1 mL plasma. The dilutions for necessary concentrations are shown in **Table 1**.

2.2. Determination of Pyrethrin/Pyrethroid Metabolites (cis-/trans-CDCA, cis-/trans-DCCA, cis-DBCA, FPBA, 3-PBA) in Urine

1. Microliter pipets, adjustable between 1 and 2000 µL (e.g., Eppendorf).
2. 20-mL tubes with Teflon-sealed screw caps.
3. Microvials (e.g., Agilent).
4. Centrifuge.
5. Block heater for hydrolysis.
6. Shaker.
7. Nitrogen evaporator.
8. GC–MS system (e.g., AutoSpec Ultima).
9. Helium 5.0.
10. Capillary column, 30 m × 0.25 mm × 0.25 µm Rtx 65.
11. HFIP (e.g., Aldrich, Poole, UK).
12. DIC (e.g., Aldrich).
13. 3-Phenoxybenzoic acid (e.g., Aldrich)
14. 2-Phenoxybenzoic acid (e.g., Aldrich) used as ISTD.
15. *cis*-3-(2,2-Dichlorovinyl)-2,2-dimethylcyclopropane carboxylic acid (e.g., Dr. Ehrenstorfer).

16. *trans*-3-(2,2-Dichlorovinyl)-2,2-dimethylcyclopropane carboxylic acid (e.g., Dr. Ehrenstorfer).
17. *cis*-3-(2,2-Dibromovinyl)-2,2-dimethylcyclopropane carboxylic acid (e.g., Roussel-Uclaf, Romainville Cedex, France).
18. 4-Fluoro-3-phenoxybenzoic acid (e.g., Bayer Industry Services, Leverkusen, Germany).
19. *cis*-CDCA (e.g., Bayer Industry Services).
20. *trans*-CDCA (e.g., Bayer Industry Services).
21. Acetonitrile (Supra-Solv).
22. Tert.-Butyl-methylether (Supra-Solv).
23. Iso-octane (Supra-Solv).
24. For preparation of the standard solutions, about 10 mg of each compound (or proportionally more if purity < 100%) is weighed into separate 10-mL flasks. Each flask is diluted to volume with acetonitrile. The concentration of these standard starting solutions is 1000 mg/L. The following dilutions are performed with these starting solutions:

    a. For stock solution 1 of 10.0 mg/L, 100 µL of each standard starting solution is added to a 10-mL flask, which is filled to volume with acetonitrile (1:100 dilution).
    b. For stock solution 2 of 1.0 mg/L, 1000 µL of dilution 1 is added to a 10-mL flask, which is filled to volume with acetonitrile (1:1000 dilution).
    c. For stock solution 3 of 0.1 mg/L, 100 µL of dilution 1 is added to a 10-mL flask, which is filled to volume with acetonitrile (1:10,000 dilution).
    d. For stock solution 4 of 0.01 mg/L, 1000 µL of dilution 3 is added to a 10-mL flask, which is filled to volume with acetonitrile (1:100,000 dilution).
    e. For the ITSD solution of 1.0 mg/L, the preparation is done with separate dilutions as for dilution 2 described in.

For the calibration experiment, defined volumes of dilution 1, 2, 3, or 4 are added to 2 mL urine. The dilutions for necessary concentrations are shown in **Table 2**.

3. Methods

3.1. Determination of Pyrethroids in Blood Plasma

3.1.1. Sample Preparation

1. Put the conditioned Oasis column on the SPE station.
2. Dilute 1 mL plasma with 1 mL ultrapure water in a test tube.
3. Add 10 µL of ITSD (e.g., Fenvalerat) (*see* **Note 5**).
4. Mix the sample slightly to get a homogeneous solution.
5. Trickle sample slowly on the Oasis column.
6. Let sample elute through the column under atmospheric pressure.
7. Rinse column with 2 mL ultrapure water.
8. Dry column 30 s under vacuum on the SPE station.
9. Dry column under nitrogen steam on the SPE station with drying option (approx 30 min) at room temperature.
10. Rinse column with 2 mL n-hexane (*see* **Note 6**).
11. Elute pyrethroids with 3 mL hexane:dichloromethane 1:1 (v/v).
12. Evaporate solution in a nitrogen evaporator (e.g., Pierce) down to approx 200 µL.
13. Transfer sample in a microvial and narrow carefully with nitrogen down to dryness (*see* **Note 7**).
14. Resolve sample in 25 µL toluene (analysis sample).

Table 2
Necessary Fortification Levels for Pyrethroid Metabolites (HFIP Method) in Urine

Concentration (μg/L)	Stock solution	Spike volume (μL) in 2 mL urine	Spike volume ISTD dilution 1 mg/L (μL)
Blank value	—	—	20
0.05	4	10	20
0.1	4	20	20
0.2	4	40	20
0.5	4	100	20
1.0	3	20	20
2.0	3	40	20
5.0	2	100	20
10.0	2	20	20
20.0	2	40	20
50.0	2	100	20
100.0	1	20	20

3.1.2. Operational Parameters for GC and MS

3.1.2.1. GC PARAMETERS

Use HP 5890II with SSL-Injector and CTC A 200S Autosampler.

DB5 (30 m × 0.25 mm × 0.1 μm) column
He 80 kPa for 1 min, 5 kPa/min, 100-kPa gas pressure
1 min off purge time
40-mL/min split
3-mL/min septum purge
60°C 1 min > 15°C/min > 320°C > 10 min
300°C injection temperature (*see* **Note 8**)
1 μL injection volume

3.1.2.2. MS PARAMETERS

Use an Micromass AutoSpec Ultima.

250°C interface
200°C source
NCI mode inner source
Ammonia CI gas, 2×10^{-5} KPa source pressure (*see* **Note 9**)
0.5-mA filament
100 eV electron energy
Maximum 8000 accelerating voltage
350-V multiplier
10,000 resolution
Perfluorokerosin calibration gas

3.1.3. Analytical Determination

Inject into the GC–MS system 1 μL of the sample (*see* **Subheading 3.1.1., step 14**). If the instrument parameters are set as described in in selected ion monitoring

Table 3
SIM Masses and Retention Time for MS Detection of Pyrethroids

Pyrethroid	Target mass (*m/z*)	Retention time (min)
Bioallethrin	167.107	10:45
Permethrin (sum of 2 isomers)	206.998	14:13/14:18
Cyfluthrin (sum of 4 isomers)	206.998	14:36/14:45
Cypermethrin (sum of 4 isomers)	206.998	14:48/14:58
Deltamethrin	296.895	15:55
Fenvalerat (ITSD)	211.053	15:25/15:34

(SIM) mode (**Table 3**), a stereoselective resolved chromatogram of each pyrethroid can be obtained as shown in **Fig. 4**.

3.1.4. Method Validation

The calibration curve and the samples for precision control are prepared with plasma of persons not exposed to pyrethroids. Necessary fortification levels for this procedure are prepared in agreement with the fortification levels in **Table 1**. The linearity of all pyrethroids is tested in a range between 5 and 1000 ng/L blood plasma, with correlation coefficients more than 0.995. If the method is working correctly, quality criteria for precision in series can be achieved as shown in **Table 4**. The average recovery of all compounds reached 90%.

3.1.5. Storage Stability

The starting solutions can be stored in a deep freezer at −18°C for at least 6 mo. Longer times were not tested. We prefer fresh (monthly) preparation of stock solutions 2, 3, and 4. It was found that pyrethroids in plasma are not stable if they are stored at +4°C *(11)*. If analysis cannot start in about a day after blood plasma sampling, the plasma samples must be stored in a deep freezer at −70°C. Then, they are stable for more than a year.

3.2. Determination of Pyrethrin/Pyrethroid Metabolites (cis/trans CDCA, cis/trans-DCCA, cis-DBCA, FPBA, 3-PBA) in Urine

3.2.1. Sample Preparation

1. Transfer 2 mL urine in a screw cap test tube.
2. Add 20 µL ITSD solution (10 µg/L 2-PBA).
3. Add 500 µL concentrated hydrochloric acid.
4. Cover test tube with screw cap.
5. Hydrolyze sample for 2 h at 100°C in a block heater.
6. Add 3 mL tert.-butyl-methylether to the cold sample.
7. Cover test tube with screw cap and shake urine sample vigorously for 5 min.
8. Centrifuge sample for 5 min at 2000 .
9. Separate organic layer in a new screw cap test tube.
10. Add 2 mL tert.-butyl-methylether to sample.
11. Cover test tube with screw cap and shake urine sample vigorously for 5 min.
12. Centrifuge sample for 5 min at 2000 .

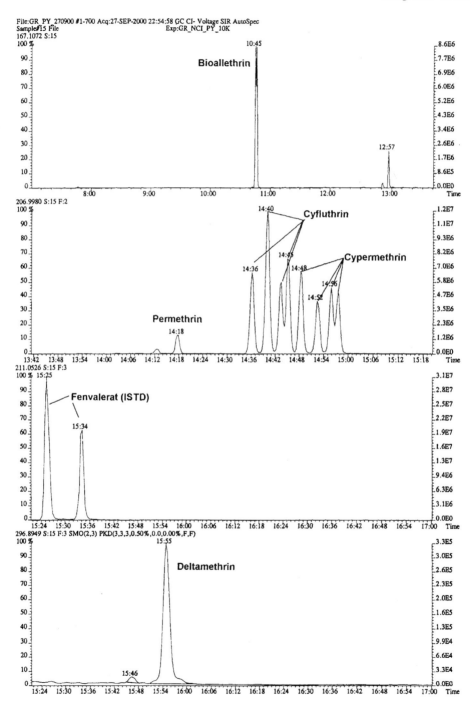

Fig. 4. High-resolution CI–MS chromatogram of pyrethroids in blood plasma separated on a 30 m × 0.25 mm × 0.1 μm DB5 capillary column.

Table 4
Precision in Series and Detection Limits of Pyrethroids

Pyrethroid	Plasma 0.1 µg/L R.S.D. (%)	Plasma 1.0 µg/L R.S.D. (%)	Detection limit (ng/L)
Bioallethrin	10.5	17.5	5
Permethrin (sum of 2 isomers)	8.4	16.9	5
Cyfluthrin (sum of 4 isomers)	9.9	9.4	5
Cypermethrin (sum of 4 isomers)	6.4	10.8	5
Deltamethrin	15.1	10.3	20

RSD, relative standard deviation.

13. Combine organic layer in the new screw cap test tube (**step 9**).
14. Discard the lower urine phase.
15. Dry organic layer under a gentle stream of nitrogen just to dryness.
16. Dissolve residue in 250 µL acetonitrile.
17. Add 30 µL of HFIP (*see* **Note 10**).
18. Add 20 µL of DIC.
19. Derivatize solution under slight mixing for 10 min at room temperature.
20. Add 1 mL 1 *M* sodium hydrogen carbonate solution.
21. Add 250 µL iso-octane.
22. Cover test tube and mix sample 10 min vigorously for extraction.
23. Centrifuge sample 5 min at 2000 for phase separation.
24. Separate iso-octane phase in a microvial.

3.2.2. Operational Parameters for GC and MS

3.2.2.1 GC PARAMETERS

Use HP 5890II with SSL-Injector and CTC A 200S Autosampler.

Rtx 65.30 m × 0.25 mm × 0.25 µm column
He 120 kPa for 1 min, 100 kPa/min, 80 kPa gas pressure
1 min off purge time
40-mL/min split
3-mL/min septum purge
60°C 1 min >8°C/min >150°C >30°C/min at 300°C for 20 min
300°C injection temperature
1 µL injection volume

3.2.2.2. MS Parameters

Use a Micromass AutoSpec Ultima.

250°C interface.
250°C source.
EI mode inner source (*see* **Note 11**).
0.3-mA filament.
70 eV electron energy.
Maximum 8000 accelerating voltage.
330-V multiplier.

Table 5
SIM Masses and Retention Time for MS Detection of Pyrethroid Metabolites (HFIP-Ester) in Urine

Pyrethroid metabolite	Target mass (*m/z*)	Retention time (min)
cis -CDCA	331.077	7:19
trans-CDCA	331.077	6:21
cis-DCCA	323.027	7:34
trans-DCCA	323.027	7:45
cis-DBCA	368.975	10:52
2-PBA (internal standard)	364.053	14:27
3-PBA	364.053	14:39
FPBA	382.044	14:05

10,000 resolution.
Perfluorokerosine calibration gas

3.2.3. Analytical Determination

Inject 1 µL of the sample (sample preparation, **step 24**) into the GC–MS system. If the instrument parameters are set as described in **Subheading 3.2.2.** in SIM EI⁺ mode (**Table 5**), a stereoselective resolved chromatogram of each pyrethroid metabolite can be obtained as shown in **Fig. 5**. This sample can optionally be analyzed in NCI mode as shown in **Fig. 6** (*see* **Note 11**).

3.2.4. Method Validation

The calibration curve and the samples for precision control are prepared with urine of persons not exposed to pyrethroids. Necessary fortification levels for this procedure were prepared in agreement with the fortification levels in **Table 2**. The linearity of all pyrethroid metabolites is tested in a range between 0.1 and 100 µg/L urine, with correlation coefficients more than 0.995. If the method is working correctly, quality criteria for precision in series can be achieved as listed in **Table 6**. The average recovery of all compounds lies between 90 and 100%.

3.2.5. Storage Stability

The starting solutions and stock solutions can be stored in a deep freezer at −18°C for at least 6 mo. Longer times were not tested. Urine samples can be stored for more than a year at −21°C in a deep freezer

4. Notes

1. This method is developed for the determination of very low pyrethroid concentrations in blood plasma, which are caused by the short pyrethroid half-lives in blood plasma. Therefore, this method is useful for the determination of pyrethroids following acute intoxication.
2. Several analytical techniques have been tested previously, but all gave poor reproducibilities or sensitivities. A direct extraction with iso-octane after precipitation with sodium chloride and ethanol also works, but only for a few samples. In such a plasma extract,

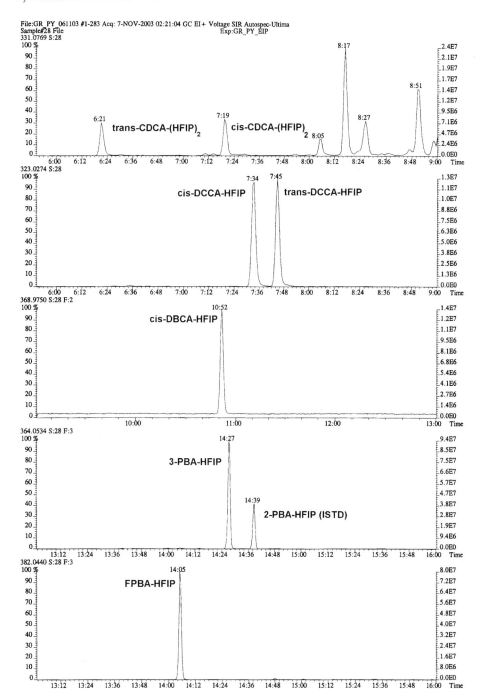

Fig. 5. High-resolution EI[+] MS chromatogram of pyrethroid metabolites in urine (as HFIP esters) separated on a 30 m × 0.25 mm × 0.25 μm Rtx 65 capillary column.

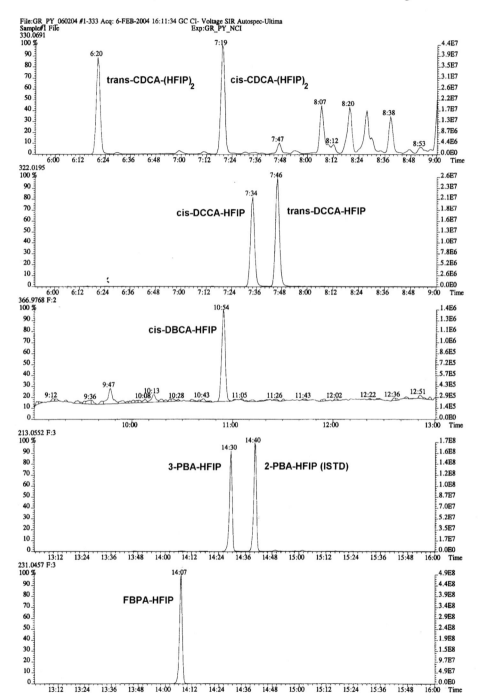

Fig. 6. High-resolution CI–MS chromatogram of pyrethroid metabolites in urine (as HFIP esters) separated on a 30 m × 0.25 mm × 0.25 µm Rtx 65 capillary column.

Table 6
Precision in Series and Detection Limits of Pyrethroid Metabolites (HFIP Ester) in Urine

Pyrethroid metabolite	Urine 1 µg/L R.S.D. (%)	Urine 10.0 µg/L R.S.D. (%)	Detection limit (µg/L)
cis-CDCA	7.4	2.4	0.1
trans-CDCA	5.3	2.7	0.1
cis-DCCA	5.2	3.4	0.1
trans-DCCA	4.4	2.1	0.1
cis-DBCA	3.5	1.9	0.1
3-PBA	3.0	2.1	0.1
FPBA	3.0	2.1	0.1

RSD, relative standard deviation.

pyrethroids and blood plasma fats are extracted together. Then, if this sample is injected into the GC, the blood plasma fat residues cannot be vaporized and build active residues in the injector, which leads to memory effects and adsorption of pyrethroids. These problems can be solved with the described SPE.

3. In contrast to analytical methods covering only pyrethroid metabolites in urine, this method has the advantage that it is possible to determine metabolites of synthetic pyrethroids (*cis/trans*-DCCA, *cis*-DBCA, 4-F-3-PBA, 3-PBA) as well as the metabolite of pyrethrins/bioallethrin (*cis/trans*-CDCA) simultaneously. Another advantage is lower cost because of shorter analysis time.

4. Several analytical techniques were tested previously, but the method chosen is the most adequate. The esterification with methanol gave poor sensitivity for *cis*- and *trans*-CDCA in the lower background level. Furthermore, it is impossible to separate the *cis*- and *trans*-CDCA–methylester on a DB5 capillary column. Esterification with ethanol enables a separation between *cis*- and *trans*-CDCA–ethylester but also shows poor sensitivity in the lower background level. Optimal sensitivity and selectivity are only found with HFIP as an excellent esterification reagent for all tested substances within a quantification level of 0.1 µg/L.

5. The spiked ISTD fenvalerat cannot compensate all different properties of the pyrethroids during the analysis and GC–MS determination. In this method, fenvalerat is used as the ISTD because this pyrethroid is not often used in Germany. Otherwise, it might be useful to work with deuterated or ^{13}C-labeled internal pyrethroid standards if they are available.

6. The washing step with n-hexane eliminates the fat residues, and the elution with hexane/dichloromethane (1:1 v/v) is done to get nearly matrix-free extracts.

7. The extract of the Oasis cartridges must be evaporated very carefully to dryness because pyrethroids with lower boiling points evaporate with nitrogen if the nitrogen steam is too high or the evaporating process is not stopped immediately after sample drying. If no fine adjustable nitrogen evaporator is available, this drying step can also be done in a vacuum centrifuge. In this case, 100 µL toluene should be added to the extract before the solvent evaporation process is started. Toluene works as a keeper and minimizes loss of pyrethroids during this step. After sample evaporation to approx 25 µL, this residue can be used for GC–MS analysis.

8. A high temperature of the GC injector is used for optimal evaporation of pyrethroids and reduction of possible memory effects based on injector temperature distribution, which can occur by condensation at cold places in the injector. Therefore, it is advantageous to

use deactivated double-gooseneck injector liners. Matrix residues of samples in injector liners or at the first centimeters of a capillary column can result in lower detection limits, especially for deltamethrin. Deltamethrin is critical to analyze because it shows the lowest response of all pyrethroids based on its unfavorable fragmentation pattern. A possible contamination source is the autosampler syringe itself because residues of pyrethroids are not washed out quantitatively when nonpolar cleaning solvent is used. The best cleaning procedure is a two-step washing process with different polar solvents (e.g., toluene and dichloromethane).

9. Ammonia is described in this method as an NCI reactant gas, but analogous results can be obtained with methane. Notice that not all GC–MS instruments and pumps are equipped for ammonia. If no high-resolution GC–MS system is available, the analyses of pyrethroids can also be done on other GC–MS systems. The only disadvantage is a detection limit that is about a factor of 10 higher.

10. The derivatization with HFIP works only in water-free samples. Therefore, it is important to separate tert.-butyl methylether (t-BME) carefully from the lower water phase. HFIP is a very powerful reagent that reacts spontaneously with carboxylic acids; DIC is used as a catalyzor and water binder (**Fig. 3**).

11. This routine method was developed for analysis in a high-resolution GC–MS system in EI⁺ mode and optional CI⁻ mode. The installation of electronegative fluorine via derivatization with HFIP also enables a very sensitive determination in CI⁻ mode (**Fig. 6**). By CI⁻ mode detection, limits in the lower nanogram-per-liter range are possible *(13)*. The advantage of high mass resolution (10,000) in both detection techniques enables the accuracy of analytical results. EI⁺ mode is used for routine analysis because it is more stable in comparison to CI⁻, which is used for verification. The stability or reproducibility in CI⁻ depends on higher influences of sample matrix to the fragmentation process, which is weaker in CI⁻ than in EI⁺ mode. This problem can be solved with deuterated or ¹³C-labeled ITSDs. A determination with low mass resolution mass spectrometers was not tested, but is known to work. Possible matrix interferences that occur on these instruments can be solved with longer columns or hydrogen as carrier gas (use caution). Of course, this leads to longer analysis time combined with higher costs.

References

1. Aldridge, W. N. (1990) An assessment of the toxicological properties of pyrethroids and their neurotoxicity. *Crit. Rev. Toxicol.* **21,** 89–104.
2. Eadsforth, C. V., Bragt, P. C., and van Sittert, N. J. (1988) Human dose-excretion studies with pyrethroid insecticides cypermethrin and alphacypermethrin: relevance for biological monitoring. *Xenobiotica* **18,** 603–614.
3. Woollen, B. H., Marsh, J. R., Laird, W. J. D., and Lesser, J. E. (1992) The metabolism of cypermethrin in man: differences in urinary metabolite profiles following oral and dermal administration. *Xenobiotica* **22,** 983–991.
4. Leng, G., Leng, A., Kühn, K.-H., Lewalter, J., and Pauluhn, J. (1997) Human dose excretion studies with the pyrethroid insecticide cyfluthrin: urinary metabolite profile following inhalation. *Xenobiotica* **27,** 1272–1283.
5. Class, T. J., Ando, T., and Casida, J. E. (1990) Pyrethroid metabolism: microsomal oxidase metabolites of (S)-bioallethrin and the six natural pyrethrins. *J. Agric. Food Chem.* **38,** 529–537.
6. Leng, G., Kuehn, K.-H., Wieseler, B., and Idel, H. (1999) Metabolism of (S)-bioallethrin and related compounds in humans. *Toxicol. Lett.* **107,** 109–121.

7. He, F., Sun, S., Han, K., et al. (1988) Effects of pyrethroid insecticides on subjects engaged in packaging pyrethroids. *Br. J. Industr. Med.* **45,** 548–551.

8. He, F., Wang, S., Liu, L., Chen, S., Zhang, Z., and Sun, J. (1989) Clinical manifestations of acute pyrethroid poisoning. *Arch. Toxicol.* **63,** 54–58.

9. Leng, G. and Lewalter, J. (1999) Role of individual susceptibility in risk assessment of pesticides. *Occup. Environ. Med.* **56,** 449–453.

10. Leng, G., Kuehn, K.-H., and Idel, H. (1996) Biological monitoring of pyrethroid metabolites in urine of pest control operators. *Toxicol. Lett.* **88,** 215–220.

11. Leng, G., Kühn, K.-H., and Idel, H. (1997) Biological monitoring of pyrethroids in blood plasma and pyrethroid metabolites in urine: applications and limitations. *Sci. Tot. Environ.* **199,** 173–181.

12. Leng, G., Ranft, U., Sugiri, D., Hadnagy, W., Berger-Preiss, E., and Idel, H. (2003) Pyrethroids used indoor—biological monitoring of exposure to pyrethroids following an indoor pest control operation. *Int. J. Hyg. Env. Health* **206,** 85–92.

13. Leng, G., Kuehn, K.-H., Leng, A., Gries, W., Lewalter, J., and Idel, H. (1997) Determination of trace levels of pyrethroid metabolites in human urine by capillary gas chromatography–high resolution mass spectrometry with negative chemical ionization. *Chromatographia* **46,** 265–274.

14. Kühn, K.-H., Leng, G., Bucholski, K. A., Dunemann, L., and Idel, H. (1996) Determination of pyrethroid metabolites in human urine by capillary gas chromatography–mass spectrometry. *Chromatographia* **43,** 285–292.

15. Berger-Preiß, E., Levsen, K., Leng, G., Idel, H., Sugiri, D., and Ranft, U. (2002) Indoor pyrethroid exposure in homes with woollen textile floor coverings. *Intl. J. Hyg. Env. Med.* **205,** 459–472.

16. Heudorf, U. and Angerer, J. (2001) Metabolites of pyrethroid insecticides in urine specimens: current exposure in an urban population in Germany. *Environ. Health Persp.* **109,** 213–217.

3

A Multianalyte Method for the Quantification of Current-Use Pesticides in Human Serum or Plasma Using Isotope Dilution Gas Chromatography–High-Resolution Mass Spectrometry

Dana B. Barr, Roberto Bravo, John R. Barr, and Larry L. Needham

Summary

We propose a sensitive and accurate analytical method for quantifying 29 current-use pesticides in human serum or plasma. These pesticides include organophosphates, carbamates, chloroacetanilides, and synthetic pyrethroids, pesticides used in both agricultural and residential settings. Our method employs simple solid-phase extraction followed by highly selective analysis using isotope dilution gas chromatography–high-resolution mass spectrometry. The method is very accurate, has limits of detection (LODs) in the low picogram/gram range, and has coefficients of variation that are typically less than 20% at the low picogram/gram end of the method linear range.

Key Words: Contemporary; gas chromatography; human; isotope dilution; mass spectrometry; pesticides; plasma; serum.

1. Introduction

Exposure assessment is an integral component of risk assessment of pesticides, but often reliable exposure assessment information lacks quantity or quality. Because human exposure to pesticides is multimedia and multiroute and varies with the use of pesticides, environmental monitoring of exposure must account for the concentration of the pesticide in all media, the time in contact with each medium, and route(s) of exposure to calculate aggregate exposure information to a given pesticide accurately *(1)*. Even when all of this information is considered, measurements of the external dose may not accurately reflect the absorbed, or internal, dose.

Because of their inherent chemical nature, current-use pesticides have short biological half-lives, usually a few hours or days *(2–7)*. Therefore, current-use pesticides are taken up rapidly into the bloodstream, metabolized, and eliminated from the body in the urine as polar metabolites, which are often conjugated as glucuronides or sulfates.

From: *Methods in Biotechnology, Vol. 19, Pesticide Protocols*
Edited by: J. L. Martínez Vidal and A. Garrido Frenich © Humana Press Inc., Totowa, NJ

Because the metabolites of pesticides are usually excreted in urine soon after exposure and because urine is usually a plentiful matrix and easy to obtain, biological monitoring of exposure to current-use pesticides has typically involved quantifying pesticide metabolites in urine *(3)*. In addition, concentrations of pesticides or their metabolites in urine are typically much higher than in blood and are detectable for a longer period of time (e.g., a few days as compared to several hours in blood). However, this approach is not without its limitations. Individual urinary metabolites can often be derived from multiple pesticides or even from exposure to the metabolite itself as an environmental degradate. Furthermore, urine volume and dilution vary among "spot" samples; thus, it may be difficult to compare concentrations among samples even when adjusting for the dilution. For most current-use pesticides, creatinine adjustment is the most common method for dilution adjustment; however, creatinine concentrations vary with demographics, so this approach loses its appeal when applied to a diverse population. Another limitation, from a practical standpoint, in measuring urinary metabolites is that authentic standards are often not commercially available because many metabolites are newly identified.

Measuring pesticides in blood has several advantages over measuring pesticide metabolites in urine. Because the parent chemical is measured, detailed metabolism information (i.e., which metabolites are formed) is not required. Also, the measurement of the intact pesticide in blood instead of a metabolite provides more specificity for the exposure. For example, 3,5,6-trichloropyridinol (3,5,6-TCPy) in urine would reflect exposure to chlorpyrifos, chlorpyrifos-methyl, or environmental degradates (e.g., TCPy itself or pesticide oxons). We cannot distinguish between these exposures with urinary TCPy. However, if these chemicals are measured in blood, we can differentiate the exposures. For example, a detectable level of chlorpyrifos in blood would unequivocally identify an exposure to chlorpyrifos. Distinguishing between exposure to each pesticide and exposure to their respective degradation products is very important in risk assessment because the toxicities, and hence the acceptable daily intake, for chlorpyrifos, chlorpyrifos-methyl, the oxons, and 3,5,6-TCPy all differ.

Because blood is a regulated fluid (i.e., the volume does not vary substantially with water intake or other factors), the blood concentrations of toxicants measured at a specified time interval after exposure will be the same as long as the absorbed amounts are constant; thus, no corrections for dilution are necessary. Furthermore, blood measurements are more likely than urine measurements to reflect the dose available for the target site *(8)* because the measured dose has not yet been eliminated from the body.

The major disadvantages of blood measurements are the venipuncture required to obtain the sample and the low toxicant concentrations. Blood collection is more costly, and the invasive sampling often limits study participation, especially for children. In addition, when samples can be obtained, the amount of blood available to perform the analysis is often limited; therefore, highly sensitive analytical techniques may be required. Analysis of blood is further complicated by the inherently low toxicant concentrations, often three orders of magnitude lower than urinary metabolite levels.

For most researchers, the disadvantages of blood measurements have far outweighed the advantages. In fact, most of the scientific literature detailing biological monitoring of current-use pesticides describes urinary measurements *(9)*. However, several methods involving blood, serum, or plasma measurements of a variety of contemporary pesticides have been published *(10–35)*. The pesticides measured using these methods include primarily organophosphate and carbamate insecticides. The majority of these methods were developed for forensic applications or for diagnosis of acute pesticide intoxication and have LOD ranges in the high nanograms per milliliter to micrograms per milliliter. In all cases, these methods lack the sensitivity or the selectivity to measure pesticides in blood or blood products resulting from low-level exposures.

We have developed a sensitive and accurate method for quantifying 29 contemporary pesticides in human serum or plasma (**Table 1**). Our method uses a simple solid-phase extraction (SPE) followed by a highly selective analysis using isotope dilution gas chromatography–high-resolution mass spectrometry (GC–HRMS).

2. Materials

1. Standards of the pesticides and labeled pesticides (**Table 2**) with purity higher than 99%.
2. Ammonium sulfate.
3. Anhydrous sodium sulfate.
4. Oasis® HLB™ SPE columns (Waters, Milford, MA).
5. Disposable empty sorbent cartridges.
6. SPE extraction manifold.
7. Pesticide quality solvents: dichloromethane, methanol, and toluene.
8. Bioanalytical grade I water.
9. SuporCap-100 filtration capsule.
10. Volumetric flasks.
11. Automatic pipets with disposable tips.
12. 15-mL glass centrifuge tubes.
13. 1-mL conical autosampler vials (silanized) with crimp tops.
14. Qorpak glass bottles with Teflon caps (30 mL).
15. TurboVap LV.
16. MAT900 magnetic sector mass spectrometer (ThermoFinnigan, Bremen, Germany).
17. GC (e.g., Agilent 6980, Palo Alto, CA) equipped with autosampler (e.g., CTC A200S) and operated using XCalibur® software
18. DB5 MS column ([14% cyanopropylphenyl]-methyl polysiloxane, 30 m, 0.25-mm id, 25-μm film) (e.g., J & W Folsom, CA).
19. Helium (>99.999%).

3. Methods *(36)*

3.1. Standards

3.1.1. Native Standards

1. Prepare individual stock solutions by dissolving 3 mg of each standard into 15 mL toluene and mixing well to obtain a concentration of 200 ng/μL.
2. Divide into 1-mL aliquots, flame seal in ampules, and store at −20°C until used.

3.1.2. Internal Standards

1. Prepare stock internal standard solutions by dissolving 5 mg of each stable isotope labeled standard into 50 mL toluene and mixing well (*see* **Note 1**).
2. An internal standard spiking solution is prepared by diluting 1 mL of each stock solution

Table 1
Pesticides Included in the Multianalyte Method

Analyte	Parent pesticide	Class	Use
Acetochlor	Acetochlor	Chloroacetanilide	Herbicide
Alachlor	Alachlor	Chloroacetanilide	Herbicide
Atrazine	Atrazine	Triazine	Herbicide
Bendiocarb	Bendiocarb	Carbamate	Insecticide
Carbofuran	Carbofuran	Carbamate	Insecticide, nematocide
Carbofuranphenol	Carbofuran, carbosulfan	Carbamate	Insecticide, nematocide
Chlorothalonil	Chlorothalonil	Miscellaneous	Fungicide
Chlorpyrifos	Chlorpyrifos	Organophosphate	Insecticide
Chlorthal-dimethyl	Chlorthal-dimethyl	Chlororterephthalate	Herbicide
Diazinon	Diazinon	Organophosphate	Insecticide, acaricide
Dichlorvos	Dichlorvos	Organophosphate	Insecticide, acaricide
Dicloran	Dicloran	Chloronitroaniline	Fungicide
Diethyltoluamide (DEET)	Diethyltoluamide (DEET)	Toluamide	Repellant
Fonophos	Fonophos	Organophosphate	Insecticide
2-Isopropoxyphenol	Propoxur	Carbamate	Insecticide
Malathion	Malathion	Organophosphate	Insecticide, acaricide
Metalaxyl	Metalaxyl	Phenylamide	Fungicide
Methyl parathion	Methyl parathion	Organophosphate	Insecticide, acaricide
Metolachlor	Metolachlor	Cloroacetanilide	Herbicide
1-Naphthol	Carbaryl, naphthalene	Carbamate, polycyclic aromatic hydrocarbon	Insecticide, plant growth regulator
Parathion	Parathion	Organophosphate	Insecticide, acaricide
cis-Permethrin	cis-Permethrin	Synthetic pyrethroid	Insecticide
trans-Permethrin	trans-Permethrin	Synthetic pyrethroid	Insecticide
Phorate	Phorate	Organophosphate	Insecticide, acaricide, nematocide
Phthalimide	Folpet	N-Trihalomethylthio	Fungicide
Propoxur	Propoxur	Carbamate	Insecticide
Terbufos	Terbufos	Organophosphate	Insecticide, nematocide
Tetrahydrophthalimide	Captan, captafol	N-Trihalomethylthio	Fungicide
Trifluralin	Trifluralin	Dinitroaniline	Herbicide

Table 2
Pesticides and Their Labeled Standards

Pesticide	Labeled pesticide
2-Isopropoxyphenol (IPP)	2-Isopropoxyphenol (IPP)
Dichlorvos (DCV)	Dimethyl-D_6-DCV
Carbofuranphenol (CFP)	Ring-$^{13}C_6$-CFP
Phthalimide (PI)	Ring/carboxyl-$^{13}C_4$-PI
Tetrahydrophthalimide (THPI)	Ring-D_6-THPI
Diethyltoluamide (DEET)	Dimethyl-D_6-DEET
1-Naphthol (1N)	Ring-$^{13}C_6$-1N
Trifluralin (TFL)	Dipropyl-D_9-TFL
Propoxur (PPX)	
Phorate (PHT)	Diethoxy-$^{13}C_4$-PHT
Bendiocarb (BCB)	
Terbufos (TBF)	Diethoxy-$^{13}C_4$-TBF
Diazinon (DZN)	Diethyl-D_{10}-DZN
Fonophos (FFS)	Ring-$^{13}C_6$-FFS
Carbofuran (CF)	Ring-$^{13}C_6$-CF
Atrazine (ATZ)	Ethylamine-D_5-ATZ
Dicloran (DCN)	Ring-$^{13}C_6$-DCN
Acetochlor (ACC)	Ring-$^{13}C_6$-ACC
Alachlor (ALC)	Ring-$^{13}C_6$-ALC
Chlorothalonil (CTNL)	
Metalaxyl (MXL)	Propionyl-D_4-MXL
Chlorpyrifos (CPF)	Diethyl-D_{10}-CPF
Methyl parathion (MP)	
Chlorthal-dimethyl (DCL)	Dimethyl-D_6-DCL
Metolachlor (MTCL)	Ring-$^{13}C_6$-MTCL
Malathion (MLTN)	D_{10}-MLTN
Parathion (PTN)	Diethyl-D_{10}-PTN
cis-Permethrin (CPM)	Phenoxy-$^{13}C_6$-CPM
trans-Permethrin (TPM)	Phenoxy-$^{13}C_6$-TPM

(including the ones purchased in methanol and nonane) with acetonitrile in a 100-mL volumetric flask to obtain a concentration of 10 pg/µL.

3. Divide into 1-mL aliquots, flame seal in ampules, and store at −20°C until used.

3.1.3. Calibration Standards

1. Create 10 working standard sets (0.25, 0.5, 2, 5, 10, 20, 50, 100, 200, and 400 pg/µL) from the native standards to encompass the entire linear range of the method. The internal standard concentration should be kept constant at 100 pg/µL.
2. Divide into 1-mL aliquots, flame seal in ampules, and store at −20°C until used.

3.2. Quality Control Materials

1. Pool serum samples from multiple donors (*see* **Note 2**) and mix well.
2. Pressure filter to 0.2 µm to remove large particles.
3. Split serum into three pools of equal volume.

4. Spike the first pool to a concentration of 15 pg/g.
5. Spike the second pool to a concentration of 50 pg/g.
6. Leave the last pool unspiked.
7. Mix for 24 h under refrigeration.
8. Divide into 4-mL aliquots. Cap, label, and store the vials at −20°C until used.
9. Determine the mean concentration and the analytic variance by the repeat measurement of at least 20 samples in different analytical runs.
10. Evaluate quality control (QC) acceptance based on the Westgard multirules *(37)*.

3.3. Laboratory Reagent Blanks

1. Pipet 4 mL bioanalytical grade I water into centrifuge tube (*see* **Note 3**).
2. Prepare as if an unknown sample as indicated vide infra **in Subheading 3.4.** (*see* **Note 4**).
3. Concentrations of the pesticides in the blank samples are required to be less than the LOD for the run to be considered acceptable.

3.4. Sample Preparation

1. Bring all samples, QCs, reagents, and standards to room temperature.
2. Weigh a 4-g aliquot of serum/plasma into a test tube.
3. Spike samples with 100 μL of the internal standard, mix, and allow to equilibrate for approx 5 min.
4. Denatured serum proteins were denatured with 4 mL saturated ammonium sulfate (*see* **Note 5**).
5. Centrifuge samples at 2140g for 5 min.
6. Meanwhile, precondition Oasis SPE columns with 2 mL methanol followed by 2 mL water.
7. Pass the supernatants from the serum samples through the SPE columns and discard.
8. Dry columns using 20 psi vacuum for 20 min.
9. Elute SPE cartridges with 4 mL dichloromethane.
10. Load empty disposable cartridges with 1 g anhydrous sodium sulfate.
11. Pass eluates through sodium sulfate cartridges and collect (*see* **Note 6**).
12. Concentrate extracts to about 500 μL using a TurboVap evaporator set at 37°C and 15 psi head pressure of nitrogen.
13. Transfer concentrated extracts to a 1-mL conical vial.
14. Add 10 μL toluene to each vial as a keeper agent.
15. Allow to evaporate to approx 10 μL at ambient temperature.
16. Cap vials and store under refrigeration until analyzed.

3.5. Instrumental Analysis

1. Perform analyses using a gas chromatograph (split/splitless injector) interfaced to a mass spectrometer equipped with an autosampler and operated using software.
2. Install the capillary column into the gas chromatograph.
3. Set helium carrier gas at a linear velocity of 35 cm/s.
4. Set the injector and transfer line temperatures at 240 and 270°C, respectively.
5. Establish a GC program as follows: 100°C initial column temperature, hold for 1 min, increase to 180°C at 15°C/min, hold for 2 min, increase to 221°C at 3°C/min, then finally increase to 280°C at 25°C/min and hold for 5 min.
6. Set up MS acquisition program as follows: selected ion monitoring (SIM) mode; 5000 K initial accelerating voltage; resolution 10,000 as defined at 10% valley; perfluorokerosene (PFK) ions used as lock and calibration masses. The monoisotopic masses for each ion

monitored for the pesticides and their respective internal standards, the ion types (i.e., fragment or molecular ion), ion composition, retention windows for analysis, and relative retention times are shown in **Table 3** (*see* **Notes 7** and **8**).

7. Inject 2 μL of each extract using a splitless injection.

3.6. Data Processing and Analysis

1. Set the detection and baseline thresholds to 40 and 4, respectively, and the minimum peak width to 1. In addition, set up processing to subtract the background signal and smooth all data (three-point smooth).
2. Process data automatically using XCalibur® software supplied with the mass spectrometer.
3. Double check peak selection and integration.
4. Download analysis data into a permanent storage database.

3.7. Method Validation

1. *LOD*. Calculate the analytical LOD for the method as $3s_0$, where s_0 can be estimated as the y-intercept of a linear regression analysis of a plot of the standard deviation vs the concentration *(30)* (**Table 4**).
2. *Extraction recovery*. Calculate extraction recovery by spiking six 4-mL blank serum samples (*see* **Subheading 3.2.**) with pesticide standards to a final concentration of 32 pg/ g. Prepare these samples simultaneously with six 4-mL unspiked samples (*see* **Subheading 3.4.**). After extraction, spike the extracts from the unspiked samples with the same amount of native pesticides as the samples spiked before extraction. Add internal standard (100 μL) to each extract. Analyze as indicated in **Subheading 3.5.** Calculate the recoveries as the ratios of spiked samples to the control samples.
3. *Relative recovery*. Determine the method of relative recovery by spiking blank serum samples (*see* **Subheading 3.2.**) with a known amount of the pesticides. Prepare and analyze samples according to the method (*see* **Subheadings 3.4.** and **3.5.**). Compare the calculated and the spike concentrations by linear regression analysis of a plot of the calculated concentrations vs the expected concentrations. With this analysis, a slope of 1.0 would be indicative of 100% relative recovery.

4. Notes

1. Exceptions to this stock preparation are carbofuran, alachlor, metolachlor, and chlorpyrifos, which are purchased as 100 μg/mL solutions in methanol or nonane.
2. We purchased expired serum from the Red Cross in Cincinnati, Ohio.
3. Because virtually all serum samples tested had detectable levels of at least one of the pesticides or metabolites of interest, water was used as a laboratory reagent blank. The blank contained the same water used in the daily preparation of reagents. They are used to ensure that contamination does not occur at any step in the preparation process.
4. Unknown serum or plasma samples, QC materials, and laboratory reagent blanks are prepared identically.
5. Acids cannot be used for denaturation because they degrade many of the compounds of interest.
6. To evaporate samples fully and extend life of the column, all residual water must be carefully removed. This step is essential.
7. If no labeled standard is available for a particular pesticide, the nearest labeled standard in the same retention time window can be used as an internal standard.
8. The total analysis time per sample is about 30 min.

Table 3
High-Resolution Mass Spectral Analysis Specifications

Analyte	Ion type	Monoisotopic mass	Ion composition	Retention window	Relative retention time
2-Isopropoxyphenol (IPP)	M	152.0837	$C_9H_{12}O_2$	1	1.00
Ring-$^{13}C_6$-2-IPPI	M + $^{13}C_6$	158.1039	$^{13}C_6{}^{12}C_3H_{12}O_2$	1	1.00
Dichlorvos (DCV)	F	184.9771	$C_4H_7ClO_4P$	1	1.20
Dimethyl-D_6-DCV	F + D_6	191.0147	$C_4D_6HClO_4P$	1	1.19
Carbofuranphenol (CFP)	M	164.0837	$C_{10}H_{12}O_2$	1	1.26
Ring-$^{13}C_6$-CFP	M + $^{13}C_6$	170.1039	$^{13}C_6{}^{12}C_4H_{12}O_2$	1	1.26
Phthalimide (PI)	M	147.0320	$C_8H_5NO_2$	1	1.90
Ring/carboxyl-$^{13}C_4$-PI	M + $^{13}C_4$	151.0454	$^{13}C_4{}^{12}C_4H_5NO_2$	1	1.90
Tetrahydrophthalimide (THPI)	M	151.0633	$8H_9NO_2$	1	2.01
Ring-D_6-THPI	M + D_6	157.1010	$C_8D_6H_3NO_2$	1	1.99
DEET	M	190.1232	$C_{12}H_{16}NO$	1	2.03
Dimethyl-D_6-DEET	M + D_6	196.1608	$C_{12}D_6H_{10}NO$	1	2.02
1-Naphthol (1N)	M	144.0575	$C_{10}H_7OH$	1	2.10
Ring-$^{13}C_6$-1N	M + $^{13}C_6$	150.0776	$^{13}C_6{}^{12}C_4H_7OH$	1	2.10
PFK	L	130.9920	n/a	1	n/a
PFK	C	180.9888	n/a	1	n/a
Trifluralin (TFL)	F	264.0232	$C_8H_5N_3O_4F_3$	2	2.27
Dipropyl-D_9-TFL	F + D_3	267.0420	$C_8D_3H_2N_3O_4F_3$	2	2.24
Propoxur (PPX)	F	152.0837	$C_9H_{12}O_2$	2	2.30
Phorate (PHT)	M	260.0128	$C_7H_{17}O_2PS_3$	2	2.32
Diethoxy-$^{13}C_4$-PHT	M + $^{13}C_4$	264.0262	$^{13}C_4{}^{12}C_3H_{17}O_2PS_3$	2	2.32
Bendiocarb (BCB)	F	166.0630	$C_9H_{10}O_3$	2	2.47
PFK	L	168.9888	n/a	2	n/a
PFK	C	268.9824	n/a	2	n/a
Terbufos (TBF)	M	288.0441	$C_9H_{21}O_2PS_3$	3	2.61
Diethoxy-$^{13}C_4$-TBF	M + $^{13}C_4$	292.0576	$^{13}C_4{}^{12}C_5H_{21}O_2PS_3$	3	12.61

Compound	Code	Mass	Formula		RT
Diazinon (DZN)	M	304.1011	$C_{12}H_{21}N_2O_3PS$ 3	2	2.65
Diethyl-D$_{10}$-DZN	M + D$_{10}$	314.1638	$C_{12}D_{10}H_{11}N_2O_3PS$	3	2.62
Fonophos (FFS)	M	246.0302	$C_{10}H_{15}OPS_2$	3	2.71
Ring-$^{13}C_6$-FFS	M + $^{13}C_6$	252.0503	$^{13}C_6{}^{12}C_4H_{15}OPS_2$ 3	2	2.71
PFK	L	230.9856	n/a	3	n/a
PFK	C	292.9824	n/a	3	n/a
Carbofuran (CF)	F	164.0837	$C_{10}H_{12}O_2$	4	2.84
Ring-$^{13}C_6$-CF	F + $^{13}C_6$	170.1039	$^{13}C_6{}^{12}C_4H_{12}O_2$	4	2.84
Atrazine (ATZ)	F	200.0703	$C_7H_{11}ClN_5$	4	2.84
Ethylamine-D$_5$-ATZ	F + D$_5$	205.1017	$C_7D_5H_6ClN_5$	4	2.83
Dicloran (DCN)	M + 2	207.9620	$C_6H_4N_2O_2{}^{35}Cl^{37}Cl$	4	2.89
Ring-$^{13}C_6$-DCN	M + $^{13}C_6$ + 2	213.9822	$^{13}C_6H_4N_2O_2{}^{35}Cl^{37}Cl$	4	2.89
Acetochlor (ACC)	F	223.0764	$C_{12}H_{14}NO_2Cl$	4	3.30
Ring-$^{13}C_6$-ACC	F + $^{13}C_6$	229.0965	$^{13}C_6{}^{12}C_6H_{14}NO_2Cl$	4	3.30
Alachlor (ALC)	F	188.1075	$C_{12}H_{14}NO$	4	3.41
Ring-$^{13}C_6$-ALC	F + $^{13}C_6$	194.1227	$^{13}C_6{}^{12}C_6H_{14}NO$ 4	4	3.41
Chlorothalonil (CTNL)	M + 2	265.8786	$C_8{}^{35}Cl^{37}Cl\ N_2$	4	3.46
PFK	L	168.9888	n/a	4	n/a
PFK	C	230.9856	n/a	4	n/a
Metalaxyl (MXL)	F	206.1181	$C_{12}H_{16}O_2N$	5	3.59
Propionyl-D$_4$-MXL	F + D$_4$	210.1432	$C_{12}D_4H_{12}O_2N$	5	3.58
Chlorpyrifos (CPF)	F	313.9574	$C_9H_{11}Cl_2NO_3PS$ 5	3	3.62
Diethyl-D$_{10}$-CPF	F + D$_{10}$	324.0202	$C_9D_{10}HCl_2NO_3PS$	5	3.58
Methyl parathion (MP)	M	263.0017	$C_8H_{10}NO_3PS$	5	3.66
Chlorthal-dimethyl (DCL)	F + 2	300.8807	$C_9H_3O_3{}^{35}Cl_3{}^{37}Cl$ 5	5	3.72

(continued)

Barr et al.

Table 3
High-Resolution Mass Spectral Analysis Specifications *(continued)*

Analyte	Ion type	Monoisotopic mass	Ion composition	Retention window	Relative retention time
Dimethyl-D_6-DCL	$F + D_3 + 2$	303.8995	$C_9D_3O_3{}^{35}Cl_3{}^{37}Cl$	5	3.70
Metolachlor (MTCL)	F	238.0999	$C_{13}H_{17}ClNO$	5	3.77
Ring-$^{13}C_6$-MTCL	$F + {}^{13}C_6$	244.1200	$^{13}C_6{}^{12}C_7H_{17}ClNO$	5	3.77
Malathion (MLTN)	F	255.9993	$C_7H_{13}O_4PS_2$	5	3.85
D_{10}-MLTN	$F + D_5$	261.0307	$C_7D_5H_8O_4PS_2$	5	3.81
Parathion (PTN)	M	291.0330	$C_{10}H_{14}NO_5PS$	5	4.08
Diethyl-D_{10}-PTN	$M + D_{10}$	301.0958	$C_{10}D_{10}H_4NO_5PS$	5	4.04
PFK	L	230.9856	n/a	5	n/a
PFK	C	292.9824	n/a	5	n/a
cis-Permethrin (CPM)	F	183.0810	$C_{13}H_{11}O$	6	5.63
Phenoxy-$^{13}C_6$-CPM	$F + {}^{13}C_6$	189.1011	$^{13}C_6{}^{12}C_7H_{11}O$	6	5.63
trans-Permethrin (TPM)	F	183.0810	$C_{13}H_{11}O$	6	5.70
Phenoxy-$^{13}C_6$-TPM	$F + {}^{13}C_6$	189.1011	$^{13}C_6{}^{12}C_7H_{11}O$	6	5.70
PFK	L	180.9888	n/a	6	n/a
PFK	C	192.9888	n/a	6	n/a

C, calibration mass; F, fragment ion; L, lock mass; M, molecular ion; n/a, not applicable; PFK, perfluorokerosine.

Table 4
Method Specifications

Analyte	LOD (pg/g)	Extraction Recovery % (N = 6)	Relative Recovery % (N = 20)	RSD (N = 6)
2-Isopropoxyphenol	3	48 ± 15	100 ± 3	17
Dichlorvos	1	15 ± 10	101 ± 4	13
Carbofuranphenol	1	80 ± 8	100 ± 3	8
Phthalimide	20	89 ± 6	98 ± 2	25
Tetrahydrophthalimide	1	91 ± 8	99 ± 5	14
Deet	10	43 ± 4	101 ± 2	10
1-Naphthol	20	12 ± 10	101 ± 4	24
Trifluralin	1	15 ± 8	98 ± 3	27
Propoxur	1	61 ± 12	99 ± 4	19
Phorate	1	21 ± 11	99 ± 4	13
Bendiocarb	5	46 ± 6	99 ± 8	20
Terbufos	1	17 ± 9	97 ± 6	17
Diazinon	0.5	27 ± 5	101 ± 5	19
Fonophos	1	20 ± 8	103 ± 6	14
Carbofuran	1	38 ± 10	98 ± 3	30
Atrazine	1	53 ± 12	101 ± 3	17
Dicloran	1	46 ± 23	100 ± 3	13
Acetochlor	1	23 ± 8	95 ± 4	13
Alachlor	1	21 ± 11	100 ± 4	14
Chlorothalonil	5	14 ± 12	101 ± 3	14
Metalaxyl	5	55 ± 9	100 ± 4	25
Chlorpyrifos	1	21 ± 14	96 ± 5	16
Methyl parathion	2	20 ± 16	100 ± 6	20
Chlorthal-dimethyl	1	18 ± 5	101 ± 3	14
Metolachlor	1	23 ± 9	101 ± 3	11
Malathion	12	22 ± 18	104 ± 8	20
Parathion	1	20 ± 18	101 ± 7	17
cis-Permethrin	1	13 ± 5	98 ± 7	31
trans-Permethrin	1	14 ± 5	100 ± 8	28

LOD, limit of detection; RSD, relative standard deviation determined from spiked serum samples.

References

1. Donaldson, D., Kiely, T., and Grube, A. (2002) *1998 And 1999 Market Estimates. Pesticides Industry Sales and Usage Report.* U.S. Environmental Protection Agency, Washington, DC.
2. Garfitt, S. J., Jones, K., Mason, H. J., and Cocker, J. (2002) Exposure to the organophosphate diazinon: data from a human volunteer study with oral and dermal doses. *Toxicol. Lett.* **134,** 105–113.
3. Barr, D. B., Barr, J. R., Driskell, W. J., et al. (1999) Strategies for biological monitoring of exposure for contemporary-use pesticides. *Toxicol. Ind. Health* **15,** 168–179.

4. Griffin, P., Mason, H., Heywood, K., and Cocker, J. (1999) Oral and dermal absorption of chlorpyrifos: a human volunteer study. *Occup. Environ. Med.* **56,** 10–13.

5. Leng, G., Kuhn, K. H., and Idel, H. (1997) Biological monitoring of pyrethroids in blood and pyrethroid metabolites in urine: applications and limitations. *Sci. Total Environ.* **199,** 173–181.

6. Leng, G., Leng, A., Kuhn, K. H., Lewalter, J., and Pauluhn, J. (1997) Human dose–excretion studies with the pyrethroid insecticide cyfluthrin: urinary metabolite profile following inhalation. *Xenobiotica* **27,** 1273–1283.

7. Nolan, R. J., Rick, D. L., Freshour, N. L., and Saunders, J. H. (1984) Chlorpyrifos: pharmacokinetics in human volunteers. *Toxicol. Appl. Pharmacol.* **73,** 8–15.

8. Needham, L. L., Ashley, D. L., and Patterson, D. G., Jr. (1995) Case studies of the use of biomarkers to assess exposures. *Toxicol. Lett.* **82–83,** 373–378.

9. Barr, D. and Needham, L. (2002) Analytical methods for biological monitoring of exposure to pesticides: a review. *J. Chromatogr. B Anal. Technol. Biomed. Life Sci.* **778,** 5.

10. Stein, V. B. and Pittman, K. A. (1976) Gas–liquid chromatographic determination of azinphos ethyl in human plasma and in mouse plasma, tissue, and fat. *J. Assoc. Off. Anal. Chem.* **59,** 1094–1096.

11. Fournier, E., Sonnier, M., and Dally, S. (1978) Detection and assay of organophosphate pesticides in human blood by gas chromatography. *Clin. Toxicol.* **12,** 457–462.

12. Nordgren, I., Bengtsson, E., Holmstedt, B., and Pettersson, B. M. (1981) Levels of metrifonate and dichlorvos in plasma and erythrocytes during treatment of schistosomiasis with bilarcil. *Acta Pharmacol. Toxicol. (Copenh.)* **49**(Suppl. 5), 79–86.

13. Michalke, P. (1984) Determination of *p*-nitrophenol in serum and urine by enzymatic and non-enzymatic conjugate hydrolysis and HPLC. Application after parathion intoxication. *Z. Rechtsmed.* **92,** 95–100.

14. Saito, I., Hisanaga, N., Takeuchi, Y., et al. (1984) Assessment of the exposure of pest control operators to organophosphorus pesticides. Organophosphorus pesticides in blood and alkyl phosphate metabolites in urine. *Sangyo Igaku* **26,** 15–21.

15. Ikebuchi, J., Yuasa, I., and Kotoku, S. (1988) A rapid and sensitive method for the determination of paraquat in plasma and urine by thin-layer chromatography with flame ionization detection. *J. Anal. Toxicol.* **12,** 80–83.

16. Gellhaus, H., Hausmann, E., and Wellhoner, H. H. (1989) Fast determination of demeton-S-methylsulfoxide (metasystox R) in blood plasma. *J. Anal. Toxicol.* **13,** 330–332.

17. Flanagan, R. J. and Ruprah, M. (1989) HPLC measurement of chlorophenoxy herbicides, bromoxynil, and ioxynil, in biological specimens to aid diagnosis of acute poisoning. *Clin. Chem.* **35,** 1342–1347.

18. Liu, J., Suzuki, O., Kumazawa, T., and Seno, H. (1989) Rapid isolation with Sep-Pak C18 cartridges and wide-bore capillary gas chromatography of organophosphate pesticides. *Forensic Sci. Int.* **41,** 67–72.

19. Sharma, V. K., Jadhav, R. K., Rao, G. J., Saraf, A. K., and Chandra, H. (1990) High performance liquid chromatographic method for the analysis of organophosphorus and carbamate pesticides. *Forensic Sci. Int.* **48,** 21–25.

20. Junting, L. and Chuichang, F. (1991) Solid phase extraction method for rapid isolation and clean-up of some synthetic pyrethroid insecticides from human urine and plasma. *Forensic Sci. Int.* **51,** 89–93.

21. Tomita, M., Okuyama, T., Watanabe, S., Uno, B., and Kawai, S. (1991) High-performance liquid chromatographic determination of glyphosate and (aminomethyl)phosphonic acid in human serum after conversion into *p*-toluenesulphonyl derivatives. *J. Chromatogr.* **566,** 239–243.

22. Kawasaki, S., Ueda, H., Itoh, H., and Tadano, J. (1992) Screening of organophosphorus pesticides using liquid chromatography–atmospheric pressure chemical ionization mass spectrometry. *J. Chromatogr.* **595,** 193–202.

23. Unni, L. K., Hannant, M. E., and Becker, R. E. (1992) High-performance liquid chromatographic method using ultraviolet detection for measuring metrifonate and dichlorvos levels in human plasma. *J. Chromatogr.* **573,** 99–103.

24. Croes, K., Martens, F., and Desmet, K. (1993) Quantitation of paraquat in serum by HPLC. *J. Anal. Toxicol.* **17,** 310–312.

25. Smith, N. B., Mathialagan, S., and Brooks, K. E. (1993) Simple sensitive solid-phase extraction of paraquat from plasma using cyanopropyl columns. *J. Anal. Toxicol.* **17,** 143–145.

26. Keller, T., Skopp, G., Wu, M., and Aderjan, R. (1994) Fatal overdose of 2,4-dichlorophenoxyacetic acid (2,4-D). *Forensic Sci. Int.* **65,** 13–18.

27. Wintersteiger, R., Ofner, B., Juan, H., and Windisch, M. (1994) Determination of traces of pyrethrins and piperonyl butoxide in biological material by high-performance liquid chromatography. *J. Chromatogr. A* **660,** 205–210.

28. Lee, X. P., Kumazawa, T., and Sato, K. (1995) Rapid extraction and capillary gas chromatography for diazine herbicides in human body fluids. *Forensic Sci. Int.* **72,** 199–207.

29. Qiu, H. and Jun, H. W. (1996) Solid-phase extraction and liquid chromatographic quantitation of insect repellent *N,N*-diethyl-*m*-toluamide in plasma. *J. Pharm. Biomed. Anal.* **15,** 241–250.

30. Cho, Y., Matsuoka, N., and Kamiya, A. (1997) Determination of organophosphorous pesticides in biological samples of acute poisoning by HPLC with diode-array detector. *Chem. Pharm. Bull.(Tokyo)* **45,** 737–740.

31. Futagami, K., Narazaki, C., Kataoka, Y., Shuto, H., and Oishi, R. (1997) Application of high-performance thin-layer chromatography for the detection of organophosphorus insecticides in human serum after acute poisoning. *J. Chromatogr. B Biomed. Sci. Appl.* **704,** 369–373.

32. Lee, H. S., Kim, K., Kim, J. H., Do, K. S., and Lee, S. K. (1998) On-line sample preparation of paraquat in human serum samples using high-performance liquid chromatography with column switching. *J. Chromatogr. B Biomed. Sci. Appl.* **716,** 371–374.

33. Sancho, J. V., Pozo, O. J., and Hernandez, F. (2000) Direct determination of chlorpyrifos and its main metabolite 3,5, 6-trichloro-2-pyridinol in human serum and urine by coupled-column liquid chromatography/electrospray–tandem mass spectrometry. *Rapid Commun. Mass Spectrom.* **14,** 1485–1490.

34. Frenzel, T., Sochor, H., Speer, K., and Uihlein, M. (2000) Rapid multimethod for verification and determination of toxic pesticides in whole blood by means of capillary GC–MS. *J. Anal. Toxicol.* **24,** 365–371.

35. Kawasaki, S., Nagumo, F., Ueda, H., Tajima, Y., Sano, M., and Tadano, J. (1993) Simple, rapid and simultaneous measurement of eight different types of carbamate pesticides in serum using liquid chromatography–atmospheric pressure chemical ionization mass spectrometry. *J. Chromatogr.* **620,** 61–71.

36. Barr, D., Barr, J., Maggio, V., et al. (2002) A multi-analyte method for the quantification of contemporary pesticides in human serum and plasma using high-resolution mass spectrometry. *J. Chromatogr. B Analyt. Technol. Biomed. Life Sci.* **778,** 99.

37. Westgard, J. O. (2002) *Basic QC practices: training in statistical quality control for health care laboratories.* Westgard QC, Madison, WI.

4

Application of Solid-Phase Disk Extraction Combined With Gas Chromatographic Techniques for Determination of Organochlorine Pesticides in Human Body Fluids

Adrian Covaci

Summary

A simple, rapid, sensitive procedure based on solid-phase disk extraction (SPDE) is described for the isolation and concentration of trace levels of selected organochlorine pesticides from human body fluids (serum, cord blood, milk, follicular and seminal fluid). Similar methodology can be used for each matrix; the only restricting factor is the viscosity of the fluid. After denaturing proteins with formic acid, an Empore™ C18-bonded silica extraction disk cartridge is used for the extraction of the analytes. Subsequent cleanup and lipid removal from the SPDE eluate is achieved by adsorption chromatography on acidified silica or Florisil, depending on the interest in acid-labile pesticides. By using the SPDE procedure, high-throughput parallel-sample processing can be achieved. Instrumental analysis is done by gas chromatography–mass spectrometry in electron-capture negative ionization mode (GC–MS/ECNI). Recoveries for selected organochlorine pesticides range from 65 to 91% (SD < 10%) for serum and from 70 to 102% (SD < 14%) for milk. Detection limits between 10 and 100 pg/mL fluid can be obtained. The method was validated through successful participation in several interlaboratory tests and through the routine analysis of human serum with various loadings of organochlorine pesticides.

Key Words: Cord blood; follicular and seminal fluid; gas chromatography; milk; organochlorine pesticides; serum; solid-phase disk extraction.

1. Introduction

Organochlorine pesticides (OCPs) are prevalent environmental contaminants with high lipophilicity and long half-lives (1). Owing to their persistence, high potential for accumulation in food chains and potential health effects (immunotoxicity, reproductive effects, endocrine disruption, and neurotoxicity), the monitoring of OCPs in humans is of high general concern (2,3). Monitoring of human exposure to OCPs is most conveniently performed by analysis of blood plasma, blood serum, or milk. The

From: *Methods in Biotechnology, Vol. 19, Pesticide Protocols*
Edited by: J. L. Martínez Vidal and A. Garrido Frenich © Humana Press Inc., Totowa, NJ

requirements for risk assessment in epidemiological studies have created the need for efficient, fast, and less-costly analytical methods.

Because of trace levels found in biological fluids and the presence of other extraneous chemicals at higher concentrations, a highly sensitive and selective multistage analytical procedure is needed. The determination of OCPs by gas chromatography (GC) usually requires preliminary purification of the extracts before instrumental analysis. Conventional methods of separating OCPs from human body fluids involve liquid–liquid extraction with nonpolar solvents *(4,5)*. They are very complex, labor intensive, and time consuming and use excessive amounts of solvents and reagents. Solid-phase extraction (SPE), using commercially available columns prepacked with various stationary phases, has been investigated as an alternative method for extraction and cleanup *(6–10)*.

The use of solid-phase disk extraction (SPDE) technology has been reported for the first time for the analysis of OCPs and polychlorinated biphenyls (PCBs) in human serum *(11–13)*. The procedure involves denaturation of serum proteins with formic acid, SPE using C_{18} Empore™ disk cartridges, followed by elimination of lipids using cleanup on acidified silica or Florisil. The use of SPDE improves the assay throughput and seems promising for its simplicity, reliability, low solvent consumption, minimal cross-contamination from high-level samples, parallel sample processing, and time reduction. The above-described method has been successfully applied to several monitoring studies *(14–18)*.

This chapter provides a reliable, simple, rapid, and sensitive methodology for the routine analysis of OCPs in various human body fluids, such as serum, plasma, cord blood serum, milk, and seminal and follicular fluids.

2. Materials

1. Analytical standards hexachlorobenzene (HCB); α-, β-, and γ-hexachlorocyclohexane (HCH) isomers; *o,p'*-DDE, *p,p'*-DDE, *o,p'*-DDD, *p,p'*-DDD, *o,p'*-DDT, *p,p'*-DDT; *trans*-chlordane; *cis*-chlordane; *trans*-nonachlor; oxychlordane; dieldrin; heptachlorepoxide; heptachlor; and mirex (e.g., from Dr. Ehrenstorfer, Augsburg, Germany) at a concentration of 10 ng/μL in iso-octane or cyclohexane. A stock solution (500 pg/μL) containing all analytes is prepared, and further dilutions to 200, 50, 10, and 1 pg/μL are made with iso-octane in volumetric flasks.
2. Internal standards (ε-HCH and PCB 143) and syringe standard (1,2,3,4-tetrachloronaphthalene [TCN]; e.g., from Dr. Ehrenstorfer) at a concentration of 10 ng/μL in iso-octane or cyclohexane. Dilutions to a concentration of 100 pg/μL for internal standard and 500 pg/μL for syringe standard are made with iso-octane in volumetric flasks.
3. Methanol, acetonitrile, hexane, dichloromethane, acetone, and iso-octane (pesticide grade; e.g. Merck, Darmstadt, Germany).
4. Formic acid 99% p.a.
5. Concentrated sulfuric acid 95 to 97% p.a. (e.g., Merck).
6. Anhydrous sodium sulfate for residue analysis, Florisil (0.15 to 0.25 mm), and silica gel 60 to 200 mesh (e.g., Merck) are washed with hexane and heated overnight at 120°C.

7. Human serum for method validation is provided by the Blood Transfusion Centre, University Hospital of Antwerp (Belgium). Blood is collected in a vacuum system tube and centrifuged (15 min, 2000*g*) within 24 h after collection. Human milk and follicular and seminal fluid samples are obtained from the Fertility Unit of the University Hospital of Antwerp, Belgium. All samples are kept frozen at –20°C until analyzed.

8. All glassware is washed with detergent, rinsed with water, soaked for 24 h in sulfochromic acid, and rinsed with distilled water, acetone, and hexane. Prior to use, the treated glassware is rinsed with the solvent with which it subsequently will contact.

9. C_{18} Empore™ disk extraction cartridges, 10 mm/6 mL (e.g., 3M Company).

10. Positive pressure manifold (part 1223-420X; e.g., 3M Company).

11. Empty polypropylene columns (3 mL) for cleanup (e.g., Supelco, Bellefonte, PA).

12. A GC (e.g., Hewlett Packard 6890, Palo Alto, CA) connected with a mass spectrometer (MS) (e.g., Hewlett Packard 5973) operated in electron-capture negative ionization (ECNI) mode.

13. A 25 m × 0.22 mm id × 0.25 mm film thickness, 8% phenyl polycarborane siloxane (HT-8) capillary column.

3. Methods

Compared to classical liquid–liquid extraction methods, the use of Empore disk technology (90% sorbent, 10% matrix–polytetrafluoroethylene [PTFE]) allows reduction of elution solvent because of small bed volume (**Fig. 1**). The C_{18} disk cartridge employed for sample cleanup and analyte enrichment has a nonpolar character, causing retention of nonpolar compounds. It has also a size exclusion function to eliminate macromolecular interference (such as serum proteins) in biological extracts.

A main disadvantage of the SPDE method is that lipid determination cannot be done on the same sample aliquot because the procedure does not allow the collection of the lipidic fraction. However, enzymatic methods are an elegant way to measure the lipid content on a very small (<150 µL) separate serum aliquot.

The method described below outlines the (1) sample preparation and loading onto the SPDE cartridge, (2) elution of analytes from the SPDE cartridge and cleanup of the eluate, (3) the GC analysis, (4) method validation, and (5) required quality control criteria.

3.1. Sample Preparation and Loading

3.1.1. Incubation With Internal Standards and Protein Disruption

To avoid column overloading and breakthrough of the analytes, the following sample volumes are recommended: 1 to 5 mL plasma, serum, or cord blood serum; 1 to 3 mL milk; 1 to 2 mL seminal fluid; and 3 to 5 mL follicular fluid.

1. Spike the appropriate amount of each type of sample with internal standards ε-HCH and PCB 143 (generally, between 5 and 10 ng/sample) (*see* **Note 1**).

2. Repeat the above procedure for procedural blanks using Milli Q water instead of samples.

3. Equilibrate the mixture in an ultrasonic bath for 30 min.

4. Incubate overnight at +4°C.

5. Prior to extraction (the next day), mix the samples with formic acid (1:1 v/v) (*see* **Note 2**).

6. Equilibrate the mixture by ultrasonic treatment for 30 min. When milk is used, it should be diluted with Milli Q water (1:1 v/v) because of higher viscosity and the possibility of clogging the SPDE cartridge.

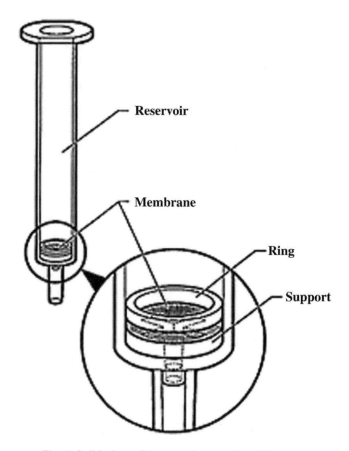

Fig. 1. Solid-phase disk extraction cartridge (3M Empore).

3.1.2. Conditioning the SPDE Cartridge

Prior to the sample application, the disk cartridges have to be rinsed with organic solvent and then conditioned with methanol and water.

1. Wash the SPDE cartridge with two 500-µL portions of dichloromethane and dry it thoroughly.
2. Add to each cartridge 250 µL methanol.
3. Add two portions of 250 µL Milli Q water. Attention must be paid that, after conditioning, the cartridges should not be allowed to dry.

3.1.3. Sample Application and Drying Step

1. Load the sample to the SPDE cartridge and apply a positive-pressure nitrogen stream (*see* **Note 3**).
2. Rinse each cartridge with two 500-µL portions of Milli Q water.
3. Dry the sorbent bed thoroughly under a nitrogen stream at 20 psi positive pressure (10 min) and then by centrifugation (15 min, 2000*g*) (*see* **Note 4**).

When applying positive pressures, low flow rates are mandatory to increase the extraction yields, allowing longer contact time between the analytes and the sorbent. However, compared to serum, the amount of milk that can be loaded on the SPDE cartridge will be less because of the presence of more lipids with similar polarity as the organochlorines and therefore higher competition for the binding sites. Furthermore, milk samples have to be diluted with Milli Q water (1:1 v/v), and higher pressures are needed for the sample loading.

3.2. Elution and Cleanup

Compared to classical SPE, lower solvent volumes are needed for the elution of analytes from the SPDE cartridge (up to 1.5 mL per cartridge). The choice of cleanup procedure is dependent on the analytes of interest. For example, if OCPs are measured together with other potentially present persistent contaminants such as PCBs, destructive cleanup using acidified silica is preferred. The resulting extracts are cleaner than those resulting from nondestructive cleanup, but some of the acid-labile OCPs, such as dieldrin, endosulfans, heptachlor, and heptachlorepoxide are completely destroyed *(19)*. Alternatively, nondestructive cleanup using deactivated silica, alumina, or Florisil may be used, but the resulting extracts may still contain traces of lipids and might hinder the GC analysis *(20)*.

3.2.1. Cleanup on Acidified Silica

The use of cleanup on acidified silica results in a lower background, which facilitates peak identification of OCPs present at very low concentrations. This cleanup method ensures better instrumental performance and longer column lifetime.

1. Fill an empty column with 500 mg acid silica (*see* **Note 5**) and 100 mg Na_2SO_4.
2. Wash the filled cleanup column with 2 mL dichloromethane:hexane (1:1 v/v), followed by 2 mL of hexane; after solvent elution, place the column under the SPDE cartridge.
3. Prepare the column setup and place the collection tube under the silica column (**Fig. 2**).
4. Elute the SPDE cartridge with two 500-µL portions of hexane and 500 µL dichloromethane:hexane (1:1 v/v).
5. Remove the SPDE cartridge.
6. Elute OCPs from the silica column with 3 mL hexane and 3 mL dichloromethane:hexane (1:1).
7. Add 50 µL iso-octane as a keeper solvent.
8. Concentrate the eluate under a gentle nitrogen stream at room temperature to near dryness.
9. Add 100 µL iso-octane and 25 µL recovery standard TCN (500 pg/µL), vortex thoroughly for 15 s, and transfer to a GC vial.

3.2.2. Cleanup on Florisil

The use of Florisil cleanup allows determination of acid-labile analytes, but results in a higher background in the chromatograms. Minor modifications, such as addition of acetonitrile in the elution step, may also increase the recovery of more polar OCPs.

1. Fill an empty column with 1 g activated Florisil and 100 mg Na_2SO_4.
2. Wash the filled cleanup column with 2 mL dichloromethane:hexane (1:1 v/v), followed by 2 mL of hexane; after solvent elution, place it under the SPDE cartridge.

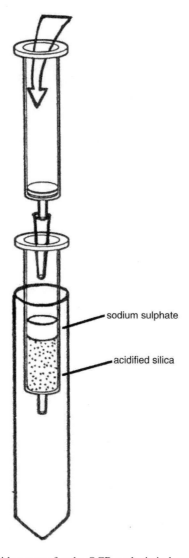

sodium sulphate

acidified silica

Fig. 2. Cartridge setup for the OCP analysis in human body fluids.

3. Prepare the column setup and place the collection tube under the Florisil column (**Fig. 2**).
4. Elute the SPDE cartridge with two 500-µL portions of hexane and 500 µL dichloromethane:hexane (1:1 v/v).
5. Remove the SPDE cartridge.
6. Elute OCPs from the Florisil column with 3 mL hexane, 3 mL dichloromethane:hexane (1:1), and 500 µL acetonitrile.
7. Add 50 µL of iso-octane as a keeper solvent.
8. Concentrate the eluate under a gentle nitrogen stream at room temperature to near dryness.

Table 1
Ion Fragments Monitored for Each Compound or Group of Compounds

Group of compounds	Quantification ion	Qualification ion
TCN (syringe standard)	266	264
PCB 143	360	362
o,p'- and p,p'-DDE	318	316
o,p'- and p,p'-DDD	71	248
o,p'- and p,p'-DDT	71	248
trans- and *cis-*Chlordane	408	410
*trans-*Nonachlor	442	444
Oxychlordane	350	424
Dieldrin	380	237
Mirex	368	370
Heptachlor	272	71
Heptachlorepoxide	388	71
ε-HCH	71	255
α-, β-, and γ-HCH	71	255
HCB	284	286

9. Add 100 µL iso-octane and 25 µL recovery standard TCN (500 pg/µL), vortex thoroughly for 15 s, and transfer to a GC vial.

3.3. GC–MS Analysis

The final cleaned extract (1 µL) is injected in pulsed splitless mode (280°C injector temperature, 30-psi pressure pulse, 1.5-min pulse time, 1.5-min splitless time) into the GC. Helium is used as carrier gas at a constant flow of 1.0 mL/min with an initial pressure of 14.40 psi. The temperature of the HT-8 column (*see* **Note 6**) is programmed from 90 (1.50 min) to 200°C (2.0 min) at a rate of 15°C/min, to 270°C (1.0 min) at a rate of 5°C/min, and finally to 290°C (10 min) at a rate of 25°C/min.

Low-resolution quadrupole MS is used in ECNI mode (methane moderating gas) to increase the sensitivity of OCP determination. The ion source, quadrupole, and interface temperatures are 150, 130, and 300°C, respectively. Two most abundant ions (if possible including the molecular ion) are monitored for OCP (**Table 1**). Dwell times are set at 50 ms.

3.4. Method Validation

3.4.1. Recoveries

To test the method performance, recoveries of internal standards and analytes should be determined from each matrix. However, because of small sample volumes of cord blood serum and follicular and seminal fluid, only recoveries of internal standards can be measured.

1. Five replicates at one spiking level and five nonspiked replicates from the same batch of pooled body fluid have to be analyzed.

Table 2
Recoveries of Internal Standards

Matrix	Mean percentage recoveries (SD)	
	ε-HCH	PCB 143
Maternal serum	68 (8)	65 (9)
Cord serum	62 (7)	61 (5)
Human milk	73 (5)	75 (7)
Follicular fluid	53 (17)	48 (12)
Seminal fluid	67 (11)	68 (14)

2. Calculate absolute recoveries of OCPs after subtracting the levels found in the nonspiked replicates from the spiked ones.

Recoveries of internal standards (ε-HCH and PCB 143) calculated from each fluid range from 48% (follicular fluid) to 76% (human milk), and relatively good reproducibility (SD < 12% for serum, cord blood, and milk and < 17% for follicular and seminal fluids) may be achieved (**Table 2**). The lower recoveries for follicular fluid can be explained by the higher speed through the extraction cartridge (thus a lower contact time) because of lower viscosity.

Recoveries of selected OCPs in serum range from 66 to 88% (SD < 9%) when the acidified silica cleanup is used. The Florisil cleanup results in recoveries between 65 and 91% (SD < 10%). For human milk, analyte recoveries range from 70 to 102% (SD < 14%) for all analytes at a fortification level similar to normal values in human milk.

3.4.2. Detection Limits

The limit of detection for each analyte is calculated as the sum between the mean of the detected signal at the analyte retention time in procedural blanks (five replicates) and 3 × SD of the signal. Typical detection limits range between 10 and 100 pg/mL for the investigated human body fluids.

3.4.3. Linearity

Linearity should be tested with a lack-of-fit test for intervals that include the normal range of pollutants in human body fluids (0.02 to 10 ng/mL in human serum, cord blood serum, and follicular and seminal fluids and 0.02 to 50 ng/mL in human milk). The correlation coefficient r^2 should not be used to assess linearity unless already verified by a lack-of-fit test *(21)*.

3.5. Quality Control

Because of extremely low levels of analytes that need to be measured in small sample volumes and the high probability of errors at these concentrations, adequate quality control needs to be applied before data reporting.

3.5.1. Internal Quality Control

Several procedures are usually applied to ensure adequate quality:

1. Review manually all peaks for proper integration.
2. Identify OCPs based on their relative retention times to the internal/recovery standards, based on the selected ion chromatograms and based on the ratios between abundance of the quantification ion and the qualifier ion. A deviation of ion ratios of less than ±15% from the theoretical value is considered acceptable.
3. For calibration, plot peak area ratios (OCPs area/internal standard area) against the amount ratios (OCPs amount/internal standard amount).
4. Run calibration curves with each sample batch and check that the correlation coefficients are kept above 0.99.
5. With each sample batch, analyze procedural blanks and inject standard solutions that include all analytes to assess variations in chromatographic and instrument performance and to check for interference.
6. Measure recoveries of internal standards in each sample to monitor their variation.
7. Analyze an in-house control serum (laboratory reference material) with every sample batch. The control serum is obtained from human serum (about 50 donations) from the Blood Transfusion Center, University Hospital of Antwerp, Belgium. The serum (approx 100 mL) is pooled and homogenized. Aliquots of 7 mL are divided into hexane-washed glass vials and kept at –20°C. Monitor the variation in concentrations of the measurable analytes (such as *p,p'*-DDE and β-HCH) and set confidence and control intervals as mean ± 2SD and mean ± 3 SD, respectively *(22)*. Plot a control chart for the variation in time of concentrations of relevant analytes measured in the in-house control serum.
8. Certified reference materials such as SRM 1588a (PCBs and OCPs in human serum, National Institute for Standards and Technology, United States) should be used at regular intervals.

3.5.2. External Quality Control

External quality control is usually done through adherence to internationally recognized interlaboratory tests. Several such studies for human serum/plasma are available throughout the year, with a larger participation at the Arctic Monitoring and Assessment Programme (AMAP) ring test, organized by the Toxicological Centre of Québec (Canada) or at the Quality Assurance Assay in Occupational and Environmental Medicine run by the University of Erlangen (Germany). Several OCPs, such as HCB, β-HCH isomers, *p,p'*-DDE, *p,p'*-DDT, and oxychlordane are requested to be measured in naturally contaminated or spiked serum samples at environmental levels.

4. Notes

1. The selection of internal standards is based on the GC elution characteristics and on their absence in the samples. PCB 143 has been shown to be a suitable internal standard for most OCPs (except HCHs), and it can also be used for the determination of PCBs, normally measured in the same extracts. Use of other internal standards, such as more expensive ^{13}C-labeled OCPs, is encouraged, but in this case the extracts cannot be analyzed by electron capture detector, which may provide better sensitivity for low traces of analytes. All samples need to be spiked with the internal standards dissolved in ethanol or acetone instead of hexane or iso-octane because the presence of nonpolar solvent was found to reduce recoveries of all compounds (especially the most lipophilic compounds).
2. Formic acid is chosen as a deproteination agent because it yields the highest analyte recoveries on the SPDE cartridge *(11)* and because it does not degrade acid-sensitive

analytes *(19)*. The serum pretreatment by denaturation without precipitation is found to release the analytes from the binding sites without the possible loss by occlusion in the precipitate *(11)*. The fact that no precipitate is formed during the denaturation makes this method compatible with the SPDE cartridge. Reduced pH (adjusted by the addition of formic acid) increases the extraction efficiency of the C_{18} sorbent. The denaturation of milk proteins and disruption of fat globules is also done by addition of formic acid and ultrasonication. Sodium oxalate or methanol (or short-chain alcohol for reduced viscosity) *(6)* may also be used for this purpose. However, the denatured samples, when not processed immediately (<1 h), sometimes form a gel-like precipitate that would prevent immediate extraction, as also reported elsewhere *(10)*. When samples are left capped and in a refrigerator for 2 to 3 h, the gel would dissipate, allowing proper SPDE column flow and analyte extraction.

3. The use of slower flow rates (2 to 4 psi) allows maximal residence time of the solvents in the sorbent bed and yields slightly improved recoveries than higher flow rates (10 to 15 psi).

4. The drying step is essential because the nonpolar eluents (hexane and dichloromethane) need to interact with all areas of the sorbent and should not be hindered by residual water trapped in the pores. Centrifugation of the cartridges before elution of OCPs is necessary because of the high compactness of the adsorbent bed, making complete drying difficult.

5. The acid silica is prepared as follows: To 50 g silica gel, 27 mL concentrated sulfuric acid (95 to 97%) were added dropwise while the mixture was stirred with a magnetic stir bar to ensure good homogeneity. After the addition of acid, the acid silica is stirred for another 30 min.

6. Although other types of stationary phase may as well be used for the determination of OCPs, the selection of the HT-8 column is driven by its excellent separation characteristics for PCBs *(23)*, which are often measured together with OCPs in routine analysis.

References

1. Simonich, S. L., and Hites, R. A. (1995) Global distribution of persistent organochlorine compounds. *Science* **269**, 1851–1854.
2. Colosio, C., Tiramani, M., and Maroni, M. (2003) Neurobehavioral effects of pesticides: state of the art. *Neurotoxicology* **24**, 577–591.
3. Fleeger, J. W., Carman, K. R., and Nisbet, R. M. (2003) Indirect effects of contaminants in aquatic ecosystems. *Sci. Total Environ.* **317**, 207–233.
4. Greizerstein, H. B., Gigliotti, P., Vena, J., Freudenheim, J., and Kostyniak, P.J. (1997) Standardization of a method for the routine analysis of PCB congeners and selected pesticides in human serum and milk. *J. Anal. Toxicol.* **21**, 558–566.
5. Najam, A. R., Korver, M. P., Williams, C. C., Burse, V. W., and Needham, L. L. (1999) Analysis of a mixture of PCBs and chlorinated pesticides in human serum by column fractionation and dual-column capillary gas chromatography with ECD. *J. AOAC Int.* **82**, 177–185.
6. Mañes, J., Font, G., and Pico, Y. (1993) Evaluation of a SPE system for determining pesticide residues in milk. *J. Chromatogr.* **642**, 195–204.
7. Dmitrovic, J., Chan, S. C., and Chan, S. H. Y. (2002) Analysis of pesticides and PCB congeners in serum by GC/MS and SPE cleanup. *Toxicol. Lett.* **134**, 253–258.
8. Barr, J. R., Maggio, V. L., Barr, D. B., et al. (2003) New high-resolution mass spectrometric approach for the measurement of polychlorinated biphenyls and organochlorine pesticides in humans serum. *J. Chromatogr. B* **794**, 137–148.

9. Pitarch, E., Serrano, R., Lopez, F. J., and Hernandez, F. (2003) Rapid multiresidue determination of organochlorine and organophosphorus compounds in human serum by solid-phase extraction and gas chromatography coupled to tandem mass spectrometry. *Anal. Bioanal. Chem.* **376**, 189–197.

10. Sandau, C. D., Sjödin, A., Davis, M. D., et al. (2003) Comprehensive solid-phase extraction method for persistent organic pollutants. Validation and application to the analysis of persistent chlorinated pesticides. *Anal. Chem.* **75**, 71–77.

11. Pauwels, A., Wells, D. A., Covaci, A., and Schepens, P. (1998) Improved sample preparation method for selected persistent organochlorine pollutants in human serum using solid-phase disk extraction with gas chromatographic analysis. *J. Chromatogr. B* **723**, 117–125.

12. Covaci, A. and Schepens, P. (2001) Improved determination of selected POPs in human serum by solid phase disk extraction and GC-MS. *Chemosphere* **43**, 439–447.

13. Covaci, A. and Schepens, P. (2001) Solid phase disk extraction method for the determination of POPs form human body fluids. *Anal. Lett.* **34**, 1449–1460.

14. Pauwels, A., Covaci, A., Delbeke, L., Punjabi, U., and Schepens, P. (1999) The relation between levels of selected PCB congeners in human serum and follicular fluid. *Chemosphere* **39**, 2433–2441.

15. Covaci, A., Hura, C., and Schepens, P. (2001) Solid phase disk extraction: an improved method for determination of organochlorine residues in milk. *Chromatographia* **54**, 247–252.

16. Covaci, A., Jorens, P., Jacquemyn, Y., and Schepens, P. (2002) Distribution of PCBs and organochlorine pesticides in umbilical cord and maternal serum. *Sci. Total Environ.* 298, 45–53.

17. Koppen, G., Covaci, A., Van Cleuvenbergen, R., et al. (2002) Persistent organochlorine pollutants in human serum of 50–65 years old women in the Flanders Environmental and Health Study (FLEHS). Part 1: Concentrations and regional differences. *Chemosphere* **48**, 811–825.

18. Covaci, A., Koppen, G., Van Cleuvenbergen, R., et al. (2002) Persistent organochlorine pollutants in human serum of 50–65 years old women in the Flanders Environmental and Health Study (FLEHS). Part 2: Correlations among PCBs, PCDD/PCDFs and the use of predictive markers. *Chemosphere* **48**, 827–832.

19. Manirakiza, P., Covaci, A., and Schepens, P. (2002) Improved analytical procedure for determination of chlorinated pesticide residues in human serum using solid phase disc extraction (SPDE), single-step clean-up and gas chromatography. *Chromatographia* **55**, 353–359.

20. Bernal, J. L., Del Nozal, M. J., and Jiménez, J. J. (1992) Some observations on clean-up procedures using sulphuric acid and Florisil. *J. Chromatogr.* **607**, 303–309.

21. Massart, D. L., Vandeginste, B. G. M., Buydens, L. M. C., De Jong, S., Lewi, P. J., and Smeyers-Verbeke, J. (1997) *Handbook of chemometrics and qualimetrics: Part A.* Elsevier, Amsterdam.

22. Luotamo, M. and Aitio, A. (1997) Quality assurance of isomer-specific analysis of PCBs in serum. *Chemosphere* **34**, 965–973.

23. Frame, G. M. (1997) A collaborative study of 209 PCB congeners and 6 Aroclors on 20 different HRGC columns. *Fresenius J. Anal. Chem.* **357**, 701–722.

5

A Comprehensive Approach for Biological Monitoring of Pesticides in Urine Using HPLC–MS/MS and GC–MS/MS

Dana B. Barr, Anders O. Olsson, Roberto Bravo, and Larry L. Needham

Summary

Many epidemiological studies have been conducted to determine if any relation exists between pesticide exposure and disease. Biological monitoring is a useful tool for establishing the presence and magnitude of exposures, which are essential parts of the exposure → disease continuum. In the past, we had almost limitless urine for biological measurements, but this has changed dramatically as the study populations continue to focus on young children, for whom urine collection is difficult, and as the number of pesticides for which exposure information is needed has increased. To accommodate the biological monitoring component of these studies, we refined our methods to allow maximum exposure information from a limited-volume urine sample. Using three separate analytical methods, each requiring only 2 mL of urine, we can successfully measure 35 different pesticides or metabolites at background levels with a high degree of selectivity and precision. We describe a comprehensive approach to biological monitoring of current-use pesticides in urine using high-performance liquid chromatography–tandem mass spectrometry (HPLC–MS/MS) and gas chromatography–MS/MS (GC–MS/MS) with quantification using isotope dilution.

Key Words: Biological monitoring; GC–MS/MS; HPLC–MS/MS; isotope dilution; pesticides; urine.

1. Introduction

In 1999, an estimated 415,000 tons of conventional pesticides were applied in the United States (1). The widespread use of pesticides and their potential relation to adverse health outcomes have increased both public and scientific interest in pesticide exposures. Consequently, many epidemiological studies are conducted to determine if any relation exists between pesticide exposure and disease.

To adequately link exposure to disease, the exposure must be accurately assessed. Biological monitoring is a useful tool for establishing the presence and magnitude of

From: *Methods in Biotechnology, Vol. 19, Pesticide Protocols*
Edited by: J. L. Martínez Vidal and A. Garrido Frenich © Humana Press Inc., Totowa, NJ

exposures *(2)*. For chemicals that do not bioaccumulate in humans, urine is the primary matrix used for biological monitoring. In the past, there has been almost-limitless urine for biological measurements, but this has changed dramatically as the study populations continue to focus on young children, for whom urine collection is more difficult, and as the number of pesticides for which exposure information is needed has increased. To accommodate the biological monitoring component of these studies, we sought to refine our methods to allow maximum exposure information from a limited-volume urine sample.

Many of the methods available in the literature focus on specific pesticide classes or individual pesticides or metabolites *(3–11)*. Few methods are available for the analysis of markers of several classes of pesticides in the same sample in the low nanogram-per-milliliter range, which is generally necessary for studies of individuals not occupationally exposed . However, in epidemiological studies, it is often important to measure biomarkers of many different pesticides to obtain an accurate representation of an individual's exposure to make adequate statistical interpretations of health outcome data. We have accomplished this in our laboratory by analyzing small aliquots of the same urine sample using a number of different methods *(8,11,12)*. This has allowed us to obtain biologically based data on 35 different chemicals using approx 6 mL of urine, an amount that can easily be obtained from adults and children. The target analytes for our methods represent pesticides from several different pesticide classes: organophosphorus insecticides, synthetic pyrethroid insecticides, triazine herbicides, chloroacetanilide herbicides, phenoxyacetic acid herbicides, carbamates, fungicides, and fumigants as well as the topical insect repellant *N,N*-diethyl-*m*-toluamide (**Table 1, Fig. 1**).

We describe a comprehensive approach to biological monitoring of current-use pesticides in urine using high-performance liquid chromatography–tandem mass spectrometry (HPLC–MS/MS) and gas chromatography (GC)–MS/MS with quantification using isotope dilution. Using three separate analytical methods, each requiring only 2 mL of urine, we can successfully measure 35 different pesticides or metabolites at background levels with a high degree of selectivity and precision. The methods we have developed are diverse in character, but all use selective detection techniques and isotope dilution quantification. Our methods are precise, reliable, and robust with low limits of detection (LODs).

2. Materials

1. Pesticide quality solvents acetonitrile, methanol, diethyl ether, hexane, and toluene.
2. Pesticides and their labeled standards with purity higher than 99% (*see* **Table 2**).
3. *n*-Butyl chloride (BuCl).
4. Glacial acetic acid, sulfuric acid, and hydrochloric acid (HCl).
5. Sodium acetate, sodium hydroxide, and sodium sulfate.
6. 0.2*M* acetate buffer.
7. Tetrabutylammonium hydrogen sulfate (TBAHSO$_4$).
8. β-Glucorinidase/sulfatase from *Helix pomatia* (G 0751, EC 3.2.1.31, type H-1).
9. Deionized water.
10. Nitrogen (>99.999% purity).
11. Zero air.

12. Argon.
13. Methane.
14. SuporCap-100 filtration capsule.
15. Volumetric flasks.
16. Automatic pipets with disposable tips.
17. 15-mL glass centrifuge tubes.
18. Qorpak glass bottles (30 mL) with Teflon caps.
19. Oven or dry bath set at 37°C.
20. Dry bath set at 60°C.
21. Oasis HLB 3-cc solid-phase extraction (SPE) cartridges.
22. Vacuum manifold for SPE.
23. Commercial lyophilizer or freeze-dryer.
24. Chem Elut (Varian, Sunnyvale, CA) sorbent-immobilized liquid extraction cartridges.
25. TurboVap LV.
26. High-performance liquid chromatograph (e.g., Agilent 1100, Palo Alto, CA) *(2)*.
27. Triple quadrupole MS (e.g., Sciex API4000, Applied Biosystems, Foster, City, CA).
28. Triple quadrupole MS (e.g., TSQ 7000 ThermoQuest, San Jose, CA) *(3)*.
29. Trace gas chromatograph with split/splitless injector *(2)*.
30. CTC A200S autosampler *(2)*.
31. Betasil C18 HPLC column (5-µm particle size, 100-A pore size, 1 mm id × 100 mm long).
32. Betasil phenyl HPLC column (5-µm particle size, 100-A pore size, 4.6 mm id × 100 mm long).
33. DB5-MS GC column (0.25-µm film, 0.25-mm id).
34. Deactivated fused silica GC column.

3. Methods

3.1. Quality Control Materials

1. For quality control (QC) materials, collect approx 20 L urine from multiple anonymous donors.
2. Combine urine collected and mix.
3. Dilute urine with water (1:1 v/v) to reduce endogenous levels of the analytes of interest.
4. Mix overnight at 20°C.
5. Pressure filter urine through a 0.45-µm SuporCap™ -100 capsule (e.g., Pall Corp., Ann Arbor, MI).
6. Divide urine into three equal-volume pools.
7. Spike first pool (low concentration; QCL) with approx 30–60 µg of each target analyte (**Table 1**). Mix overnight at 20°C to homogenize (*see* **Note 1**).
8. Spike second pool with 100–120 µg of each target analyte (**Table 1**). Mix overnight at 20°C to homogenize (*see* **Note 2**).
9. Do not spike last pool (*see* **Note 3**).
10. Aliquot urine pools into 30-mL Qorpak glass bottles with Teflon caps. Label and freeze at –70°C until needed.

3.2. Standard Preparation

3.2.1. Multiclass Method

3.2.1.1. LABELED STANDARDS

1. Weigh 0.5 mg of each labeled standard (**Table 3**) into a 25-mL volumetric flask.
2. Dilute to mark with acetonitrile.
3. Dilute mixture 1:20 with acetonitrile (*see* **Note 4**).

Table 1
The Target Analytes, Abbreviations, Their Metabolic Status, and Parent Pesticide and Class

Analyte name	Abbreviation	Type of marker	Indicator of exposure to (pesticide class[a])
Atrazine mercapturate	ATZ	Conjugate	Atrazine (TAH)
Acetochlor mercapturate	ACE	Conjugate	Acetochlor (CAH)
Metolachlor mercapturate	MET	Conjugate	Metolachlor (CAH)
Alachlor mercapturate	ALA	Conjugate	Alachlor (CAH)
2,4,5-Trichlorophenoxyacetic acid	2,4,5-T	Parent	2,4,5-T (PH)
2,4-Dichlorophenoxyacetic acid	2,4-D	Parent	2,4-D (PH)
4-Nitro-phenol	PNP	Metabolite	Parathion[a] (OPI)
5-Chloro-1-isopropyl-[3H]-1,2,4-triazol-3-one	CIT	Metabolite	Isazophos[a] (OPI)
3-Chloro-4-methyl-7-hydroxycoumarin	CMHC	Metabolite	Coumaphos[a] (OPI)
2-Diethylamino-6-methyl pyrimidin-4-ol	DEAMPY	Metabolite	Pirimiphos[a] (OPI)
2-[(Dimethoxyphosphorothioyl)sulfanyl]succinic acid	MDA	Metabolite	Malathion[a] (OPI)
3,5,6-Trichloro-2-pyridinol	TCPY	Metabolite	Chlorpyriphos[a] (OPI)
2-Isopropyl-6-methyl-4-pyrimidiol	IMPY	Metabolite	Diazinon[a] (OPI)
Dimethylphosphate	DMP organophosphorus pesticides (OPI)	Metabolite	O,O-Dimethyl-substituted
Dimethylthiophosphate	DMTP organophosphorus pesticides (OPI)	Metabolite	O,O-Dimethyl-substituted thio
Dimethyldithiophosphate	DMDTP organophosphorus pesticides (OPI)	Metabolite	O,O-Dimethyl-substituted dithio
Diethylphosphate	DEP organophosphorus pesticides (OPI)	Metabolite	O,O-Diethyl-substituted
Diethylthiophosphate	DETP organophosphorus pesticides (OPI)	Metabolite	O,O-Diethyl-substituted thio

(Continued)

Diethyldithiophosphate	DEDTP	Metabolite	O,O-Diethyl-substituted dithio organophosphorus pesticides (OPI)
4-Fluoro-3-phenoxy benzoic acid	4-F-3-PBA	Metabolite	Cyfluthrin (PI)
3-Phenoxy benzoic acid	3-PBA	Metabolite	10 of 18 commercially available pyrethroids in the United States (PI)
cis-3-(2,2-Dibromovinyl)-2,2-dimethylcyclopropane-1-carboxylic acid	DBCA	Metabolite	deltamethrin (PI)
cis and trans-3-(2,2-Dichlorovinyl)-2,2-dimethylcyclopropane-1-carboxylic acids	cis-/trans-DCCA	Metabolite	Cyfluthrin, permethrin, cypermethrin (PI)
N,N-Diethyl-m-toluamide	DEET	Parent	DEET (IR)
2-Isopropoxyphenol	IPP	Metabolite	Propoxur (CI)
2,3-Dihydro-2,2-dimethyl-7-hydroxybenzofuran (carbofuranphenol)	CFP	Metabolite	Carbofuran, benfuracarb, carbosulfan, furathiocarb (CI)
1-Naphthol	1N	Metabolite	Carbaryl (CI); naphthalene (FG, PAH)
2-Naphthol	2N	Metabolite	Naphthalene (FG, PAH)
ortho-Phenylphenol	OPP	Parent	OPP (F)
2,5-Dichlorophenol	25DCP	Metabolite	para-Dichlorobenzene (FG, CH)
2,4-Dichlorophenol	24DCP	Metabolite	meta-Dichlorobenzene (CH)
2,4,5-Trichlorophenol	245TCP	Metabolite/parent	245TCP, trichlorobenzene, pentachlorophenol, lindane (CH)
2,4,6-Trichlorophenol	246TCP	Metabolite/parent	246TCP, trichlorobenzene, pentachlorophenol, lindane (CH)
Pentachlorophenol	PCP	Parent	Pentachlorophenol (F)

CAH, chloroacetoanilides herbicides; CH, chlorinated hydrocarbon; CI, carbamate insecticide; F, fungicide; FG, fumigant; IR, insect repellant; OPI, organophosphate insecticides; PAH, polycyclic aromatic hydrocarbon; PH, phenoxyacetic acid herbicides; PI, pyrethroid insecticides; TAH, triazine herbicides.

[a]Or their methyl counterparts.

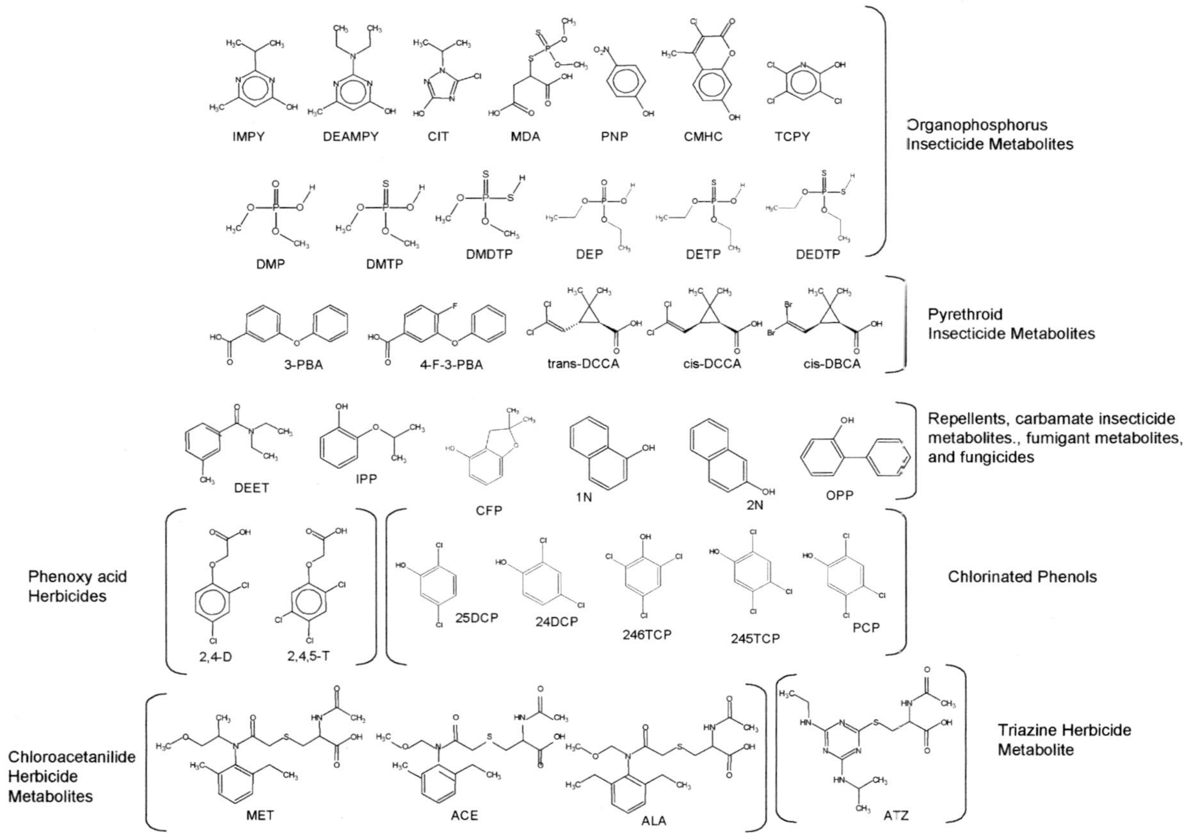

Fig. 1. Structures of target analytes (abbreviations according to **Table 1**).

66

Table 2
Optimized Precursor/Product Ion Pairs and the Collision Offset Energy (CE)
for the Dialkylphosphate Method Target Analytes on the TSQ 7000 Mass Spectrometer

Analyte	Precursor ion		Product ion		CE	Ion
(label pattern)	Native	Labeled	Native	Labeled	(V)	mode
DMP (d_6)	203	209	127	133	−12	+
DMTP (d_{10})	219	225	143	149	−13	+
DMDTP (d_6)	235	241	125	131	−10	+
DEP (d_6)	231	241	127	133	−13	+
DETP (d_{10})	247	257	191	193	−12	+
DEDTP ($^{13}C_4$)	263	267	153	157	−12	+

4. Aliquot into 1-mL ampules and flame seal.
5. Store at −20°C until used.

3.2.1.2. NATIVE STANDARDS

1. Weigh 5 mg of each target analyte (**Table 4**) into a 25-mL flask (*see* **Note 5**).
2. Dilute to marks with acetonitrile to create individual stock solutions.
3. Prepare eight calibration standard spiking solutions at the following concentrations: 0.020, 0.040, 0.080, 0.20, 0.40, 0.80, 2.0, and 4.0 µg/mL (*see* **Notes 6** and **7**).
4. Divide into 1-mL aliquots in glass ampules and flame seal.
5. Store at −20°C until used.

3.2.2. Dialkylphosphate Method

3.2.2.1. LABELED STANDARDS

1. Weigh 0.5 mg of each isotopically labeled analyte into a 100-mL volumetric flask.
2. Dilute to mark with acetonitrile (*see* **Notes 8** and **9**).
3. Divide solution into 1-mL aliquots in glass ampules and flame seal.
4. Store at −20°C until used.

3.2.2.2. NATIVE STANDARDS

1. Seven sets of calibration spiking standards in acetonitrile with analyte concentrations ranging from 0.0145–1.7 ng/µL are prepared under contract by Battelle Memorial Institute (Bel Air, MD) (*see* **Note 10**).

3.2.3. Phenols Method

3.2.3.1. LABELED STANDARDS

1. Prepare individual stock-labeled internal standard (ISTD) solutions by weighing approx 0.5 mg of each of the isotopically labeled analytes (**Table 4**) into a 2.5-mL volumetric flask (*see* **Note 11**).
2. Dilute to marks with acetonitrile to yield 200-µg/mL solutions.
3. Prepare a multiple-analyte standard solution by adding 250 µL of each individual stock solution into a 5-mL volumetric flask.
4. Dilute to mark with acetonitrile to obtain a final concentration of 10 ng/µL (*see* **Note 12**).
5. Divide all solutions into 1-mL aliquots in ampules and flame seal.
6. Store at −20°C until used.

Table 3
Pesticides and Their Labeled Standards

Pesticide	Labeled standard
Acetochlor mercapturate (*N*-acetyl-*S*-[2-(2-ethyl-6-methylphenyl)(ethoxymethyl)amino]-2-oxoethyl-L-cysteine) (ACE)	$^{13}C_6$ ACE
Alachlor mercapturate (*N*-acetyl-*S*-[2-(2,6-diethylphenyl)(methoxymethyl)amino]-2-oxoethyl-L-cysteine) (ALA)	$^{13}C_6$ ALA
Metolachlor mercapturate (*N*-acetyl-*S*-[2-(2-ethyl-6-methylphenyl)(2-methoxy-1-methylethyl)amino]-2-oxoethyl-L-cysteine) (MET)	$^{13}C_6$ MET
5-Chloro-1,2-dihydro-1-isopropyl-[3H]-1,2,4-triazol-3-one (CIT)	d_7 CIT
3-Chloro-4-methyl-7-hydroxycoumarin (CMHC)	
2-Diethylamino-6-methyl pyrimidin-4-ol (DEAMPY)	$^{13}C_4$ CMHC
2-Isopropyl-6-methyl-4-pyrimidiol (IMPY)	d_6 DEAMPY
3,5,6-Trichloro-2-pyridinol (TCPY)	$^{13}C_4$ IMPY
N,N-Diethyl-*m*-toluamide (DEET)	$^{13}C_5$ ^{15}N TCPY
2-[(Dimethoxyphosphorothioyl)sulfanyl]succinic acid) (MDA)	d_6 DEET
Atrazine mercapturate (*N*-acetyl-*S*-[4-(ethylamino)-6-[(1-methylethyl)amino]-1,3,5-triazin-2-yl]-L-cysteine) (ATZ)	d_7 MDA
2,4-Dichlorophenoxyacetic acid (2,4-D)	$^{13}C_3$ ATZ
2,4,5-Trichlorophenoxyacetic acid (2,4,5-T)	$^{13}C_6$ 24D
2, 5-Dichlorophenol (25DCP)	$^{13}C_6$ 245T
2,4-Dichlorophenol (24DCP)	$^{13}C_6$ 25DCP
2,4,5-Trichlorophenol (245TCP)	$^{13}C_6$ 2,4DCP
ortho-Phenyl phenol (OPP)	$^{13}C_6$ 245TCP
Pentachlorophenol (PCP)	$^{13}C_6$ OPP
4-Nitrophenol (PNP)	$^{13}C_6$ PCP
2,4,6-Trichlorophenol (246TCP)	$^{13}C_6$ PNP
1-Naphthol (1N)	$^{13}C_6$ 246TCP
2-Naphthol (2N)	$^{13}C_6$ IN
cis- and *trans*-3-(2,2-Dichlorovinyl)-2,2-dimethylcyclopropane-1-carboxylic acids (*cis*-DCCA, *trans*-DCCA)	Carbofuranphenol (CFP)
4-Fluoro-3-phenoxybenzoic acid (4F3PBA)	$^{13}C_3$ *trans*-DCCA
cis-3-(2,2-Dibromovinyl)-2,2-dimethylcyclopropane-1-carboxylic acid (DBCA)	
3-Phenoxybenzoic acid (3PBA)	
Dimethylphosphate (DMP)	$^{13}C_6$ 3PBA
Diethylphosphate (DEP)	d_6DMP
Diethylthiophosphate (DETP)	$^{d}_{10}$ DEP
Diethyldithiophosphate (DEDTP)	$^{d}_{10}$DETP
Dimethylthiophosphate (DMTP)	$^{13}C_4$ DEDTP
Dimethyldithiophosphate (DMDTP)	$^{d}_6$ DMTP
	$^{d}_6$ DMDTP

Optimized Precursor/Product Ion Pairs, Declustering Potential (DP), Collision Cell Exit Potential (CXP), and the Collision Offset Energy (CE) for the Multianalyte Method Target Analytes on the Sciex API4000 and TSQ 7000 Mass Spectrometers

Analyte (label pattern)	Precursor ion		Product ion		Ion Instrument	DP Mode	CE (V)	CXP (V)	(V)
	Native	Labeled	Native	Labeled					
trans-DCCA ($^{13}C_2$)	207	210	35	35	Sciex	–	–50	30	–5
cis-DCCA[a]	207	[a]	35	[a]	Sciex	–	–50	30	–5
DBCA[a]	294	[a]	79	[a]	Sciex	–	–35	18	–3
3PBA ($^{13}C_6$)	213	219	93	99	Sciex	–	–55	28	–7
4F3PBA[b]	231	[b]	93	[b]	Sciex	–	–60	36	–17
IMPY ($^{13}C_4$)	153	157	84	88	TSQ	+	NA	–22	NA
DEAMPY (d$_6$)	182	188	154	158	TSQ	+	NA	–22	NA
CIT (d$_7$)	203	210	120	121	TSQ	+	NA	–24	NA
ATZ ($^{13}C_3$)	343	346	214	217	TSQ	+	NA	–23	NA
ACE ($^{13}C_6$)	351	357	130	130	TSQ	+	NA	–15	NA
ALA ($^{13}C_6$)	365	371	162	168	TSQ	+	NA	–25	NA
DEET (d$_6$)	192	198	119	119	TSQ	+	NA	–22	NA
MDA (d$_7$)	273	280	141	147	TSQ	–	NA	13	NA
PNP ($^{13}C_6$)	138	144	108	114	TSQ	–	NA	22	NA
CMHC ($^{13}C_4$)	209	213	145	148	TSQ	–	NA	24	NA
MET ($^{13}C_6$)	409	415	280	286	TSQ	–	NA	21	NA
2,4-D ($^{13}C_6$)	219	225	161	167	TSQ	–	NA	18	NA
TCPY 1 ($^{13}C_5$ ^{15}N)	198	204	198	204	TSQ	–	NA	15	NA
TCPY 2[c] ($^{13}C_5$ ^{15}N)	196	202	196	202	–	15		NA	
2,4,5-T ($^{13}C_6$)	255	261	197	203	TSQ	NA	16	NA	NA
3-PBA ($^{13}C_6$)	213	219	93	99	TSQ	–	NA	25	NA

NA, parameter not applicable to instrument used.
[a] Labeled *trans*-DCCA used as internal standard.
[b] Labeled 3-PBA used as internal standard.
[c] Confirmation ion.

3.2.3.2. NATIVE STANDARDS

1. Prepare a native standard stock solution by weighing approx 5 mg of the native standard into a 25-mL volumetric flask.
2. Dilute with acetonitrile to yield a 200-ng/µL solution.
3. Prepare a set of seven standard solutions at the following concentrations by diluting stock solutions with acetonitrile: 0.05, 0.1, 0.25, 0.5, 1, 2.5, and 5 µg/mL (*see* **Note 13**).
4. Divide into 1-mL aliquots in glass ampules and flame seal.
5. Store at –20°C until used.

3.3. Sample Preparation

3.3.1. Multiclass Method

1. Pipette 2-mL aliquots (2 mL) of urine into centrifuge tubes.
2. Spike with 25 µL isotopically labeled ISTD giving an approx 25 ng/mL concentration of the standards in the urine *(12)*.
3. Add 1.6 mg β-glucuronidase/sulfatase dissolved in 1.5 mL of a 0.2 *M* acetate buffer to each sample (*see* **Notes 14** and **15**).
4. Incubate in a dry bath or oven at 37°C for 17 h (*see* **Note 16**).
5. Place Oasis SPE cartridges on vacuum manifold.
6. Condition Oasis SPE cartridges with 1 mL methanol followed by 1 mL 1% acetic acid.
7. Pass urine hydrolysates through the preconditioned cartridges.
8. Wash cartridges with 1 mL 5% methanol in 1% acetic acid.
9. Dry cartridges for approx 30 s using a vacuum.
10. Elute SPE cartridges with 1.5 mL methanol and collect.
11. Add 2 mL acetonitrile to the methanol eluates (*see* **Note 17**).
12. Evaporate to dryness using a Turbovap LV at 40°C and 10 psi of nitrogen as the evaporating gas.
13. Reconstitute the dried residue in 50-µL acetonitrile.
14. Separate extract into 10 and 40 µL fractions for analysis using HPLC with turbo ionspray atmospheric pressure ionization (TSI)–MS/MS and HPLC with atmospheric pressure chemical ionization (APCI)–MS/MS, respectively.

3.3.2. Dialkylphosphate Method

1. Pipet 2 mL urine into 15-mL centrifuge tubes.
2. Spike with 10 µL of the ISTD solution and mix to give urinary ISTD concentrations of 25 µg/L for each analyte *(8,15)*.
3. Place in a commercial lyophilizer system.
4. Operate lypophilizer overnight in the program mode without further manual manipulation: Freeze samples for 4 h at –34°C and atmospheric pressure. Apply vacuum to 25.5 mTorr at –34°C for 4 h. Samples are taken to –20°C for 2 h, 0°C for 1 h, and finally 20°C for 1 h.
5. Add 2 mL acetonitrile and 2 mL diethyl ether to dried residues and vortex mix for 1 min.
6. Pour supernatants into fresh 15-mL centrifuge tubes to separate them from the undissolved residue.
7. Rinse tubes with the undissolved residue with another 1 mL acetonitrile, then add to supernatant.
8. Evaporate supernatants to approx 1 mL using a Turbovap LV at 30°C and 10 psi of nitrogen (*see* **Note 18**).
9. Pour into 15-mL test tubes containing a few grains of potassium carbonate.

10. Add 50 μL CIP and mix.
11. Place in a dry bath set at 60°C for 3 h.
12. Transfer supernatants to clean tubes and evaporate to dryness using the TurboVap.
13. Reconstitute samples in 75 μL of toluene for GC–MS/MS analysis.

3.3.3. Phenols Method

1. Pipet 2 mL urine into 15-mL centrifuge tubes.
2. Spike with 5 μL of the ISTD solution to give approx 25 μg/L concentration of the ISTD in the urine *(11)*.
3. Add 1.8 mg β-glucuronidase/sulfatase in 2 mL 0.1 *M* acetate buffer to each sample (*see* **Notes 14** and **15**).
4. Incubate in oven or dry bath for 17 h at 37°C (*see* **Note 16**).
5. Precondition Oasis SPE cartridges with 1 mL of 20% diethyl ether/*n*-butyl chloride, 1 mL methanol, and 1 mL 0.05*N* HCl.
6. Acidify urine hydrolysates with 250 μL 2*M* H$_2$SO$_4$.
7. Pass urine through SPE cartridges.
8. Wash SPE cartridges with 5% methanol.
9. Elute cartridges with 4 mL of 20% ethyl ether/*n*-butyl chloride.
10. Add 1 mL of 3*N* NaOH to eluates and mix (*see* **Note 19**).
11. Discard organic phase.
12. Add 0.5 mL 0.4*M* TBAHSO$_4$ and 0.5 mL 10% CIP in *n*-butyl chloride to each sample.
13. Incubate in a 60°C dry bath for 1 h to form the chloropropyl ethers of the target analytes.
14. Apply reaction mixtures to 3 cc ChemElut sorbent-immobilized liquid extraction 3-cc cartridges.
15. Elute twice with 2 mL hexane.
16. Evaporate eluates to dryness using a Turbovap LV evaporator at 30°C and 10 psi of nitrogen for approx 30 min.
17. Reconstitute samples with 75 μL of toluene for analysis using GC–MS/MS.

3.4. Instrumental Analysis

Perform all analyses on a TSQ 7000 triple quadrupole MS (ThermoQuest) coupled to a GC or HPLC or a Sciex API4000 triple quadrupole mass spectrometer (Applied Biosystems/MDS Sciex) coupled to an HPLC. The HPLC apparatus should be equipped with a binary pump, a degasser, an autosampler, and a temperature-stable column compartment. For the APCI application, use a TSQ 7000. For the TSI application, use a Sciex API4000.

3.4.1. Multianalyte Method

1. Install a Betasil phenyl column on the HPLC connected to the TSQ 7000.
2. Set up HPLC connected to the TSQ 7000 to perform an isocratic elution with a mobile phase mixture of 36% acetonitrile in water with 0.1% acetic acid.
3. Set the flow rate at 1.0 mL/min and the injection volume at 10 μL.
4. Keep the column temperature at 25°C during the analysis.
5. On the TSQ 7000, set the heated capillary to 450°C, the corona discharge to 4.0 kV, and the capillary temperature to 250°C.
6. Set instrument to use 50 psi sheath gas (N$_2$) and 2 mT collision gas (Ar).
7. Set up analysis program in the multiple reaction monitoring (MRM) mode using the parameters described in **Table 3** for the TSQ 7000 target analytes.

8. Inject each sample twice. One injection should be in the positive mode and the other in the negative mode.
9. For the first injection, acquire data in positive ionization mode. Divide the acquisition into two distinct timed segments, 0–3.5 and 3.5–7.25 min. The total runtime is 7.25 min.
10. For the second injection, acquire data in the negative ionization mode. Divide the acquisition into five distinct timed segments: 0–3.2, 3.2–4.3, 4.3–6.8, 6.8–9, and 9–13 min. The total runtime is 13 min.
11. Install a Betasil C18 HPLC column to the HPLC connected to the Sciex.
12. Set up the HPLC to run an isocratic elution using a mobile phase of 51% acetonitrile in water with 0.1% acetic acid.
13. Set the flow rate to 0.05 mL/min and the injection volume to 2 µL (*see* **Note 20**).
14. Keep the column at 35°C.
15. Set nitrogen pressure for ion source and curtain gases to 16 and 20 psi, respectively. Set zero air for the collision-activated dissociation and heater gases to 6 and 16 psi, respectively. The heater gas is at 450°C.
16. Set the ionspray current to –4.5 kV and the entrance potential to –10 V.
17. Set up acquisition program in the MRM mode using negative ion TSI using the parameters listed in **Table 3** for the Sciex target analytes.
18. Inject samples and acquire data in one segment. The total runtime is less than 9 min.

3.4.2. Dialkylphosphate Method

1. Set up GC to inject 1-µL samples via an autosampler by splitless injection with an injection purge delay of 60 s.
2. Install a DB5 MS column on the GC.
3. Ensure that a deactivated silica guard column is installed in-line preceding the GC column (*see* **Note 20**).
4. Set the temperatures of the injector and transfer line to 250°C.
5. Set up a GC program as follows: 80°C for 2 min with linear ramp to 250°C at 17°C/min. The final temperature of 250°C is held for 2 min.
6. Set the TSQ 7000 to acquire data in the positive chemical ionization (CI$^+$) MRM mode.
7. Set the methane (CI reagent gas) pressure to 1500 mT.
8. Set argon (collision-induced dissociation gas) pressure to 2 mT.
9. Perform a full autotune of the mass spectrometer.
10. Set TSQ parameters as follows: 150°C source temperature, 200-eV electron energy, and the potential for the continuous dynode electron multiplier will vary depending on multiplier lifetime.
11. Set up an MRM acquisition program using the parameters outlined in **Table 2** using a mass window of 0.4 amu and a scan rate of 0.03 s^{-1}. Divide the run into five distinct timed segments: 7.0–7.5, 7.5–8.5, 8.5–9.2, 9.2–9.9, and 9.9–10.4 min.

3.4.3. Phenols Method

1. Set up GC to inject 1-µL samples via an autosampler by splitless injection with an injection purge delay of 60 s.
2. Install a DB5 MS column on the GC.
3. Ensure that a deactivated silica guard column is installed in-line preceding the GC column (*see* **Note 20**).
4. Set the temperatures of the injector and transfer line to 250°C.
5. Set up a GC program as follows: 80°C for 2 min, heat linearly 160°C at 10°C/min, and

then heat to 260°C at 4°C/min. The final temperature of 260°C is held for 2 min.
6. Set TSQ 7000 to acquire data in the CI⁺ MRM mode.
7. Set the methane (CI reagent gas) pressure to 1500 mT.
8. Set argon (collision-induced dissociation gas) pressure to 2 mT.
9. Perform a full autotune of the MS.
10. Set TSQ parameters as follows: 150°C source temperature, 200-eV electron energy, and vary the potential for the continuous dynode electron multiplier depending on multiplier lifetime.
11. Set up and MRM acquisition program using the parameters outlined in **Table 5** using a mass window of 0.4 amu and a scan rate of 0.06 s⁻¹. Divide the run into six distinct timed segments: 9.5–11, 11–12.8, 12.8–14, 14–15.4, 15.4–16.4, and 16.4–21 min.

3.5. Data Processing

1. Automatically integrate peaks using appropriate software (e.g., Xcalibur® version 2.1 from ThermoQuest provided with the TSQ 7000 or Analyst® software from Applied Biosystems/MDX Sciex provided with the Sciex API4000).
2. Subtract the background signal and smooth data.
3. Check and correct any discrepancies in peak selection to provide an accurate integration.
4. Export peak areas and other pertinent data associated with the analysis into a Microsoft Excel® file and load into a Microsoft Access® database for permanent storage.
5. Perform all statistical analyses using SAS software (SAS Institute Inc., Cary, NC).

3.6. Quantification and Quality Control of Analytical Runs

1. Prepare a seven- or eight-point calibration plot for quantification by spiking 2 mL blank urine (*see* **Subheading 3.1.**) with the amount of native standard indicated *vide supra* (*see* **Subheading 3.2.**; 25 μL for multiclass method, 100 μL for dialkylphosphate method, 20 μL for phenols method) (*see* **Notes 22** and **23**).
2. Spike with ISTD solution (10 μL for multiclass and dialkylphosphate methods, 5 μL for phenols method) and mix.
3. Prepare calibration samples in parallel with unknown ($N = 36$) and QC (one QCL and one QC high) samples according to methods outlined in **Subheading 3.3.**
4. Analyze samples according to methods outlined in **Subheading 3.4.**
5. Derive an equation from a linear regression analysis of the best-fit line of a plot of the calibration standard concentrations against the area$_{native}$/area$_{ISTD}$.
6. Use equation to derive concentrations of unknown samples by using each analyte area$_{native}$/area$_{ISTD}$.
7. Evaluate QC of the analytical runs using Westgard multirules for quality control *(16)*.

3.7. Method Validation

3.7.1. Limits of Detection

1. Calculate the (LODs) as three times the standard deviation of the noise at zero concentration *(17)* (**Table 6**). The estimate of the noise should be based on the variation in precision at concentrations close to the LOD. This can be calculated using the four lowest calibration standards from available validation and analytical runs that will provide an integrated LOD value over several runs (*see* **Note 24**).
2. Verify the calculated LODs by comparing with the results of the calibration standard samples to ensure that the calculated values are in agreement with the peaks observed in the lowest calibration standards.

Table 5
Optimized Precursor/Product Ion Pairs and the Collision Offset Energy (CE)
for the Phenols Method Target Analytes on the TSQ 7000 Mass Spectrometer

Analyte (label pattern)	Precursor ion		Product ion		CE (V)	Ion Mode
	Native	Labeled	Native	Labeled		
IPP[a]	229	235	187	193	5.9	+
25DCP ($^{13}C_6$)	239	245	163	169	7.2	+
24DCP ($^{13}C_6$)	239	245	163	169	7.2	+
CFP[b]	247	247	199	205	7.2	+
246TCP ($^{13}C_6$)	273	279	197	203	7.5	+
TCPY ($^{13}C_5^{15}N$)	274	280	198	204	7.5	+
PNP ($^{13}C_6$)	216	219	140	143	10.1	+
245TCP ($^{13}C_6$)	273	279	197	203	7.5	+
1N ($^{13}C_6$)	221	227	145	151	10.1	+
2N[c]	221	227	145	151	10.1	+
OPP ($^{13}C_6$)	247	253	171	177	13	+
PCP[d] ($^{13}C_6$)	228	234	NA	NA	NA	–

NA, not applicable.
[a]Labeled 25DCP and used as internal standard.
[b]Labeled 246TCP and used as internal standard.
[c]Labeled 1N and used as internal standard.
[d]MS analysis only.

3.7.2. Extraction Efficiency

1. Determine the extraction recovery at two concentrations by spiking five blank urine samples (*see* **Subheading 3.1.**) with the appropriate standard concentration (**Table 6**).
2. Extract (and derivatize, if required) samples according to the methods in **Subheading 3.3.**; however, do not spike with ISTD or perform the evaporation step (*see* **Note 25**).
3. After extraction is completed, spike extract with ISTD.
4. Extract (and derivatize, if required) five additional blank urine samples (*see* **Subheading 3.1.**) according to the method in **Subheading 3.3.**, but do not spike these samples with native or ISTD and do not perform the evaporation step (*see* **Note 26**).
5. After the extraction is completed, spike with native standards and ISTD.
6. Evaporate samples and reconstitute in solvent according to the methods in **Subheading 3.3.**
7. Analyze all samples according to the methods in **Subheading 3.4.**
8. Calculate the recovery by comparing the responses of the blank urine samples spiked before extraction to the responses of the blank urine samples spiked after the extraction. These values can be expressed as a percentage.

3.7.3. Precision

Determine the precision of the method by calculating the relative standard deviation (RSD) of repeat measurements of samples from the QC pools. Multiple instruments and individuals can be used to calculate the RSDs to incorporate all sources of analytical variation (**Table 6**).

Table 6
Analytical Specifications of the Target Analytes

Method	Analyte	Extraction recovery Low level[a] Mean (%)	N	Extraction recovery High level[b] Mean (%)	N	RSD (%) Low pool (N)	RSD (%) High pool (N)	LOD (ng/mL)
Multiclass	MDA	75 ± 6	19	68 ± 5	10	6 (25)	5 (25)	0.3
	PNP	95 ± 3	19	93 ± 4	10	6 (83)	5 (83)	0.1
	CMHC	95 ± 7	19	96 ± 13	10	10 (85)	9 (82)	0.2
	MET	91 ± 9	19	N/A	NA	9 (85)	9 (81)	0.2
	24D	96 ± 9	19	87 ± 4	10	6 (82)	6 (79)	0.2
	TCPY 1	88 ± 8	19	93 ± 9	10	9 (84)	9 (81)	0.3
	TCPY 2[c]	93 ± 10	19	94 ± 8	10	9 (25)	6 (25)	0.4
	245T	97 ± 5	19	90 ± 2	10	6 (85)	5 (82)	0.1
	3PBA (APCI)	95 ± 5	19	90 ± 4	10	5 (25)	5 (25)	0.2
	IMPY	99 ± 12	19	81 ± 9	10	11 (83)	10 (82)	0.7
	DEAMPY	98 ± 7	19	95 ± 3	10	9 (80)	8 (82)	0.2
	CIT	98 ± 20	19	90 ± 11	10	14 (25)	12 (25)	1.5
	ATZ	96 ± 4	19	94 ± 2	10	8 (82)	6 (80)	0.3
	ACE	98 ± 5	19	94 ± 3	10	8 (81)	7 (79)	0.1
	DEET	96 ± 4	19	93 ± 3	10	8 (83)	8 (82)	0.1
	3PBA (TSI)	94 ± 4	19	92 ± 2	10	7 (86)	5 (84)	0.1
	4F3PBA	106 ± 13	19	104 ± 9	10	7 (87)	6 (87)	0.2
	cis-DCCA	108 ± 15	19	101 ± 15	10	14 (87)	10 (86)	0.2
	trans-DCCA	95 ± 4	19	92 ± 2	10	7 (87)	5 (86)	0.4
	DBCA	114 ± 18	19	N/A	NA	15 (87)	15 (87)	0.1
Dialkyl phosphate	DMP	94 ± 8	6	95 ± 10	6	15 (84)	11 (83)	0.6
	DEP	99 ± 4	6	99 ± 4	6	13 (84)	10 (83)	0.2
	DMTP	100 ± 11	6	82 ± 10	6	13 (84)	10 (83)	0.2
	DMDTP	100 ± 4	6	82 ± 11	6	15 (84)	14 (83)	0.1
	DETP	82 ± 6	6	87 ± 9	6	15 (84)	14 (83)	0.1
	DEDTP	75 ± 3	6	85 ± 3	6	13 (84)	11 (83)	0.1
Phenols	IPP	84 ± 9	3	89 ± 9	3	13 (83)	12 (83)	0.4
	25DCP	93 ± 3	3	94 ± 1	3	14 (83)	10 (83)	0.1
	24DCP	94 ± 3	3	92 ± 2	3	10 (83)	12 (83)	0.3
	CFP	92 ± 7	3	95 ± 6	3	11 (84)	11 (84)	0.4
	TCPY	95 ± 3	3	94 ± 2	3	10 (83)	9 (83)	0.4
	245TCP	80 ± 2	3	84 ± 2	3	17 (84)	14 (83)	0.9
	246TCP	91 ± 6	3	95 ± 3	3	13 (83)	14 (83)	1.3
	PNP	97 ± 3	3	94 ± 3	3	8 (83)	9 (83)	0.8
	1N	88 ± 2	3	93 ± 4	3	11 (84)	9 (84)	0.3
	2N	97 ± 3	3	99 ± 4	3	11 (83)	10 (83)	0.2
	OPP	94 ± 5	3	93 ± 2	3	10 (84)	8 (84)	0.3
	PCP	64 ± 3	3	66 ± 1	3	10 (84)	8 (84)	0.5

LOD, limit of detection; *N*, number of samples tested; NA, not applicable; RSD, relative standard deviation.
[a]5 ng/mL for multiclass method; 10 ng/mL for dialkylphosphate method; 25 ng/mL for phenols method.
[b]50 ng/mL for multiclass and dialkylphosphate methods; 100 ng/mL for phenols method.
[c]Confirmation ion.

3.7.4. Intra- and Intermethod Comparison

1. Determine comparability of analytes measured in multiple methods (i.e., TCPY and 4-nitro-phenol [PNP] measured in the multiclass and phenols methods) by measuring a minimum of 25 samples using both methods.
2. Compare results from both methods. Ideally, the analyses will demonstrate agreement within ±10%.
3. Determine comparability of 3-phenoxy benzoic acid (3PBA) concentrations using the Sciex with TSI ionization and TSQ 7000 using APCI ionization by comparing results from both instruments. Ideally, the analyses will demonstrate agreement within ±5%.

4. Notes

1. Target concentration of the QCL pool is 5–10 µg/L.
2. Target concentration of the HQCL pool is 15–20 µg/L.
3. Unspiked pool will serve as matrix for calibration standards and blanks.
4. Solution concentration is 20 µg/mL. The diluted solution is 1 µg/mL. This solution is used as an ISTD spiked (10 µL) in all unknown samples, QC materials, and calibration standards for the multiclass method.
5. *cis*-3-(2,2-Dibromovinyl)-2,2-dimethylcyclopropane-1-carboxylic acid (DBCA) is purchased as a 10-µg/mL acetonitrile solution so a stock solution is not made.
6. DBCA is not added to the two highest standard solutions because of the diluted stock solution.
7. To prepare a calibration curve for the multiclass method, 25 µL of each standard solution are added to each 2 mL blank urine sample.
8. Solution concentration is 5 ng/µL.
9. This solution is used as an ISTD spiked (10 µL) in all unknown samples, QC materials, and calibration standards for the dialkylphosphate method.
10. To prepare a calibration curve for the dialkylphosphate method, 100 µL of each standard solution are added to each 2 mL blank urine sample.
11. All analytes except 2-isopropoxyphenol (IPP), 2,3-dihydro-2,2-dimethyl-7-hydroxy-benzofuran (carbofuranphenol, CFP), and 2-naphthol (2N) have analogous isotopically labeled standards. For these, the closest eluting labeled analyte is used as an ISTD. Thus, for IPP, labeled 2,5-dichlorophenol (25DCP) is used as an ISTD; for CFP, labeled 2,4,6-trichlorophenol (246TCP) is used as an ISTD; and for 2N, 1-naphthol (1N) is used as an ISTD.
12. This solution is used as an ISTD spiked (5 µL) in all unknown samples, QC materials, and calibration standards for the phenols method.
13. To prepare a calibration curve for the phenols method, 20 µL of each standard solution are added to each 2 mL blank urine sample.
14. β-Glucuronidase/sulfatase type H-1 from *Helix pomatia* with a specific activity of approx 500 units/mg is used to liberate glucuronide- and sulfate-bound conjugates of metabolites in the multiclass and phenols methods.
15. 0.2M acetate buffer is prepared by mixing 3.1 mL glacial acetic acid, 9.7 g sodium acetate, and 1 L water. The pH is 4.5. A 0.1M acetate buffer is made by doubling the water used.
16. From 1 to 6 h are required for complete hydrolysis; 17 h is a convenient time to use because it represents the time from the end of a working day to the beginning of another working day.
17. Acetonitrile forms azeotropes with residual water from SPE extraction and promotes more efficient evaporation of the methanol eluate.
18. This takes about 10 min.

19. Phenols are back-extracted into base, then extracted again into solvent to remove interfering components from the sample.
20. A small-diameter column and low flow is crucial for obtaining maximum sensitivity on the Sciex application.
21. The guard column extends the useful lifetime of the GC column. When the analysis sensitivity is reduced, especially for DMP, the guard column can be easily, and inexpensively, changed.
22. The concentrations of the calibration standards should range from 0.25 to 100 ng/mL for most analytes.
23. Best calibration results are obtained if an entire calibration sample set is generated for each analytical run.
24. We consider the integrated LOD to be a better estimate of the method LOD because interperson, interday, and, if applicable, interinstrument variation have all been incorporated into the resulting LODs.
25. ISTD will be added after extraction to all samples to account for instrument variability only in the recovery calculations.
26. These samples will be reference samples indicative of 100% recovery.

References

1. Donaldson, D., Kiely, T., and Grube, A. (2002) *1998 and 1999 Market Estimates. Pesticides Industry Sales and Usage Report.* US Environmental Protection Agency, Washington, DC.
2. Barr, D. B., Barr, J. R., Driskell, W. J., et al. (1999) Strategies for biological monitoring of exposure for contemporary-use pesticides. *Toxicol. Ind. Health* **15,** 168–179.
3. Aprea, C., Sciarra, G., and Lunghini, L. (1996) Analytical method for the determination of urinary alkylphosphates in subjects occupationally exposed to organophosphorus pesticides and in the general population. *J. Anal. Toxicol.* **20,** 559–563.
4. Baker, S. E., Barr, D. B., Driskell, W. J., Beeson, M. D., and Needham, L. L. (2000) Quantification of selected pesticide metabolites in human urine using isotope dilution high-performance liquid chromatography/tandem mass spectrometry. *J. Exp. Anal. Environ. Epidemiol.* **10,** 789–798.
5. Beeson, M. D., Driskell, W. J., and Barr, D. B. (1999) Isotope dilution high-performance liquid chromatography/tandem mass spectrometry method for quantifying urinary metabolites of atrazine, malathion, and 2,4-dichlorophenoxyacetic acid. *Anal. Chem.* **71,** 3526–3530.
6. Biagini, R. E., Tolos, W., Sanderson, W. T., Henningsen, G. M., and MacKenzie, B. (1995) Urinary biomonitoring for alachlor exposure in commercial pesticide applicators by immunoassay. *Bull. Environ. Contam. Toxicol.* **54,** 245–250.
7. Cho, Y., Matsuoka, N., and Kamiya, A. (1997) Determination of organophosphorous pesticides in biological samples of acute poisoning by HPLC with diode-array detector. *Chem. Pharm. Bull. (Tokyo)* **45,** 737–740.
8. Bravo, R., Driskell, W. J., Whitehead, R. D., Needham, L. L., and Barr, D. B. (2002) Quantification of dialkyl phosphate metabolites of organophosphate pesticides in human urine using GC–MS/MS with isotope dilution method. *J. Anal. Toxicol.* **26,** 245–252.
9. Cocker, J., Mason, H. J., Garfitt, S. J., and Jones, K. (2002) Biological monitoring of exposure to organophosphate pesticides. *Toxicol. Lett.* **134,** 97–103.
10. Drevenkar, V., Stengl, B., Tkalcevic, B., and Vasilic, Z. (1983) Occupational exposure control by simultaneous determination of N-methylcarbamates and organophosphorus pesticide residues in human urine. *Int. J. Environ. Anal. Chem.* **14,** 215–230.

11. Hill, R. H., Jr., Shealy, D. B., Head, S. L., et al. (1995) Determination of pesticide metabolites in human urine using an isotope dilution technique and tandem mass spectrometry. *J. Anal. Toxicol.* **19,** 323–329.

12. Olsson, A., Baker, S. E., Nguyen, J. V., et al. (2004) A liquid chromatography–tandem mass spectrometry multiresidue method for quantification of specific metabolites of organophosphorus pesticides, synthetic pyrethroids, selected herbicides, and DEET in human urine. *Anal. Chem.* **76,** 2453–2461.

13. Centers for Disease Control and Prevention. *National Health and Nutrition Examination Survey.* National Center for Health Statistics, Hyattsville, MD. Available at: http://www.cdc.gov/nchs/nhanes.htm.

14. Barr, D. B., Bravo, R., Weerasekera, G., et al. (2004) Concentrations of dialkyl phosphate metabolites of organophosphorus pesticides in the US population. *Environ. Health Perspect.* **112,** 186–200.

15. Bravo, R., Caltabiano, L. M., Weerasekera, G., et al. (2004) Measurement of dialkyl phosphate metabolites of organophosphorus pesticides in human urine using lyophilization with gas chromatography–tandem mass spectrometry and isotope dilution quantification. *J. Exp. Anal. Environ. Epidemiol.* **15,** 271–281.

16. Westgard, J. O. (2002) *Basic QC Practices: Training in Statistical Quality Control for Health Care Laboratories.* Westgard QC, Madison, WI.

17. Taylor, J. K. (1987) *Quality Assurance of Chemical Measurements.* CRC Press, Boca Raton, FL.

18. Centers for Disease Control and Prevention. (2003) *Second National Report on Human Exposure to Environmental Chemicals.* National Center for Environmental Health, Atlanta, GA. Available at: www.cdc.gov/exposurereport.

6

Urinary Ethylenethiourea as a Biomarker of Exposure to Ethylenebisdithiocarbamates

Silvia Fustinoni, Laura Campo, Sarah Birindelli, and Claudio Colosio

Summary

A method for the determination or urinary ethylenethiourea (ETU), a major metabolite of ethylenebisdithiocarbamates (EBDTCs), is described. ETU is extracted from human urine, in the presence of ethylenethiourea-d_4 as the internal standard, using dichloromethane. The residue is reacted to form the *bis*-(*tert*-butyldimethyilsilyl) derivative and analyzed using gas chromatography–mass spectrometry (GC–MS) in the electron impact/single-ion monitoring (EI/SIM) mode. The linearity and precision of the assay are good over the entire investigated range (0–200 µg/L). The detection limit of the assay after correction for urinary creatinine is 0.5 µg/g creatinine. A protocol for biological monitoring through determination of urinary ETU is provided. The protocol was applied to study 47 agricultural workers exposed to EBDTCs in the vineyards and 33 controls. In workers postexposure samples, median level of urinary ETU was 8.8 (from less than 0.5 to 126.3) µg/g creatinine. This level is significantly higher than those found in worker preexposure samples and in controls, used as reference values.

Key Words: Agricultural workers; biological monitoring; EBDTC; EBDTC production workers; ethylenethiourea; ethylenebisdithiocarbamates; ETU; gas chromatography–mass spectrometry; general population; occupational exposure in humans; urine; vineyard workers.

1. Introduction

Ethylenebisdithiocarbamates (EBDTCs) are widely used in agriculture as fungicides, mainly on fruits, vegetables, and ornamental plants. Because of low acute toxicity and short environmental persistence, their use is increasing worldwide. The metabolic pathway of EBDTCs in mammals is complex and gives rise to the production of several metabolites, among which is ethylenethiourea (ETU; **Fig. 1**) *(1)*. ETU can also be produced by the spontaneous degradation of EBDTCs in the treated crops or during their production, formulation, and storage *(2,3)*. Although the acute toxicity of EBDTCs and ETU is low, possible long-term effects consequent to low-dose prolonged exposures have not been fully elucidated yet *(4,5)*.

From: *Methods in Biotechnology, Vol. 19, Pesticide Protocols*
Edited by: J. L. Martínez Vidal and A. Garrido Frenich © Humana Press Inc., Totowa, NJ

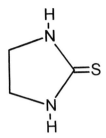

Fig. 1. Molecular structure of ETU.

Human occupational exposure to EBDTCs or ETU takes place in industrial settings and in agriculture. The general population may be exposed through the ingestion of contaminated food and water *(6,7)*. To assess this exposure, the determination of urinary ETU is proposed *(8–13)*, and here is reported a sensitive analytical method suitable for this aim. A protocol for the biological monitoring of exposure to EBDTCs or ETU is outlined. The results obtained by investigating a group of agricultural workers and a group of healthy subjects belonging to the general population, chosen as controls, are reported.

2. Materials

1. ETU or 2-imidazolidinethione (>98%).
2. Propylenethiourea (PTU; 94%, e.g., Dr. Ehrenstofen, Ausburg, Germany).
3. Ethylene-d_4-diamine (98%, e.g., Aldrich, Poole, UK)
4. Carbon disulfide (CS_2; >99.5%).
5. Potassium hydroxide (KOH).
6. Ethanol (EtOH; >99.8%).
7. Hydrochloric acid in water (HCl; 37% w/v).
8. Acetone (>99.8%).
9. Silica gel for flash chromatography (200–400 mesh).
10. Methanol (MeOH; >99.9%).
11. Ammonium chloride (NH_4Cl; >99%).
12. Potassium fluoride (KF; >99%).
13. Dichloromethane (>99.9%).
14. Water (high-performance liquid chromatographic grade).
15. Derivatization mixture: anhydrous acetonitrile, *N*-(*tert*-butyldimethylsilyl)-*N*-methyltrifluoroacetamide (BSTFA; derivatization grade in glass-sealed ampoule, e.g. Aldrich), and *tert*-butyldimethylsilyl chloride (*t*-$BuMe_2Si$-Cl; >97%) at the ratio 5:4:1 v/v/v.
16. Anhydrous acetonitrile: acetonitrile (CH_3CN; >99.9%) on molecular sieve (1.6-mm pellets, 4-Å pore).
17. Silica gel thin-layer chromatographic plates, 60-Å F_{254} with fluorescent indicator 250-μm layer thickness (e.g., Merck, Darmstadt, Germany).
18. Iodine (I_2; 99.8%).
19. 3-mL diatomaceous earth column (e.g., Chem Elut 1003, Varian, Sunnyvale, CA).
20. Picric acid in water (1% w/v).
21. Sodium hydroxide in water (NaOH; 2.5*M*).

22. Chromatographic column CPSil 19 CB, 30 m long, 0.25 mm id, 0.25-µm film thickness (e.g., Varian).
23. Gas chromatograph equipped with split–splitless injector operating in the splitless mode and a mass detector operating in the electron impact (EI) mode.

3. Methods

The method described below outlines (1) the preparation of ethylenethiourea-d_4 (ETU-d_4) for use as an internal standard; (2) the chromatographic method for the determination of ETU in human urine; (3) a protocol suitable for the field studies with typical levels of ETU excreted in Italian vineyard workers and general population.

3.1. Ethylenethiourea-d_4

The synthesis, purification, and characterization of ETU-d_4 used as an internal standard in the determination of urinary ETU are described in **Subheadings 3.1.1.–3.1.3.**, respectively.

3.1.1. Synthesis

Prepare ETU-d_4 accordingly to a previous publication, with some modification (**14**). Briefly, introduce ethylene-d_4-diamine (135 µL, 2 mmol) in a two-neck flask containing 300 µL H_2O. Add a solution of KOH (225 mg, 2 mmol) in H_2O (300 µL) and heat the mixture at 40°C for 20 min. Cool the solution at room temperature and add EtOH (600 µL) and CS_2 (300 µL, 5 mmol). Reflux the reaction mixture at 60°C for 45 min and then at 100°C for 30 min. Cool the reaction mixture at room temperature again, add aqueous HCl (300 µL), then heat at 100°C for 7 h. Finally, cool the reaction mixture overnight at room temperature to obtain a whitish solid. Filter the solid and wash it with few milliliters of cold acetone to yield raw ETU-d_4 purified as described below.

3.1.2. Purification

Purify the raw product by flash chromatography. Fill a glass column (50 cm long, 2 cm id) with 200–400 mesh silica gel to obtain a 20-cm dry bed. Prepare about 500 mL of a mixture of dichloromethane:methanol (95:5 v/v) to use as eluting solvent. Condition the column by passing the eluting solvent through the bed until it is uniformly wet and free of air bubbles. To speed up the operation and to obtain better packing, pump air at the top of the column using a manual pump. Load raw ETU-d_4 dissolved in a minimum amount of eluting solvent (about 5 mL) at the top of the silica bed (*see* **Note 1**). Elute ETU-d_4 with the eluting mixture and collect single fractions (about 7 mL each) in glass tubes using the manual pump. Check the chromatographic fractions for the presence of ETU-d_4 using silica gel thin-layer chromatography plates (*see* **Subheading 3.1.3.**). Pool the appropriate fractions and evaporate the solvent under vacuum to obtain chromatographically pure ETU-d_4 (>99%) as a white solid.

3.1.3. Characterization

Perform characterization of ETU-d_4 by thin-layer chromatography and GC–mass spectrometry (MS). For thin-layer chromatography characterization, use silica gel plates and elute ETU-d_4 vs ETU with the dichloromethane:methanol 95:5 mixture.

Develop the chemicals as yellow-brown spots in the presence of I_2 vapors. In this condition, ETU and ETU-d_4 show a retention index of about 0.3. For the GC–MS characterization, react a solution of ETU-d_4 in anhydrous acetonitrile (1 mg/mL) with 100 µL of derivatization mixture at 60°C for 30 min. Analyze the reaction mixture by GC–MS using the eluting conditions described in **Subheading 3.2.2.** For peak detection, select the dynamic range *m/z* 40–400 and acquire the mass spectra in the full-scan mode. Under this condition, a chromatogram with three peaks, corresponding to the unreacted ETU-d_4 (*tr* = 5.60 min), the monosilanized derivative (*t*-BuMe$_2$Si)-ETU-d_4 (*tr* = 6.24 min), and the *bis*-silanized derivative *bis*(*t*-BuMe$_2$Si)-ETU-d_4 (*tr* = 6.46 min), is obtained. The principal peak ions in the mass spectra are as follows:

ETU-d_4: *m/z* 106 $[M]^{+\bullet}$ (100%) (for mass spectra, *see* **Fig. 2**)
(t-BuMe$_2$Si)-ETU-d_4: *m/z* 220 $[M]^{+\bullet}$ (4%), 163 $[M^{+\bullet}–C(CH_3)_3^\bullet]^+$ (100%), 205 $[M^{+\bullet}–CH_3^\bullet]^+$ (4%)
bis(*t*-BuMe$_2$Si)-ETU-d_4: *m/z* 334 $[M]^{+\bullet}$ (0.5%), 277 $[M^{+\bullet}–C(CH_3)_3^\bullet]^+$ (100%), 319 $[M^{+\bullet}–CH_3^\bullet]^+$ (7%)

3.2. Determination of Urinary ETU

The extraction of ETU from urine and its derivatization, the gas chromatography–mass spectrometry (GC–MS) analysis of the derivative, the preparation of the solutions for the calibration curve, the preparation of the internal standard solution, and the calculation of urinary ETU concentration are described in **Subheadings 3.2.1.–3.2.5. Subheading 3.2.6.** describes the determination of urinary creatinine.

3.2.1. Extraction and Derivatization of ETU

Leave urine samples at room temperature until completely thawed. Mix the sample, wait a few minutes, then transfer 3 mL of the specimen's supernatant (*see* **Note 2**) in a glass vial containing 0.1 g NH$_4$Cl and 1.5 g KF (*see* **Note 3**). Add 0.1 mL of internal standard solution to the final concentration of 83.3 µg/L of ETU-d_4 in urine (*see* **Note 4**). Vigorously stir the mixture to facilitate dissolution of salts and pour the solution onto a diatomaceous earth column (*see* **Note 5**). After urine percolation (about 5 min), add 12 mL dichloromethane through the column and collect the organic solvent in a 20-mL glass vial. Evaporate the extract at 25–35°C using a stream of nitrogen. Dissolve the residue with 1 mL dichloromethane (0.5 mL × 2) and transfer the solution in a 1.8-mL glass vial. Gently evaporate the solvent at 25–35°C using a stream of nitrogen and add the residue with 0.1 mL of derivatization mixture (*see* **Note 6**). Seal the vial with a plastic screw cap lined with polyperfluoroethylene gasket and react the mixture at 60°C overnight. Under these conditions ETU and ETU-d_4 react to give the *bis*-silanized derivatives *bis*(*t*-BuMe$_2$Si)-ETU and *bis*(*t*-BuMe$_2$Si)-ETU-d_4 (*see* **Note 7**). Transfer the residue, cooled at room temperature, in a conical insert and analyze it as described below.

3.2.2. Gas Chromatography–Mass Spectrometry

Inject 1 µL of the derivatized mixture containing *bis*(*t*-BuMe$_2$Si)-ETU and *bis*(*t*-BuMe$_2$Si)-ETU-d_4 in acetonitrile into the chromatographic column through the injector liner kept at 250°C. Use helium as a carrier gas at constant flow of 1 mL/min. Keep the oven temperature at 150°C for 1 min, then increase the temperature to 240°C at the rate of 20°C/min. Finally, keep the oven at 240°C for 5 min. Under these conditions,

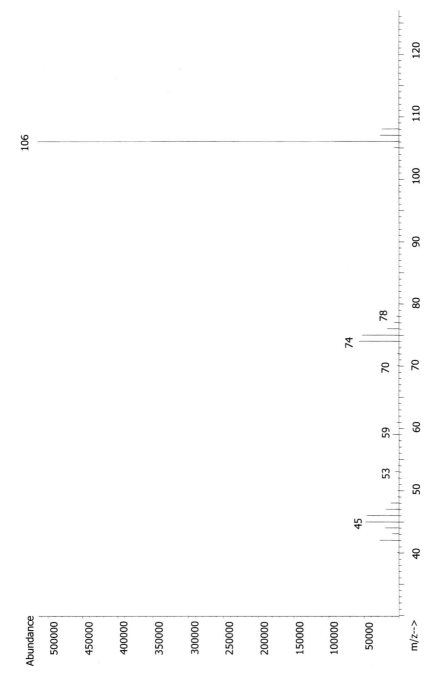

Fig. 2. Mass spectra of ETU-d$_4$ acquired in the EI mode.

the ETU and ETU-d_4 derivatives [*bis*(*t*-BuMe$_2$Si)-ETU and *bis*(*t*-BuMe$_2$Si)-ETU-d_4] are eluted. The total run time is 10.5 min.

Set the MS detector with the electron impact (EI) source (70 eV) kept at 230°C in the selected ion monitoring (SIM) mode. Select a 4-min delay time and 100-ms dwell time. From 4 to 9 min, focus the spectrometer at ions m/z 273 and 277 [M$^{+\cdot}$–C(CH$_3$)$_3$]$^+$ for *bis*(*t*-BuMe$_2$Si)-ETU and *bis*(*t*-BuMe$_2$Si)-ETU-d_4, respectively. Under the described conditions, approximate retention time for both *bis*(*t*-BuMe$_2$Si)-ETU and *bis*(*t*-BuMe$_2$Si)-ETU-d_4 is 6.42 min (*see* **Fig. 3**).

3.2.3. Preparation of the Calibration Solutions

Use calibration solutions of ETU in urine at concentrations of 200, 100, 50, 25, 12.5, and 2.5 µg/L to obtain the calibration curve. Prepare the calibration solutions for dilution of an aqueous ETU solution at the initial concentration of 10 mg/L with a pool of urine obtained from nonsmoking, nonoccupationally exposed donors. Use an unspiked sample of the same urine as blank. Divide the calibration solution and the urine blank in small portions (about 10 mL) and store at −20°C in the dark. In these conditions, the solutions are stable for at least 6 mo.

3.2.4. Preparation of Internal Standard Solution

Prepare the internal standard solution by diluting ETU-d_4 in water at the concentration of 2.5 mg/L. The internal standard solution, stored at −20°C in the dark, is stable for at least 6 mo.

3.2.5. Calibration Curve and Calculation

Obtain the calibration curve analyzing the above-mentioned calibration solutions in the presence of ETU-d_4 as an internal standard following the procedure of extraction, derivatization, and analysis outlined in **Subheadings 3.2.1.** and **3.2.2.** Use the least-square linear regression analysis to calculate the slope m of the function $Y = mX$, where Y is the ratio between the chromatographic peak area of *bis*(*t*-BuMe$_2$Si)-ETU and *bis*(*t*-BuMe$_2$Si)-ETU-d_4 at the different concentrations subtracted by the same ratio in the blank, and X is the ETU concentration (*see* **Note 8**).

Use the calibration curve to calculate the ETU concentration in unknown urine samples. Divide the ETU concentration, expressed in micrograms per liter, by urinary creatinine concentration, expressed as grams creatinine per liter and determined as described in **Subheading 3.2.6.**, to obtain the ETU concentration expressed as micrograms per gram creatinine (*see* **Note 9**). The limit of detection for the entire assay is 0.5 µg/g creatinine (*see* **Note 10**).

3.2.6. Urinary Creatinine

Determine the concentration of urinary creatinine by Jaffe's colorimetric assay. In brief, react 20 µL urine with 5 mL aqueous solution of picric acid (1% w/v) in NaOH (2.5 M). Wait 10 min and read the absorption of the complex creatinine–picrate at 512 nm using an ultraviolet–visible (UV–Vis) spectrophotometer.

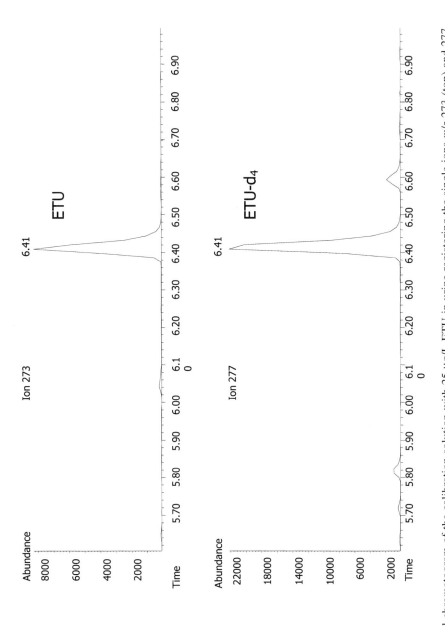

Fig. 3. The typical chromatogram of the calibration solution with 25 μg/L ETU in urine registering the single ions m/z 273 (top) and 277 (bottom) for ETU and ETU-d_4 derivatives, respectively.

3.3. Biological Monitoring of Exposure Through the Determination of Urinary ETU

A protocol suitable for the biological monitoring of exposure through the measurement of urinary ETU is provided. Sampling strategy, sample handling, and delivery and storage conditions are outlined in **Subheadings 3.3.1.** and **3.3.2.** In **Subheading 3.3.3.**, an example of field study with urinary ETU levels measured in agricultural workers and in the general population and suggestions for the interpretation of results are given.

3.3.1. Sampling Strategy

3.3.1.1. Workers

To perform the field study in subjects exposed to several active ingredients, typically agricultural workers, carefully consider the scheduled applications and select a period in which only EBDTCs are used or multiple exposures are minimized.

Submit a personal questionnaire to gather general information (e.g., health status, gender, age, race) and specific information concerning work (e.g., job title and description, name and amount of pesticide formulation handled, time and kind of exposure, use of protective devices).

Choose the best time for urine collection based on the pattern of exposure over time: continuous (typical for industrial workers) or intermittent (typical for agricultural workers).

For preexposure sampling (*see* **Note 11**) for industrial workers, collect a spot urine sample before shift the first working day after the weekend or a rest period. For agricultural workers, collect a spot urine sample before the seasonal applications. Preferentially collect the second urine of the morning (*see* **Note 12**) *(15)*.

For postexposure sampling for industrial workers, collect a spot urine sample at the end of the shift or prior to the next shift (preferentially the second urine of the morning). For agricultural workers, collect a spot urine sample at the end of the exposure or the day after, prior to the next shift (*see* **Note 13**). When a significant variation in exposure levels is anticipated among different working days, repeat specimen collection.

3.3.1.2. Controls

Select controls among the general population without known exposure to EBDTCs/ ETU matched with workers for health status, age, gender, race, and geographical area (*see* **Note 11**). For these subjects, assuming constant low-level exposure because of traces of EBDTCs or ETU ingested with diet, the sampling period is not critical; however, to achieve a better comparison with workers, choose the seasonal period in which the field study is performed.

For control sampling, collect a spot urine sample. To achieve a better comparison, choose the same moment used for the collection of worker's specimens.

3.3.2. Sample Handling, Delivery, and Storage

Pour a urine sample (10–13 mL) in a plastic tube equipped with a plastic screw cap suitable for cryo storage. Leave some empty space in the tube (*see* **Note 14**). Label the tube with an identification number using a water-resistant pen or a water-resistant

Table 1
Summary of Statistics for Urinary ETU Excretion in 47 Vineyard Workers
(Pre- and Postexposure Samples) and in 33 Controls

Subjects (N)	Sampling time	Urinary ETU (μg/g creatinine)				
		Mean	SD	Median	Minimum	Maximum
Vineyard	Preexposure	1.3	1.9	<0.5[a]	<0.5	8.2
workers (47)	Postexposure	21.5	29.8	8.8	<0.5	126.3
Controls (33)	—		1.7	2.2	0.9	<0.5	11.6

[a]0.5 μg/g creatinine is the detection limit of the assay.

adhesive label. Shield the tube from light with aluminum foil (*see* **Note 15**). Refrigerate the sample at 4°C as soon as possible and deliver it chilled to the laboratory within 24 h of sampling. If the delivery is delayed, store urine at –20°C. ETU in a specimen kept a –20°C in the dark is stable for at least 6 mo.

3.3.3. The Field Study: An Example

Table 1 indicates the urinary ETU levels measured by applying the protocol to 47 agricultural workers exposed to EBDTCs in the vineyards during mixing, loading, application, and reentry and 33 controls. Regard the ETU concentration in preexposure specimens as an internal reference level for workers. Moreover, regard the ETU concentration in controls as a local reference value (*see* **Note 11**). Compare the ETU concentration determined in workers' postexposure specimens with the reference values. The comparison shows significant EBDTCs exposure in the investigated vineyard workers. Furthermore, the ETU levels found in the absence of occupational exposure confirm a slight ubiquitous exposure to EBDTCs/ETU presumably attributable to diet.

Finally, because of the absence of biological exposure indices for urinary ETU, no conclusion can be drawn regarding the health risk associated with this exposure. However, all the workers studied were in good health and had no sign of health impairment attributable to exposure.

4. Notes

1. In this condition, part of the raw ETU-d_4 may not dissolve because of the presence of inorganic salts. Load the dissolved fraction on the chromatographic column.
2. Take only the urine supernatant to avoid loading sediment onto the liquid–liquid extraction column.
3. Add NH_4Cl to adjust the pH. Add KF to increase ion strength and facilitate liquid–liquid extraction of ETU. KF is highly hydroscopic, so quickly weigh it and keep it in a closed vial before use.
4. The use of PTU as an internal standard for the determination of urinary ETU has been alternatively investigated. The performances of the entire procedure regarding linearity, repeatability, and sensitivity were similar to those obtained using ETU-d_4. Major differences were in the registration of single-ion m/z 287 for the internal standard derivative of PTU and in its retention time, which is 6.27 min under the chromatographic conditions described in **Subheading 3.2.2**. The advantage of using PTU instead of ETU-d_4 is that the former is commercially available. A major drawback is that PTU is itself the metabolite

of pesticides such as Propineb, so it may not be employed if this exposure takes place.

5. Vigorously stir the mixture as soon as possible to avoid the formation of clots.

6. Prepare the derivatization mixture daily. Preferentially, purchase the derivatization reagent BSTFA in small glass-sealed ampules.

7. The kinetic of the derivatization of ETU to form the $bis(t\text{-BuMe}_2\text{Si})$-ETU derivative is slow, and the reaction is not completed under the conditions described in the present assay. In fact, from each reactant two major chemicals are formed: the mono- and the bis-tBuMe$_2$ derivatives in a ratio of about 1:1. To complete the formation of $bis(t\text{-BuMe}_2\text{Si})$-ETU, more than 60 h are required. This is a long time, and 16 h or overnight is chosen instead for convenience. Nevertheless, this procedure provides good results because the assay is performed in the presence of the internal standard, which ensures that the analyte-vs-internal standard signal ratio is constant over a wide range of time (i.e., from 5 to 60 h). Of course, higher sensitivity could be achieved if the complete conversion to bis-tBuMe$_2$ derivative was performed.

8. The linearity of the regression curve is good over the entire investigated range, with a correlation coefficient typically higher than 0.99. The repeatability of the assay, as coefficient of variation percentage, at a urinary ETU concentration of 25 µg/L, is <5%.

9. In a healthy adult, the daily urine volume may vary substantially (from 600 to 2500 mL). Therefore, urine dilution may be a significant confounding factor in the determination of ETU in a spot urine sample. In particular, this is the case for agricultural workers, who often perform difficult jobs in the presence of such adverse weather conditions as high temperature or humidity. Moreover, the activity of the kidney in the excretion of xenobiotics or their metabolites is not constant during the day. An obvious, but not a practical, solution would be the collection of a 24-h urine sample. A more popular approach, which we also propose for the determination of urinary ETU, is the correction for creatinine. Creatinine in urine (normal values range from 0.3 to 3.4 g/L) is both an indicator of urine dilution and an index of kidney activity. Support for the use of creatinine for the correction of ETU excretion is the similarity between the two molecules, so that an analogous mechanism of kidney excretion may be postulated.

10. An analytical limit of detection of 0.6 µg/L is calculated as five times the signal obtained by submitting a water sample to the procedure of extraction, derivatization, and analysis outlined in **Subheadings 3.2.1.** and **3.2.2.** The limit of detection of the assay of 0.5 µg/g creatinine is estimated by dividing the analytical limit of detection by the mean level of creatinine excreted in healthy subjects (1.2 g creatinine/L).

11. The collection of a preexposure urine sample in workers and of urine sample in controls allows us to obtain ETU levels in the absence of occupational exposure to be used as internal and local reference values, respectively. Because of the lack of biological exposure indices, these values are necessary terms of reference for the comparison with urinary ETU levels in workers' postexposure samples.

12. For practical purposes, we suggest the collection of the second urine of the morning *(15)*. From workers, this specimen is easily obtained just before the beginning of the shift. For controls, we suggest the same sampling time to achieve a better comparison with workers.

13. To better define the best time for sample collection, we performed a study to investigate the kinetic of urinary ETU excretion in agricultural workers. The results of this study showed that the ETU levels in urine collected at the end of the exposure or prior to the next shift (about 16 h after the end of exposure) are comparable (unpublished results).

14. Space is left in the tube to compensate for expansion of urine volume caused by freezing.

15. Aluminum foil shields the tube from light to avoid photodegradation of ETU *(16)*.

References

1. Somerville, L. (1986) The metabolism of fungicides. *Xenobiotica* **16**, 1017–1030.
2. Blazquez, C. H. (1973) Residue determination of ethylene-thiourea (2-imidazol-idinethione) from tomato foliage, soil, and water. *J. Agric. Food Chem.* **21**, 330–332.
3. Bontoyan, W. R., Looker, J. B., Kaiser, T. E., Giang, P., and Olive, B. M. (1972) Survey of ethylenethiourea in commercial ethylenebis-dithiocarbamate formulations. *J. AOAC Int.* **55**, 923–925.
4. Houeto, P., Bindoula, G., and Hoffman, J. R. (1995) Ethylenebisdithiocarbamate and ethylenethiourea: possible human health hazards. *Environ. Health Perspect.* **103**, 568–573.
5. International Agency for Research on Cancer. (2001) *Monographs, vol. 79, Some Thyrotropic Agents.* IARC, Lyon, France, pp. 659–701.
6. Newsome, W. H. (1976) Residues of four ethylenebisdiocarbammate and their decomposition products on field-sprayed tomatoes. *J. Agric. Food Chem.* **24**, 999–1001.
7. Nitz, S., Moza, P. N., Kokabi, J., Freitag, D., Behechti, A., and Kote, F. (1984) Fate of ethylenebis(dithiocarbamates) and their metabolites during the brew process. *J. Agric. Food Chem.* **32**, 600–603.
8. Aprea, C., Sciarra, G., and Sartorelli, P. (1998) Environmental and biological monitoring of exposure to mancozeb, ethylenethiourea, and dimethoate during industrial formulation. *J. Toxicol. Environ. Health A* **53**, 263–281.
9. Aprea, C., Betta, A., Catenacci, G., et al. (1996) Reference values of urinary ethylenethiourea in four regions of Italy (multicentric study). *Sci. Total Environ.* **192**, 83–93.
10. Aprea, C., Betta, A., Catenacci, G., et al. (1997) Urinary excretion of ethylenethiourea in five volunteers on a controlled diet (multicentric study). *Sci. Total Environ.* **203**, 167–179.
11. Canossa, E., Angiuli, G., Garasto, G., Buzzoni, A., and De Rosa, E. (1993) Indicatori di dose in agricoltori esposti a mancozeb. *Med. Lav.* **84**, 42–50.
12. Colosio, C., Fustinoni, S., Birindelli, S., et al. (2002) Ethylenethiourea in urine as an indicator of exposure to mancozeb in vineyard workers. *Toxicol. Lett.* **134**, 133–140.
13. Kurttio, P., and Savolainen, K. (1990) Ethylenethiourea in air and urine as an indicator of exposure to ethylenebisdiocarbammate fungicides. *Scan. J. Work Environ. Health* **16**, 203–207.
14. Doerge, D. R., Cooray, N. M., Yee, A. B. K., and Niemczura, W. P. (1989) Synthesis of isotopically-labelled ethylenethiourea. *J. Label. Comp. Radiopharm.* **28**, 739–742.
15. Hoet, P. (1996) *Biological Monitoring of Chemical Exposure in the Workplace.* World Health Organization, Geneva, pp. 1–19.
16. Ross, R. D., and Crosby, D. G. (1973) Photolysis of ethylenethiourea. *J. Agric. Food Chem.* **21**, 335–337.

7

Analysis of 2,4-Dichlorophenoxyacetic Acid and 2-Methyl-4-Chloro-Phenoxyacetic Acid in Human Urine

Cristina Aprea, Gianfranco Sciarra, Nanda Bozzi, and Liana Lunghini

Summary

Two methods for the quantitative analysis of 2,4-dichlorophenoxyacetic acid (2,4-D) and 2-methyl-4-chlorophenoxyacetic acid (MCPA) in urine are reported. Hydrochloric acid, sodium chloride, and internal standards (2,3-dichlorophenoxyacetic acid [2,3-D] or 4-chlorophenoxyacetic acid [4-CPA]) are added to the urine. The mixture is extracted with dichloromethane. For the high-performance liquid chromatographic (HPLC) method, the concentrated extract is drawn into a previously conditioned silica cartridge in which 2,4-D and MCPA are enriched. For the gas chromatographic (GC) method, after extraction the two compounds are converted to their pentafluorobenzyl esters and purified in a previously conditioned silica cartridge. Calibration is carried out using standard solutions in urine, which are processed in the same way as the urine samples and are determined by HPLC with diode array detector (DAD) or GC with electron capture detection (ECD). The peak areas of the chlorophenoxycarboxylic acids obtained are divided by the peak areas of the respective internal standard. The resulting quotients are plotted as a function of the concentration of the chlorophenoxycarboxylic acids to obtain a calibration curve. The two methods have detection limits of about 15 µg/L and 1 µg/L for 2,4-D and MCPA, respectively.

Key Words: Chlorophenoxycarboxylic acids; general population; GC analysis; herbicides; HPLC analysis; occupational exposure; pentafluorobenzylbromide derivatization.

1. Introduction

2,4-Dichlorophenoxyacetic acid (2,4-D) and 2-methyl-4-chlorophenoxyacetic acid (MCPA) are selective, hormone-type pesticides widely used as postemergence herbicides to eliminate weeds from cereal and grass crops.

From: *Methods in Biotechnology, Vol. 19, Pesticide Protocols*
Edited by: J. L. Martínez Vidal and A. Garrido Frenich © Humana Press Inc., Totowa, NJ

Intake of 2,4-D and MCPA may occur by inhalation, ingestion, or absorption through intact skin. In humans, the two compounds are mainly eliminated unchanged by urinary excretion. A small proportion of conjugates has been detected in a few cases *(1–3)*. Urinary levels of 2,4-D and MCPA as free acids can therefore be used as indicators of exposure to these compounds and their salts *(4)*.

Exposure to mixtures of phenoxyacetic, chlorophenol, and chlorinated dibenzodioxin herbicides seems to be associated with an increase in the incidence of malignant lymphoma and sarcoma of the soft tissues *(5)*. Because many tumors have been associated with exposure to phenoxy herbicides and their contaminants or other chemical compounds, it is uncertain whether exposure to 2,4-D and MCPA is specifically related to the development of sarcoma of the soft tissues.

Because of widespread use, principally of 2,4-D, methods are required to monitor exposure of workers and the general population. Various methods of determining 2,4-D and MCPA in urine have been developed. High-performance liquid chromatographic (HPLC) procedures *(1,6,7)* involve assay of nonderivatized acids but have quite high detection limits. The high polarity of these compounds makes it impracticable to analyze them directly by gas chromatography (GC), and they must first be derivatized to stable and more volatile compounds. The derivatives most commonly prepared are alkyl esters: methylation with dimethylsulfate was proposed by Vural and Burgaz *(8)*; other authors have used diazomethane *(9,10)* and diazoethane *(11)*. Pentafluorobenzylbromide is another derivatizing agent used to increase the response of electron capture detectors (ECDs) in GC analysis of the two herbicides *(12)*. These procedures rarely have detection limits below 20–30 µg/L, except those of Holler et al. *(11)* and Hughes et al. *(10)*, which have limits of 1 µg/L and 5 µg/L, respectively. A radioimmunoassay method with a detection limit of 1 µg/L was used for the analysis of 2,4-D in urine *(13,14)*.

The two methods described here enable simple and reliable determination of 2,4-D and MCPA in urine using HPLC with diode array detection (DAD) and GC with ECD. The excellent reproducibility, recovery, and operational simplicity of the methods make them suitable for routine use in most laboratories. The detection limit of 15 µg/L means that the HPLC method can be used to determine 2,4-D and MCPA in occupationally exposed subjects. The GC method with a detection limit of 1 µg/L can also be used to assay the two compounds in the general population. The GC method and the HPLC method are equivalent for determining 2,4-D and MCPA in urine of occupationally exposed subjects *(15)*.

2. Materials

2.1. Equipment

1. Apparatus for sample evaporation (e.g., rotating evaporator with water bath or automated system for several samples).
2. Apparatus for Ultrapure water production (e.g., Milli-RO/Milli Q system, Millipore, Bedford, MA).
3. Workstation for vacuum extraction (e.g., Supelco, Bellefonte, PA) or automated system (e.g., Gilson, OH).

4. Vortex mixer.
5. Equipment for liquid–liquid extraction, preferably mechanical automated system for several samples.
6. Centrifuge.
7. Solid-phase extraction (SPE) silica cartridges (3-mL column reservoir, 500 mg sorbent) (e.g., Supelco).
8. Sampler vials (approx 1.5 mL) with crimp caps and crimping tongs.
9. 10-, 100-, 1000-mL volumetric flasks with a ground glass stopper.
10. 1-, 2-, 5-, and 10-mL glass pipets.
11. Microliter pipets, adjustable between 100 and 1000 µL (e.g., Eppendorf, Hamburg, Germany).
12. 50-mL glass vials with plastic or glass stoppers.
13. HPLC system, diode array detector (DAD) and computer software or integrator.
14. HPLC column that is LC8, 25 cm long, 4.6 mm id, 5-µm grains (e.g., Supelco).
15. Filters for samples before injection into the HPLC column (e.g., Millex HV 13, Millipore).
16. 20-µL syringe for HPLC, preferably an autosampler.
17. Gas chromatograph with split–splitless injector, ECD, and computer software or integrator.
18. GC column 50 m long, 0.32 mm id, 0.4-µm thick film; dimethyl (95%), diphenyl (5%), polysiloxane stationary phase (e.g., CP Sil 8, Chromopack, Middleburg, The Netherlands).
19. 2-µL syringe for GC, preferably an autosampler.

2.2. Chemicals (see Note 1)

1. Pentafluorobenzylbromide (e.g., from Aldrich, Poole, UK).
2. Acetone.
3. Methanol.
4. *n*-Hexane.
5. Dichloromethane.
6. LiChrosolv-type acetonitrile.
7. Analytical-grade sodium chloride.
8. Analytical-grade anhydrous potassium carbonate.
9. Pure anhydrous sodium sulfate (fine powder).
10. Analytical-grade monobasic potassium phosphate.
11. Pure 85% orthophosphoric acid.
12. Analytical-grade 37% hydrochloric acid.
13. Pure 100% glacial acetic acid.
14. 2,4-D (99.77% purity) (e.g., Dr. Ehrenstorfer, Ausburg, Germany).
15. MCPA (99.7% purity) (e.g., Dr. Ehrenstorfer).
16. 2,3-Dichlorophenoxyacetic acid (2,3-D) (98.4% purity) (e.g., Dr. Ehrenstorfer).
17. 4-Chlorophenoxyacetic acid (4-CPA) (98% purity) (e.g., Dr. Ehrenstorfer).
18. Ultrapure water (equivalent to ASTM type 1).
19. Purified nitrogen.
20. Helium for GC.
21. Argon–methane (5% methane) for GC.

2.3. Solutions

1. Internal standard solution for HPLC (*see* **Note 2**). About 10 mg 4-CPA is weighed exactly in a 100-mL volumetric flask, which is then filled to the mark with methanol (contents

100 mg/L). Of this solution, 20 mL are pipeted into a 100-mL volumetric flask, which is then filled to the mark with Ultrapure water (contents 20 mg/L). This solution can be stored in the refrigerator at 4°C for about 4 wk.

2. For the internal standard solution for GC (*see* **Note 3**), about 10 mg 2,3-D and 10 mg 4-CPA are weighed exactly in a 100-mL volumetric flask, which is then filled to the mark with methanol (contents 100 mg/L). Of this solution, 5 mL is pipeted into a 100-mL volumetric flask, which is then filled to the mark with Ultrapure water (contents 5 mg/L). This solution can be stored in the refrigerator at 4°C for about 4 wk.

3. For dichloromethane solution acidified with acetic acid, 1 mL acetic acid is diluted to 100 mL with dichloromethane in a 100-mL volumetric flask.

4. For pentafluorobenzylbromide solution, 100 μL pure pentafluorobenzylbromide is diluted to 10 mL with acetone in a 100-mL volumetric flask. This solution can be stored in the refrigerator at 4°C for about 1 wk.

5. For potassium carbonate solution, about 6 g anhydrous potassium carbonate is weighed in a 10-mL volumetric flask, which is then filled to the mark with Ultrapure water (content about 60% w/v). This solution can be stored in the refrigerator at 4°C for about 4 wk.

6. *n*-Hexane/dichloromethane solution (1:70 mL), *n*-hexane is transferred to a 100-mL volumetric flask, which is then filled to the mark with dichloromethane. This solution must be prepared daily.

7. For the *n*-hexane/dichloromethane solution (2:60 mL), *n*-hexane is transferred to a 100-mL volumetric flask, which is then filled to the mark with dichloromethane. This solution must be prepared daily.

8. For the phosphate buffer, 1.36 g monobasic potassium phosphate and 500 μL glacial acetic acid are transferred to a 1000-mL volumetric flask, which is then filled to the mark with Ultrapure water. The pH is adjusted to 3.2 by adding a few drops of orthophosphoric acid. This solution must be prepared daily.

2.4. Calibration Standards

2.4.1. Starting Solution

About 10 mg 2,4-D and 10 mg MCPA are weighed exactly in a 10-mL volumetric flask, which is then filled to the mark with methanol (content 1 mg/mL). This solution can be stored in the refrigerator at 4°C for about 4 wk.

2.4.2. Stock Solution

For the stock solution, 2 mL of the starting solution is pipeted into a 10-mL volumetric flask, which is then filled to the mark with Ultrapure water (contents 200 μg/mL). This solution can be stored in the refrigerator at 4°C or about 4 wk.

2.4.3. Working Solutions

Working solutions (WSs) to construct calibration curves in urine are prepared from stock solution by diluting with Ultrapure water. They contain between 0.78 and 100 mg chlorophenoxycarboxylic acid per milliliter (**Table 1**). These solutions are prepared daily.

Table 1
Scheme for the Preparation of the WSs

Identification of the working solution	Volume of stock solution or WS	Final volume of working solutions (mL)	Concentration of WS (μg/mL)
WS1	5 of the stock solution	10	100
WS2	5 of the WS1	10	50
WS3	5 of the WS2	10	25
WS4	5 of the WS3	10	12.5
WS5	5 of the WS4	10	6.25
WS6	5 of the WS5	10	3.125
WS7	5 of the WS6	10	1.5625
WS8	5 of the WS7	10	0.78125

2.5. Preparation of Silica Cartridges

2.5.1. HPLC Method

A silica cartridge is conditioned with 6 mL dichloromethane. The prepared cartridge must still be moist when the sample is introduced.

2.5.2. GC Method

A silica cartridge is conditioned with 3 mL dichloromethane and 6 mL n-hexane in that order. The prepared cartridge must still be moist when the sample is introduced.

2.6. Specimen Collection

Specimens are collected in sealable plastic bottles without preservatives or stabilizers.

3. Methods

3.1. Sample Preparation

A 20-mL urine sample, transferred to a 50-mL glass vial, is spiked with 400 μL internal standard solution (4-CPA solution for HPLC and 4-CPA/2,3-D solution for GC), 200 μL concentrated HCl, and about 4 g sodium chloride (saturation of sample) (*see* **Note 4**). Thus prepared, the vial is sealed with a plastic or glass stopper, and the urine sample is extracted twice with 12 mL dichloromethane. Each extraction step lasts 10 min and is followed by centrifugation at 2100g for 5 min. For sample preparation by liquid–liquid extraction, it is convenient to use equipment that can mechanically extract several urine samples at the same time.

The organic extracts are pooled in a new 50-mL glass vial, dehydrated with anhydrous sodium sulfate, and evaporated to dryness in a rotating vacuum evaporator at 30°C or in an automated system designed to process several samples at the same time.

3.1.1. HPLC Method (see **Note 5**)

After liquid–liquid extraction, the residue, made up with 1 mL dichloromethane solution acidified with acetic acid, is slowly drawn through the silica cartridge prepared as described and still moist. The eluate is not collected. The column is rinsed with 1 mL dichloromethane solution acidified with acetic acid, and the eluate is collected with the subsequent 3 mL of dichloromethane. Only at this point is the cartridge sucked dry. The organic phase obtained is evaporated to dryness under a gentle nitrogen stream. The residue is made up with 0.5 mL methanol and filtered in a sampler vial (approx 1.5 mL). After the vial is sealed with a crimp cap, the sample is injected into the HPLC column. For sample preparation by liquid–solid extraction, it is convenient to use equipment that can handle several cartridges at the same time.

3.1.2. GC Method (see **Note 6**)

After liquid–liquid extraction, the residue is spiked with 200 µL pentafluorobenzylbromide solution, 15 µL potassium carbonate solution, and 4 mL acetone. The mixture is shaken in a vortex mixer and left to react at ambient temperature for at least 5 h (the time for maximum reaction yield is 5 h, but no changes or degradation occur after that time).

After the reaction, 3 mL Ultrapure water and 10 mL dichloromethane are added. The mixture is shaken for 5 min in a mechanical shaker and then centrifuged at 1369g for 5 min. The organic phase is transferred to a new 50-mL glass vial, dried on anhydrous sodium sulfate, and evaporated to dryness in a rotating vacuum evaporator at 30°C or in an automated system equipped to process several samples at the same time. The residue, made up with 1 mL hexane, is drawn slowly through the silica cartridge, prepared as described and still moist. The eluate is not collected. The column is rinsed with 2 mL n-hexane/dichloromethane solution 1 (the eluate is not collected) and then eluted with 4 mL n-hexane/dichloromethane solution 2. Only at this point is the cartridge sucked dry. The organic phase is evaporated to about 0.5 mL under a gentle nitrogen stream and transferred to a sampler vial (approx 1.5 mL). Once the vial is sealed with a crimp cap, the sample is injected into the gas chromatograph (*see* **Note 7**). For sample preparation by liquid–solid extraction, it is convenient to use equipment that can deal with several cartridges at the same time.

3.2. Operational Parameters for HPLC

Mobile phase acetonitrile/phosphate buffer with the elution gradient are reported in **Table 2**. Elution flow is 2 mL/min; 10-µL injection volume; 30°C column temperature; and 230-nm detector wavelength. Under these conditions, retention times are about 6 min for 4-CPA (internal standard), about 13 min for 2,4-D, and about 14 min for MCPA.

3.3. Operational Parameters for GC (see **Note 8**)

Detector temperature is 300°C; for carrier gas, the flow rate is 0.6 mL/min, and linear velocity is 27.9 cm/s. Make-up gas flow rate is 65 mL/min. Injection volume is 1 µL by the splitless technique (splitless time 1 min). Injector temperature is 250°C.

Table 2
Elution Gradient for HPLC Analysis

Step	% Acetonitrile	% Phosphate buffer	Time (min)
Equilibrium	25	75	10
1	25	75	5
2	33	67	Ramp 7, hold 5
3	25	75	Ramp 3

Table 3
Column Temperature Program for GC Analysis

Step	Temperature	Time (min)
Equilibrium	50	2
1	50	1
2	270	Ramp 35°C/min, hold 30

The column temperature program is reported in **Table 3**. Under these conditions, retention times are about 22 min for 4-CPA (internal standard), about 24 min for MCPA, about 26 min for 2,4-D, and about 27 min for 2,3-D (internal standard).

3.4. Calibration

2,4-D and MCPA are determined by urine calibration curves constructed by adding 100 µL of all WSs to 20-mL aliquots of urine pooled from subjects not occupationally exposed to chlorophenoxycarboxylic acids. A blank of the same urine (not spiked with standard 2,4-D or MCPA) is also prepared. The additions are in the range 0.78 to 10 µg for each of the two compounds in 20-mL urine. The concentrations of 2,4-D and MCPA in urine are reported in **Table 4**.

The UCSs (urine calibration standards) are processed in the same way as the urine samples (HPLC method or GC method) and analyzed as described by GC or HPLC. A calibration curve is obtained by plotting the quotient of peak areas of chlorophenoxycarboxylic acids with internal standard as a function of concentration. In the HPLC method, 4-CPA is suitable as an internal standard for the assay of 2,4-D and MCPA. In the GC method, 4-CPA is used as the internal standard for MCPA, and 2,3-D is used for 2,4-D.

The calibration curves are linear ($r > 0.990$) between the detection limit and 500 µg of chlorophenoxycarboxylic acids per liter urine.

3.5. Calculation of Analytical Result

Recorded peak areas of chlorophenoxycarboxylic acids are divided by the peak area of the respective internal standard. The quotients thus obtained are used to read the appropriate concentrations, in micrograms of chlorophenoxycarboxylic acids per liter of urine, from the calibration curve. It may be necessary to take a reagent blank

Table 4
Scheme for the Preparation of the UCSs

Identification of the standard solution	Volume of the WS (μL)	Volume of urine (mL)	Concentration added to urine (μg/L)
Blank	—	20	0
UCS1	100 of WS8	20	3.90625
UCS2	100 of WS7	20	7.8125
UCS3	100 of WS6	20	15.625
UCS4	100 of WS5	20	31.25
UCS5	100 of WS4	20	62.5
UCS6	100 of WS3	20	125
UCS7	100 of WS2	20	250
UCS8	100 of WS1	20	500

value into account because the purity of chemicals varies from one manufacturer to another and can be different from one lot to another. In addition, a reagent blank can indicate that a contamination eventually occurred during the analysis. If the reagent blank shows a peak that can interfere with the determination of 2,4-D or MCPA, it is necessary to investigate for the causes or subtract the signal observed from the analytical result.

3.6. Standardization and Quality Control

To determine the precision of the methods, a sample containing a constant concentration of 2,4-D and MCPA is analyzed. As material for quality control is not commercially available, it is prepared in the laboratory by adding a defined amount of the two chlorophenoxycarboxylic acids to urine. Aliquots of this solution can be stored in the freezer for up to 1 yr and used for quality control. The mean expected value and tolerance range of this material is obtained in a preanalytical period (one determination of the control material in 10 different analytical series).

3.7. Reliability of the Methods

3.7.1. Precision

To determine precision in the series, two pools of urine from subjects not occupationally exposed to 2,4-D and MCPA are spiked with WSs of the two compounds to have concentrations of 125 μg/L (HPLC method) and 30 μg/L (GC method). Ten 20-mL aliquots are obtained from each pool and analyzed by the two methods: Relative standard deviation values between 6.2 and 6.8% in the HPLC method and between 5.5 and 8.0% (**Table 5**) in the GC method are easily obtained.

The day-to-day precision is evaluated on the urine pools used for determination of precision in the series spiked with 2,4-D and MCPA. Ten 20-mL aliquots from each pool are analyzed by the two methods on 10 different days: A relative standard deviation between 7.2 and 8.3% is found for the HPLC method and between 5.8 and 9.0% for the GC method (**Table 6**).

Table 5
Precision in the Series for the HPLC and GC Determination of 2,4-D and MCPA ($N = 10$)

Compound	HPLC method		GC method	
	Concentration added (μg/L)	RSD (%)	Concentration added (μg/L)	RSD (%)
2,4-D	125	6.2	30	8.0
MCPA	125	6.8	30	5.5

RSD, relative standard deviation.

Table 6
Day-to-Day Precision for the HPLC and GC Determination of 2,4-D and MCPA ($N = 10$)

Compound	HPLC method		GC method	
	Concentration added (μg/L)	RSD (%)	Concentration added (μg/L)	RSD (%)
2,4-D	125	7.2	30	9.0
MCPA	125	8.3	30	5.8

3.7.2. Recovery of Sample Processing

To check for losses occurring during sample preparation in the HPLC method, recovery is evaluated by comparing the results obtained analyzing the solutions used for evaluating precision in the series with those of standard solutions in methanol at the same concentrations without further processing. Losses caused by sample processing are between 14.7 and 18.6% (**Table 7**). For the GC method, recovery is evaluated by comparing the results obtained analyzing the solutions used for evaluating precision in the series with those of standard solutions at the same concentrations after pentafluorobenzylbromide derivatization but without further processing. Losses caused by sample processing are between 5.7 and 12.9% (**Table 7**).

3.7.3. Detection Limit

Under these conditions of sample preparation and determination, the detection limits are 15 and 1 μg of either compound per liter of urine for the HPLC and GC methods, respectively. As no reagent blank value is detectable, the detection limits are calculated as three times the signal-to-noise ratio.

4. Notes

1. Solvents used for sample preparation and extraction must be of the highest purity as this varies greatly from one manufacturer to another.

Table 7
Recovery of Sample Processing ($N = 10$)

	HPLC method		GC method	
Compound	Concentration added (μg/L)	Mean recovery (%)	Concentration added (μg/L)	Mean recovery (%)
2,4-D	125	81.4	30	87.1
MCPA	125	85.3	30	94.3

2. In the HPLC procedure, analysis is carried out with the addition of the internal standard 4-CPA, which is suitable for the assay of both compounds.
3. In the GC method, two internal standards, 4-CPA and 2,3-D, are used because they behave identically to MCPA and 2,4-D, respectively, during extraction and chromatography. This increases the reproducibility of the assay.
4. Before the procedure of extracting 2,4-D and MCPA from urine, the addition of concentrated HCl transforms the two compounds in the not-dissociated form and increases recovery in dichloromethane. Saturation of the urine with NaCl gives fast separation of the organic phase from the aqueous phase and practically complete recovery of the—organic solvent.
5. In the HPLC method, purification of the urine extract on an SPE silica cartridges is necessary because, without this preliminary phase, chromatograms obtained would show much interference. Dichloromethane used as an eluant during the purification gives selective migration of the two compounds and the internal standard (4-CPA) through the silica column, leaving behind other components of the urine extract that could interfere with the analysis. The addition of acetic acid to dichloromethane dissolves the two phenoxycarboxylic acids present in the residue of the urine extract and favors subsequent extraction of the two compounds from the silica, increasing the precision of the procedure.
6. In the GC method, purification on SPE silica cartridges eliminates interference by urinary and derivatization products. The first step of washing with a hexane/dichloromethane (70:30) mixture eliminates certain apolar components from the samples so the eluate is not collected. Subsequent elution with hexane/dichloromethane (60:40) enables quantitative and selective recovery of the pentafluorobenzyl esters of the two compounds and the internal standards (2,3-D and 4-CPA), leaving the more polar components, which are not eluted, on the silica.
7. Interference by other substances is not observed: No HPLC or gas chromatographic peak occurred at or near the characteristic retention times of 2,4-D, MCPA and the internal standards (4-CPA and 2,3-D) for urine samples from persons occupationally and not occupationally exposed to the two compounds. **Figure 1** reports a chromatogram of a real sample obtained with the HPLC method, and **Fig. 2** is a chromatogram of another sample obtained with the GC method.
8. A modified operating procedure is developed as an alternative to the gas chromatographic method with ECD detection described above. Sample preparation is the same as above but with the following changes in GC operational parameters:

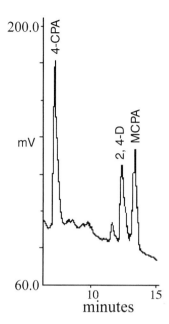

Fig. 1. Chromatogram of urine extract obtained by the HPLC method on a real sample containing 2,4-D and MCPA.

Capillary column	Low polarity, 30 m long, 0.25-mm id, 0.25-µm film thickness (e.g., Meridian MDN12 from Supelco)
Detector	Mass selective detector in the selective-ion monitoring mode; 70-eV electronic impact; 1.20-kV detector
Temperatures	Injector 280°C
	Interface 280°C
	Column: 2 min at 50°C, then increase 7°C per minute to 250°C; 15 minutes at the final temperature Equilibrium time 2 min
Carrier gas	Helium
	Flow rate 0.6 mL/min, linear velocity 27.9 cm/s
Splitless time	1 min
Sample volume	1 µL
Retention times; *m/z*	4-CPA 29.44 min; 366.00 (internal standard)
	MCPA 30.32 min, 141.00
	2,4-D 31.51 min, 181.00
	2,3-D 32.00 min, 175.00 (internal standard)

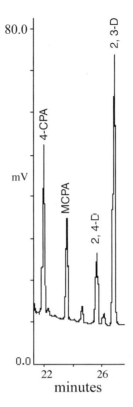

Fig. 2. Chromatogram of urine extract obtained by the GC method on a real sample containing 2,4-D and MCPA.

The results obtained using this modified procedure are comparable to those achieved by the previously described method. Using negative chemical ionization with methane gas (m/z 366 for 4-CPA, 380 for 4-CPA, 400 for 2,4-D, 400 for 2,3-D), we obtained detection limits of about 0.1 µg/L for both 2,4-D and MCPA.

References

1. Fjellstad, P. and Wannag, A. (1977) Human urinary excretion of the herbicide 2-methylchlorophenoxyacetic acid. *Scand. J. Work Environ. Health* **3,** 100–103.
2. Sauerhoff, M. W., Braun, W. H., Blau, G. E., and Gehering, P. J. (1977) The fate of 2,4-dichlorophenoxyacetic acid (2,4 D) following oral administration to man. *Toxicology* **8,** 8–11.
3. Kolmodin-Hedman, B., Sverker, H., Ake, S., and Malin, A. (1983) Studies on phenoxy acid herbicides. II. Oral and dermal uptake and elimination in urine of MCPA in humans. *Arch. Toxicol.* **54,** 267–273.
4. Kanna, S. and Fang, S. C. (1966) Metabolism of C^{14}-labeled 2,4-dichlorophenoxyacetic acid in rats. *J. Agric. Food Chem.* **14,** 500–503.
5. World Health Organization. (1984) *Environmental Health Criteria 29. 2,4-Dichlorophenoxyacetic Acid (2,4-D)*. WHO, Geneva.

6. Nidasio, G. F., Burzi, F., and Sozzé, T. (1984) Determination of 2-methyl-4-chlorophenoxyacetic acid (MCPA) in urine using combined TLC and HPLC methods. *G. Ital. Med. Lav.* **6,** 261–264.

7. Hutta, M., Kaniansky, D., Kovalcikova, E., et al. (1995) Preparative capillary isotachophoresis as a sample pretreatment technique for complex ionic matrices in high performance liquid chromatography. *J. Chromatogr. A* **689,** 123–133.

8. Vural, N. and Burgaz, S. (1984) A gas chromatographic method for determination of 2,4-D residues in urine after occupational exposure. *Bull. Environ. Contam. Toxicol.* **35,** 518–524.

9. Smith, A. E. and Hayden, B. J. (1979) Method for the determination of 2,4-dichlorophenoxyacetic acid residues in urine. *J. Chromatogr.* **171,** 482–485.

10. Hughes, D. L., Ritter, D. J., and Wilson, R. D. (2001) Determination of 2,4-dichlorophenoxyacetic acid (2,4-D) in human urine with mass selective detection. *J. Environ. Sci. Health* **B36,** 755–764.

11. Holler, J. S., Fast, D. F., Hill, R. H., et al. (1989) Quantification of selected herbicides and chlorinated phenols in urine by using gas chromatography/mass spectrometry/mass spectrometry. *J. Anal. Toxicol.* **13,** 152–157.

12. Manninen, A., Kangas, J., Klen, T., and Savolainen H. (1986) Exposure of Finnish farm workers to phenoxy acid herbicides. *Arch. Environ. Contam. Toxicol.* **15,** 107–111.

13. Knopp, D. and Glass, S. (1991) Biological monitoring of 2,4-dichlorophenoxyacetic acid-exposed workers in agriculture and forestry. *Int. Arch. Occup. Environ. Health* **63,** 329–333.

14. Knopp, D. (1994) Assessment of exposure to 2,4-dichlorophenoxyacetic acid in the chemical industry: results of a 5 year biological monitoring study. *Occup. Environ. Med.* **51,** 152–159.

15. Aprea, C., Sciarra, G., and Bozzi, N. (1997) Analytical methods for the determination of urinary 2,4-dichlorophenoxyacetic acid and 2-methyl-4-chlorophenoxyacetic acid in occupationally exposed subjects and in the general population. *J. Anal. Toxicol.* **21,** 262–267.

8

Determination of Herbicides in Human Urine by Liquid Chromatography–Mass Spectrometry With Electrospray Ionization

Isabel C. S. F. Jardim, Joseane M. Pozzebon, and Sonia C. N. Queiroz

Summary

A method for the determination of triazines (simazine, atrazine) and their metabolite 2-chloro-4,6-diamino-1,3,5-triazine by liquid chromatography–mass spectrometry with electrospray ionization (LC–MS/ESI) in human urine is described. The method outlines the sample preparation, which involves protein precipitation with acetonitrile and solid-phase extraction using C18 cartridges, and the qualitative and quantitative chromatographic analyses. The method may be used to assess occupational exposure to triazine herbicides following the urinary excretion of low levels of both the parent compounds and their metabolites.

Key Words: Atrazine; 2-chloro-4,6-diamino-1,3,5-triazine; human urine; liquid chromatography–mass spectrometry with electrospray ionization; simazine; solid-phase extraction; triazines.

1. Introduction

Occupational exposure to triazine herbicides results in urinary excretion of low levels of both the parent compounds and their metabolites *(1–6)*. Quantitative and qualitative analyses require highly sensitive and specific purification and determination procedures.

The increasing use of liquid chromatographic (LC) methods for pesticide analysis is because of their applicability in determining thermally labile compounds and polar compounds, which require derivatization before gas chromatographic (GC) analysis *(7)*. The good compatibility of aqueous samples with reversed-phase liquid chromatography contributes to the fact that LC is frequently applied for analyses of environmental and biological fluid samples. Although ultraviolet (UV) detectors are the most common choice for LC, on-line liquid chromatography with mass spectrometry (LC–MS) is an interesting approach because conventional LC detectors can cause false-

From: *Methods in Biotechnology, Vol. 19, Pesticide Protocols*
Edited by: J. L. Martínez Vidal and A. Garrido Frenich © Humana Press Inc., Totowa, NJ

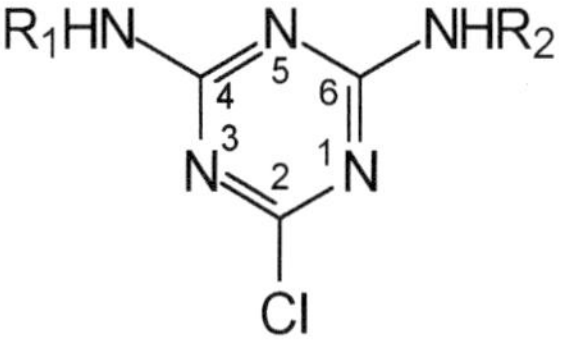

$$R_1HN \diagdown \diagup N \diagdown \diagup NHR_2$$

Fig. 1. Chemical structures of the herbicides studied.

Herbicide	Substituents in positions 4 and 6	
	R_1	R_2
2-chloro-4,6-diamino-1,3,5-triazine (metabolite)	-H	-H
Simazine	$-C_2H_5$	$-C_2H_5$
Atrazine	$-C_2H_5$	$-CH(CH_3)_2$

Fig. 1. Chemical structures of the herbicides studied.

positive results. This technique combines the advantages of coupling the separation power of LC to the unequivocal identification potential of MS for the determination of pesticides in biological samples *(8,9)*.

This work describes a method for the determination by liquid chromatography–mass spectrometry with electrospray ionization (LC–MS/ESI) of the triazines simazine and atrazine and their metabolite 2-chloro-4,6-diamino-1,3,5-triazine (**Fig. 1**), present in human urine.

2. Materials

1. Standard grade simazine (>98%), atrazine (>97%), and the metabolite 2-chloro-4,6-diamino-1,3,5-triazine (>96%).
2. Methanol, pesticide grade.
3. Acetonitrile, HPLC grade.
4. Chloroform, pesticide grade.
5. Water purified by a Milli-Q Plus System (Millipore, Bedford, MA).
6. Envi C18 Supelclean 3-mL extraction tubes (e.g., Supelco, Bellefonte, PA).
7. Urine samples. These must be kept in the freezer at –20°C until use. The samples are stable for 15 d (*see* **Note 1**).
8. LC–MS system equipped with an injector with a 10-µL loop and a UV detector (e.g., Waters, Milford, MA).
9. Chromatographic column (150 × 3.9 mm id) and guard column (20 × 3.9 mm id) C18 (e.g., Waters Nova-Pak C18, 4 µm).
10. Vacuum extraction system (manifold) (e.g., Supelco).

3. Methods

The steps described below outline (1) sample preparation and (2) chromatographic analysis.

3.1. Sample Preparation

Urine samples are frozen at –20°C until use. After the urine samples are thawed, they must be shaken for homogenization. The required volume must be sampled as quickly as possible to avoid sedimentation of solids.

3.1.1. Removal of Proteins

1. Take a 2-mL volume of urine.
2. Basify the sample by addition of 200 µL of NH$_4$OH (0.01%) (pH ~9.0) and add 4 mL of acetonitrile at room temperature.
3. Centrifuge the mixture (5 min, 3000g).
4. Take a 3-mL aliquot of the supernatant containing urine–acetonitrile (1:2 v/v) and dilute with 20 mL Milli-Q water (*see* **Note 2**).

3.1.2. Cleanup Using Solid-Phase Extraction

The cleanup is made by solid-phase extraction using Envi C18 Supelclean extraction tubes (3 mL) and a vacuum extraction system (manifold). The steps that follow are involved.

3.1.2.1. SORBENT CONDITIONING

Rinse the cartridge with 10 mL methanol and then with 5 mL Milli-Q water for equilibration. Avoid allowing the sorbent to dry; otherwise, the recovery will be decreased.

3.1.2.2. SAMPLE APPLICATION

Apply all the diluted sample (**Subheading 3.1.1.**) to the cartridge, under vacuum, at a flow rate of 3 mL/min.

3.1.2.3. WASHING/REMOVAL OF INTERFERENCES

Remove undesired matrix components by passing 5 mL Milli-Q water through the cartridge. This eluate is discarded, and the sorbent bed is then dried under vacuum for 1 min.

3.1.2.4. ELUTION AND CONCENTRATION OF THE SAMPLE

Wipe the delivery needles of the manifold and place labeled collection tubes under the cartridges. With the vacuum off, add the elution solvent, 3 mL chloroform, to each cartridge (*see* **Note 3**). Turn on vacuum and carefully open the tap of the manifold to initiate elution of the analyte. The elution must be carried out slowly to obtain suitable recovery of the analyte.

The organic layer is evaporated to dryness under a stream of nitrogen, and the residue is dissolved in 200 µL of acetonitrile. The preconcentration factor is fivefold because the volume applied to the cartridge corresponds to 1 mL of initial urine sample, and the final volume is 200 µL.

3.2. Chromatographic Analysis

3.2.1. Standard Solution Preparation

Stock solutions of simazine, atrazine, and the metabolite 2-chloro-4,6-diamino-1,3,5-triazine are prepared in methanol at concentrations of 100 µg/mL. Intermediate concentrations are prepared in mobile phase at concentrations of 1000 µg/L of each herbicide (*see* **Note 4**). These solutions are stored in the refrigerator at 4°C, at which they are stable for at least 60 d. Further solutions at five different concentrations (5, 30, 60, 100, and 200 µg/L) of each analyte are prepared daily for construction of the analytical curve.

3.2.2. Spiked Samples

Because certified reference material is not available and there are insufficient samples for intercomparison assays, method accuracy is determined using spiked samples. Three different concentrations (40, 60, and 80 µg/mL) of each pesticide are spiked into known volumes of pesticide-free urine (blank) to calculate recovery. After spiking the samples, carry out procedures as in **Subheadings 3.1.1.** and **3.2.1.**

3.2.3. Chromatographic Conditions

The mobile phase is acetonitrile:H_2O (40:60 v/v) with the pH of the mobile phase adjusted to 3.0 with 0.1% CH_3COOH (see **Notes 5** and **6**). The mobile phase flow rate is set at 0.3 mL/min. The column is directly coupled to the inlet of a chromatograph with a quadrupole MS system operated using the ESI source. Measurements are carried out using the positive ESI mode. All measurements are carried out at room temperature (see **Note 7**).

For optimization of the MS parameters, each compound is dissolved in pH 3.0 mobile phase and injected separately. A source temperature of 150°C is used. Nitrogen is used as both nebulizer and drying gas at flow rates of 30 and 300 L/h, respectively.

The capillary voltage is 25 V for the determination of these herbicides and their metabolite. For identification, the instrument is operated in the total ion mode; for quantification, acquisition is in the selected ion monitoring (SIM) mode. The selected ions for quantification are 145, 202, and 216 for 2-chloro-4,6-diamino-1,3,5-triazine, simazine, and atrazine, respectively (see **Note 8**).

3.2.4. Qualitative and Quantitative Analyses (see **Note 9**)

1. Inject the following sequence: solvent (mobile phase), matrix blank (urine without pesticide), standards, solvent, spiked samples (urine with spiked pesticides), and samples. Matrix blank, spiked samples, and samples are injected after processing according to the steps in **Subheadings 3.1.1.** and **3.1.2.** The standards, spiked samples, and samples are all injected in triplicate.
2. Compare the chromatograms. No peak should be detected at the retention times of the pesticides in the chromatograms of solvent and matrix blank.
3. Determine the limit of detection (LOD) and the limit of quantification (LOQ).
4. Construct the analytical curve with the areas obtained from the chromatograms vs concentration. Linear analytical curves ($r^2 \geq 0.99$) must be obtained in the range from LOQ to 200 µg/L.
5. Calculate the recoveries by comparing the replicate responses of extracted spiked samples with those of standards that represent 100% recovery. Recovery is calculated using the following equation:

    Recovery (%) = (Mass of analyte after extraction × 100)/Mass of analyte added

6. Calculate the precision in terms of repeatability using the results obtained from the recovery samples.
7. If peaks are detected in the chromatograms of the samples, confirm the presence of the herbicide by comparing both retention time and the mass spectra using full-scan mode.
8. Quantify the corresponding amount of the herbicide by the external standard method using the linear equation obtained from the analytical curve.

4. Notes

1. The stabilities of the herbicides in urine with storage time are evaluated by monitoring aliquots of urine having two different concentrations of simazine and atrazine (80, 150 µg/L) and submitting them to replicate freeze/thaw cycles *(10)*. The concentrations of the herbicides are determined after the initial thawing (zero time) and after thawing at successive prefixed times (once each week). Each concentration is compared with the mean of the zero time (initial concentration). Degradation values of <10% are obtained over a period of 15 d.

2. The sample is diluted to decrease solution strength before the extraction procedure.

3. The C18 Supelclean extraction tubes are eluted with methanol and compared with those eluted with chloroform. Recovery is generally good for both solvent systems, but backgrounds are cleaner using chloroform. The evaporation time for chloroform eluate is also shorter than that for methanol. Therefore, the use of chloroform as an elution solvent is recommended for the triazines under study.

4. An analytical curve prepared using mobile phase as solvent is preferable for dissolving the components of the sample, which is a prerequisite for using LC.

5. The pH of the mobile phase is adjusted using a calibrated pH meter with glass and thermal compensation electrodes.

6. A prerequisite in LC–MS is that the analyte must be ionizable in solution, so the mobile phase often contains a small amount of a volatile acid or base. If such additives impair the chromatographic separation, they can be added after the separation, before the eluent enters the ESI interface. The separation of the triazine herbicides and their metabolite is tested at various compositions of the eluent and at different pH values. Because of their polar character, the triazines do not interact strongly with the C18 reversed phase, the most utilized stationary phase in LC. Thus, it is necessary to add acetic acid to the mobile phase to increase the molecular ionization of the compounds, which also improves the detection in analysis by MS.

7. Quantification can also be performed by LC with UV detection. The UV detector is set at 220 nm as a compromise between the maximum absorbance of the analytes and the reduced background of the eluents at this wavelength *(11)*. Acetonitrile is chosen as an organic modifier in the mobile phase owing to its low absorbance background in the UV region.

8. After defining the conditions for SIM of the triazines, the product ions are recorded with a single quadrupole set at a fixed m/z value representing $[M + H]^+$. A urine sample is a complex matrix consisting of various components; therefore, it is essential to use an extraction procedure to remove these interferents. However, using SIM, no problem is encountered in the quantification of the analytes because only the ion corresponding to each compound is selectively monitored.

9. The method must be validated using the instrumentation with which the analyses will be made. According to our results for validation of this method *(12)*, analytical curves are obtained using standard solutions of the triazines, and they showed good linearity in the range from LOQ to 800 µg/L with correlation coefficients > 0.999. The following values are easily obtained for the validation parameters: average recoveries ranging from 82 to 114%; of 0.4 µg/L LOD; 1.3 µg/L LOQ; and precision values of 0.5% (repeatability) and 2.4% (intermediate precision). These values are considered acceptable for biological samples *(11)*.

Acknowledgments

We acknowledge FAPESP for financial support and a fellowship (to J. M. P.). We also thank C. H. Collins and W. Vilegas for helpful discussions and suggestions.

References

1. Mendas, G., Drevenkar, V., and Zupancic-Kralj, L. (2001) Solid-phase extraction with styrene-divinylbenzene sorbent for high-performance liquid or gas chromatographic determination of urinary chloro- and methylthiotriazines. *J. Chromatogr. A* **918,** 351–359.
2. Catenacci, G., Barbieri, F., Bersani, M., Fereoli, A., Cottica, D., and Maroni, M. (1993) Biological monitoring of human exposure to atrazine. *Toxicol. Lett.* **69,** 217–222.
3. Mendas, G., Tkalcevic, B., and Drevenkar, V. (2000) Determination of chloro- and methylthiotriazine compounds in human urine: extraction for gas chromatographic analysis with nitrogen-selective and electron capture detection. *Anal. Chim. Acta* **424,** 7–18.
4. Perry, M. J., Christiani, D. C., Mathew, J., Degenhart, D., Tortorelli, J., Strauss, J., and Sonzogni, W. C. (2000) Urinalysis of atrazine exposure in farm pesticide applicators. *Toxicol. Ind. Health* **16,** 285–290.
5. Catenacci, G., Maroni, M., Cottica, D., and Pozzoli, L. Assessment of human exposure to atrazine through the determination of free atrazine in urine. *Bull. Environ. Contam. Toxicol.* **44,** 1–7.
6. Barr, D. A. and Needham, L. (2002) Analytical methods for biological monitoring of exposure to pesticides: a review. *J. Chromatogr. B* **778,** 5–29.
7. Hogendoorn, E. and van Zoonen, P. (2000) Recent and future developments of liquid chromatography in pesticide trace analysis. *J. Chromatogr. A* **892,** 435–453.
8. Careri, M., Mangia, A., and Musci, M. (1996) Applications of liquid chromatography-mass spectrometry interfacing systems in food analysis: pesticide, drug and toxic substance residues. *J. Chromatogr. A* **727,** 153–184.
9. Henion, J., Brewer, E., and Rule, G. (1998) Sample preparation for LC/MS/MS: analyzing biological and environmental samples. *Anal. Chem.* **70,** 650A–656A.
10. Causon, R. (1997) Validation of chromatographic methods in biomedical analysis. Viewpoint and discussion. *J. Chromatogr. B* **689,** 175–180.
11. Sacchero, G., Apone, S., Sarzanini, C., and Mentasti, E. (1994) Chromatographic behaviour of triazine compounds. *J. Chromatogr. A* **668,** 365–370.
12. Pozzebon, J. M., Vilegas, W., and Jardim, I. C. S. F. (2003) Determination of herbicides and a metabolite in human urine by liquid chromatography–electrospray ionization mass spectrometry. *J. Chromatogr. A* **987,** 375–380.

9

Analysis of Pentachlorophenol and Other Chlorinated Phenols in Biological Samples by Gas Chromatography or Liquid Chromatography–Mass Spectrometry

Ji Y. Zhang

Summary

Gas chromatographic (GC) and liquid chromatographic (LC)–mass spectrometric (MS) methods have been described to analyze pentachlorophenol and other chlorinated phenols in biological samples. After addition of internal standard (ISTD), the samples are hydrolyzed with sulfuric acid to release free phenols. The mixtures are extracted using liquid–liquid extraction or solid-phase extraction, including solid-phase microextraction. The resulting samples are injected onto GC–MS or LC–MS for analysis. Selected ion monitoring with a mass spectrometer is used to detect chlorinated phenols. The linearity is obtained in a wide range, from 0.1 to 100 ng/mL, with limit of detection at low nanograms/milliliter for the GC–MS method. The LC–MS technique in negative ion detection provides good linearity and reproducibility for chlorinated phenols using atmospheric pressure chemical ionization (APCI) ion sources. However, detection limits for the LC–MS method are higher than for the GC–MS method.

Key Words: Chlorinated phenols; gas chromatography–mass spectrometry; liquid chromatography–mass spectrometry; liquid–liquid extraction; pentachlorophenol; solid-phase extraction; solid-phase microextraction.

1. Introduction

Pentachlorophenol (PCP) and related chloronated phenols (CPs) are commonly used chemicals that have extensively distributed throughout the environment (1). CPs can be detected in human tissues following both occupational and ambient exposures (2–4). These compounds are considered to be procarcinogen in rodents (5) and possibly in humans (6,7), for whom the liver is the target organ of toxicity and carcinogenicity. PCP and some of the CPs have been included in the US Environmental Protection Agency list of 11 priority pollutant phenols in water (8). Gas chromatography (GC) and liquid chromatography (LC) coupled with mass spectrometry (MS) have

From: *Methods in Biotechnology, Vol. 19, Pesticide Protocols*
Edited by: J. L. Martínez Vidal and A. Garrido Frenich © Humana Press Inc., Totowa, NJ

been widely used to detect PCP and other CPs in environmental *(9–12)* and biological samples *(13–18)*. With liquid–liquid extraction or solid-phase extraction (SPE) for purifying and concentrating samples, GC–MS and LC–MS methods can provide the sensitivity and specificity needed for measurement of PCP or CPs in biological samples. This chapter discusses several extraction procedures and GC–MS and LC–MS methods for determination of PCP or CPs in biological samples. Suitable extraction and detection methods can be selected for combination use based on interest and capability of the laboratory.

2. Materials

1. Pesticide-quality solvents ethanol, ethyl acetate, toluene, isopropanol, and acetone.
2. Acetic anhydride and pyridine.
3. Standards of the CPs, including PCP and the ISTD 2,3,4,6-tetrachlorophenol (2,3,4,6-TeCP) with purity higher than 98% (e.g., Sigma-Aldrich, Milwaukee, WI).
4. Analytical-grade sulfuric acid, potassium carbonate, and sodium hydroxide.
5. High-purity water is obtained using a Millipore system (Milford, MA).
6. Human urine and plasma blank acquired from healthy human volunteers who did not take any medications for at least 1 mo.
7. Ultrahigh-purity helium (minimum purity 99.999%).
8. Centrifuge.
9. C18 SPE cartridges (Bond Elut SPE 300 mg, 5-cc reservoir; e.g. Varian, Harbor City, CA).
10. C18 SPE disk (47 mm id, 3M, MN).
11. Solid-phase microextractor (SPME) fiber coated with an 85-μm film thickness polyacrylate (e.g., Supelco, Bellefonte, PA).
12. Gas chromatograph (e.g., Hewlett-Packard MS Engine GC–MS, Palo Alto, CA) equipped with a split/splitless programmed temperature injector and an autosampler.
13. A 3-m fused-silica capillary column for GC with 0.25-mm id, 0.5-μm thickness (e.g., DB-5.635, J&W Scientific, Folsom, CA).
14. Liquid chromatograph with an autosampler and an MSD mass spectrometer (e.g., Agilent 1100 series HPLC, Agilent Technologies, Palo Alto, CA).
15. A 150 × 2.1 mm id, 3-μm particle size column for LC (e.g., Thermo Hypersil, Bellefonte, PA).

3. Methods (see Note 1)

3.1. Preparation of Standard and Quality Control Samples

1. For primary solutions (200 μg/mL), weigh 20 mg of each standard, as well as of the ISTD, into a 100-mL volumetric flask and fill the flask with isopropanol (*see* **Note 2**).
2. For the secondary solutions (2 μg/mL), transfer 1 mL each of the primary solution for all standards into a 100-mL volumetric flask and fill the flask with isopropanol:water (50:50 v/v) to obtain a work solution containing all the target compounds (*see* **Note 2**).
3. For the ISTD solution (500 ng/mL), transfer 250 μL of the ISTD primary solution into a 100-mL volumetric flask and fill the flask with acetone (*see* **Note 2**).
4. Spike urine or plasma with aliquots of secondary solution to obtain 0.1, 0.5, 1, 5, 10, 50, 75, and 100 ng/mL concentrations of the calibration standards.
5. Urine and plasma pools spiked with CPs at 0.1, 5, 10, and 100 ng/mL are prepared for quality control (QC) (*see* **Note 3**).
6. The calibration standards and QC standards are aliquoted and stored in a freezer at approx −20°C.

3.2. Liquid–Liquid Extraction

1. Thaw the frozen urine or plasma samples in a water bath at room temperature and centrifuge at 2000*g* for 5 min at 4°C.
2. Place a 5-mL aliquot of each urine sample into a screw-capped test tube.
3. Add 50 μL of a solution of ISTD (500 ng/mL) in acetone.
4. Hydrolyze the samples with 120 μL of concentrated sulfuric acid at 100°C for 60 min.
5. Cool to room temperature and extract the urine samples with 4 mL of toluene by agitating the mixture for 2 min.
6. Centrifuge the samples for 5 min at 3000*g* to break up the emulsions.
7. Remove the upper organic layer, acetylate with 100 mL of acetic anhydride-pyridine (5:2 v/v), and then agitate for 2 min.
8. Remove the excess acetylation reagent by shaking with 2 mL of 100 m*M* potassium carbonate buffer for 2 min.
9. Centrifuge, then take the organic layer, and dry with sodium sulfate.
10. For plasma samples, dilute 1 mL of the plasma with the same volume of water before addition of ISTD.
11. Extract the plasma samples in the same way as for urine but without hydrolysis (*see* **Note 4**).
12. Concentrate the extracts to approx 0.5 mL and directly inject onto GC–MS for analysis (*see* **Note 5**).
13. The samples are underivatized for LC–MS analysis.
14. The extracts need to be dried under nitrogen and resolubilized in 20 μL of acetonitrile before LC–MS analysis.

3.3. Solid-Phase Extraction

1. Place cartridges or disks on a vacuum manifold and condition with 2 mL of methanol and 2 mL of water.
2. Pass the urine, after hydrolysis, or the diluted plasma through the cartridge or disk.
3. Elute with 2 mL of ethyl acetate, concentrate by evaporation under a steam of nitrogen, and then resolubilize in 20 μL of acetonitrile before GC–MS or LC–MS analysis.

3.4. Solid-Phase Microextraction

1. Add HCl to adjust the pH of the urine samples to pH 1.0 (*see* **Note 6**).
2. Condition the fibers under helium in the hot injector of the gas chromatograph at 300°C for 2–3 h prior to use.
3. Perform SPME in a 40-mL vial containing 25 mL of urine and equipped with a 1-in. stir bar and a stirring plate. During the extraction, the urine samples are continuously agitated at around 700*g*.
4. Continue the extraction of PCPs and CPs in urine samples with SPME for 50 min (*see* **Note 7**).
5. Set the needle on the SPME manual holder at its maximum length (4 cm) in the GC injector.
6. Use a desorption temperature of 290°C for 2 min, which produces the highest sensitivity for detection of PCPs and CPs.

3.5. Gas Chromatography–Mass Spectrometry

1. Perform GC–MS analyses using injector in the splitless mode.
2. Keep helium carrier gas at a rate of 1 mL/min. For fiber injection, the injector is held isothermally at 290°C. The temperature of the transfer line is held at 310°C.

3. Maintain the ion source of the mass spectrometer at 250°C and 150°C for electronic impact and chemical ionization (CI), respectively.
4. Set the column initially at 60°C, ramped at 30°C/min to 190°C and from 190 to 310°C at a rate of 10°C/min.
5. Use electron impact ionization with a 70°C electron energy, positive chemical ionization (PCI) and negative chemical ionization (NCI) with methane as reagent gas to trace the optimum ionization mode for analysis of PCPs and CPs.
6. Use single-ion monitoring (SIM) to detect the phenols and ISTD at the nominal molecular mass and isotopic molecular mass at a constant ratio of CPs (**Table 1**). The ions used for quantitation are the molecular ions of CPs (*see* **Note 8**).

3.6. Liquid Chromatography–Mass Spectrometry

1. Use a mobile phase of (A) aqueous ammonium acetate/acetic acid (5 m*M*, pH 4.5), (B) acetonitrile, and (C) methanol at a flow rate of 0.2 mL/min. The elution program starts at an initial composition of A:B:C (60:30:10 v/v/v), and the isocratic step is maintained up to 29 min; then, the composition is programmed to A:B:C (12:86:2 v/v/v) in 5 min and held at this composition for 6 min. Finally, the mobile phase is returned to the initial composition in 2 min and reequilibrated for 10 min before the next injection.
2. Atmospheric pressure chemical ionization (APCI) in negative mode is used with a 400°C vaporizer temperature; −4000-V needle voltage; 70-V fragmentor potential; nebulizer and dry gas (nitrogen) at 60 psi and 12 L/min, respectively. The APCI interface and MS parameters are optimized to obtain maximum sensitivity at unit resolution.
3. Use SIM to enhance sensitivity by monitoring [M-H]⁻ ions (*see* **Note 8**).

3.7. Quantitation and Performance Characteristics

1. Use SIM mode analysis to generate the peak areas of CPs and ISTD.
2. Calculate the ratios of the peak areas of CPs to ISTD.
3. Build calibration curves by a weighted (1/Concentration^2) least-squares linear regression analysis.
4. Estimate concentrations of CPs in the samples using the equations from the appropriate calibration curves.
5. For GC–MS analysis, linearity of the standard curves is between 0.1 and 100 ng/mL. The lowest limit of quantitation (LLOQ) is 0.1 ng/mL (*see* **Note 9**) for all CPs.
6. For LC–MS analysis, standard curves are linear from 5 to 100 ng/mL, with a 5 ng/mL LLOQ.
7. Acceptable precision (≤15%) and accuracy (100 ± 15%) are obtained for concentrations above the sensitivity limit and within the standard curve range for both GC–MS and LC–MS analysis.

4. Notes

1. All glassware should be silanized prior to use by socking the glassware overnight in toluene solution at a concentration of 10% dichlorodimethylsilane. The glassware is rinsed with toluene and methanol, then dried in an oven for 2 h. Commercially available silanized glassware can be used without further silanization.
2. The stock solutions were kept refrigerated (4°C) and discarded 1 mo after preparation.
3. The calibration standards and QC samples are prepared with appropriate volumes of stock solutions in 100-mL volumetric flasks by diluting to the volumes with human urine or plasma to achieve the desired concentrations.

Table 1
Analytical Conditions of CPs as Determined by GC–MS

No.	Compound	t_R (min)	Selected ion/confirmed ion (isotope ratio)		
			EI	PCI	NCI
1	2-Chlorophenol (2-CP)	3.35	128/130 (3:1)	129/131 (3:1)	128/130 (3:1)
2	2,4-Dichlorophenol (2, 4-DCP)	4.34	162/164 (3:2)	163/165 (3:2)	162/164 (3:2)
3	2,4,6-Trichlorophenol (2,4,6-TCP)	5.34	196/198 (1:1)	197/199 (3:1)	196/198 (1:1)
4	2,3,4,6-Tetrachlorophenol (2,3,4,5-TeCP)	6.64	232/230 (4:3)	233/231 (4:3)	232/230 (4:3)
5	Pentachlorphenol (PCP)	8.17	266/264/268 (15:9:10)	267/269 (3:2)	230/232 (3:2)

Reprinted with permission from **ref. 14**. Copyright 1998 Elsevier.

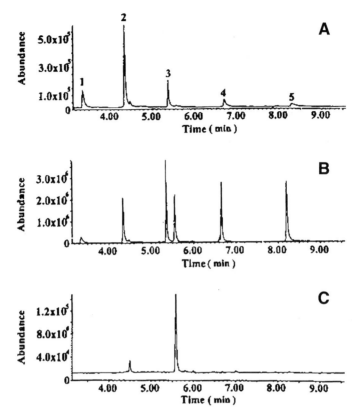

Fig. 1. Mass ion chromatogram produced by SPME–GC–MS of (**A**) 25 µg/L CPs in urine at pH 7.0 and (**B**) at pH 1.0; (**C**) blank urine. Numbers on the peaks correspond to those in **Table 1**. Reprinted with permission from **ref. *14***. Copyright 1998 Elsevier.

4. CPs are excreted in urine as free forms and sulfate and glucuronide conjugates, with the amount of conjugation depending on the particular CP and its concentrations in urine. However, CPs exist as free forms in plasma.

5. Acylation has been used in derivatization of CPs for GC–MS analysis for several reasons. First, acylation reduces the polarity of the phenol groups, which in the underivatized forms may not chromatograph very well because of nonspecific absorption effects, such as tailing and ghost peaks. Second, acylation improves volatility on CPs, making them more accessible to GC analysis. Third, acylation may help to separate closely related CPs, which in the underivatized state may be difficult to resolve.

6. Extraction efficiency of CPs can be enhanced by decreasing the solubility of CPs in biological fluids. It is found that the best recovery is achieved by adjusting pH to approx 1.0 (**Fig. 1**). Because the extraction enhancement is related to the K values of CPs, a higher increased factor is obtained with higher K value compounds. The extraction recovery for PCP at pH 1.0 is approx 9 greater than that obtained in the original solution (pH 8.0). Addition of saturated salts such as sodium chloride and potassium chloride combined with acidification did not improve the extraction efficiency of CPs.

7. The efficiency of the extraction depends on the time. Good recoveries can be obtained with time extraction longer than 40 min.

8. The use of the SIM technique in GC–MS and LC–MS analysis offers a second dimension for compound identification. In addition to retention time, the ratios of ion clusters caused by chlorine isotopes are used for identifying the CPs. The following criteria are used in the identification (otherwise, no compounds are considered detected): at the retention times of analytes deviation between the isotopic ratios obtained from the analysis, and the theoretical values should be less than 13%. Also, for quantitation, the peak will be calculated only when the signal-to-noise ratio is greater than 3.

9. LLOQ is the lowest concentration of QCs that meets validation criteria with a precision of $\leq 15\%$ and accuracy of $\leq 100 \pm 15\%$.

References

1. Jensen, J. (1996) Chlorophenols in the terrestrial environment. *Rev. Environ. Contam. Toxicol.* **146**, 25–51.

2. Ahlborg, U. G., Lindgren, J. E., and Mercier, M. (1974) Metabolism of pentachlorophenol. *Arch. Toxicol.* **32**, 271–281.

3. Ahlborg, U. G. and Thunberg, T. (1980) Chlorinated phenols: occurrence, toxicity, metabolism and environmental impact. *CRC Crit. Rev. Toxicol.* **7**, 1–35.

4. Kutz, F. W. and Cook, B. T. (1992) Selected pesticide residue and metabolites in urine from a survey of the US general population. *J. Toxicol. Environ. Health* **37**, 277–291.

5. Schwetz, B. A., Quast, J. F., Keeler, P. A., Humiston, C. G., and Kociba, R. J.(1978) Results of 2-year toxicity and reproduction studies on pentachlorophenol in rats, in *Pentachlorophenol: Chemistry, Pharmacology, and Environmental Toxicology* (Rao, K. R., ed.), Plenum Press, New York, pp. 301–309.

6. Hardell, L. and Sandstorm, A. (1979) Case–control study: soft-tissue sarcomas and exposure to phenoxyacetic acid or chlorophenols. *Br. J. Cancer* **39**, 711–717.

7. Pearce, N. E., Smith, A. H., Howard, J. K., Sheppard, R. A., Giles, H. J., and Teague, C. A. (1986) Non-Hodgkin's lymphoma and exposure to phenoxyherbicides, chlorophenols, fencing work, and meat work employment: a case–control study. *Br. J. Ind. Med.* 43, 75–83.

8. (1997) *Sampling and Analysis Procedure for Screening of Industrial Effluents for Priority Pollutants.* U.S. Environment Monitoring and Support Laboratory, Cincinnati, OH.

9. Jauregui, O., Moyano, E., and Galceran, M. T. (1997) Liquid chromatography–atmospheric pressure ionization mass spectrometry for the determination of chloro- and nitrophenolic compounds in tap water and sea water. *J. Chromatogr. A* **787**, 79–89.

10. Ribeiro, A. Neves, M. H., Almeida, M. F., Alves, A., and Santos, L. (2002) Direct determination of chlorophenols in landfill leachates by solid-phase micro-extraction-gas chromatography-mass spectrometry. *J. Chromatogr. A* **975**, 267–274.

11. Hanada, Y., Imaizumi, I., Kido, K., et al. (2002) Application of a pentaflurobenzyl bromide derivatization method in gas chromatography/mass spectrometry of trace levels of halogenated phenols in air, water and sediment samples. *Anal. Sci.* **18**, 655–659.

12. Sarrion, M. N., Santos, F. J., Moyano, E., and Galceran, M. T. (2003) Solid-phase microextraction liquid chromatography/tandem mass spectrometry for the analysis of chlorophenols in environmental samples. *Rapid Commun. Mass Spectrom.* **17**, 39–48.

13. Hargesheimer, E. E. and Coutts, R. T. (1983) Selected ion mass spectrometric identification of chlorophenol residues in human urine. *J. Assoc. Off. Anal. Chem.* **66**, 13–21.

14. Lee, M., Yeh, Y, Hsiang, W., and Chen, C. (1998) Application of solid-phase microextraction and gas chromatography–mass spectrometry for the determination of chlorophenols in urine. *J. Chromatogr. B* **707**, 91–97.

15. Crespin, M. A., Gallego, M., and Valcarcel, M. (2002) Solid-phase extraction method for the determination of free and conjugated phenol compounds in human urine. *J. Chromatogr. B* **773,** 89–96.
16. Kontsas, H., Rosenberg C., Pfaffli, P., and Jappinen, P. (1995) Gas chromatographic–mass spectrometric determination of chlorophenols in the urine of sawmill workers with past use of chlorophenol-containing anti-stain agents. *Analyst* **120,** 1745–1749.
17. Treble, R. G., and Thompson, T. S. (1996) Normal values for pentachlorophenol in urine samples collected from a general population. *J. Anal. Toxicol.* **20,** 313–317.
18. Hovander, L., Malmberg, T., Athanasiadou, M., et al. (2002) Identification of hydroxylated PCB metabolites and other phenolic halogenated pollutants in human blood plasma. *Arch. Environ. Contam. Toxicol.* **42,** 105–117.

10

Analysis of 2,4-Dichlorophenoxyacetic Acid in Body Fluids of Exposed Subjects Using Radioimmunoassay

Dietmar Knopp

Summary

The development of a radioimmunological method (radioimmunoassay, RIA) for the determination of the herbicide 2,4-dichlorophenoxyacetic acid (2,4-D) in human body fluids such as blood serum and urine is described. It comprises three major parts. The first part deals with the generation of a polyclonal anti-2,4-D antiserum in rabbits. This includes the preparation, purification, and characterization of 2,4-D–protein conjugates (immunogens), immunization of rabbits, and evaluation of antiserum quality. A guide is presented that with high probability will lead to useful antibodies. The second part outlines the synthesis and purification of the tracer (i.e., a tritiated 2,4-D with high specific activity) that is selectively labeled in the 6-position of the aromatic moiety. The third part describes carrying out the RIA, setting up the calibration curve, cross-reactivity testing, and evaluation of matrix interferences.

Key Words: Cross-reactivity; 2,4-dichlorophenoxyacetic acid; exposed subjects; herbicide; immunization; immunoassay; immunogens; matrix interferences; polyclonal antibodies; radioimmunoassay; serum; tracer synthesis; urine.

1. Introduction

Immunoassay technology has been used successfully for decades in the clinical laboratory as an extraordinarily effective analytical approach for both low and high molecular analytes of clinical importance *(1)*. The use of immunoassay has further expanded to application for the determination of chemical pollutants in environmental monitoring and human exposure assessment *(2)*.

The new technique opened its road to success with the use of radioisotopes as labels. Radioactivity can be measured very easily and precisely; therefore, radioimmunoassays (RIAs) are known to be robust and less affected by matrix interferences. As a disadvantage, instrumentation is rather complex, and dealing with radioactive materials is subject to strict regulations governing use and disposal. As a consequence, development and use of RIA is restricted to special laboratories only.

From: *Methods in Biotechnology, Vol. 19, Pesticide Protocols*
Edited by: J. L. Martínez Vidal and A. Garrido Frenich © Humana Press Inc., Totowa, NJ

The search for alternative labels led to the enzymeimmunoassay (EIA), that is, the use of enzymes as labels in conjunction with a suitable substrate to produce an assay signal. As radioisotopes, enzymes offer a wide range of formats as well. RIA appears to have diminished in its significance as judged by corresponding publications. However, radioisotopes are widely used, particularly in research.

In this chapter, the development of an RIA for the herbicide 2,4-dichlorophenoxy-acetic acid (2,4-D) is described *(3)*. It involves the generation of a polyclonal anti-2,4-D antiserum in rabbits, the synthesis and purification of a tritiated 2,4-D that is selectively labeled in the 6-position of the aromatic moiety, and carrying out the RIA. Corresponding research was initiated in the early 1980s when the World Health Organization asked for data on the individual body burden of occupationally exposed workers to enable better assessment of the potential health hazard of this chemical *(4)*. The RIA was successfully used to determine 2,4-D in different samples, such as blood and urine samples of (1) employees in the chemical industry, (2) herbicide sprayers in agriculture and forestry industries, and (3) accidentally exposed persons *(5–7)*. The assay was further applied to study herbicide resorption and excretion in rats and the behavior of 2,4-D, if present in the dialysate solution, during hemodialysis treatment *(8,9)*.

According to the described schedule, RIA development is possible also with other chlorinated phenoxyalcanoic acids. As an example, the immunoassay for 2-(2,4-dichlorophenoxy) propionic acid should be mentioned *(10)*.

2. Materials

1. 2,4-D.
2. Research-grade crystalline bovine serum albumin (BSA).
3. Tri-*n*-butylamine.
4. Isobutyl chloroformate.
5. 2,4,6-Trinitrobenzenesulfonic acid (TNBS).
6. Ultraviolet/visible (UV/Vis) spectrophotometer.
7. Sodium azide.
8. Dialysis tubing.
9. Freund's complete and incomplete adjuvants (e.g., Difco Laboratories).
10. 10% Pd/CaCO$_3$.
11. Liquid scintillation counter (LSC)/beta scintillation counter (LSC).
12. Liquid scintillator cocktail such as Ultima cocktails (e.g., Perkin Elmer, Boston, MA).
13. Silica gel thin-layer chromatographic (TLC) plates.
14. Thin-layer scanner.
15. RIA buffer (dilution buffer for antiserum, tracer, and samples): 50 mM sodium phosphate buffer, pH 7.4, 150 mM NaCl, 0.1% sodium azide, 0.1% gelatin (stable at 4°C for up to 1 mo).
16. UV lamp.
17. Human γ-globulin (HGG).
18. Polyethylene glycol (PEG) solution: 36.7 mM PEG 6000 in RIA buffer (stable at 4°C for up to 1 mo).
19. Dialysis buffer: 10 mM glycine buffer, pH 9.0 (stable at 4°C up to 1 mo).
20. Freeze-dryer.
21. Vortex mixer.

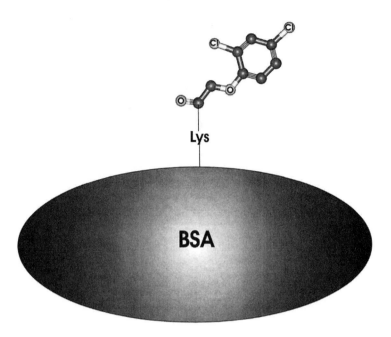

Fig. 1. Schematic presentation of the 2,4-D–BSA conjugate (inmunogen).

3. Methods

The methods described outline (1) the synthesis and characterization of a 2,4-D–protein conjugate, (2) the generation of polyclonal 2,4-D antibodies in rabbits and their characterization, (3) the synthesis and purification of [6-^{3}H]2,4-D, and (4) the performance of the RIA.

3.1. 2,4-D–Protein Conjugate

Small molecules (haptens) do not alone elicit an immune response. They have to be conjugated covalently to a larger carrier molecule before immunization (reviewed in **ref. 11**; *see* **Fig. 1**). The steps described in **Subheadings 3.1.1.–3.1.3.** outline the procedure for the preparation, purification, and characterization of the immunogen used for the subsequent immunization.

3.1.1. Preparation of 2,4-D–Protein Conjugates

Conjugates are prepared by the mixed anhydride method *(12)*.
1. Dissolve 0.1 mmol 2,4-D in 3 mL dry dioxane and add an equimolar amount of tri-*n*-butylamine.
2. Cool the solution to about 5°C (put the solution into a refrigerator).
3. Add an equimolar amount of isobutyl chloroformate, vortex the solution, and put it again into the refrigerator for about 20 min (solution A).

4. Dissolve 40 mg BSA in 6 mL distilled water and 2 mL dioxane (solution B) (*see* **Note 1**).
5. Add solution A drop by drop to solution B and keep the mixture at room temperature for at least 4 h.

3.1.2. Purification of the 2,4-D–Protein Conjugate

Methods of purification involve dialysis and gel filtration chromatography. Although more time consuming, dialysis is reliable and very simple to perform.

1. Fill the coupling mixture into a dialysis tubing and close both ends of the tubing carefully (*see* **Note 2**).
2. Take a 1-L beaker and fill it with dialysis buffer. Put the filled tubing into the buffer and dialyze for 2 d against glycine buffer (dialysis buffer) and 1 d against distilled water. Stirring with a magnetic stirrer and frequent changes of the dialysate will accelerate dialysis.
3. After finishing dialysis, open one end of the tubing and put the liquid into a glass vessel for subsequent freeze-drying (*see* **Note 3**).
4. Put the lyophilized conjugate into a tube and store it in the refrigerator.

3.1.3. Characterization of the 2,4-D–Protein Conjugate

Determination of the extent of incorporation of a hapten into the conjugate is strongly recommended. Optimum coupling density depends on the selected carrier protein, although there is no consensus regarding which ratio is best. What is important is not the molar ratio, but the packing density; that is larger proteins should have higher substitution ratios. If available, a small amount of radio-labeled hapten of known specific activity could be included in the conjugation reaction. Another possibility available is matrix-assisted laser desorption ionization time-of-flight mass spectrometry (MALDI-TOF-MS) *(13)*. As a limitation, it can only be applied for smaller immunogens such as serum albumins. A more general indirect method for conjugates that are produced by reaction of ε-lysine-amino groups is measurement of the remaining free amino groups in the conjugate. The method of Habeeb involves titration of free amino groups with TNBS *(14)*.

1. Dissolve protein conjugate at a concentration of 2.4 mg/mL in 1N NaOH.
2. To 0.3 mL of conjugate solution, add the subsequent same amount of 1N HCl, 1.4 mL 4% sodium bicarbonate solution at pH 8.5, and 1 mL of 0.1% aqueous TNBS solution. Mix and incubate for 2 h at 40°C.
3. Add 1 mL 10% sodium dodecyl sulfate solution and 0.5 mL of 1N HCl. Mix and measure absorbance at 335 nm. As a reference, use an analogously treated solution that contains no protein.
4. Calculate the number of free amino groups n according to the formula (*see* **Note 4**)

$$n = (\Delta E \ MW_{prot})/(\varepsilon_{NH2} \ C_{prot})$$

where ΔE is the difference of extinction between conjugate and reference sample, MW_{prot} is the molecular weight of protein, C_{prot} is the molar concentration of protein, and ε_{NH2} is the molar extinction coefficient of a single free amino group of the carrier protein.

3.2. Generation of Antibodies

Antibodies can be obtained from a number of sources (reviewed in **ref. *15***). The generation of polyclonal 2,4-D antibodies is described in **Subheadings 3.2.1.–3.2.4.** This includes the immunization of rabbits and the evaluation of antiserum quality.

3.2.1. Immunization of Rabbits

Immunize two or more adult rabbits (weight depends on breed) intradermally at 10 multiple sites by injecting 0.1-mL aliquots of the emulsified 2,4-D–BSA conjugate, which is prepared by dissolving 2 mg conjugate in 1 mL sterilized physiological saline solution and emulsifying with 1 mL Freund's adjuvant to give a stable water-in-oil emulsion (*see* **Note 5**). For the initial immunization (priming), Freund's complete adjuvant is used. Administer further immunizations (booster injections) 8 wk after priming and then about five times at 4-wk intervals with Freund's incomplete adjuvant (*see* **Note 6**). Seven to 10 d after each boost, bleed approx 2 mL from a marginal ear vein of each rabbit to assess the progress of the immunization by performing the 2,4-D RIA (*see* **Note 7**).

3.2.2. Evaluation of Antiserum Quality

During the course of immunization, the quantity (titer; this is the dilution of antiserum that will bind typically 50% of the mass of tracer added to the assay tube) and quality (specificity, affinity, and sensitivity to matrix interferences) of the generated 2,4-D antibodies should be checked. Therefore, from the blood taken after each boost either serum or plasma is separated from blood cells and submitted to testing by the RIA (*see* **Note 8**). For antiserum assessment, perform the RIA as described in **Subheading 3.4.1.** with some modifications.

1. Prepare an antibody dilution curve without analyte but with different antiserum dilutions (e.g., $1/10$ to $1/10^5$). Use RIA buffer as the sample matrix.
2. Plot percentage of bond tracer against antiserum dilution to obtain a sigmoidal dilution curve.
3. Repeat **steps 1** and **2** with the addition of analyte concentration equivalent to the required limit of detection. Use RIA buffer as the sample matrix. The resulting curve is termed the *displacement curve* (*see* **Note 9**).
4. Repeat **steps 1** and **2** with a serially diluted blank serum (from a nonimmunized animal) instead of antiserum to assess nonspecific binding across the antiserum dilution range. Use RIA buffer as the sample matrix.
5. Repeat **steps 1** and **2** with the addition of test compound at a concentration of about 100-fold above that of the analyte of interest, for which lack of cross-reactivity in the assay is critical. Use RIA buffer as the sample matrix.
6. Establish two calibration curves as described in **Subheading 3.4.2.** using standard dilutions prepared (1) with RIA buffer and (2) with sample medium/RIA buffer (10:90 v/v) (*see* **Note 10**).

When titer and quality are satisfactory, greater amounts of blood are collected from the ear vein (do not collect more than 50 mL of blood at one time). Separate serum and divide it into 1-mL aliquots for storage at lower temperatures (*see* **Note 7**).

3.3. Tracer Synthesis

Tracer synthesis is described in **Subheadings 3.3.1** and **3.3.2** and includes (1) the synthesis of 6-bromo-2,4-D, (2) the radio-labeling with tritium, and (3) the purification of [6-^{3}H]2,4-D (*see* **Fig. 2**).

3.3.1. Synthesis of 6-Bromo-2,4-Dichlorophenoxyacetic Acid

1. Dissolve 50 g 2,4-dichlorophenol in a mixture of 250 mL water and 250 mL acetic acid.

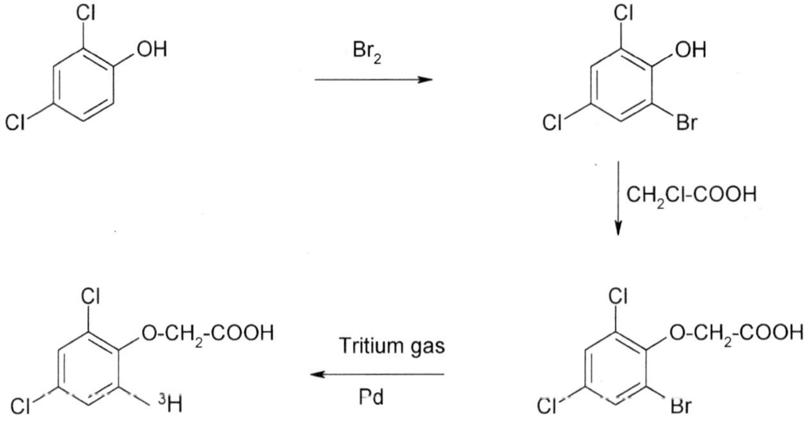

Fig. 2. Scheme of tracer synthesis.

2. Dissolve 16.5 mL bromine in 90 mL acetic acid and add to dichlorophenol solution under stirring with a glass agitator (*see* **Note 11**). After cooling, the solution grows stiff to form a pulp.
3. Add an excess of water and filtrate the formed precipitate over a paper filter.
4. Wash the precipitate extensively with about 500 mL of boiling water.
5. Recrystallize the solid precipitate in a water/ethanol mixture.
6. Check the purity of the 6-bromo-2,4-dichlorophenol, for example, by determination of its melting point (68–69°C) or by TLC using a silica gel TLC plate and cyclohexane/benzene/acetic acid/paraffin (100:15:10:5 v/v/v/v) as mobile phase over a distance of 10 cm beyond the origin. For detection, spray the dried TLC plate with silver nitrate reagent solution (2.12 g silver nitrate, 12.5 mL distilled water, 3.75 mL 5*N* ammonia solution; fill to 250 mL with acetone) and expose to UV light (e.g., CAMAG UV lamp) until the halogen-containing compounds become visible as darkish spots (*see* **Note 12**).
7. Dissolve 5 g 6-bromo-2,4-dichlorophenol in 30 mL 70% aqueous monochloroacetic acid and adjust the final pH between 10.0 and 12.0 by adding 30% sodium hydroxide solution.
8. Boil the reaction mixture under reflux and stirring for 4 h.
9. Cool the solution to room temperature, collect the precipitate on a ceramic frit, and redissolve it in diethyl ether.
10. Shake the ether solution intensively with about the same volume of a 0.5% aqueous sodium hydrogencarbonate solution in a separation funnel.
11. Discard the organic phase and acidify the aqueous phase with concentrated sulfuric acid.
12. Add about the same volume of diethyl ether to dissolve the precipitate in the organic phase.
13. Separate the ether phase in a separation funnel
14. Repeat **steps 10–13**.
15. Remove the organic solvent by vacuum evaporation.
16. Check the purity of the 6-bromo-2,4-dichlorophenoxyacetic acid, for example, by determination of its melting point (181–184°C) or by TLC as described in **step 6** (*see* **Note 13**).

3.3.2. Preparation of [6-^{3}H]2,4-Dichlorophenoxyacetic Acid

Low-level radioactive materials are often used as tracers and labels. Dealings with radioactive materials are subject to strict regulations governing use and disposal. Only trained laboratory personnel should handle radioactive materials. The following catalytic hydrogenation with ^{3}H gas is reserved for special laboratories only. Generally, the radiochemical should be put at a customer's disposal from corresponding suppliers (e.g., GE Healthcare, formerly Amersham Biosciences, Uppsala, Sweden).

1. Dissolve 0.05 mmol 6-bromo-2,4-dichlorophenoxyacetic acid in 2 mL dry dioxane in a round-bottom glass piston.
2. Add 300 mg catalyst (10% Pd/CaCO$_3$).
3. Freeze the solution with liquid nitrogen and append the piston to the vacuum-handling apparatus.
4. Exhaust the apparatus and thaw the solution by warming.
5. Induce tritium gas to start the hydrogenation. Stir the solution gently with a magnetic stirrer for the entire reaction time at room temperature.
6. Absorption of tritium is finished after about 75 min.
7. Stop the reaction after another 60 min, freeze the solution with liquid nitrogen, and take down the piston from the installation.
8. Thaw the solution and separate the catalyst by centrifugation at 1500*g* for 10 min.
9. Remove the solvent from the clear supernatant by freeze-drying.
10. Redissolve the residue in ethanol and store the solution for 48 h at room temperature (*see* **Note 14**).
11. Remove the solvent by lyophilization and redissolve the residue in 10 mL ethanol.
12. Measure total radioactivity of the solution (e.g., on an LSC).
13. Purify the [6-^{3}H]2,4-D by preparative TLC: 10-µL fractions of the ethanolic solution are applied at a distance of 3 cm on TLC plates (silica gel G 60/kieselgur G, 2:3). On a separate trace, apply nonradioactive 2,4-D as a control. Plates are developed with cyclohexane/benzene/acetic acid/paraffin (100:15:10:5 v/v/v/v) as the mobile phase over a distance of 18 cm beyond the origin. Plates are air dried. Detect nonradioactive 2,4-D by spraying the separate trace with silver nitrate solution, followed by exposure to UV light. If a radio-TLC scanner is available, it can be used to scan the radioactive traces completely. Scrape off the silica gel layer of the radioactive traces from the TLC plate in the area (distance from origin) of the nonradioactive 2,4-D spot (*see* **Note 15**).
14. Extract the silica gel with ice-cold methanol and repeat TLC (**step 13**) to confirm radiochemical purity of the compound. Store the solution at −20°C. Radiochemical purity higher than 95% can be obtained easily (*see* **Note 16**).

3.4. Radioimmunoassay

3.4.1. Procedure

The immunoassay is performed in 3-mL throwaway polyethylene tubes (RIA tubes), which fit into commercial 20-mL LSC vials. As diluent, use RIA buffer. Equilibrate all reagents to room temperature before use (*see* **Fig. 3**).

1. Add 100 µL 1:40 diluted normal rabbit serum (*see* **Note 17**).
2. Add 100 µL sample (*see* **Note 18**).
3. Add 100 µL diluted tracer [6-^{3}H]2,4-D (*see* **Note 19**).
4. Add 100 µL antiserum dilution and close the tube (*see* **Note 20**).
5. Vortex for 10 s and incubate for 24 h at 4°C (*see* **Note 21**).
6. Open the tube and add 100 µL 0.33% HGG solution (*see* **Note 22**).
7. Add 1 mL PEG solution and close the tube.

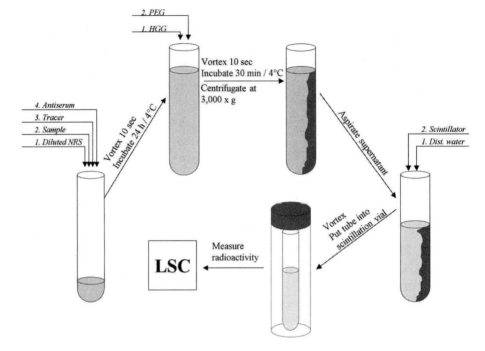

Fig. 3. Flow chart of the 2,4-D radioimmunoassay. NRS, normal rabbitt serum, from nonimmunized animal.

8. Vortex for 10 s and incubate an additional 30 min at 4°C.
9. Centrifuge at 3000g for 10 min (*see* **Note 23**).
10. Open the tube and aspirate the supernatant carefully with a Pasteur pipet (*see* **Note 24**).
11. Redissolve the pellet with 200 µL distilled water.
12. Add 1 mL scintillator cocktail and close the tube (*see* **Note 25**).
13. Vortex the tube vigorously and put it into a 20-mL glass scintillation vial (*see* **Note 26**).
14. Screw down the scintillation vial and measure the radioactivity for at least 10 min on an LSC. The measured radioactivity (e.g., given as cpm) represents the antibody bond tracer B.
15. Interpolate analyte amount of unknown sample from the calibration curve.

3.4.2. Calibration Curve

Required assay sensitivity for human biomonitoring studies is in the nanogram- or picogram-per-milliliter range. Performance must strictly be controlled for each matrix type, which is under investigation.

1. Prepare stock solution by dissolving 25 mg 2,4-D in 100 mL methanol.
2. Prepare standard solutions (10, 50, 100, 500, 1000, 2500, 5000, 10,000 ng/mL) by appropriate dilution of the methanolic stock (from 1:25 to 1:25,000; v/v) with sample medium.
3. Prepare 1:10 (v/v) dilutions of standards with dilution buffer.
4. To measure total counts T, add 100 µL diluted tracer into an empty RIA tube and mix with 1 mL scintillator cocktail.
5. Prepare all samples in triplicate and measure radioactivity.

6. Calculate data, for example, with weighted linear regression analysis of logit B/B_0 vs log dose or with spline function using appropriate software (e.g., GraphPad Prism 4 or RiaCalc evaluation program) (*see* **Note 27**). Use normal rabbit serum dilution of a nonimmunized animal as a blank and subtract corresponding counts as background (nonspecific binding) from each data point (*see* **Note 20**).
7. Determine intra- and interassay precision by repetitive analysis of replicates of different standard samples over the working range of the calibration curve (*see* **Note 28**).

3.4.3. Cross-Reactivity Testing

For the user of an immunoassay, it is very important to know which analytes can be trapped in the assay. The ability of the antiserum to discriminate between structurally related chemicals, called *specificity,* can be tested by incubating a fixed amount of antiserum and tracer with varying amounts of test chemicals. The corresponding procedure is known as *cross-reactivity testing (16).* First, other chlorinated phenoxyacid herbicides and related phenols (possible metabolites) should be included (*see* **Fig. 4** and **Note 29**). Use the same procedure as described in **Subheading 3.4.1.** but prepare higher competitor concentrations of 200 to 8000 ng/mL of sample medium. Calculate percentage cross-reactivity according to the following formula:

$$\%CR = [(2,4\text{-D Concentration at } 50\% \text{ antibody binding})/(\text{Concentration of the test chemical at } 50\% \text{ antibody binding})] \times 100$$

3.4.4. Matrix Interferences

Antibodies are the key reagents in immunoassays. These proteins are sensitive to nonphysiological conditions to a different extent. Concluding from this, all physical and chemical factors that can interfere with the protein structure can also adversely affect the assay. Considering biological samples such as blood serum/plasma and urine, these are sample pH and ionic strength mainly. The effect of these parameters can be checked very easily by simple modification of the RIA buffer regarding pH, molarity, and even buffer type.

4. Notes

1. In general, 2,4-D should be coupled to a soluble carrier. Besides BSA, other proteins such as keyhole limpet hemocyanin and thyroglobulin are known to be excellent carrier proteins.
2. Most easily, plastic clips should be used. One end of the tube should be closed with a magnetic clip to allow comfortable moving of the tubing on a magnetic stirrer during dialysis.
3. Depending on the freeze-dryer, either glass pistons or glass dishes can be used.
4. Generally, we use a second carrier protein sample that is treated in the same way but without addition of the analyte (2,4-D) as a reference for the maximal absorbance (i.e., for the total number of measurable free amino groups).
5. For successful immunization, it is most important to prepare a stable emulsion. It can be tested by expelling a small droplet of the mixture onto the surface of some water in a beaker. The droplet should form a stable bead.
6. Repeated immunization with Freund's complete adjuvant will lead to severe lesions at the site of injection, which should be avoided. Besides Freund's adjuvant, a number of other immunological adjuvants such as Ribi Adjuvant, TiterMax, AdjuPrime, Adjuvax,

Compound	Structure
2,4-D	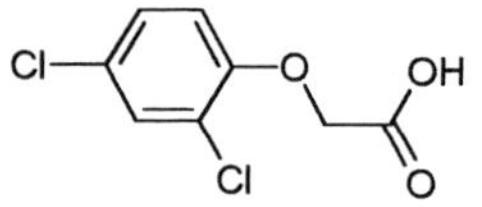
2,4-DME	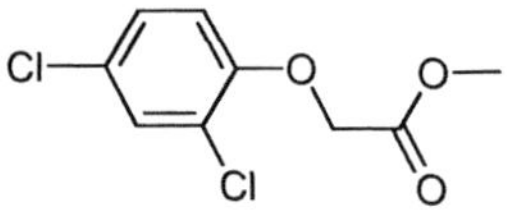
2,4,5-T	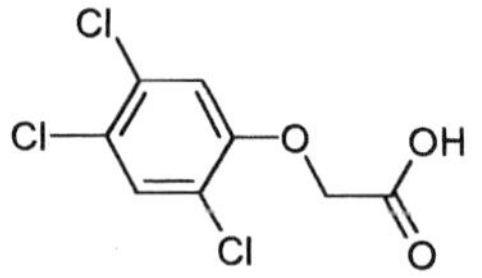
MCPA	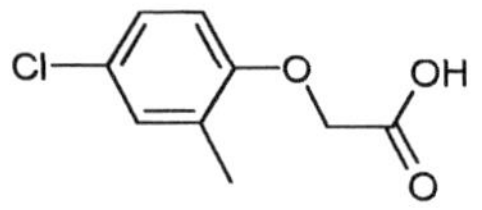
2,4-DP	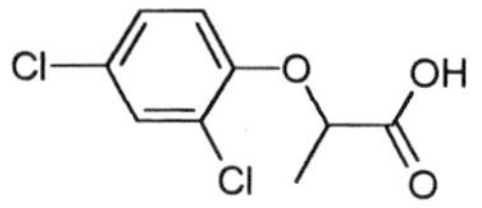
MCPP	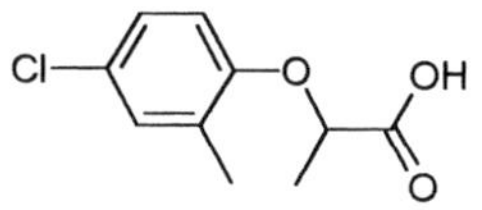
2,4-DPME	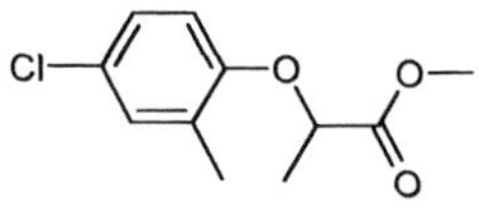
2,4-Dichlorophenol	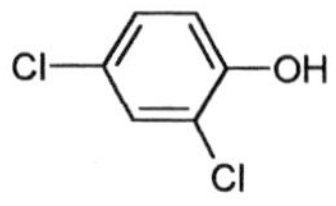
2,4-DB	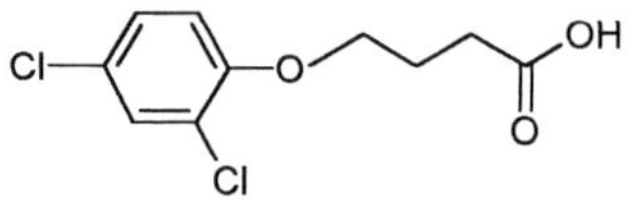
MCPB	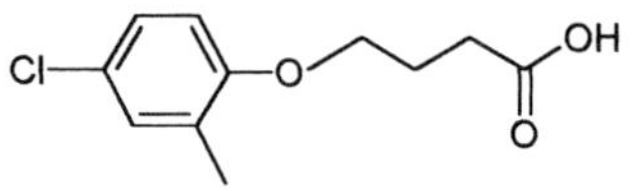

Fig. 4. Phenoxyalcanoic acid herbicides that should be included in cross-reactivity testing (2,4-D: 2,4-dichlorophenoxyacetic acid; 2,4-DME: (2,4-dichlorophenoxy)acetic acid methyl ester; 2,4,5-T: (2,4,5-trichlorophenoxy)acetic acid; MCPA: 2-methyl-4-chlorophenoxyacetic acid; 2,4-DP: 2,4-dichlorophenoxypropionic acid; MCPP: 2-(2-methyl-4-chlorophenoxy)propionic acid; 2,4-DPME: 2,4-dichlorophenoxypropionic acid methyl ester; 2,4-DB: 4-(2,4-dichlorophenoxy)butyric acid; MCPB: 4-(2-methyl-4-chlorophenoxy)butyric acid.

and others are known to enhance or modify the immune response to coadministered antigens. Generally, the examiner's experience and preference seem to be most important for the choice of adjuvant. The number of necessary booster injections varies and depends on several factors, such as type of immunogen, the animal, route of immunization, dose of antigen, choice of adjuvant, and so on.

7. Bleeding of rabbits requires a little more skill than immunizing them. The easiest method is to cut the vein with a clean razor blade. Collect the blood by allowing it to drip into a clean test tube. For preparation of serum, blood should be allowed to clot at 4°C overnight. Serum should be removed carefully with a Pasteur pipet. After preparation, serum can be stored for many years at −20°C or in liquid nitrogen.

8. Although many different techniques are available for evaluating antibodies, the method used should reflect the ultimate use of the antibody. In the present case, therefore, the 2,4-D RIA should be used. This means that the tracer and other necessary reagents must be prepared or made available very early during assay development. However, the sensitivity of the assay, which is ultimately related to the affinity of the antibodies, is dependent on the assay conditions. Work at this stage will only give an indication of what is ultimately attainable.

9. The greater the shift of the displacement curve compared to the antibody dilution curve, the higher is the affinity of the generated antibodies (i.e., the higher is the sensitivity of the assay).

10. In the case of the absence of matrix effects, both calibration curves are almost identical. Generally, matrix effect disappears using higher sample dilutions. However, this may interfere with the required sensitivity.

11. Bromine vapor is highly toxic and may lead to irritation of eyes and mucous membranes of the respiratory organs. Direct contact with the chemical should be avoided, and the experiment should be performed under a fume hood.

12. Besides the synthesized compound, the starting compound (2,4-dichlorophenol) should be applied to the plate as a reference. The retardation factor *Rf*, which is defined as the ratio of the migration distance of the analyte to the migration distance of the mobile phase, is clearly higher for the 6-bromo-2,4-dichlorophenol (compared to the nonbrominated counterpart). A yield of about 25% will be obtained.

13. A yield of about 75% will be obtained. For TLC, as reference, 2,4-D should be applied. Similar to the phenols (*see* **Note 12**), *Rf* value is higher for the brominated acid compared to the nonbrominated compound. If a radio-TLC scanner is available, it can be used to scan the radioactive traces completely. From the resulting chromatogram, the radiochemical purity of the labeled 2,4-D can easily be evaluated and its position on the TLC plate accurately be localized. In default of a scanner, the position of the non-radio-labeled 2,4-D is used as a reference for its radio-labeled counterpart.

14. During this time, tritium (which is unstable bound to the phenoxyacetic acid) is removed.

15. Purification is possible readily also by preparative high-performance liquid chromatography, if available.

16. Radiolyses (i.e., the disintegration of radio-labeled chemical compounds by radiation of nuclides during storage of the compounds) is a well-known phenomenon. It is recommended to store the radiochemical in the absence of light and at low temperature. Further, the radiochemical concentration should be low (i.e., as high a dilution as possible) should be used. The latter will be specified by the intended use. Check radiochemical purity at 12-wk intervals. If necessary, repeat purification.

17. Normal rabbit serum can be obtained from a nonimmunized rabbit. Working dilution should be prepared fresh.

18. No or only minimal sample preparation is one of the outstanding properties of immunological methods. This depends mainly on the claimed sensitivity and selectivity, the robustness of the assay, and sample matrix type. According to the 2,4-D concentration, the sample should be analyzed by the RIA at an appropriate dilution. Generally, serum and urine sample dilutions lower than 1:10 interfere with the immunological method.

19. Prepare a tracer dilution to achieve about 12,000 cpm in a volume of 100 µL. For a [6-^{3}H]2,4-D tracer with a specific radioactivity of 465 GBq/mmol, as prepared in our study, this corresponds to about 380 pg tracer per assay tube. Working dilution should be prepared fresh.

20. Ideally, a preimmune serum as a reference should be collected from each animal selected for the production of 2,4-D antibodies to estimate the progress of antibody generation in the course of the immunization (i.e., for the evaluation of different bleedings). Antiserum working dilution should be prepared fresh. Use an antiserum dilution that binds about 50% of the tracer added to the test tube.

21. During assay optimization, other conditions such as room temperature and shorter incubation times should be tested. Higher temperature may lead to faster attainment of equilibration, hence quicker assays.

22. A small amount of HGG has to be added as a bulking agent to obtain a visible white precipitate in the antibody precipitation (**step 9**) to prevent its exhaustion during aspiration (**step 10**).

23. Use a fixed-angle rotor.

24. Aspirate supernatant from the bottom of the tube opposite the precipitate.

25. Use of a commercial scintillator is recommended, which gives high counting efficiency for tritium and forms a stable microemulsion with up to 20% (v/v) of aqueous sample. If possible, use newer safety cocktails that are less hazardous (e.g., Ultima-Flo M for serum samples and Ultima Gold LLT for urine samples, both from Perkin Elmer). In any case, follow hazard warnings and safety phrases clearly visible on the packing.

26. Use of plastic throwaway tubes is more economic and comfortable because much less liquid scintillator cocktail is necessary (only approx 7% compared to the use of 20-mL glass vials), and careful repetitive cleaning of vials can be avoided.

27. B and B_0 represent counts bound by the antiserum in the presence or absence of non-radio-labeled 2,4-D (sample analyte).

28. Delete calibration standards within the quantitation range if they are not within ±20% of the nominal value and the coefficient of variation of the triplicates is not ≤15%. The curve fit can be considered acceptable if five or more points remain.

29. Be aware of interference by structurally related pharmaceuticals. As an example, in our experiments false-positive results (2,4-D present in the sample) have been caused by meclofenoxate [p-chlorophenoxyacetic acid-2-(dimethylamino)ethyl ester]. As a metabolite, 4-chlorophenoxyacetic acid is formed in the body that reacted with the 2,4-D antibodies.

References

1. Diamandis, E. P. and Christopoulos, T. K. (eds.). (1997) *Immunoassay.* Academic Press, New York.

2. Aga, D. S. and Thurman, E. M. (eds.). (1996) *Immunochemical Technology for Environmental Applications.* ACS Symposium Series 657, ACS, Washington, DC.

3. Knopp, D., Nuhn, P., and Dobberkau, H.-J. (1985) Radioimmunoassay for 2,4-dichlorophenoxyacetic acid. *Arch. Toxicol.* **58,** 27–32.

4. World Health Organization. (1984) *2,4-Dichlorophenoxyacetic acid (2,4-D).* Environmental Health Criteria 29, WHO, Geneva.

5. Knopp, D. (1994) Assessment of exposure to 2,4-dichlorophenoxyacetic acid in the chemical industry: results of a five year biological monitoring. *Occup. Environ. Med.* **51,** 152–159.
6. Knopp, D. and Glass, S. (1991) Biological monitoring of 2,4-dichlorophenoxyacetic acid-exposed workers in agriculture and forestry. *Int. Arch. Occup. Environ. Health* **63,** 329–333.
7. Knopp, D., Schmid, M., and Niessner, R. (1993) Accidental bystander overexposure to the herbicide 2,4-dichlorophenoxyacetic acid. *Fresen. Environ. Bull.* **2,** 148–150.
8. Knopp, D. and Schiller, F. (1992) Oral and dermal application of 2,4-dichlorophenoxy-acetic acid sodium and dimethylamine salts to male rats: investigations on absorption and excretion as well as induction of hepatic mixed-function oxidase activities. *Arch. Toxicol.* **66,** 170–174.
9. Knopp, D., Unger, G., and Niessner, R. (1994) Organic trace contaminants in water—a potential health hazard for chronic dialysis patients. *Zbl. Hyg.* **195,** 509–515.
10. Knopp, D., Skerswetat, M., Schmid, M., and Niessner, R. (1993) Preparation and characterization of polyclonal antibodies against the herbicide 2-(2,4-dichlorophenoxy)propionic acid (dichlorprop). *Fresenius Environ. Bull.* **2,** 274–280.
11. Hermanson, G. T. (ed.). (1996) *Bioconjugate Techniques.* Academic Press, New York.
12. Erlanger, B. F. (1973) Principles and methods for the preparation of drug protein conjugates for immunological studies. *Pharmacol. Rev.* **25** , 271–280.
13. Wengatz, I., Schmid, R., Kreissig, S., et al. (1992) Determination of the hapten density of immunoconjugates by matrix-assisted UV laser desorption ionization mass-spectrometry. *Anal. Lett.* **25,** 1983–1997.
14. Habeeb, A. F. S. A. (1966) Determination of free amino groups in proteins by trinitrobenzenesulfonic acid. *Anal. Biochem.* **14,** 328–336.
15. Harlow, E. and Lane D. (eds.) (1988) *Antibodies: A Laboratory Manual.* Cold Spring Harbor Laboratory (www.cshlpress.com).
16. Abraham, G. E. (1969) Solid-phase radioimmunoassay of estradiol-17β. *J. Clin. Endocrinol. Metab.* **29,** 866–870.

11

A High-Throughput Screening Immunochemical Protocol for Biological Exposure Assessment of Chlorophenols in Urine Samples

Mikaela Nichkova and M.-Pilar Marco

Summary

This chapter presents a high-throughput screening (HTS) immunochemical proce-dure suitable for processing and analyzing simultaneously multiple urine samples. The method presented here is addressed to assess the level of exposure of the population to certain organochlorine substances (i.e., dioxins, lindane, hexachlorobenzene, chlorophenols, etc.) by analyzing the concentration of chlorophenols in urine. The pro-cedure consists mainly of three steps. First, the urine samples are treated in basic media to hydrolyze the glucuronide and sulfate conjugates of the chlorophenols. Next, the free chlorophenol fractions are selectively isolated by immunosorbents (ISs), specially pre-pared for this purpose. Finally, this fraction is quantified with a microplate ELISA (en-zyme-linked immunosorbent assay). All solid-phase extraction (SPE) and analytical methods can be performed on a 96-plate configuration. The HTS–IS–ELISA procedure has been proved to be efficient, consistent, and reliable. The main advantages are the simplicity, the high sample throughput capabilities, and the small urine sample volumes required for each assay. With the present analytical protocol, quantitation can be accu-rately performed within 0.3 and 30 µg/L 2,4,6-trichlorophenol. About 100 samples per day can be processed with inter- and intraassay precision (%CV) below 20%, except when measurements take place at the level of the ELISA limit of detection. A protocol such as that presented here may be generally applied in environmental or biological monitoring programs for which many complex samples have to be screened as far as antibodies with the necessary specificity and affinity are available.

Key Words: Biomarker; chlorophenols; dioxins; ELISA; exposure; hexachlorobenzene; HTS; immunoaffinity extraction; lindane; 96-well immunosorbent SPE; urine.

1. Introduction

The monitoring of chlorophenols in human urine is normally used as an indication of exposure to chlorophenols or to compounds having chlorinated phenols as products of metabolism, such as chlorobenzenes, hexachlorobenzene, hexachlorocyclohexanes,

From: *Methods in Biotechnology, Vol. 19, Pesticide Protocols*
Edited by: J. L. Martínez Vidal and A. Garrido Frenich © Humana Press Inc., Totowa, NJ

and chlorophenoxycarboxylic acids *(1–3)*, which are widely used as solvents, deodorants, disinfectants, and plant-protecting agents. A continuous excretion of trichlorophenols may also indicate a risk of dioxin exposure *(4,5)*. Determination of the actual chlorophenol intake by the population would provide a reliable estimation of the individual health risk because of environmental *(6–10)* and occupational exposure *(11–16)*.

Immunochemical techniques are gaining application in the area of human exposure assessment *(17,18)*, offering rapid, simple, and cost-effective alternative approaches for analytical screening. High sample throughput is a feature of immunoassays that makes them particularly suited to large-scale biomonitoring studies. Numerous urinary immunoassays (enzyme-linked immunosorbent assay, ELISA) have been developed as screening tools to a variety of industrial chemicals, trace contaminants, and pesticides (for review, *see* **refs.** *19–21*). Mostly, immunoassay interferences are avoided by simple dilution of the biological sample, which means simply reducing, but not removing, the interfering component. However, the interferences caused by matrix components may vary from sample to sample, especially in complex biological matrices from a large number of individuals. Thus, removing the interfering compounds would result in more controlled analysis for human biomonitoring purposes.

Immunoaffinity extraction, using immunosorbents (ISs) as solid-phase extraction (SPE) stationary phases, of low molecular weight analytes from complex environmental and biological matrices provides highly selective extraction based on specific molecular recognition and minimizes reliance on organic solvents to achieve efficient separation *(22–24)*. Furthermore, antibody cross-reactivity allows multiresidue analysis targeting of a parent compound and its metabolites or a class of structurally related analytes.

Here, we present an application of extraction performed in 96-column format coupled to ELISA (high-throughput screening–immunosorbent–ELISA, HTS-IS-ELISA) for the urinary detection of chlorophenols as an analytical tool for biological exposure assessment. The detailed evaluation and optimization of the HTS–IS–ELISA method for chlorophenol urinary determination was reported in our previous work *(25)*. The optimized HTS-IS-ELISA protocol presented here consists of three main steps: urine hydrolysis, chlorophenol extraction by HTS-IS-SPE, and ELISA quantification. The method was validated by gas chromatography coupled to mass spectrometry (GC–MS) analysis of more than 100 urine samples from different individuals, and an excellent correlation between both methods was observed *(25)*.

The methodology described here could be applied to the immunochemical determination of other urinary metabolites if specific antibodies against them are available. In this case, it is necessary to optimize the HTS–IS–SPE conditions (sample volume, washing and elution, loading level effects, etc.) to achieve quantitative extraction and to eliminate possible matrix effects on the ELISA. Furthermore, the HTS–IS–SPE of chlorophenols can be used as a cleanup method prior to their chromatographic identification and determination by GC-MS and gas chromatography with electron capture detection (GC/ECD).

2. Materials

2.1. Standards and Quality Controls

1. 2,4,6-Trichlorophenol (2,4,6-TCP).
2. Ethanol (EtOH).
3. Phosphate-buffered saline (PBS) buffer: 10 mM phosphate buffer, 0.8% saline solution, pH 7.5.
4. Dimethylsulfoxide (DMSO).

2.2. Urine Samples

1. Urine samples (*see* **Note 1** regarding their collection). Keep urine samples aliquoted (12 mL) at –30°C. Volume needed for two replicate analyses is 12 mL.
2. Pooled urine sample from different individuals (who tested negative for specific drugs of abuse; e.g., Bio-Rad Laboratories) is used as a control matrix.
3. 15M KOH.
4. Concentrated H_2SO_4.
5. Sand bath.

2.3. High-Throughput Screening–Immunosorbent–Solid-Phase Extraction

1. Immunosorbent based on immunoglobulin G (IgG) fraction of polyclonal antisera As45 against 2,4,6-TCP *(26,27)* coupled to NHS (*N*-hydroxysuccinimide) activated Sepharose® 4 Fast Flow (e.g., Pharmacia Biotech). About 25 mL of gel suspension is needed to fill 96 columns (*see* **Note 2**).
2. VersaPlate 96-Well SPE System consisting of a 96-well baseplate, 96 removable empty columns, and a vacuum manifold set (e.g., Varian, Sunnyvale, CA).
3. Vacuum controller (mechanical gages).
4. VersaPlate accessories (disposable waste reservoir, cartridge removal tool, 20-mm pore frits, 96 glass vials (0.75 mL) in a collection rack, 96-well microplate Teflon-coated silicone rubber seal, sealing tape pads, and sealing caps) (e.g., Varian).
5. Multichannel electronic pipet (50–1200 mL) (e.g., Eppendorf, Hamburg, Germany).

2.4. Enzyme-Linked Immunosorbent Assay

1. Polyclonal antisera As43 against 2,4,6-TCP *(26,27)*. A working aliquot is stored at 4°C and remains stable for about 6 mo.
2. Coating antigen (8-BSA) is 3-(2-hydroxy-3,6-dichlorophenyl) propanoic acid coupled to bovine serum albumin (BSA) by the active ester method *(27)*. A stock solution of 1 mg mL^{-1} is prepared in 10 mM PBS and kept at 4°C. The solution is stable for 1 mo.
3. 2,4,6-TCP. *See* **Subheading 3.1.** regarding preparation of the stock solution.
4. Goat antirabbit IgG (Sigma, A-8275). Store at –20°C in working aliquots. For continuous use, store at 0–5°C.
5. PBS buffer: 10 mM phosphate buffer, 0.8 % saline solution, pH 7.5.
6. 7% EtOH/PBS: 10 mM PBS with 7% EtOH.
7. PBST buffer: PBS with 0.05% Tween 20.
8. Coating buffer: 50 mM carbonate–bicarbonate buffer, pH 9.6; store at 4°C.
9. Citrate buffer: 40 mM solution of sodium citrate at pH 5.5; store at 4°C.
10. 0.6% TMB (3,3',5,5'-tetramethylbenzidine) is prepared by dissolving 6 mg of TMB base in 1 mL DMSO (dimethylsulfoxide). This stock solution should be stored at room temperature and protected from light.

11. For 1% H_2O_2, dilute 100 μL of 30% H_2O_2 in 3 mL water. This solution is stored in the refrigerator on a plastic recipient for about 1 mo.
12. ELISA substrate solution: 0.01% TMB and 0.004% H_2O_2 in citrate buffer. Immediately before use, for 25 mL of citrate buffer mix 400 μL TMB (0.6% DMSO) and 100 μL 1% H_2O_2.
13. Polystyrene microtiter 96-well plates (e.g., Nunc).
14. Microplate washer.
15. Microplate spectrometer.
16. Software: ELISA competitive curves are analyzed using SoftmaxPro (Molecular Devices).

3. Methods

The methods described outline (1) the preparation of quality controls (QCs), (2) the urine hydrolysis (deconjugation step), (3) the HTS–IS–SPE extraction of chlorophenols, and (4) their ELISA quantification.

3.1. Preparation of Standards and QCs

1. The stock solution of 2,4,6-TCP needed for the preparation of the ELISA standard curve is 1 mM solution prepared in DMSO, and it should be stored at 4°C.
2. For QCs, a stock solution of 2,4,6-TCP at a concentration of 1 g/L is prepared in ethanol. Standards of 2,4,6-TCP (80, 16, 8, and 1 mg/L concentrations) in PBS buffer (buffer QCs) are prepared by further dilution with PBS. Urine QCs of the same concentrations are prepared in a hydrolyzed pooled urine sample (*see* **Subheading 3.2.** for hydrolysis). The urine QCs (6 mL each) are prepared by spiking with the corresponding volume of 500 mg/L 2,4,6-TCP standard solution prepared in PBS. All stock and QC solutions are stored at 4°C.

3.2. Urine Hydrolysis (Deconjugation of Chlorophenol–Glucuronides and Sulfates)

In biological monitoring studies it is important to determine the total urinary chlorophenol concentration (free and conjugated) (*see* **Note 3**). Therefore, the first step in the urine analysis is the hydrolysis of the chlorophenol conjugates. Here, we suggest alkaline hydrolysis (*see* **Note 4**).

1. Add 2 mL 15M KOH to 10 mL urine sample in a glass vial, seal it, and heat it on a sand bath for 30 min at 100°C.
2. Cool the sample to room temperature and neutralize it to pH 7.5 by adding dropwise concentrated H_2SO_4 and mixing very well (*see* **Note 5**).
3. Centrifuge (10 min at 1520g) if a precipitation of salts or proteins is formed and use the supernatant.

3.3. High-Throughput Screening–Immunosorbent–Solid-Phase Extraction

The detailed optimization, evaluation, and validation of the high-throughput–immunoaffinity–solid-phase extraction (HTS–IS–SPE) (*see* **Note 6**) of chlorophenols from urine samples were described in our previous work *(25)*. We have also demonstrated that the extracts obtained after the HTS–IS–SPE have no matrix effects on the ELISA measurements *(28,29)*. Here, we present the optimized HTS–IS–SPE procedure that might be followed for efficient chlorophenol extraction from urine samples.

3.3.1. HTS–IS–SPE Procedure

The HTS–IS–SPE procedure is performed with the VersaPlate 96-Well SPE System, which consists of a vacuum manifold set equipped with a vacuum controller and water pump. All steps of the HTS–IS cleanup cycle are performed under gentle low vacuum, maintaining a flow rate in the range 1–2 mL/min. The vacuum is manually controlled to allow the different solvents (buffers, samples) to pass through the IS extraction columns: The columns should be kept wet after both conditioning and loading steps and dried after the washing and elution steps. To protect the immunosorbent from drying in the conditioning and loading steps, the vacuum should be stopped as soon as the first columns are drained. All liquid loadings are done manually using an eight-channel electronic pipet. During sample loading, washing, and regeneration steps, waste is collected in the disposable reservoir. Before elution, the waste reservoir should be replaced with the collection rack of 96 glass vials (0.75 mL).

The HTS–IS–SPE cycle consists of conditioning, sample loading, washing of the unbound material, eluting of the specifically retained analyte, and regenerating of the column for a next cycle (*see* **Note 7**).

1. Bring the columns to room temperature and wash them with 5 mL PBS.
2. Condition the columns by washing with 1.2 mL 70% EtOH followed by 1.2 mL PBS.
3. Load each column with 6 mL sample (spiked PBS, urine samples, and QCs).
4. Wash the columns with 1.2 mL PBS (if you want to elute the chlorophenol family) or with 20% EtOH for more specific extraction (*see* **Note 8**).
5. Elute the bound chlorophenols with 0.6 mL 70% EtOH. For ELISA analysis, 0.1 mL of the eluted fractions is diluted 10 times with 10 mM PBS to a 7% EtOH content (*see* **Note 9**).
6. Regenerate the columns with 1.2 mL PBS.
7. Dilute the collected extract 10 times with PBS and keep at 4°C sealed with a 96-well microplate silicone rubber seal. Analyze the extracts by ELISA curve run in 7% EtOH/ PBS or by chromatographic analysis.
8. When not in use, store the VersaPlate 96-Well IS–SPE assembly sealed with caps at 4°C in PBS containing 0.1% NaN$_3$.

3.3.2. Quality Assurance

3.3.2.1. CONTROL OF COLUMN PREPARATION

As the ISs are packed manually, it is important to evaluate the variation in the capacity of different columns. An experiment can be designed in which different columns placed in different positions of the rack holder are loaded with different amounts of 2,4,6-TCP. **Table 1** presents the recoveries obtained for HTS–IS–SPE of 2,4,6-TCP from PBS standards of different concentrations (buffer QCs).

It should be noted that at the maximum capacity the columns are very reproducible (%CV = 4.69, $N = 10$). The highest CV of 27.19% ($N = 16$) is observed for the lowest loading level (0.72%). In general, in spite of the manual packing, the reproducibility within columns is very good. If any of the prepared columns does not satisfy the above characteristics, it should be discarded and replaced by a new column.

3.3.2.2. CONTROL OF IMMUNOSORBENT BINDING CAPACITY (*SEE* **NOTE 10**)

We strongly recommend the use of buffer and/or urine QCs in each HTS–IS–SPE run to control the performance of the extraction procedure. A matrix blank (nonspiked

Table 1
Recovery Obtained for HTS–IS–SPE of 2,4,6-TCP From PBS Standards
(PBS Washing, 70% EtOH Elution in 0.6 mL)

Loading level[a] (%)	2,4,6-TCP (ng)	N[b]	Recovery ± SD (%)
100	500	10	101.05 ± 4.74
20	100	3	84.81 ± 13.25
2.4	12	16	97.75 ± 16.84
0.72	3.6	16	95.77 ± 26.04

[a]*See* **Note 2**.
[b]*N* is the number of mini-ISs used.

Table 2
Features of the 2,4,6-TCP Immunoassay (As43/8-BSA)

Parameter	Values[a]
A_{max}	0.796 ± 0.174
A_{min}	0.018 ± 0.021
IC_{50} (µg/L)	1.132 ± 0.361
Dynamic range	0.288 ± 0.045 to 3.117 ± 0.679
Slope	1.22 ± 0.23
LOD (µg/L)	0.175 ± 0.027
r^2	0.991 ± 0.006

LOD, limit of detection.

[a]The parameters are extracted from the four-parameter equation used to fit the standard curve. The data presented correspond to the average of 34 calibration curves run in different plates in a 3-mo period. Each curve is built using two-well replicates.

pooled urine) can be used as a control for possible contamination. In addition, buffer QCs (500, 100, 50, and 10 ng 2,4,6-TCP) corresponding to 100, 20, 10, and 2% loading levels, respectively, should be used regularly as QC of the immunosorbent-binding capacity within several applications of urine samples. Quantitative recovery (in the range 80–100%) should be maintained for up to 35 analyses for all loading levels according to our experience. If the binding capacity of some column decreases, the column should be replaced by a new one. We observed that the recovery of 6% of the 96 IS (at 2.4% loading level) became lower than 40% after 30 cycles. This drawback of the HTS–IS format can be improved with better packing of the columns or using more resistant sorbent.

3.4. Enzyme-Linked Immunosorbent Assay

The ELISA (As43/8-BSA) for detection of 2,4,6-TCP was developed by Galve et al. *(27)* (for immunoassay description, *see* **Note 11**). The assay parameters of the immunoassay are summarized in **Table 2**.

1. Coat the microtiter plates with 8-BSA in coating buffer (0.625 µg/mL, 100 µL/well) overnight at 4°C covered with adhesive plate sealers.
2. The following day, wash the plates four times with PBST.
3. Prepare a standard curve of 2,4,6-TCP standards (1000 to 0.625 n*M* in 7% EtOH/PBS) in a separate mixing plate. The stock solution of 2,4,6-TCP is 1 m*M* in DMSO. The stock solution is diluted 1/1000 in PBS to make a concentration of 1000 n*M* corresponding to the highest concentration of the standard curve. Pipet 300 mL of this solution to well A1. Pipet 270 mL of 7% EtOH/PBS into well B1 and 240 mL of 7% EtOH/PBS into well H1. Pipet 150 mL of 7% EtOH/PBS into wells C1 through G1 of the mixing plate. Transfer 30 mL from well A1 to B1. Mix by gently drawing up and expelling the solution back into the well three or four times. Transfer 150 mL from well B1 to C1 and mix. Continue to transfer and mix. The last transfer of 150 mL is from well F1 to G1. Transfer 60 mL from well G1 to H1 and mix. Pipet 150 mL of 7% EtOH/PBS into wells A2, A3, and A4 (zero analyte concentration). This makes enough solution to run the calibration curve in duplicate on the coated ELISA plate.
4. Prepare serial dilutions in 7% EtOH/PBS of the HTS–IS–SPE extracts of the unknown urine samples. For example, pipet 300 mL of one sample into well A4. Pipet 150 mL of 7% EtOH/PBS into wells B4 and C4. Transfer 150 mL from well A4 to B4 and mix. Transfer 150 mL from well B4 to C4 and mix. This results in a sample that will be tested neat, diluted by two and four. Larger dilutions may be made by adjusting volumes of sample, diluent, and transfer volume appropriately. Each dilution can be analyzed in duplicate on the coated ELISA plate.
5. Add the 2,4,6-TCP standards and the samples to the coated plates (50 µL/well), followed by the sera As43 (1/2000 in PBST, 50 µL/well), and incubate for 30 min at room temperature.
6. Wash the plates again four times with PBST.
7. Add a solution of goat antirabbit IgG coupled to horseradish peroxidase in PBST (1/6000) to the wells (100 mL/well) and incubate for 30 min at room temperature.
8. Wash the plates four times with PBST.
9. Add the substrate solution (100 mL/well).
10. Stop color development after 30 min at room temperature with 4*N* H_2SO_4 (50 mL/well) and read the absorbance at 450 nm.
11. Calculation of sample concentration using the standard curve. The concentration should be reported in 2,4,6-TCP immunoreactivity equivalents (For ELISA crossreactivity see **Note 11**). The standard curve should be fitted to a four-parameter logistic equation according to the following formula:

$$y = (A - B/[1 - (x/C)D]) + B$$

where *A* is the maximum absorbance, *B* is the minimum absorbance, *C* is the concentration producing 50% of the maximum absorbance, and *D* is the slope at the inflection point of the sigmoidal curve (*see* **Table 2**).
12. Quality assurance and quality control of the HTS–IS–ELISA protocol. In each analysis, it is necessary to run matrix spikes and blank sample through the entire procedure. We suggest the analysis of unknown urine samples be performed in duplicate. The main features of the urinary analysis of 2,4,6-TCP by the HTS–IS–ELISA method are summarized in **Table 3**.

Table 3
Features of the Urinary Analysis of 2,4,6-TCP by HTS–IS–ELISA

Parameters	HTS–IS–ELISA
Sample volume	6 mL
Speed of IS cleanup	96 samples/h
Speed of total analysis	96 samples/d
LOD^a (mg/L)	0.3
LOQ^b (mg/L)	0.55
MDC^c (mg/L)	30
Interday precision (%CV)^d	17.4–22.9% for 0.7–8 mg/L ($N = 24$)
Intraday precision (%CV)^e	6.2–11.3% for 0.7–8 mg/L ($N = 72$)
Number of false positives	0
Number of false negatives	0

^{a}The limit of detection (LOD) is evaluated according to the LOD of the ELISA ($LOD_{7\%EtOH/PBS} = 0.2$ mg/L, 90% of the signal at zero analyte concentration) and the corresponding recoveries of the IS-SPE and of the hydrolysis.

^{b}The limit of quantification (LOQ) is evaluated according to the LOQ of the ELISA ($LOQ_{7\%EtOH/PBS} = 0.37$ mg/L, 80% of the signal at zero analyte concentration) and the corresponding recoveries of the IS-SPE and of the hydrolysis.

^{c}MDC (maximum detectable concentration with recovery > 70%)

^{d}The %CV is the average of the %CVs for each concentration for each day (within one 96-well SPE procedure).

^{e}The %CV corresponds to the recovery obtained for each concentration for 3 d (between three 96-well SPEs). N is the number of ISs used for each concentration.

4. Notes

1. A study carried out with sawmill factory workers demonstrated that tri- and tetrachlorophenols are excreted totally conjugated (97–92.9% for 24-h urine, 80.5–79.1% for morning urine, and 86.4–81.6% for afternoon urine) and the extent of conjugation of PCP (pentachlorophenol) is lower (76.2% for 24-h urine, 69% for afternoon urine) *(29)*. The urinary half-times for tri-, tetra-, and pentachlorophenol are 18 h, 4.3 d, and 16 d, respectively.

2. The IgG fraction of As45 is isolated by 35% ammonium sulfate precipitation to remove serum albumins according to a standard protocol *(30)*. The obtained IgG is immobilized to the NHS-activated Sepharose 4 Fast Flow gel by covalent coupling via the amino groups as recommended by the manufacturer (Pharmacia Biotech). The NHS-activated Sepharose 4 Fast Flow is highly cross-linked 4% agarose matrix with a 16–23 mmol NHS/mL drained medium ligand density, 90-mm mean particle size, and 3.0–13.0 pH stability. The antibody coupling can be performed at different scales (using 1, 5, 12, and 24 mL Sepharose suspension) with a coupling efficiency of about 97% in all cases *(25)*. The drained gel bed volume of each IS is 0.2 mL. This corresponds to maximum theoretical binding capacity for each column of approx 1 mg (5.1 nmol) 2,4,6-TCP; based on the amount of IgG coupled (9.7 mg) and the assumption that bivalent binding takes place, 10% of the polyclonal IgG is specific, and 100% of the immobilized IgG is accessible. If 50% steric hindrance or no efficient antibody orientation is assumed, the theoretical binding capacity would be 0.5 mg (2.22 nmol) 2,4,6-TCP.

3. Chlorophenols are excreted to the urine as such or in the form of glucuronide and sulfate conjugates, with the amount of conjugation depending on the particular chlorophenol and its concentration in the urine *(29,31)*. At low concentration, sulfate conjugation is dominant, but when chlorophenol concentration increases, acid conjugation becomes more important.

4. Alternatively, chlorophenol glucuronides and sulfates can be cleaved by acid *(13)* or enzymatic hydrolysis *(29)*. However, we have demonstrated that for alkaline hydrolysis, quantitative analysis (extraction recovery higher than 70%) can be performed in a broader range (1–20 mg/L 2,4,6-TCP urinary concentration) than for acid/enzymatic hydrolysis *(28)*.

5. The neutral pH of the urine sample is very important to ensure effective antigen–antibody interaction in the IS, resulting in an efficient immunoaffinity extraction.

6. SPE devices in a 96-well plate format were introduced in 1996, and they enjoyed widespread application and rapid acceptance in biotechnology and pharmaceutical laboratories, in which HTS is sought *(32)*. The 96-well SPE sorbents *(33)* afford rapid development and automation of SPE methods to eliminate traditional time-consuming and labor-intensive sample preparation steps for environmental *(34)* and biological samples *(35–39)*. All these applications of 96-SPE formats are based on nonselective SPE sorbents. However, the trends in SPE research are oriented not only toward reduction of the SPE format and the automation for a high throughput, but also toward the development of more selective extraction procedures, such as those using immunoextraction sorbents *(40)*. Immunoaffinity extraction provides highly selective extraction of low molecular weight analytes from complex matrices based on the specific molecular recognition *(22–24,41,42)*. Antibodies are covalently bonded onto an appropriate sorbent to form the so-called immunosorbent. Single analytes can be targeted, but thanks to the antibody cross-reactivity, immunoextraction sorbents have also been designed to target a group of structurally related analytes. Because of antibody specificity, the problem of the coextraction of matrix interferences is circumvented.

7. Detailed information about the basic principles and the factors affecting the IS–SPE procedures can be found in **refs.** *22*, *23*, and *43*.

8. Halogenated phenols structurally related to 2,4,6-TCP are usually present in urine samples. In previous work *(28)*, we evaluated the selectivity of the IS–SPE procedure described here and used IS with 1-mL bed size. The specificity studies were carried out by GC/ECD analysis of the eluted fractions after toluene extraction and derivatization of the chlorophenols with *N,O–bis*(trimethylsilyl) trifluoroacetamide. When a mixture of penta-, tetra-, tri-, and dichlorophenol isomers and dibromophenols are loaded to the column (16 compounds, total loading level of the mixture is 20% [248 ng 2,4,6-TCP]), individual loading level is 1.7% (21 ng 2,4,6-TCP), and PBS washing is applied in the cleanup; all penta-, tetrachlorophenols (2,3,4,6-TtCP, 2,3,4,5-TtCP, and 2,3,5,6-TtCP), trichlorophenols (2,4,6-TCP, 2,4,5-TCP, 2,3,4-TCP, 2,3,5-TCP, and 2,3,6-TCP), 2,4,6-tribromophenol, and 2,4-dibromophenol are retained in more than 80%. The dichlorophenols are not detected in the eluted fractions. When the same mixture of halogenated phenols is loaded but 20% EtOH washing is used, the immunoaffinity extraction is more specific: Only 2,4,6-TCP, 2,4,6-TBP, 2,3,4,6-TtCP, PCP, and 2,4-DBP were retained by the IS. Thus, when PBS washing is applied, the IS–SPE procedure can be used as a class-selective IS–SPE of chlorophenols from urine samples that can be later analyzed by chromatographic or immunochemical methods. A more selective extraction can be performed by washing with 20% ethanol buffer. If ELISA is used as a detection method after both types of IS–SPE procedure, only the halogenated phenols with significant immunoassay cross-reactivity would be detected.

9. Chlorophenols can be identified and determined by GC/ECD or GC–MS. The 7% EtOH solutions obtained after the HTS–IS–SPE are extracted with toluene. Then, the chlorophenols are derivatized with silylating agent [*N,O-bis*(trimethylsilyl)trifluoroacetamide] and directly analyzed *(25,28)*.

10. An important issue to be considered when the HTS–IS–SPE protocol is used is the control of the immunosorbent stability (i.e., be sure that it keeps its efficient binding capacity). When water–organic modifier mixture is used for elution, the presence of nonpolar solvents reduces the hydrophobic binding component of the antibody–antigen interaction. However, it also affects the stability of the hydrophobic bonds that maintain the antibody tertiary structure, resulting in the release of the antigen. These harsh eluting conditions can irreversibly denature antibodies, but because small volumes are required, contact times can be minimized. In our studies, the regeneration of the IS is performed by passing 10 bed volumes of PBS through the column. However, in the HTS–IS–SPE procedure, the washing and elution steps are performed until the columns are dried (to avoid error in the collected volume), which can have a negative effect on the immunosorbent stability. In addition, the backpressure formed is not equal for all the columns because different resistance is created by the manually placed frits (different packing). The applied pressure to all the columns in the 96-well rack is not homogeneous. Some columns get drier under elution. All these can create problems with column stability.

11. The antisera is raised against 3-(3-hydroxy-2,4,6-trichlorophenyl)propanoic acid covalently coupled by the mixed anhydride method to keyhole limpet hemocyanin. The indirect ELISA uses a heterologous coating antigen prepared by conjugation of 3-(2-hydroxy-3,6-dichlorophenyl)propanoic acid to BSA using the active ester method. The assay performs well between pH 7.5 and 9.5, and it is inhibited at pH lower than 6.0. The immunoassay detectabilities do not change significantly when the ionic strength of the media is in the range 12–25 mS/cm. The ELISA for 2,4,6-TCP is quite specific, but some cross-reactivity with other chlorinated phenols, such as 2,3,4,6-TtCP (21%), 2,4,5-TCP (12%), and 2,3,5-TCP (15%), is observed. Brominated phenols are even more recognized than the corresponding chlorinated analogues (e.g., 2,4,6-TBP, 710%; 2,4-DBP, 119%).

Acknowledgments

This work was supported by MCyT (AGL 2002-04653-C04-03) and EC (QLRT-2000-01670). Mikaela Nichkova thanks the Spanish Ministry of Education for her fellowship to the FPU program.

References

1. Angerer, J., Maaß, R., and Heinrich, R. (1983) Occupational exposure to hexachlorocyclohexane. VI. Metabolism of HCH in man. *Int. Arch. Occup. Environ. Health* **52,** 59–67.

2. Hill, R. H., Jr., Head, S. L., Baker, S., et al. (1995) Pesticide residues in urine of adults living in the United States: reference range concentrations. *Environ. Res.* **71,** 99–108.

3. Guidotti, M., Ravaioli, G., and Vitali, M. (1999) Total *p*-nitrophenol determination in urine samples of subjects exposed to parathion and methyl-parathion by SPME and GC/MS. *J. High Resolut. Chromatogr.* **22,** 628–630.

4. Wrbitzky, R., Angerer, J., and Lehnert, G. (1994) Chlorophenols in urine as an environmental medicine monitoring parameter. *Gesundheitswesen* **56,** 629–635.

5. Wittsiepe, J., Kullmann, Y., Schrey, P., Selenka, F., and Wilhelm, M. (2000) Myeloperoxidase-catalyzed formation of PCDD/F from chlorophenols. *Chemosphere* **40,** 963–968.

6. Angerer, J., Heinzow, B., Schaller, K. H., Weltle, D., and Lehnert, G. (1992) Determination of environmental caused chlorophenol levels in urine of the general population. *Fresenius J. Anal. Chem.* **342**, 433–438.

7. Hill, R. H., Jr., Ashley, D. L., Head, S. L., Needham, L. L., and Pirkle, J. L. (1995) *p*-Dichlorobenzene exposure among 1000 adults in the United States. *Arch. Environ. Health* **50**, 277–280.

8. Schmid, K., Lederer, P., Goen, T., et al. (1997) Internal exposure to hazardous substances of persons from various continents. Investigations on exposure to different organochlorine compounds. *Int. Arch. Occup. Environ. Health* **69**, 399–406.

9. Bartels, P., Ebeling, E., Kramer, B., et al. (1999) Determination of chlorophenols in urine of children and suggestion of reference values. *Fresenius J. Anal. Chem.* **365**, 458–464.

10. Lampi, P., Vohlonen, I., Tuomisto, J., and Heinonen, O. P. (2000) Increase of specific symptoms after long-term use of chlorophenol polluted drinking water in a community. *Eur. J. Epidemiol.* **16**, 245–251.

11. Kauppinen, T., Kogevinas, M., Johnson, E., et al. (1993) Chemical exposure in manufacture of phenoxy herbicides and chlorophenols and in spraying of phenoxy herbicides. *Am. J. Ind. Med.* **23**, 903–920.

12. Wrbitzky, R., Goen, T., Letzel, S., Frank, F., and Angerer, J. (1995) Internal exposure of waste incineration workers to organic and inorganic substances. *Int. Arch. Occup. Environ. Health* **68**, 13–21.

13. Kontsas, H., Rosenberg, C., Pfäffli, P., and Jäppinen, P. (1995) Gas chromatographic-mass spectrometric determination of chlorophenols in the urine of sawmill workers with past use of chlorophenol-containing anti-stain agents. *Analyst* **120**, 1745–1749.

14. Kontsas, H., Rosenberg, C., Tornaeus, J., Mutanen, P., and Jappinen, P. (1998) Exposure of workers to 2,3,7,8-substituted polychlorinated dibenzo-*p*-dioxin (PCDD) and dibenzofuran (PCDF) compounds in sawmills previously using chlorophenol-containing antistain agents. *Arch. Environ. Health* **53**, 99–108.

15. Rosenberg, C., Kontsas, H., Tornaeus, J., et al. (1995) Chlorinated dioxin and dibenzofuran levels in plasma of sawmill workers exposed to chlorophenol-containing anti-stain agents. *Organohalogen Comp.* **26**, 81–84.

16. Wrbitzky, R., Beyer, B., Thoma, H., et al. (2001) Internal exposure to polychlorinated dibenzo-*p*-dioxins and polychlorinated dibenzofurans (PCDDs/PCDFs) of Bavarian chimney sweeps. *Arch. Environ. Contam. Toxicol.* **40**, 136–140.

17. Van Emon, J. M. (2001) Immunochemical applications in environmental science. *J. AOAC Int.* **84**, 125–133.

18. Draper, W. M. (2001) Biological monitoring: exquisite research probes, risk assessment, and routine exposure measurement. *Anal. Chem.* **73**, 2745–2760.

19. Knopp, D. (1995) Application of immunological methods for the determination of environmental pollutants in human biomonitoring. A review. *Anal. Chim. Acta* **311**, 383–392.

20. Biagini, R. E., Hull, R., Striley, C. A., et al. (1996) Biomonitoring for occupational exposure using immunoassay. In *Environmental Immunochemical Methods. Perspectives and Applications* (Van Emon, J. M., Gerlach, C. L., and Johnson, J. C., eds.) ACS, Washington, DC, pp. 286–296.

21. Oubiña, A., Ballesteros, B., Carrasco, P. B., et al. (2000) Immunoassays for environmental analysis. In *Sample Handling and Trace Analysis of Pollutants. Techniques, Applications and Quality Assurance* (Barceló, D., ed.), Elsevier, Amsterdam, The Netherlands, Vol. 21, pp. 289–340.

22. Delaunay, N., Pichon, V., and Hennion, M. C. (2000) Immunoaffinity solid-phase extraction for the trace-analysis of low-molecular-mass analytes in complex sample matrices. *J. Chromatogr. B* **745**, 15–37.

23. Stevenson, D. (2000) Immuno-affinity solid-phase extraction. *J. Chromatogr. B* **745**, 39–48.

24. Weller, M. G. (2000) Immunochromatographic techniques—a critical review. *Fresenius J. Anal. Chem.* **366**, 635–645.

25. Nichkova, M. and Marco, M.-P. Development of HTS-IS-ELISA for urinary detection of chlorophenols. *Environ. Health Perspect.*, submitted.

26. Galve, R., Camps, F., Sanchez-Baeza, F., and Marco, M.-P. (2000) Development of an immunochemical technique for the analysis of trichlorophenols using theoretical models. *Anal. Chem.* **72**, 2237–2246.

27. Galve, R., Nichkova, M., Camps, F., Sanchez-Baeza, F., and Marco, M.-P. (2002) Development and evaluation of an immunoassay for biological monitoring of chlorophenols in urine as potential indicators of occupational exposure. *Anal. Chem.* **74**, 468–478.

28. Nichkova, M. and Marco, M.-P. (2005) Development and evaluation of C18 and immunosorbent solid-phase extraction methods prior immunochemical analysis of chlorophenols in human urine *Anal. Chim. Acta* **533**, 67–82.

29. Pekari, K., Luotamo, M., Järvisalo, J., Lindroos, L., and Aitio, A. (1991) Urinary excretion of chlorinated phenols in saw-mill workers. *Int. Arch. Occup. Environ. Health* **63**, 57–62.

30. Baines, M.G. and Thorpe R. (1992) Purification of Immunoglobulin G (IgG) In *Immunochemical Protocols, Methods in Molecular Biology* (Manson, M. M., ed.), Humana Press, Totowa, NJ., Vol.10, pp.79–104

31. Drummond, I., van Roosmalen, P. B., and Kornicki, M. (1982) Determination of total PCP in the urine of workers. A method incorporating hydrolysis, an internal standard and measurement by LC. *Int. Arch. Occup. Environ. Health* **50**, 321–327.

32. Rossi, D. T. and Zhang, N. (2000) Automating solid-phase extraction: current aspects and future prospects. *J. Chromatogr. A* **885**, 97–113.

33. Wells, D. A. (1999) 96-well plate products for solid-phase extraction. *LC GC North America* **17**, 600-610.

34. Quayle, W. C., Jepson, I., and Fowlis, I. A. (1997) Simultaneous quantitation of 16 organochlorine pesticides in drinking waters using automated solid-phase extraction, high-volume injection, high-resolution gas chromatography. *J. Chromatogr. A* **773**, 271–276.

35. Janiszewski, J., Schneider, R. A., Hoffmaster, K., Swyden, M., Wells, D., and Fouda, H. (1997) Automated sample preparation using membrane microtiter extraction for bioanalytical mass spectrometry. *Rapid Commun. Mass Spectrom.* **11**, 1033–1037.

36. Souppart, C., Decherf, M., Humbert, H., and Maurer, G. (2001) Development of a high throughput 96-well plate sample preparation method for the determination of trileptal (oxcarbazepine) and its metabolites in human plasma. *J. Chromatogr. B: Biomed. Sci. Appl.* **762**, 9–15.

37. Shou, W. Z., Jiang, X., Beato, B. D., and Naidong, W. (2001) A highly automated 96-well solid phase extraction and liquid chromatography/tandem mass spectrometry method for the determination of fentanyl in human plasma. *Rapid Commun. Mass Spectrom.* **15**, 466–476.

38. Rule, G. and Henion, J. (1999) High-throughput sample preparation and analysis using 96-well membrane solid-phase extraction and liquid chromatography–tandem mass spectrometry for the determination of steroids in human urine. *J. Am. Soc. Mass Spectrom.* **10**, 1322–1327.

39. Zhang, H. and Henion, J. (1999) Quantitative and qualitative determination of estrogen sulfates in human urine by liquid chromatography/tandem mass spectrometry using 96-well technology. *Anal. Chem.* **71,** 3955–3964.

40. Hennion, M.-C. (1999) Solid-phase extraction: method development, sorbents, and coupling with liquid chromatography. *J. Chromatogr. A* **856,** 3–54.

41. Pichon, V., Bouzigc, M., Micgc, C., and Hcnnion, M. C. (1999) Immunosorbcnts: natural molecular recognition materials for sample preparation of complex environmental matrices. *TrAC Trends Anal. Chem.* **18,** 219–235.

42. Rhemrev-Boom, M. M., Yates, M., Rudolph, M., and Raedts, M. (2001) (Immuno)affinity chromatography: a versatile tool for fast and selective purification, concentration, isolation and analysis. *J. Pharm. Biomed. Anal.* **24,** 825–833.

43. Ballesteros, B. and Marco, M.-P. (1998) Basic principles of the use of immunoaffinity chromatography for environmental analysis. *Food Technol. Biotechnol.* **36,** 145–155.

II

ASSESSMENT OF INHALATORY AND POTENTIAL DERMAL EXPOSURE

12

Assessment of Postapplication Exposure to Pesticides in Agriculture

Joop J. van Hemmen, Katinka E. van der Jagt, and Derk H. Brouwer

Summary

Occupational exposure to pesticides may occur not only during the actual application to crops and enclosed spaces, but also after the actual application when the crops are handled (e.g., harvesting) or when treated spaces are reentered. This postapplication (reentry) exposure may occur on a daily basis (e.g., for harvesting of ornamentals or vegetables in greenhouses) and may have the duration of a full work shift. An overview is given for the methodology that can be used for assessing the levels of exposure via skin and inhalation. Such data are used for the risk assessment of the use scenarios relevant for registration purposes throughout the world and form the basis for predictive exposure modeling. For Europe, such a predictive postapplication exposure model is developed in the EUROPOEM project funded by the European Union. Because exposure may have to be reduced with various techniques in cases of anticipated unacceptable health risks, the use of control measures comes into play, which are described for postapplication exposure. The assessment of internal exposure levels using biological monitoring methodology is also described.

Key Words: Biological monitoring; exposure modeling; fluorescent techniques; gloves; hand washes; occupational exposure; personal protective equipment (PPE); postapplication; reentry; tape stripping; whole body measurement; wiping.

1. Introduction

Agricultural crops, grown either indoors or outdoors, are frequently treated with pesticides to prevent or rid pests of all kinds, such as weeds, insects, and fungi. In doing so, either the ambient atmosphere (indoors) or the foliage or soil is treated with amounts of pesticide that vary for effectivity, nature of the pest, degree of infestation, and in practice the requirements posed by (inter)national regulations for export, such as "no tolerance." This means that the foliage and fruits of the crop become contaminated with pesticide residues that may or may not be removed or lost before workers come into contact with the crop during various crop-related activities, such as harvest-

From: *Methods in Biotechnology, Vol. 19, Pesticide Protocols*
Edited by: J. L. Martínez Vidal and A. Garrido Frenich © Humana Press Inc., Totowa, NJ

ing. In practice, the applied amount and the time between application of the crop and postapplication activities in the crop determines the level of exposure during these activities. These levels are regulated by pesticide laws throughout the world. In Europe, the Plant Protection Products Directive (91/414/EC) requires a human health risk assessment of exposure during application and postapplication (reentry) and of exposure to bystanders present during application and postapplication activities.

In this chapter, the methodology for assessing exposure during postapplication is described, as is the assessment of exposure for registration purposes. In Europe, the EUROPOEM project has attempted to gather publicly available data for the risk assessment process required by European legislation for exposure during reentry of treated crops and treated premises, such as greenhouses (*1*). Methodology is described to assess internal exposure to pesticides, as well as methods to reduce exposure. An overview of many exposure measurements carried out mainly in greenhouses is presented in **ref. 2.**

2. Materials and Methodology

The methodology for sampling the amount of exposure during postapplication should in principle cover inhalation, skin exposure, and oral exposure. It might cover either external exposure or internal exposure as assessed by biological monitoring. Note, however, that oral exposure cannot yet be quantitatively estimated and is therefore either neglected or estimated together with exposures via the other routes of uptake by biological monitoring.

These different methodologies are discussed to the extent needed for the present purpose. The basic approach for exposure assessment to agricultural pesticides is presented in **ref. 3.** This is relevant for operators and reentry workers. A similar guidance document is currently in preparation by the Organization for Economic Cooperation and Development specifically for postapplication agricultural scenarios.

2.1. Inhalation Exposure Assessment

Aspiration efficiency and retention of the captured aerosols are key issues in the performance of sampling devices. Aspiration efficiency is a device-depending property that varies with the aerodynamic diameter of the aerosol. Conventions have been established to define aerosol size fraction and aspiration efficiencies by nose and mouth breathing (*4*). The inhalable fraction (i.e., all material capable of being drawn into the nose and mouth) is the most relevant fraction to measure.

Aerosols generated by sprayers or misters usually are mixed-phase aerosols in which both vapor and the liquid or solid phase are present. Traditional particle-sampling devices or vapor-trapping devices (impingers, adsorbent tubes) have been used, or these are combined to sample concurrently for particles and vapors in so-called sampling trains. These approaches are now well documented in basic textbooks and need no further discussion here.

Because particle sampling requires much higher flow rates, usually approx 2 L/min, compared to vapor sampling (50 to 200 mL/min), retention of the vapors in the vapor absorption/adsorption part may be inaccurate because of loss of material by stripping from the adsorbent by the relatively high flow rate. To overcome such a problem, an existing aerosol sampler that meets the criterion of inhalability has been modified to retain vapors as well *(5)*.

2.2. Dermal Exposure Assessment

Improved understanding of the process of dermal exposure has been achieved through a conceptual model of dermal exposure *(6)* that systematically describes the transport of contaminant mass from exposure sources to the surface of the skin. The conceptual model describes the dermal exposure process as an event-based mass transport process resulting in "loading" of the skin (i.e., the skin contaminant layer compartment that is formed by sweat, skin oil, dead cells, and contaminants/dirt). To assess skin loading, two major groups of methods are distinguished: direct methods, which indicate directly how much pesticide is on the skin, and indirect methods, which provide indirect indication of potential for skin exposure.

2.2.1. Direct Methods for Dermal Exposure Sampling

Direct methods for assessing dermal exposure include methods that indicate the mass of a contaminant or analyte that has deposited onto the skin surface over a period of exposure. Direct methods can be grouped into three major sampling principles.

- Collection of agent mass using collection media placed at the skin surface or replacing work clothing during the measurement period (i.e., surrogate skin techniques).
- Removal of the agent mass from the skin surface at the end of the measurement period (i.e., removal techniques).
- *In situ* detection of the agent or a tracer at the skin surface (e.g., through image acquisition and processing systems).

2.2.1.1. SURROGATE SKIN TECHNIQUES

Surrogate skin techniques (more precisely, interception techniques) are widely used methods to assess dermal exposure *(7,8)*. Basically, all methods use a collection medium onto which chemicals of interest are deposited on or transferred to by direct contact. Reports in the literature show a variety of collection media, such as cotton, gauze, paper, polyester, and charcoal. The ideal collection medium should mimic the skin in terms of both collection from the environment and retention vs subsequent loss.

The collection medium is located against the skin of body parts during exposure. After sampling, the medium is removed from the body part and transferred to the laboratory, in which the relevant component is extracted from the medium and quantified by chemical analysis.

The size of the collection medium varies, from relatively small-size collection media located at different body parts (e.g., 10-cm^2 patches) to a collection medium of the same type covering a complete body part (i.e., whole-body garment sampling).

Surrogate sampling techniques have the advantages of relative ease of use in the field, low capital cost, applicability for all body parts, and potential for high resolution of exposure because the collection medium can be divided in small subsamples to be analyzed separately and the ability to perform repeated sampling during an exposure interval.

A general disadvantage is the need to assume that surrogate skin techniques mimic the capture and retention properties of the skin. An additional disadvantage of small-size samplers (i.e., patches) is that the results of individual patches need to be extrapolated to the body part the patch represents.

2.2.1.2. REMOVAL TECHNIQUES

Removal techniques (i.e., removal of chemicals deposited on the skin by washing, wiping, tape stripping) and subsequent chemical analysis of the amount of chemical recovered from the washing solution, the wiping medium, or the adhesive strip are used to assess dermal exposure *(8)*. The techniques have the clear advantage of low capital costs and ease of use; however, the use of solvents may disrupt skin barrier function and enhance percutaneous absorption of the chemical *(3)*. Removal of contaminants from the skin surface is accomplished by providing an external force that equals or exceeds the force of adhesion.

For (hand) washing generally, two basic methods can be identified *(9)*: washing and rinsing. *(Hand) washing* can be defined as scrubbing the skin by mechanical agitation exercised by movement and pressure of both hands in liquid in a routine washing fashion. The contaminant is detached from the skin by a combination of mechanical force and wet chemical action (dissolution). Tap water/soap flow or water/soap in bags (500 mL) are commonly used methods. *(Hand) rinsing* or pouring can be defined as liquid–skin contact by which the contaminant is removed by a combination of hydrodynamic drag and wet chemical action (dissolution).

Clearly, the basic distinction between both methods is the presence or absence of mechanical forces in the process of detachment. Often, detergents are introduced in the process to enhance the detachment of insoluble particles. Bags (250 mL for one hand or 500 mL for two hands) are used and contain a variety of solvent with mild irritating effects pure or in a water solution.

Identified sampling protocols for hand washing/rinsing show a reasonable similarity of procedures. However, they deviate at possible key issues, such as amount of liquid and duration of rinsing (bag rinsing), amount of liquid, amount of soap, duration of washing (water/soap methods). In general, removal efficiency varies between 40 and 90% *(10)*. Because of the limited data set on removal efficiencies and large differences in components (related to physical properties), wash methods, and levels of loadings, no general conclusions can be drawn on the strengths of the variables distinguished.

Skin wiping can be defined as the removal of contaminants from skin by providing a manual external force to a medium that equals or exceeds the force of adhesion over a defined surface area. Similar to hand washing, the contaminant is detached from the skin by a combination of mechanical forces and wet chemical action (dissolution). Identified sampling protocols deviate at possible key issues, such as wipe materials and type of wiping. A wide variety of sampling media (cotton, gauzes, and sponges) is used, dry or wet, for different surface areas of the skin. In addition, the number of passes (i.e., the number of contacts of the wiped area with a single wipe) may vary between 1 and 15, and the actual wiping movement may vary from a circular movement starting in the center to rectangular movements with or without use of templates.

Sampling efficiency usually is between 40 and 90% *(10)* and will be determined not only by the sampling method, but also by the chemophysical properties of the compound and wipe solvent.

Because of their ease of use and their low capital costs, the application of removal techniques is widespread to assess dermal exposure. In spite of its potential for use for all body parts, mostly the uncovered parts of the body are monitored. Especially for wipe sampling, relatively high resolution of exposure per surface area can be achieved; however, for hand washing this is not the case. Repeated sampling is possible, but the exposure process is disturbed, and skin surfaces may be affected.

There is clear evidence that wipe sampling is less effective for removing contaminants from the skin despite the high removal efficiencies of wipe sampling reported in one study *(11)*. In a pesticide reentry study, Fenske et al. *(12)* compared hand exposure rates determined by hand wash sampling and wipe sampling. They observed on average a sixfold lower hand exposure rate for wipe sampling compared to hand wash sampling.

Tape stripping can be defined as the removal of stratum corneum cell layers by (repeated) application of an adhesive tape to the skin. Tape stripping has been used for dermatopharmacokinetic characterization of topical drug product movement into different layers of the horny layer or to assess the penetration of chemicals without the intact barrier function of the skin. Commercially available adhesive tapes are used. The surface area of the strips (3.8 to 10 cm^2) as well as the number of strips (1 to 30) varies between different studies.

Limited data are available that enable an evaluation of the precision, within- and between-operator variability, and the influence of some physical sampling parameters, such as applied pressure, adhesion time, or removal speed and angle. The data from **ref.** *13* indicate a moderate variation of removal efficiency over different exposure sites and different volunteers.

The application of the stripping method for field evaluation of dermal exposure may be limited by analytical limits of sensitivity, but extraction of series of tapes may overcome this problem. The need to sample a large quantity of (first) tape strips could also be important for reasons of sampling strategy if there are heterogeneous surface concentrations. The relatively small surface area of the tapes (typically less than 10 cm^2) compared to larger surface areas that could have been contaminated may result in similar problems in sampling strategy as for patch sampling and skin wiping.

2.2.1.3. In Situ Detection Techniques

A fluorescent tracer technique to assess dermal exposure quantitatively known as VITAE (video imaging technique for assessing dermal exposure) was introduced in the late 1980s *(14)*. A second-generation type of this technique was adopted by other laboratories *(15,16)*, whereas the basics of the technique have also been explored to develop a novel lighting system, known as the fluorescent interactive video exposure system (FIVES), to overcome significant problems related to quantification *(17,18)*.

Basically, the techniques combine the introduction of a fluorescent tracer compound in the application process with image acquisition techniques. A fluorescent tracer, selected from a wide variety of fluorescent whitening agents, deposited on the skin is exposed to long-wave ultraviolet A (UV-A) light (320–400 nm), which activates the emission of fluorescent light by the tracer molecules. A (video) camera is used to take and record images of the body part exposed. The amount of light emitted is detected by digitizing the analogue camera signal. The images consist of discrete area units known as picture elements (pixels), which may have a value between 0 and 256 (gray value). Gray values of the pixels of pre- and postimages are compared to calculate the increase of gray values resulting from exposure. A known relationship between pixel gray levels and the amount of tracer enables calculation of exposure to the tracer. Assuming a fixed tracer–substance ratio, the amount of tracer deposited on the skin surface can be extrapolated to the amount of the substance of interest for dermal exposure.

Application of fluorescent tracer techniques for dermal exposure assessment has clear advantages compared to other direct techniques. Major strengths of these techniques are their ability to spot *in situ* dermal contamination on the skin surface. Absorption and retention processes that influence the results of surrogate skin sampling or parameters affecting removal efficiency for removal techniques do not bias the measurement. Because image acquisition is a noninvasive measurement, it does not disturb these loading and unloading processes on the skin and enables repeated sampling within a work shift. Exposure processes can be studied relatively easily. For risk assessment purposes, the high-resolution properties of the fluorescent tracer techniques have a clear advantage. Because the true area exposed is detected by the system, no estimates have to be made for the surface area exposed or the distribution of exposure over the body part sampled. The possibility for a visual check on the distribution of exposure over the body part that has been evaluated is helpful.

Major drawbacks of these techniques are related to the introduction of a fluorescent tracer into the process of exposure. However, the most important limitation of these techniques is that a tracer is detected and not the substance relevant for dermal uptake. The premise is the similar behavior of tracer and relevant compounds during the entire process of exposure.

Other relevant *in situ* determination techniques are available but hardly used routinely for pesticides; these include portable X-ray fluorescence (PXRF) *(19)* and dirichlet tesselation *(20)*.

2.2.2. Indirect Methods

2.2.2.1. SURFACE SAMPLING

Indirect methods include surface-sampling techniques and biological monitoring. Surface sampling is relevant for an identified skin-surface contact. For reentry exposure, pesticide residues are sampled from surfaces to which workers come into contact.

The concept of dislodgeable foliar residue (DFR) *(21,22)*, used in agriculture reentry exposure scenarios, apparently partly circumvents some of the problems related to surface-sampling variability. This approach consists of a protocol to sample a discrete surface area by removing (parts of) the surface (i.e., leaf portions or leaf punches are taken from the foliage). Leaf punch samplers with punch diameters of 0.64, 1.27, and 2.54 cm are currently available that provide double-sided leaf areas of 2.5, 5.0, and 10 cm^2, respectively. After sampling, the leaf portions or disks are extracted twice by shaking at 200 strokes per min for 30 min with 100 mL distilled water per 100 cm^2 leaf surface area, containing 4 drops of a 1:50 dilution of a surfactant (e.g., Triton-X100 solution). Then, the bottle containing the leaves is rinsed with 25 mL liquid per 100 cm^2 leaf surface area, and after removal of the leaves, it is rinsed again twice with 10 mL methanol/100 cm^2 leaf surface area. When leaf samples are taken (instead of punches), the leaf surface area should be determined afterward (e.g., by a light-detection-based surface area meter). Advantages of this approach compared to surface wipe sampling are the standardized extraction procedure and person independency.

Although it is unclear how the removal procedure mimics the transfer from the foliage surface to the worker's skin or clothing, DFR has been used to predict dermal exposure resulting from workers' contact with foliage (*see* **Subheading 3.**). In **refs. 23–25**, pesticide residues monitored in treated fields were related to hourly dermal exposure. The transfer coefficient (TC) has been introduced as an empirical multiplier and usually is expressed in units of hourly dermal exposure (grams per hour) per unit of DFR (grams per square centimeter).

2.2.2.2. Biological Monitoring

Biological monitoring is a method of evaluating the absorption of chemicals by measuring the chemicals or their metabolites in body fluids, usually urine, blood, or exhaled air. This is perceived as the principle advantage of biological monitoring over methods of ambient exposure monitoring because the total mass of biological marker represents the individual's exposure from all routes of entry: inhalation, dermal, and primary and secondary ingestion. The method requires detailed human metabolism and pharmacokinetics data for the chemical involved for quantification (*26*) for an appropriate selection of the metabolite to sample, excretion medium, and duration of collection. Urine sampling, as a noninvasive method, is considered an ideal sampling matrix (*3*), and urine collection has been practicable up to several days. Completeness of urine voids over the full period of sampling is essential (knowledge of half-lives needed), and no additional exposure should occur during the sampling period.

The absorbed dose, determined by biological monitoring, may be difficult to relate to external exposure for multiple-route exposure pathways because these are very likely to occur in pesticide exposure scenarios. However, by subtracting or eliminating other routes of exposure, the contribution of one of the routes can be estimated in theory.

As stated, in using biological monitoring in field research of pesticide exposures, it is important to understand the relationship between skin exposure and the biological monitoring results. Generally, chemicals are absorbed through the skin more slowly than through inhalation or the oral route. Also, the skin can act as a dynamic reservoir of contaminants of past exposures, ready for mobilization and absorption under suitable conditions. Examples of these phenomena can be found in a study on exposure to propoxur (*27*) (as described in **Subheading 4.**).

In field pesticide exposure studies, biological monitoring has been used to evaluate and model reentry exposure (e.g., **refs. 25** and **28**) or to evaluate exposure reduction by protective measures (e.g., **refs. 27** and **29–31**).

2.2.3. Summary and Discussion

Diverse measurement methods, partly based on different sampling principles, can be observed. An overview for dermal exposure sampling techniques is given in **Table 1**.

Table 1
Overview of Measurement Methods for Dermal Exposure

Method	Sampling principle	Measured compartment
UV fluorescence of agent or added tracer	*In situ* detection	Skin, surface
Portable x-ray fluorescence monitor	*In situ* detection	Surface, skin
Wet wipe	Removal (manual wiping)	Surface, skin
Wet wipe	Removal (mechanized wiping)	Surface
Fixed pressure dislodgeable residue sampler	Removal (mechanical transfer in situ)	Surface
Dislodgeable foliar residue sampling	Removal (surface removal)	Surface
Adhesive tape	Removal (skin stripping)	Skin
Hand wash	Removal (wash with water or alcohol)	Skin
Patch	Surrogate skin (passive)	Skin
Whole body	Surrogate skin (passive)	Skin

It should be emphasized that all sampling methods have fundamental problems:

- Removal methods (e.g., skin stripping and solvent washing) influence the characteristics of the skin, limiting use for repeated sampling.
- Removal techniques (e.g., skin washing) are not appropriate for all body parts.
- Interception and retention characteristics of surrogate skin techniques differ from real skin and might differ from the clothing. The amount recovered from the surrogate sampler does not represent the loading of the skin surface.
- Extrapolation from small areas sampled (e.g., patches, skin tape strips, or *in situ* detected spots) to the entire exposed area can introduce substantial errors.
- The behavior of a (fluorescent) tracer introduced in the mass transport when using *in situ* techniques may differ from the behavior of the substances of interest.

Therefore, it is recommended that these limitations be taken into account when interpreting the sampling results, especially for risk assessment processes.

3. Exposure Modeling

On the basis of the available data on dermal exposure, attempts have been made to develop a general approach for exposure modeling relevant for registration purposes. This approach is based on the following steps in the process of dermal exposure: It starts with the application of the pesticide, leading to coverage of the foliage with pesticide residue that may or may not disappear in time because of various reasons, such as uptake in the foliage or hydrolysis of some kind. What remains on the foliage (DFR) may be transferred to clothing or skin of a worker who comes into contact with the foliage. The transfer (via the transfer coefficient) will depend on the nature of the contact and the degree of contact between body and foliage and the duration of the work. The resulting generic model has the following algorithm:

$$\text{Potential dermal exposure (DE)} = \text{DFR} \times \text{TC} \times \text{T} \tag{1}$$

where DFR is the dislodgeable foliar residue (typically micrograms per square centimeter), TC is the transfer coefficient (typically square centimeters per hour), and T is the time of contact (typically hours) *(32)*. The DFR can be considered the applied amount divided by the leaf area index (LAI):

$$\text{DFR} = \text{AR/LAI} \tag{2}$$

where AR is the application rate. The LAI is the ratio between the (one-sided) foliage surface area and the ground surface area on which it grows. In these formulas, one factor is not yet included: the dissipation (decay) of the active substance on the foliage. This may be introduced as a factor or as a formula if the exact nature of the dissipation over time is known. If no data are available on the degree of dissipation, the conservative approach is to assume no dissipation between application and time of reentry. In that case, DFR_0 (at time zero) is used for calculations, that is, the residue available directly after application (when dry).

In practice, this would mean that if no dermal exposure measurements are available, the exposure can be calculated using the relevant application rate or data or assumptions on DFR, the duration of the work activity, and information on TCs. This requires a database on TCs, with special emphasis on the relevant scenario.

The various factors are discussed in detail in the **ref. *1*** on reentry exposure of the EUROPOEM project, but the major issues involved are discussed here.

3.1. Dislodgeable Foliar Residue

The amount of residue on foliage depends on several factors, not only the application rate and droplet sizes, but also the crop type and the amount of foliage (LAI). Moreover, dissipation of residues on crop foliage over time depends on the physical and chemical properties of the applied active substance as well as on environmental conditions.

Common methodologies for determination of foliar residues have been described. Usually, a diluted surfactant in water is used for rinsing a certain leaf area, resulting (after analysis) in an expression of residue amount per area: the DFR. It is important to note whether the area given refers to one side or to both sides of the leaves (*see* **Subheading 2.**).

However, experimentally determined DFR data are not available in all cases. In these cases, an estimation of the amount of DFR immediately after application can be made by taking into account the application rate, the crop habitat (LAI), and the (possible) extent of residues remaining on foliage from previous applications *(1)*.

3.2. Transfer Coefficient

The transfer of residues from the crop foliage to the clothes or skin of the worker can be regarded as more or less independent of the kind of product applied, and the level of worker exposure will depend only on the intensity of contact with the foliage. This again is determined by the nature and duration of the maintenance activity to be carried out during reentry.

Therefore, it is advisable to group the various crop habitats and maintenance activities to reentry scenarios. Investigations to this end have been carried out, primarily in the United States. These data are, however, of a proprietary nature. Especially, generic transfer coefficients have been developed for a number of scenarios. Because the nature of the transfer coefficient used may depend on the data at hand (data for potential or actual exposure, full body or only body parts), it is essential to make clear the type of TC meant.

3.3. Exposure and Dermal Absorption

Dermal exposure (and, concomitantly, inhalation exposure) is by no means the ultimate goal of the assessment because, next to possible local effects on the skin and in the airways, the active substance must enter the body for systemic health effects. This requires absorption through the skin. Although the end point for the current report is exposure assessment and not risk assessment, it is worthwhile to indicate the relevance of knowledge on absorption and the possible validation of the use of data on dermal exposure and dermal absorption. In the next section, the use of biological monitoring for estimation of uptake in the human body is discussed in more detail.

3.3.1. Using the Generic Model

Equation (1) is applicable, using the database on TCs, when measured DFR values are available. In most cases, especially when developing a new product, these data are not available at an early stage. For the estimation of worker exposure at that stage, an extended version of the formula together with a tiered approach can be used. The TC is assumed to be relatively pesticide independent and crop and task specific. However, between-crops and task variances of the TC may be substantial *(33,34)*. In **ref. 1**, several generic values for TCs are given related to specific use scenarios.

3.3.2. Tiered Approach to Risk Assessment for Reentry Workers

If use conditions are relevant to reentry exposure, a tiered approach to risk assessment is proposed *(1)*. Adopting a tiered approach allows flexibility in the assessment procedure. Although tier 1 uses only generic data and assumptions, the demand for further and more specific information increases with each successive tier. Accordingly, information and assessments become less general (i.e., more refined and specific to the situations under consideration, as described below).

Comparing the estimated exposure value at any tier level with the AOEL (acceptable operator exposure level, which is applicable also to the reentry worker) may demonstrate an acceptable risk, leading to a regulatory decision to authorize the product. On the other hand, failure to demonstrate an acceptable risk takes the assessment to the next tier, which demands more exact input data. The general form of this tiered approach is depicted in words in **Table 2**.

3.4. Estimation of Inhalation Exposure

The description of the generic model and the tiered approach reflects only dermal exposure, not inhalation exposure. Although in many cases inhalation exposure will be less important for the risk assessment than dermal exposure, some emphasis must

Table 2
Tiered Approach for Assessment of Reentry Exposure

Tier 1	Uses the generic assumption on initial DFR and database for transfer factors to give single conservative point estimates (surrogate values) for total potential exposure, fully exploiting the capacity of the database applicable to a broad range of reentry scenarios common to European conditions.
	If the estimated reentry exposure is within the AOEL, no further action is required and approval can be granted.
Tier 2	Uses the generic database plus additional information relating to exposure-mitigating factors (i.e., exposure reduction coefficients for personal protective equipment [PPE]) pertinent to the case. This offers a middle course in which supplementary use-specific information is used to refine the exposure estimation, thus reducing uncertainty.
	If the estimated reentry exposure, including defined specific instructions on worker exposure, is within the AOEL, no further action is required, and approval can be granted.
Tier 3	Uses additional data on product-specific percutaneous absorption and on DFRs and their dissipation curves from foliar dislodgeable residue studies under *actual conditions* of use.
	If the estimated reentry exposure, including the redefined specific instructions on worker exposure (if necessary), is within the AOEL, no further action is required, and approval can be granted.
Tier 4	Uses product-specific data from biological monitoring studies or reentry exposure studies on the active substance under consideration and the actual reentry conditions. This provides absolute exposure data and places the greatest demands on the quality and relevance of data required.
	If the measured reentry exposure, including the redefined specific instructions (if necessary) on worker exposure, is within the AOEL, no further action is required, and approval can be granted.
	If the measured reentry exposure exceeds the AOEL, reentry restrictions have to be established.

be given to inhalation exposure. For the few available data, an algorithm is given for some reentry scenarios based on the active substance (as):

$$\text{mg as/h inhaled} = \text{kg/as/ha applied} \times \text{Task-specific factor}$$

The task-specific factors, which can be used in the first tier of the exposure and risk assessment, have been estimated for a small set of exposure data on harvesting of ornamentals and reentry of greenhouses about 8–16 h after specific applications. Some task-specific factors are given in **ref. *1***.

In many cases, inhalation exposure is expected to be quite low in comparison with dermal exposures, of course with exceptions for situations for which aerosols and volatile pesticides are of concern.

Inhalation exposure may be not only to vapors, but also to dusts. The relevance of soil exposure to inhalation contamination with pesticides is also covered in **ref. *1***, with a possible approach to estimating this whenever considered relevant. Generally, contaminated soil exposure will be relatively low compared to other exposures. The case of possible dermal exposure to soil containing pesticide residues is treated using the concept of dermal adherence.

4. Relevance of Methodology for Internal Exposure Assessment

The TC concept (as described in **Subheading 3.**) and the acceptance of its validity are essential for the credibility and acceptance of a database of reentry exposure and generic TCs for predicting reentry worker exposure. The concept has never been validated in terms of its ability to predict dermal exposure when used in conjunction with compound-specific DFR data. Biological monitoring is recognized for giving the most accurate estimate of the absorbed dose of a pesticide, particularly if studies are designed and interpreted with the aid of human metabolism and pharmacokinetic data. A direct comparison of the passive dosimetry and biological monitoring approaches to the estimation of the absorbed dose would go a long way to providing the necessary confidence in the TC concept's validity.

Biological monitoring can also provide a good estimate of the uptake of a compound over a day's work, considering the work process, use scenario, any measures used for mitigation of exposure (control by engineering measures, personal protective equipment [PPE], personal hygiene, etc.). In field practice, it therefore has the advantage of including all exposure pathways. Furthermore, using biological monitoring has the advantage that additional factors in skin penetration under specific conditions of protective clothing can be included in the interpretation of results, a scenario not allowed by measurement of external potential or actual exposure.

The evaluation of protective clothing is a major reason for doing intervention type of studies ("as is" and with a specific PPE regime) for which biological monitoring is the gold standard for assessing the internal exposure. The main reason for studying the internal exposure levels is that these levels are the most relevant for risk assessments; that is, they are much better than external exposure levels corrected for clothing protection and taking percutaneous absorption into account, as is currently done for registration procedures.

De Vreede et al. *(35)* reported large variations in penetration of work clothing (from a few percentage up to 30% for methomyl in operators, depending on location on the body and on exposure levels). This indicates the importance of more detailed studies, which have been carried out for some specific conditions *(2)*. An intervention study has been carried out for pest control operators using custom personal protection for the pesticide chlorpyrifos *(36)*.

For reentry conditions, the intervention type of study, using biological monitoring, is seldom used, mainly because the study design is difficult, and the costs for such a study are very high. In the Netherlands, such a study has been carried out for the harvesting of carnations in greenhouses.

4.1. Greenhouse Reentry Example

An intervention study was carried out to evaluate the effectiveness of protective clothing on the reduction of dermal exposure for the pesticide propoxur under field conditions during application and postapplication harvesting of carnations *(37)*. The study was carried out in different greenhouses in the Netherlands (a relatively controlled environment). Both exposure of the hands and inhalation were measured for applicators and harvesters for different protective clothing scenarios. The study was carried out as an intervention study (normal working conditions), with a normal clothing scenario prior to the intervention and with additional protective clothing after the intervention.

Both potential and actual exposure were assessed using the whole body technique *(3)*. Potential exposure to the hands was measured using monitoring gloves. Postintervention, actual exposure to the hands was assessed for 18 harvesters following reentry. Hand exposure was assessed using hand washes; the rinse-off water was collected and analyzed after two hand washes. Respiratory exposure was assessed using an Institute of Occupational Medicine, Edinburgh, UK (IOM) sampler. To assess propoxur absorption, biological monitoring was carried out. A dose excretion study *(38)* using volunteers indicated a significant increase in the dermal uptake of the active ingredient under occlusion conditions, signifying increased blood flow, a rise in skin temperature, and skin moisture. The relevance of skin moisture was identified *(39)*. In the study by Brouwer et al. *(37)*, skin moisture was monitored on various locations on the body.

Biological monitoring was interpreted by assessing the total amount of 2-isopropoxy-phenol (IPP) (metabolite of propoxur) excreted in the urine. Volunteer studies revealed a one-to-one relationship to absorbed propoxur and excreted IPP. A pulmonary retention of 40% was found *(40)* and used to calculate the relative contribution of respiratory exposure to the internal dose. For dermal exposure, the calculated respiratory portion was subtracted from the total amount of IPP. The study found that the amount of IPP excreted after working with normal clothing was 83–2189 nmol propoxur and with protective clothing was significantly reduced from 16 to 917 nmol for harvesters for similar external exposure patterns. It was also shown that all body parts except the palms of the hands revealed higher skin moisture during the use of protective clothing.

To enable the interpretation of biological monitoring, some very specific information had to be available. Information on the excretion pattern (e.g., to allow for proper data collection), the absorption rate through the different exposure pathways (respiratory, oral, and dermal) and the relationship between the excreted amount and the initial dose must be known. PBPK studies provide some of this information, but they are not readily available for most chemicals. Also, as shown in **refs.** *37* and *39*, the excretion of metabolite can be affected by the influence of other factors on absorption into the body (e.g., occlusion), complicating the interpretation of the contribution from the different exposure routes and requiring more data.

The data obtained in the study compared reasonably well with other similar studies (as indicated in **ref.** *37*). This may indicate the value of the data for reentry exposure. The apparent relatively small reduction in internal exposures, even after correction for the small contribution of inhalation exposure seems to indicate that the typical modeling approach for registration purposes may be overly conservative when using only external exposure data.

Acknowledgments

We thank the Dutch Ministry of Social Affairs and Employment for their financial support of the experimental work in assessing exposure during pesticide application and postapplication activities, which made the writing of this chapter possible. We would also like to thank our colleagues at TNO Chemistry for their support and, most important, the colleagues in the EUROPOEM project, with whom the many discussions on the issues involved have sharpened our approaches and views and made this work possible. Many discussions with international colleagues from competent authorities and agrochemical industry at workshops and conferences have also largely contributed to the present state of the art of the postapplication exposure assessment and modeling and their use in risk assessment for registration purposes.

References

1. EUROPOEM. (2003) *Post-application Exposure of Workers to Pesticides in Agriculture.* Report of the Re-entry Working group, EUROPOEM II Project (FAIR3-CT96-1406), TNO-BIBRA, Carshalton, UK.
2. Van Hemmen, J. J., Brouwer, D. H., and De Cock, J. S. (2001) Greenhouse and mushroom house exposure, in *Handbook of Pesticide Toxicology.* Volume I. *Principles* (Krieger, R. I., ed.), Academic Press, New York, pp. 457–478.
3. Organisation for Economic Co-operation and Development. (1997) *Guidance Document for the Conduct of Studies of Occupational Exposure to Pesticides During Agricultural Application,* OECD Environmental Health and Safety Publications Series on Testing and Assessment, no. 9, Organisation for Economic Co-operation and Development, Paris, France.
4. CEN EN 481. (1993) *Workplace Atmospheres. Size Fraction Definitions for Measurement of Airborne Particles,* European Committee for Standardization, Brussels, Belgium.
5. Brouwer, D. H., Ravensberg, J. C., De Kort, W. L. A. M., and Van Hemmen, J. J. (1994). A personal sampler for inhalable mixed-phase aerosols. Modification to an existing sampler and validation test with three pesticides, *Chemosphere* **28,** 1135–1146.
6. Schneider, T., Vermeulen, R., Brouwer, D. H., Cherrie, J. W., Kromhout, H., and Fogh, C. L. (1999) Conceptual model for assessment of dermal exposure. *Occup. Environ. Med.* **56,** 765–773.
7. Soutar, A., Semple, S., Aitken, R. J., amd Robertson, A. (200) Use of patches and whole body sampling for the assessment of dermal exposure. *Ann. Occup. Hyg.* **44,** 511–518.
8. Brouwer, D. H. (2002) *Assessment of Occupational Exposure to Pesticides in Dutch Bulb Culture and Glasshouse Horticulture,* PhD thesis, Utrecht University, Zeist, The Netherlands.
9. Brouwer, D. H., Boeniger, M. F., and Van Hemmen, J. J. (2000) Hand wash and manual skin wipes. *Ann. Occup. Hyg.* **44,** 501–510.
10. Marquart, J., Brouwer, D. H., and Van Hemmen, J. J. (2002) Removing pesticides from the hands with a simple washing procedure using soap and water. *J. Occup. Environ. Med.* **44,** 1075–1082.
11. Geno, P. W., Camann, D. E., Harding, H. J., Villalobos, K., and Lewis, R. G. (1996) Handwipe sampling and analysis procedure for the measurement of dermal contact with pesticides. *Arch. Environ. Contam. Toxicol.* **30,** 132–138.
12. Fenske, R. A., Simcox, N. J., Camp, J. E., and Hines, C. J. (1999) Comparison of three methods for assessment of hand exposure to Azinphos-Methyl (Guthion) during apple tinning. *Appl. Occup. Environ. Hyg.* **14,** 618–623.

13. Surakka, J., Johnsson, S., Lindh, T., Rosén, G., and Fischer, T. (1999) A method for measuring dermal exposure to multifunctional acrylates. *J. Environ. Monit.* **1**, 533–540.

14. Fenske, R. A., Leffingwell, J. T., and Spear, R. C. (1986) A video imaging technique for assessing dermal exposure. I. Instrument design and testing. *Am. Ind. Hyg. Assoc. J.* **47**, 764–770.

15. Archibald, B. A., Solomon, K. R., and Stephenson, G. R. (1994) A new procedure for calibration the video imaging technique for assessing dermal exposure to pesticides. *Arch. Environ. Contam. Toxicol.* **26**, 398–402.

16. Bierman, E. P. B., Brouwer, D. H., and Van Hemmen, J. J. (1998) Implementation and evaluation of the fluorescent tracer technique in greenhouse exposure studies. *Ann. Occup. Hyg.* **42**, 467–475.

17. Roff, M. W. (1994) A novel illumination system for the measurement of dermal exposure using a fluorescent dye and image processor. *Ann. Occup. Hyg.* **38**, 903–919.

18. Roff, M. W. (1997) Accuracy and reproducibility of calibrations on the skin using the FIVES fluorescence monitor. *Ann. Occup. Hyg.* **41**, 313–324.

19. Dost, A. A. (1995) Monitoring surface and airborne inorganic contamination in the workplace by a field portable x-ray fluorescence spectrometer. *Ann. Occup. Hyg.* **40**, 589–610.

20. Wheeler, J. P. and Warren, N. D. (2002) A dirichlet tessellation-based sampling scheme for measuring whole-body exposure. *Ann. Occup. Hyg.* **46**, 209–217.

21. Gunther, F. A., Barkley, J. H., and Westlake, W. E. (1974) Work environment research II. Sampling and processing techniques for determining dislodgeable pesticides residues on leaf surfaces. *Bull. Environ. Contam. Toxicol.* **12**, 641–644.

22. Iwata, Y., Knaak, J. B., Spear, R. C., and Foster, R. J. (1977) Worker re-entry into pesticide-treated crops: I Procedure for the determination of dislodgeable residues on foliage. *Bull. Environ. Contam. Toxicol.* **10**, 649–655.

23. Nigg, H. N., Stamper, J. H., and Queen, R. M. (1984) The development and use of a universal model to predict tree crop harvesters pesticide exposure. *Am. Ind. Hyg. Assoc. J.* **45**, 182–186.

24. Zweig, G., Leffingwell, J. T., and Popendorf, W. J. (1985) The relationship between dermal pesticide exposure by fruit harvesters and dislodgeable residues. *J. Environ. Sci. Health* **B20**, 25–59.

25. Popendorf, W. J. and Leffingwell, J. T. (1982) Regulating organophosphate residues for worker protection. *Residue Rev.* **82**, 125–201.

26, Woollen, B. H. (1993) Biological monitoring for pesticide absorption. *Ann. Occup. Hyg.* **37**, 525–540.

27. Brouwer, R., Van Maarleveld, K., Ravensberg, L., Meuling, W., De Kort, W. L. A. M., and Van Hemmen, J. J. (1993) Skin contamination, airborne concentrations, and urinary metabolite excretion of propoxur during harvesting of flowers in greenhouses. *Am. J. Indust. Med.* **24**, 593–603.

28. Popendorf, W. J. (1992) Re-entry field data and conclusions. *Rev. Environ. Contam. Toxicol.* **128**, 71–117.

29. Maddy, K. T., Krieger, R. I., O'Connel, L., et al. (1989) Use of biological monitoring data from pesticide users in making pesticide regulatory decisions in California. Study of captan exposure of strawberry picker, in *Biological Monitoring for Pesticide Exposure* (Wang, R. G. M., Franklin, R. C., Honeycutt, R. C., and Reinert, J. C., eds.), ACS Symposium Series no. 282, American Chemical Society, Washington, DC, pp. 338–353.

30. Chester, G., Dick, J., Loftus, N. J., Woollen, B. H., and Anema, B. H. (1990) The effectiveness of protective gloves in reducing dermal exposure to, and absorption of, the herbicide fluazifop-*p*-butyl by mixer-loader-applicators using tractor sprayers, in *Proceedings of the Seventh International Congress of Pesticide Chemistry*, IUPAC, New York, Vol. 3, p. 378.

31. Aprea, C., Sciarra, A., Sartorelli, P., Desideri, E., Amati, R., and Satorelli, E. (1994) Biological monitoring of exposure to organophosphorous insecticides by assay of urinary alkylphosphates: Influence of protective measures during manual operations with treated plants. *Int. Arch. Occup. Environ. Health* **66**, 333–338.

32. Van Hemmen, J. J., Van Golstein Brouwers, Y. G. C., and Brouwer, D. H. (1995) Pesticide exposure and re-entry in agriculture, in *Methods of Pesticide Exposure Assessment* (Curry, P. B., Iyengar, S., Maloney, P. A., and Maroni, M., eds.), Plenum Press, New York, pp. 9–19.

33. Krieger, R. I., Blewett, C., Edmiston, S., et al. (1991) Gauging pesticide exposure of handlers (mixer/loaders/applicators) and harvesters in California agriculture. *Med. Lavoro* **81,** 474–479.

34. Brouwer, D. H., De Haan, M., and Van Hemmen, J. J. (2000) Modelling re-entry exposure estimates. Application techniques and –rates, in *Worker Exposure to Agrochemicals. Methods for Monitoring and Assessment* (Honeycutt, R. C., and Day, E. W., Jr., eds.), Lewis, Washington, DC, pp. 119–138.

35. De Vreede, J. A. F., De Haan, M., Brouwer, D. H., et al. (1996) *Exposure to Pesticides. Part IV. Application to Chrysanthemums in Greenhouses*, Report S131-4, Ministry of Social Affairs and Employment, The Hague, The Netherlands.

36. Van der Jagt, K. E., Tielemans, E., Links, I., Brouwer, D., and Van Hemmen, J. (2004) Effectiveness of personal protective equipment: relevance of dermal and inhalation exposure to chlorpyrifos among pest control operators. *Am. Ind. Hyg. Assoc. J.* in press.

37. Brouwer, D. H., De Vreede, J. A. F., Meuling, W. J. A., and Van Hemmen, J. J. (2000) Determination of the efficiency for pesticide exposure reduction with protective clothing: a field study using biological monitoring, in *Worker Exposure to Agrochemicals. Methods for Monitoring and Assessment* (Honeycutt, R. C., and Day, E. W., Jr., eds.), Lewis, Washington, DC, pp. 63–84.

38. Meuling, W. J. A., Bragt, P. C., Leenheers, L. H., and De Kort, W. L. A. M. (1991) Dose-excretion study with the insecticide propoxur in volunteers, in *Prediction of Percutaneous Penetration Methods. Methods, Measurements and Modelling* (Scott, R. C., Guy, R. H., Hagraft, J., Bodde, H. E., eds.), IBC Technical Services, London, Vol. 2, pp. 13–19.

39. Meuling, W. J. A., Franssen, A. C., Brouwer, D. H., and Van Hemmen, J. J. (1997) The influence of skin moisture on the dermal absorption of propoxur in human volunteers: a consideration for biological monitoring practices. *Sci. Total Environ.* **199**, 165–172.

40. Machemer, L., Eben, A., and Kimmerle, G. (1982) Monitoring of propoxur exposure, *Stud. Environ. Sci.* **18**, 255–262.

13

Field Study Methods for the Determination of Bystander Exposure to Pesticides

C. Richard Glass

Summary

Techniques to estimate bystander exposure are described. Passive sample media such as filter paper are used to collect spray drift. Air-sampling devices are used to determine the airborne concentration of pesticides. The use of a mannequin or volunteer dressed in a disposable coverall standing downwind of the treated zone gives the most accurate indication of the potential dermal exposure of a bystander.

Key Words: Bystander; exposure; field study; pesticides.

1. Introduction

The proximity of many rural populations to agriculture has brought about increased awareness of the potential for exposure of bystanders to pesticides as particulates and vapor fractions following the application of pesticides in both the open air and enclosed areas. In this context, a *bystander* can be described as someone who may be at risk of exposure to pesticide drift but who is not involved with the application process itself. Therefore, the bystander is not protected from dermal or inhalation exposure and is often not even aware of the pesticide application.

Techniques to estimate bystander exposure are similar to those used to measure spray drift *(1,2)* and involve the location of sampling media at distances between the application area itself and sites downwind of this area here there is the potential for bystanders to be present. Passive sample media such as filter paper collect spray drift that sediments onto the ground, with cylindrical lines such as fishing line or fine polythene tubing collecting the airborne fraction. Air sampling devices can also be used to determine the airborne concentration of pesticides. The use of a mannequin or volunteer dressed in a disposable coverall standing downwind of the treated zone gives the most accurate indication of the potential dermal exposure of a bystander because the collection of airborne droplets is determined by factors such as the airflow around

From: *Methods in Biotechnology, Vol. 19, Pesticide Protocols*
Edited by: J. L. Martínez Vidal and A. Garrido Frenich © Humana Press Inc., Totowa, NJ

the collector and the momentum of the droplets. Such mathematical reasoning is well documented (*3*) and does not require further explanation here.

2. Materials

1. Petri dishes with lids.
2. Filter paper.
3. Chromatography paper (5-cm diameter).
4. Fishing line, strings, or Portex 2-mm diameter polythene tubing.
5. Disposable coveralls (e.g., Tyvek, Kimberly-Clark, Sontara).
6. Air-sampling pumps (2-L/min flow rate).
7. Institute of Occupational Medicine (IOM) sampling head/sorbent tubes (e.g. XAD-2).
8. Anemometer.
9. Wind vane.
10. Thermometer (wet and dry bulb).
11. Mannequins or human volunteers.
12. Suitable crop and sprayer.
13. Pesticide or tracer.
14. Measuring cylinders.
15. Stopwatch.

3. Methods

The methods described below outline (1) the selection of the field site for the study, (2) the selection and calibration of spraying equipment and pesticide, (3) the location of sampling media and devices, (4) measurements during the application of the pesticide, (5) labeling and storage of the sample media, and (6) expression of the data.

3.1. Field Site Selection

There is a wide range of field studies that could be done to determine potential bystander exposure. Field studies can be set up in fields that allow easy sampling of the spray drift, although these may not be situations in which bystander exposure is likely to occur. Alternatively, studies can be done at specific locations that may be representative of actual bystander exposure scenarios, such as sites where there are dwellings or footpaths adjacent to fields or crops that are sprayed with pesticides. Typical scenarios could be as follows:

1. Areas adjacent to an arable field of winter wheat with a height of 70–100 cm, for which a late fungicide or insecticide treatment is made with a boom sprayer fitted with hydraulic nozzles.
2. Areas adjacent to a top fruit orchard, such as for apples, with a tree height of 3m and for which a fungicide or insecticide treatment is made with an axial fan orchard air blast sprayer fitted with hydraulic nozzles.
3. Areas adjacent to a semiopen greenhouse with a tomato crop 2 m high for which a fungicide or insecticide treatment is made with a handheld lance with hydraulic nozzles.

There are many more scenarios that could involve the application of granules to the soil surface or gases injected into the soil. In some cases, the vapor and particulates from the pesticide application may travel several kilometers (*4,5*).

However, unless a specific situation needs to be investigated, it is advised that initial studies be done in areas where the topography is uniform and flat and there are few obstructions such as hedges, trees, or buildings. Such structures can interfere with both wind speed and wind direction and cause eddies in the air around the field. Such changes to the air movement may result in lack of deposition of pesticide drift in the areas where collection media have been placed. If there are doubts about the suitability of a field site, the use of a smoke generator or pellets can give a useful indication of the air currents. The field and surrounding area need to be large enough to allow the study to be set up as described in **Subheading 3.3.4.**

The type of crop used for the field study is important because this will determine the boom height for typical arable sprayers and as such the release height of the pesticide from the nozzle. The filtering effects of the crop will also influence the amount of drift. For many temperate regions of the world, a mature crop such as wheat with a height of approx 0.7 m represents a typical scenario of agricultural land that may be adjacent to dwellings. Such crops need to be treated with a pesticide in the form of a liquid using a boom sprayer. However, taller crops such as apples or oilseed rape (canola) are likely to result in higher levels of bystander exposure because of greater crop height or upward application technique in the case of orchard crops. For tree and bush crops, the height of the crop and the density of the foliage will affect the amount of spray drift.

3.2. Spraying Equipment and Pesticide

When a field study is planned specifically to generate data for bystander exposure, it is important to consider the type of pesticide application equipment used. For the typical bystander exposure scenario referred to in **Subheading 3.1.**, this would be the arable situation, with the pesticide applied to the crop using a mounted, trailed, or self-propelled boom sprayer with hydraulic nozzles. The amount of bystander exposure is considered closely related to the amount of airborne and ground-deposited spray drift *(6)*. The application technique, nozzles used, and operating pressure will affect the amount of spray drift, which has been documented by a number of authors *(7,8)*. The pesticide used for the study also needs to be considered carefully because the volatility of the active substance in the pesticide product will affect the airborne levels in the period following the application. The selected pesticide must also be known to be recoverable from the sampling media used for the study, as described in **Subheading 3.3.**

An alternative to using a pesticide is to use a tracer such as a food color (e.g., Green S) or a fluorescent tracer (e.g., fluorescein). A tracer cannot provide data for exposure to the volatile fraction but can be used for exposure to the particulates (droplets) for short-term bystander exposure studies. Tracers are usually much cheaper and easier to analyze than pesticides, although some fluorescent tracers can be unstable in strong sunlight. Tracers also are advantageous because they are nonhazardous, making the fieldwork easier by reducing the need for protective clothing. When tracers are used, they should be tank mixed with an appropriate surfactant to give the spray liquid a surface tension similar to diluted pesticide formulations. Surface tension is known to affect the droplet production by nozzles, especially twin fluid nozzles.

3.2.1. Selection of Application Equipment

Once the crop and application technique have been selected, the type of application equipment (sprayer) should be chosen. This is not as critical as the selection of the application technique and is often determined by the availability of sprayers on the farms available for the field study. Some of the key factors to consider are the following:

1. Boom width, which should be either 12 or 24 m to be representative of the most common widths found on arable farms.
2. The boom suspension type because this will affect the boom stability. This is a key factor because on uneven or rutted fields the movement of the boom causes uneven application because of yawing. Rolling of the boom results in nozzles on the boom that are not at the correct height above the crop.
3. Boom height. Arable field studies tend to be done with the boom 50 cm above the crop if 110° flat-fan nozzles are used, as described in **Subheading 3.2.2.**
4. If orchard crops are used for the study, an appropriate sprayer should be used that is typically used locally.

3.2.2. Nozzle Selection

In northern Europe, most arable sprayers are fitted with 110° flat-fan hydraulic nozzles. In some countries, such as Argentina *(9)*, 80° flat-fan nozzles are more commonly used. It is advisable to use nozzles that are typical for the region where the study is done or for the region where the data are to be used. During the late 1990s, a number of new nozzle types were introduced onto the market in Europe. Twin fluid nozzles have been available for some time; air and water are mixed under pressure to produce droplets containing air. There are now low-drift nozzles that have a Venturi system and others that have a preorifice mixing chamber. Such nozzles result in less drift during the application by reducing the number of droplets smaller than 100 μm. Therefore, it is important to select a nozzle appropriate to the local conditions under study or the conditions for which the data are generated.

Alternative nozzle types used on application equipment include hollow-cone and solid-cone nozzles. There are also spinning disks, often described as controlled droplet application.

3.2.3. Calibration of Application Equipment

The application equipment used in field studies must be calibrated to ensure that the rate of application of the pesticide is known. The application equipment will have user manuals that describe how the calibration should be done. This is often a simple process that for arable sprayers involves measuring the output of liquid from selected nozzles at the pressure intended to be used for the application (usually between 2 and 4 bar pressure). The pressure indicated on the pressure gage may not be accurate, so it is advisable to consult the nozzle manufacturer's recommendations. It is better to set the pressure to give the nozzle output recommended in the manufacturer's literature because the pressure gages on farm equipment can have large errors. Record all details of the calibration in the field record book for referral when the equipment is prepared for the application in the field, as described in **Subheading 3.3.**

Fig. 1. Field deployment for two types of ground deposit media.

3.3. Preparation and Deployment of Sampling Media

All media used for the study must be evaluated in the laboratory prior to the commencement of the fieldwork. Simple triplicate fortifications (spiking) with diluted samples of the selected pesticide formulation for each media type to be used can determine the recovery of the active substance. This ensures that the field samples will allow the pesticide to be extracted for analysis. Certain pesticides are difficult to extract from material such as polythene or cotton. The stability and recovery of the pesticide should be determined over the period of time the samples are expected to be stored between the fieldwork and analysis.

3.3.1. Sampling Media for Ground Fallout of Spray Drift

One of the methods used to estimate bystander exposure is to measure the amount of pesticide spray drift deposited on the ground. This type of media is usually filter paper or similar media (e.g., Benchkote), which is laid out on the ground or on supports such as wood or petri dishes. Typical media used with tracer studies are shown in **Fig. 1**.

3.3.2. Sampling Media for Airborne Pesticides

There are both active and passive collectors that can be used to measure airborne pesticides. The vertical drift profile should be measured at two distances from the edge of the swath using collection media such as 2-mm diameter Portex fine-bore polythene line. The lines can be suspended horizontally or vertically across 0.5-m

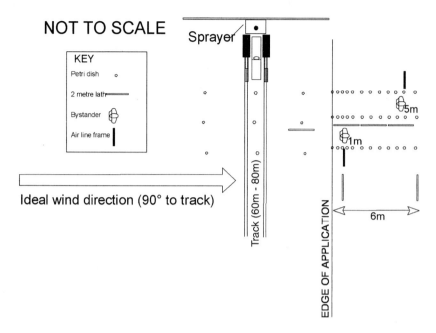

Fig. 2. A field layout for a bystander exposure study.

wooden frames at 0.25-m intervals from 0.25 to 1.75 m above ground level. Two by-standers, wearing hooded absorbent coveralls as collection media, stand at the same positions as the line frames. A diagrammatic representation of the field layout is given in **Fig. 2**. When vertical lines are deployed, it is normal practice to collect the spray drift up to a height of 10 m above the ground because this complies with the latest International Organization for Standardization draft protocol for spray drift measurement in the field *(10,11)*.

For active sampling of airborne pesticides, personal air samplers are used. These can be connected to samplers such as the IOM head or sorbent tubes such as XAD-2 *(12)*. These samples should be placed at heights above the ground representing the normal breathing zone (i.e., 1.5–2 m for adults, although this could be 0.5 m for children). It is normal practice to attach the sample head to the lapel of the mannequin or human volunteer. The flow rate of air through the sampler should be set according to manufacturer's recommendations for the particular sample head or absorbent tube used.

3.3.3. Sampling Using Mannequins or Volunteers

Life-size mannequins can be dressed in disposable coveralls such as Tyvck Classic or Kimberly-Clark Kleenguard and positioned downwind of the application area close to other sampling media such as the polythene lines. Alternatively, if the study is done with nonhazardous tracer, human volunteer bystanders wearing the hooded coveralls can stand at the same positions as the frames with polythene lines as described in **Subheading 3.3.2.** and illustrated in **Fig. 3**. This method of sampling will give an

Fig. 3. Typical field study showing bystander volunteer downwind of tracer application.

indication of the pesticide likely to be deposited on the whole body surface of a person. The coverall can be cut into sections to identify which areas of the body received the most contamination.

3.3.4. Location of Sampling Media in the Field

It is advisable to position a range of sampling media to cover an area of crop and adjacent land that includes sampling media for deposited pesticide within the application area and an area downwind of the application to a distance of at least one boom width from the edge of the treated crop. A typical field layout for an arable crop situation is shown in **Fig. 2**. The distances sampled from the application area depend on the type of the study and the expected levels of spray or vapor drift.

3.4. Monitoring of the Pesticide Application

Once the application equipment has been calibrated and set up for the field study, there are a number of critical observations and recordings that should be made. The application of the pesticide (or tracer) and the interaction with the prevailing wind conditions are the factors that have greatest influence on the bystander exposure data generated.

3.4.1. Pesticide Application Parameters

Details of the pesticide application are observed and recorded with a photographic record advisable to allow rapid recording of visual information. Digital photographs or videos provide a useful source of information that can be used to provide answers to questions that may arise during the analysis of the data. The following is a list of essential data that should be recorded.

1. Start time for application.
2. End time for application.
3. Area of crop treated (particularly the number of sprayer passes for the sampling media).
4. Observations during the mixing and loading procedure to ensure that the volume of pesticide formulation added to the spray tank is correct and recorded.
5. The ground speed of the sprayer. This can be done by recording the time taken to travel a measured distance (e.g., 100 m). For example,

$$Ground\ speed = Distance/Time$$

If it takes 44 s to travel 100 m, then the speed is 100/44 = 2.3 m/s. To convert from ms^{-1} to km/h, multiply by 3.6. To convert from m/s to mph multiply, by 5.11.

6. Record all details for the application equipment and settings. The list should include the following:

    a. Manufacturer and model of the application equipment.
    b. Spray tank size.
    c. Boom width and height above crop and ground.
    d. Nozzle number and type (manufacturer's marking).
    e. Working pressure (taken from pressure gage),
    f. Flow rate for a minimum of three nozzles with clean water.
    g. Volumes of water and pesticide added to spray tank.
    h. Estimate of volume left following application.
    i. Application rate (liters of water per hectare or gallons per acre).

7. Details of the crop need to be recorded, including crop type, height, and growth stage (with estimate of leaf area index if appropriate). Light detection and ranging (LIDAR) is becoming commonly used now for tree and bush crops to determine dose rates based on amount of foliage *(13)*.

3.4.2. Meteorological Conditions

It is essential to record the ambient conditions for the duration of the field study. It is recommended that data are collected at two heights above ground, with one of the heights boom (spray release) height in the case of arable sprayers. The following data should be collected:

1. Temperature.
2. Relative humidity.
3. Wind speed.
4. Wind direction.

The use of a data logger can allow these measurements to be taken at frequent intervals over a long period.

3.5. Collection and Storage of Media

3.5.1. Time of Collection

The media placed in the field for the field trial can be collected shortly after the end of the pesticide application because the liquid droplets dry quickly on all media types.

In warm and sunny conditions, the pesticide may begin to degrade, so the media should be collected as soon as possible. This also minimizes problems with contamination from other sources in the field.

3.5.2. Method of Collection

Scientific staff collecting the media should wear disposable gloves to avoid cross contamination of samples. Scientists who have been involved with handling of the pesticide, either diluted or the concentrate, should not be involved with handling field samples because of the risks of cross contamination. Samples should always be collected such that those expected to have the lowest residues (i.e., those furthest away from the application area) are collected first. The samples can be collected in a number of ways, depending on the nature of the sample.

The information can be written onto the label in the field or can be printed on self-adhesive labels, which can be taken to the field. The field samples should be collected in the following order: air samplers, mannequin coveralls, Benchkote, and petri dishes.

3.5.2.1. AIR SAMPLERS

1. Switch off the sampling pumps.
2. Remove IOM sample heads or absorbent tubes (e.g., XAD-2) from the mannequins.
3. Remove the cassette containing the filter and secure with seal.
4. Place in a labeled polythene bag.

3.5.2.2. MANNEQUIN COVERALLS

1. Remove the mannequin (or volunteer) to a clean area of the field and place on a sheet of polythene on the ground.
2. Remove the coverall from the mannequin, avoiding cross contamination between areas of the coverall.
3. If the coverall is to be sectioned, this can be done in the field with a clean pair of scissors.
4. Place each of the sections or the whole coverall in a labeled polythene bag.

3.5.2.3. BENCHKOTE (OR SIMILAR FLAT MEDIA PLACED ON GROUND)

1. Starting with the media samples furthest from the application area, carefully roll or fold the sections of Benchkote and place in labeled polythene bags.
2. When the collection media is fixed by staples or a similar mechanism, care needs to be taken when removing the media from the support so the media does not tear, which may result in lost pieces.
    a. Collect the media from within the application area last because this is the most heavily contaminated.

3.5.2.4. PETRI DISHES

1. Starting with the petri dishes furthest from the application area, place clean lids on each and fix the lid with adhesive tape.
2. Label each petri dish.
3. Collect the petri dishes in sets of 5 or 10 and bind together with adhesive tape so that they can be kept in order.
4. Place in clean polythene bags.
5. Collect the petri dishes from within the application area last because this is the most heavily contaminated.

3.5.3. Labeling of Media

Each individual piece of collection media needs to have a unique label, so that it can be readily identified, should it be separated from the rest of the samples during transit. The label is the only way that the item can be correctly identified and as such needs to have, as a minimum, the following information:

1. Study number.
2. Replicate number.
3. Date.
4. Sample location.

3.5.4. Fortified and Blank Samples

For each day of the field study, there needs to be fortified and blank media samples from the field. The volume selected for the fortification should be representative of the residue expected on the media in the field study. However, the volume fortified needs to be accurate, so very small volumes are usually avoided. The steps required for field fortification are as follows:

1. Select an area of the field or building similar to that where other media samples for the study are prepared. This needs to be free from contamination and should be an area up-wind of the application area.
2. Samples of each media type (minimum three replicates for fortified samples and three replicates for blank sample) should be laid out on polythene sheeting or aluminum foil in an area not too distant (e.g., less than 500 m) from the area where the pesticide application is taking place.
3. Fortify (spike) the media samples with the sample of pesticide solution taken from the nozzle of the application equipment. Suggested fortification volumes are as follows:

   a. 0.1 mL for ground deposit media and coveralls.
   b. 0.05 mL for the media used for the airborne pesticide sampling.

   A gas chromatographic syringe can be used for the 0.05-mL volumes, and an Eppendorf-type pipet can be used for the 0.1-mL volumes.
4. Leave all samples for a period equivalent to the duration of the field study to ensure that the media are exposed to the same ambient conditions as the media used for the experimental samples from the field.
5. At the end of the exposure period, the samples are treated in the same manner as the field samples, as detailed in **Subheading 3.5.2.**

3.5.5. Storage of Samples

Once all of the field samples together with field blanks and fortified samples have been collected and labeled, they need to be stored together in a container to protect them from extreme temperatures and sunlight. This is sufficient for most pesticides (analytes). The analytical validation work done in the laboratory by the analytical service prior to commencing the fieldwork will indicate the stability of the analyte used in the study. If the analyte is unstable in ambient conditions (e.g., 20°C), then samples will need to be transported to the laboratory immediately or carried in transit in a cool box with ice blocks. In extreme cases, dry ice (solid carbon dioxide) may be required for long-distance transport and care should be taken to avoid a buildup of carbon dioxide gas in vehicles during transport.

3.6. Expression of Data

3.6.1. Analytical Data

The media samples from the field can be analyzed by any competent analytical laboratory. The analytical protocol needs to stipulate the nature of the data that will be returned. It is normal for the raw data to be returned to express the mass of pesticide or volume of tracer solution on each individual media sample. The raw data need to be manipulated to allow the data to be presented in terms of potential bystander exposure. The ground fallout data should be presented as the mass of pesticide or volume of tracer per unit area or as a proportion of the pesticide or tracer applied to the crop *(14)*.

3.6.2. Ground Fallout Data

The data for ground fallout should be presented as the mass of pesticide or volume of tracer per unit area (e.g., square meter). These data can also be represented as a proportion of the pesticide or tracer applied to a unit area of the crop (hectare or acre). A typical example of a calculation for a pesticide study is as follows:

$$\text{Area of ground sample media} = 60 \text{ cm}^2$$

$$\text{Mass of pesticide on sample media} = 0.005 \text{ mg}$$

$$\text{Mass of pesticide per square meter} = 0.833 \text{ mg/m}^2 \,(10{,}000 \text{ cm}^2 = 1 \text{ m}^2)$$

From the dose rate of the pesticide, the unit dose can be calculated. For example, a product applied at 500 g per hectare would have, per square meter of crop area,

$$500/10{,}000 = 0.05 \text{ gm}^{-2} \,(50 \text{ mg/m}^2)$$

In the example, the proportion of the applied dose drifting would be

$$(0.833/50) \times 100 = 1.67\%$$

Such values can be used as generic data to estimate exposure in similar field conditions with pesticides used at different dose rates. The current trend in Europe is to use data for the drift deposition on a 2-m^2 area of ground to be equivalent to the potential dermal exposure of a bystander at the same distance from the application area *(6)*.

3.6.3. Airborne Drift Data

For passive samples such as the polythene lines, these data can be presented as the mass of pesticide or volume of tracer passing through a unit area. If 2-mm diameter lines are used as suggested in **Subheading 3.3.2.**, then the mass of pesticide can be estimated for a surface area equivalent to a bystander. The default value for a bystander is often taken as 2 m^2, so it could be assumed that half of this area would be exposed to any airborne drift.

Again, this can be related to the output from the sprayer by calculating the output of spray per meter traveled for a single pass. For example, if for every meter traveled by the sprayer, 150 mL of liquid were applied to the crop, this is calculated as follows:

Forward speed of sprayer calculated as 2 m/s (0.5 s to travel 1 m)

Flow rate measured as 0.75 L/min for each of 24 nozzles (18 L/min)

Flow rate per 0.5 s (1 m) would be

$$18/120 = 0.15 \text{ L (150 mL)}$$

If the volume of spray deposited on the lines was 10 mL for a frame 1-m wide, then the proportion of spray passing through his frame would be

$$(10/150) \times 100 = 0.67\%$$

For active sampling using the personal air samplers, then the calculation is simpler. The mass of pesticide sampled is related to the volume of air sampled. For example, if 15 mg of pesticide are found to be on the sampling device, and the volume of air sampled was 10 L, then the concentration of pesticide in the air would be

$$15/10 = 1.5 \text{ µg/L}$$

The concentration of the pesticide in the air can be related to inhalation exposure of the bystander by using an appropriate breathing rate for an adult, such as 3.6 m^3/h.

3.6.4. Mannequin or Volunteer Data

The mannequin or volunteer data are likely to be the most realistic because the amount of pesticide or tracer deposited on the coverall of the mannequin or volunteer is the value for the potential dermal exposure of the bystander. The data can be presented for regions of the body (e.g., the mass of pesticide found on the hood of the coverall or on the arms). However, it is normal practice to use the value for the entire body and express these data as the potential dermal bystander exposure. For most studies, the duration of the study will be short, so relating the exposure to a period of exposure is not appropriate. For inhalation exposure data, the study may last longer. The data can be related to duration of exposure.

If a number of collection types have been used in the field, the data from these can be compared. This will give an indication of the relative merits of each type of collection device.

Acknowledgments

I wish to acknowledge the financial support of the UK Department for Environment, Food, and Rural Affairs (DEFRA), formerly the Ministry of Agriculture, Fisheries, and Food (MAFF), which has allowed many of the methods described here to be developed and validated. The financial support of the European Union SMT program is also acknowledged, through project SMT4-CT96-2048.

References

1. Gilbert, A. J. and Bell, G. J. (1988) Evaluation of the drift hazards arising from pesticide spray application. *Aspects Appl. Biol.* **17,** 363–367.
2. Mathers, J. J., Wild, S. A., and Glass, C. R. (2000) Comparison of ground deposit collection media in field drift studies. *Aspects Appl. Biol.* **57,** 242–248.
3. May, K. R. and Clifford, R. (1967). The impaction of aerosols on cylinders, spheres, ribbons and discs. *J. Occup. Hygiene* **10,** 83–95.
4. EPPO (2003) Environmental risk assessment scheme for plant protection products. *EPPO Bull.,* **33,** 115–129.
5. Landers, A. (2000) Drift reduction in the vineyards of New York and Pennsylvania. *Aspects Appl. Biol.* **57,** 67–73.
6. Gilbert, A. EUROPOEM Bystander Working Group Report. (2002, December). Project FAIR CT96-1406, European Commission, Brussels, Belgium.
7. Zande, J. C. van de, Porskamp, H. A. J., Michielsen, J. M. P. G., Holterman, H. J., and Huijsmans, J. F. M. (2000) Classification of spray application for driftability to protect surface water. *Aspects Appl. Biol.* **57,** 57–65.
8. Birchfield, N. (2004) Pesticide spray drift and ecological risk assessment in the US EPA: a comparison between current default spray drift deposition levels and AgDRIFT predictions in screening-level risk assessment. *Aspects Appl. Biol.* **71,** 125–131.
9. Martínez Peck, R. (2004) Spraying techniques used in Argentina. *Aspects Appl. Biol.* **71,** 475–480.
10. BSI. *Equipment for Crop Protection. Methods for the Field Measurement of Spray Drift*, Draft British Standard (ISO/CD 12057), BSI, London.
11. Moreira, J. F., Santos, J., Glass, C. R., Wild, S. A., and Sykes, D. P. (2000) Measurement of spray drift with hand held orchard spray applications. *Aspects Appl Biol.* **57,** 399–404.
12. Capri, E., Alberci, R., Glass, C. R., Minuto, G., and Trevisan, M. (1999) Potential operator exposure to procymidone in greenhouses. *J. Agric. Food Chem.* **47,** 4443–4449.
13. Cross, J. V., Murray, R. A., Walklate, P. J., and Richardson, G. M. (2004) Pesticide Dose Adjustment to the Crop Environment (PACE): efficacy evaluations in UK apple orchards 2002–2003. *Aspects Appl. Biol.* **71,** 287–294.
14. Matthews, G. and Hamey, P. Y. (2003). Exposure of bystanders to pesticides. *Pesticide Outlook* **14,** 210–212.

14

Determination of Household Insecticides in Indoor Air by Gas Chromatography–Mass Spectrometry

Edith Berger-Preiss and Lutz Elflein

Summary

An analytical method for the determination of commonly used insecticides and acaricides (pyrethroids, organophosphates, carbamates, organochlorine pesticides) in indoor air is described. Air samples are collected with a sampling train consisting of a glass fiber filter (GFF) and two polyurethane foam (PUF) plugs, followed by a high-volume air pump. This combination is used to sample particle-bound compounds (on the GFF) as well as gaseous compounds (on the PUF plugs). GFFs and PUF plugs are extracted separately with ethyl acetate as solvent in an ultrasonic bath subsequent to the sampling. The extracted insecticides and acaricides are identified and quantified by gas chromatography–mass spectrometry with electron impact ionization in the selected ion monitoring mode (GC–MS/EI/SIM).

Key Words: Air sampling; analysis; carbamates; gas chromatography; mass spectrometry; glass fiber filter; organochlorine pesticides; organophosphorus compounds; polyurethane foam; pyrethroids.

1. Introduction

A wide variety of biocidal products for indoor use is available on the market for both consumers and professionals. These products contain active ingredients such as pyrethroids, organophosphates, organochlorines, and carbamates. Some of these compounds may persist in the indoor environment over a long period of time. To assess possible hazards to human health that may result from the indoor use of insecticides/acaracides, multicomponent analytical methods to determine the active ingredients of these biocides, especially in indoor air, are a prerequisite. Several methods for the analysis of single components or compound classes have been reported previously for different air-sampling techniques (using, e.g., Tenax®, Chromosorb®, XAD®, extraction disks, polyurethane foam, air containers, impingers, impactors, glass fiber filters (GFFs) or activated carbon as sampling media) and analytical methods for analyte determination (e.g., liquid chromatography with ultraviolet detection and gas chroma-

From: *Methods in Biotechnology, Vol. 19, Pesticide Protocols*
Edited by: J. L. Martínez Vidal and A. Garrido Frenich © Humana Press Inc., Totowa, NJ

tography [GC] with electron capture, nitrogen–phosphorus-sensitive, or mass spectrometric [MS] detection) *(1–19)*.

The method described in this chapter permits sensitive and simultaneous determination of active ingredients of biocide products such as carbamates, pyrethroids, organophosphorus, and organochlorine compounds in indoor air down to the low nanogram-per-square-meter range *(20)*. A high-volume pump and a sampling unit with a GFF and two polyurethane foam (PUF) plugs are used for air sampling, allowing the collection of large volumes of air and separate determination of particle-bound and gaseous compounds. Gas chromatography–mass spectrometry with electron impact ionization and in selected ion monitoring (GC–MS/EI/SIM) mode is applied to provide high sensitivity and selectivity in multicomponent analysis.

2. Materials

2.1. Equipment

2.1.1. Air-Sampling Equipment

1. Air-sampling unit (*see* **Fig. 1**) consisting of a stainless steel cylinder carrying two PUF plugs, a filter holder carrying a GFF, and a tube adapter.
2. GFF GF 10, 50-mm diameter.
3. PUF plugs, 60-mm diameter × 50 mm (*see* **Note 1**).
4. High-volume pump (flow rate ≥ 0.05 m^3/h).
5. Gas meter for sample volume determination.
6. Flow meter with restrictor for flow rate adjustment.
7. Timer.
8. Thermometer.
9. Hygrometer.
10. Barometer.
11. A pair of tweezers (filter handling).
12. Crucible tongs (PUF handling).
13. Petri dishes with caps for storage of filter samples.
14. 250-mL Amber bottles with a wide opening and screw tops for storage of PUF plugs.
15. Aluminum foil.

2.1.2. Equipment for Extraction

1. Ultrasonic bath for extraction.
2. Glass ware: beakers (50, 300, 600 mL), bulb flasks (250 mL), bulb or tubular flasks with graduated stem (50 mL, graduation 0–5 mL), Pasteur pipets, volumetric flasks (1, 10, and 100 mL), and GC vials (2 mL) with inserts.
3. Plastic (polypropylene) microliter pipet tips plugged with silanized glass wool.

2.1.3. Analytical Equipment

1. Gas chromatograph (e.g., Agilent 6890, Palo Alto, CA) with split/splitless injector and an autosampler (e.g., Agilent 7683).
2. Mass selective detector (MSD; e.g., Agilent 5973N).
3. Software for data acquisition and data analysis (e.g., HP Chemstation, Avondale, PA), including a mass spectral database (e.g., National Institute of Standards and Technology [NIST]).
4. GC capillary column 60 m long, 250 μm id, 0.25 μm *df*; bonded phase of 5% diphenyl and 95% dimethylpolysiloxane on fused silica (e.g., Agilent HP-5 MS).

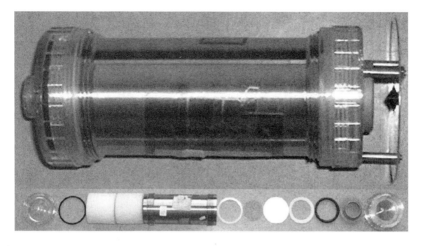

Fig. 1. Air-sampling unit for the collection of particle-bound and gaseous compounds in indoor air (top: assembled; bottom: disassembled). The parts from right to left are air plate, air streamer, rubber sealing, Teflon O-ring, GFF, wire mesh support, Teflon O-ring, cylinder, PUF plugs, rubber sealing, tube adapter.

2.2. Chemicals

1. Ethyl acetate (pesticide grade).
2. Standard substances: allethrin, chlorodecone, chlorpyrifos, cyfluthrin, cyhalothrin, cypermethrin, deltamethrin, diazinon, dichlorvos, fenitrothion, fenthion, fenvalerate, lindane, malathion, permethrin, phenothrin, piperonyl butoxide, propoxur, resmethrin, tetramethrin.

2.3. Calibration Solutions (see Note 2)

2.3.1. Preparation of Standard Solutions

1. Weigh approx 20 mg of each standard substance into separate 10-mL volumetric flasks. Dissolve in 5–8 mL ethyl acetate. If necessary, utilize ultrasonic irradiation to facilitate dissolution. Fill up to the index mark with ethyl acetate (individual standard solutions, concentration 2000 µg/mL each).
2. Transfer 0.5 mL of each individual standard solution to a 100-mL volumetric flask and fill up to the index mark with ethyl acetate (standard mix solution, concentration 10 µg/mL).

2.3.2. Preparation of Matrix-Matched Calibration Solutions

1. Prepare extracts of GFF and PUF plugs after sampling pesticide-free air according to **Subheadings 3.1.** and **3.2.** (extracts of five samples). Collect the individual extracts of the GFF and PUF plugs in separate vessels.
2. Prepare matrix-matched calibration solutions of 0.1, 0.5, 1.0, 1.5, and 2.0 µg/mL by pipeting 10, 50, 100, 150, or 200 µL of the standard mix solution (10 µg/mL) into a 1-mL volumetric flask and by filling up to the index mark with the extracts obtained according to **step 1**. Prepare matrix-matched calibration solutions for GFF and PUF matrix calibration separately.

3. Methods

The method outlines (1) the sampling of selected active ingredients of biocidal products in air (i.e., the pyrethroids allethrin, cyfluthrin, cyhalothrin, cypermethrin, deltamethrin, fenvalerate, permethrin, phenothrin, resmethrin, and tetramethrin; the organophosphates chlorpyrifos, dichlorvos, diazinon, fenitrothion, fenthion, and malathion; diflubenzuron [benzoyl-phenyl urea]; propoxur [carbamate]; the chlorinated pesticides chlorodecone and lindane; and insecticide synergist piperonyl butoxide; (2) the extraction methods used for GFFs and polyurethane foams; and (3) the analytical determination by GC–MS.

3.1. Air Sampling

1. Insert the two PUF plugs into the cylinder of the air-sampling unit (**Fig. 1**) using clean crucible tongs. Screw the tube adapter with the rubber sealing onto the cylinder to close this side of the cylinder.
2. Place the GFF together with the Teflon O-rings and the wire mesh support into the filter holder on the other side of the cylinder (according to the order shown in **Fig. 1**) using a clean pair of tweezers. Screw the air plate and steamer with the rubber sealing onto the cylinder.
3. Install the sampling unit about 1–1.5 m above the floor and at least 1 m away from the walls. Connect the sampling unit, flow meter, gas meter, and pump in a series. Record time, temperature, relative humidity, and air pressure. Start the pump and collect a total sample volume of about 10 m^3 air (flow rate of about 0.05 m^3/min).
4. Stop the pump, record the time and sample volume, and remove the filter and PUF plugs from the air-sampling unit; transfer the filter into a petri dish protected with aluminum foil (*see* **Note 3**) and the PUF plugs into a wide-opening amber bottle (use a clean pair of tweezers and clean crucible tongs, respectively, to avoid contamination).

3.2. Sample Preparation (see Note 4)

3.2.1. Glass Fiber Filter

1. Cut the GFF (containing the particle fraction of the sample) with a clean pair of scissors and transfer it into a 50-mL glass beaker.
2. Add approx 10 mL of ethyl acetate.
3. Place the glass beaker into an ultrasonic bath and extract the filter for 5 min.
4. Transfer the extract into a bulb or tube flask with graduated stem and repeat this ultrasonic extraction procedure two more times.
5. Reduce the combined extracts to about 0.5 mL with a gentle flow of nitrogen (*see* **Note 5**).
6. Take up the reduced extract solution with a Pasteur pipet and filtrate through a 1-mL microliter pipet tip plugged with silanized glass wool into a 1-mL volumetric flask. Rinse the flask with approx 0.4 mL ethyl acetate and wash the pipet tip used for filtration with the solvent.
7. Adjust the final volume in the volumetric flask to 1 mL with ethyl acetate.

3.2.2. PUF Plugs

1. Place each PUF plug (containing the gaseous fractions of the sample) separately into a 600-mL glass beaker.
2. Add approx 50 mL ethyl acetate onto the top of the plug.
3. Put the glass beaker into an ultrasonic bath and squeeze the PUF plug periodically with the bottom of a 300-mL glass beaker during a 2-min extraction time.

4. Squeeze the plug between the two beakers and pour the extract into a bulb flask. Repeat this extraction procedure three more times.
5. Reduce the combined extracts to 2–3 mL using a rotary vacuum evaporator. Transfer the extract with a Pasteur pipet into a bulb or tube flask with graduated stem. Rinse the bulb flask two times with approx 1 mL ethyl acetate and transfer the rinses into the graduated flask. Reduce the combined solutions to a volume of about 0.5 mL with a gentle flow of nitrogen (*see* **Note 5**).
6. Take up the reduced extract solution with a Pasteur pipet and filtrate through a 1-mL microliter pipet tip plugged with silanized glass wool into a 1-mL volumetric flask. Rinse the flask with approx 0.4 mL ethyl acetate and wash the pipet tip used for filtration with the solvent.
7. Adjust the final volume in the volumetric flask with ethyl acetate to 1 mL.

3.3. Analytical Determination

3.3.1. Measurement Parameters

1. Column: *see* **Subheading 2.1.3.**; carrier gas is helium 1.4 mL/min (constant-flow mode).
2. GC temperatures are as follows: 250°C injection port; oven program 60°C (1 min), 60–170°C (10°C/min), 170–280°C (4°C/min), 280°C (25 min); 280°C transfer line.
3. MSD: positive EI ionization mode, 70-eV ion potential, 230°C ion source, 150°C quadrupole.
4. Data acquisition: Select the SCAN mode for compound identification. Use SIM mode for quantification; select two ions per analyte.
   Target and qualifier ions (*m/z*): allethrin (<u>123</u>/79), chlordecone (<u>272</u>/237), chlorpyrifos (<u>197</u>/314), cyfluthrin (<u>163</u>/215), cyhalothrin (<u>181</u>/141), cypermethrin (<u>163</u>/127), deltamethrin (<u>181</u>/93), diazinon (<u>179</u>/137), dichlorvos (<u>109</u>/185), fenitrothion (<u>277</u>/125), fenthion (<u>278</u>/125), fenvalerate (<u>167</u>/125), lindane (<u>181</u>/219), malathion (<u>173</u>/125), permethrin (<u>183</u>/127), phenothrin (<u>123</u>/183), piperonyl butoxide (<u>176</u>/119), propoxur (<u>110</u>/152), resmethrin (<u>123</u>/171), tetramethrin (<u>164</u>/123) (*see* **Notes 6–8**).

3.3.2. Matrix-Matched Calibration and Sample Measurement

3.3.2.1. GFF SAMPLES

1. Use the SCAN mode of the MSD and inject 1 μL of a calibration sample (2 μg/mL) into the GC system.
2. Identify each compound by comparison of the measured mass spectra with the search results of the mass spectral database.
3. Use the SIM mode (target and qualifier ions [*m/z*]; *see* **Subheading 3.3.1.**) and subsequently inject (splitless) 1 μL of each matrix-matched calibration solution. Measure each level twice (*see* **Note 9**).
4. Plot the concentration of each compound of the calibration solution as a function of the peak areas (obtained after peak integration) and calculate the regression function of the calibration curve for each compound (*see* **Note 10**). Examples of calibration curves for chlorpyrifos and permethrin are shown in **Fig. 2**.
5. Inject 1 μL (splitless) of a GFF sample extract (*see* **Subheading 3.2.1.**) and determine the peak area after peak integration. A chromatogram of a filter sample extract is shown in **Fig. 3**.

3.3.2.2. PUF PLUG SAMPLES

1. Apply the same procedure as described for the GFF samples, but now use the matrix-matched PUF calibration solutions (*see* **Note 9**).

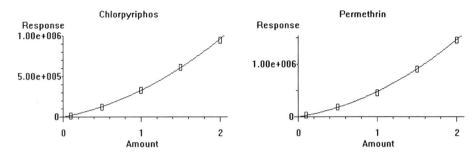

Fig. 2. Calibration curves (*y*-axis: counts; *x*-axis: concentration in micrograms per millili-ter) of chlorpyriphos and permethrin (GFF matrix).

2. Inject 1 µL (splitless) of a PUF sample extract (*see* **Subheading 3.2.2.**) and determine the peak area after peak integration. Sample chromatograms are shown in **Figs. 4** and **5**.

3.3.3. Analyte Identification and Quantification

1. Identify each compound by its retention time and target/qualifier ion response ratio. Dif-ferences in retention times should not exceed those of the calibration solutions by more than 0.5% (usually about 0.1%). Differences in the ion peak ratios used for analyte identi-fication should not exceed those determined for the calibration solutions by more than 20%.
2. Determine the total amount of each analyte detected in the sample solution using the corresponding calibration function. If necessary, take dilution factors into account (*see* **Note 11**).
3. Calculate the amount of analyte in the air as follows:

$$C_{air} = (C_{sample} \times V_{sample} \times d)V_{air}$$

where C_{air} is the concentration of the active ingredient in the air ($\mu g/m^3$), C_{sample} is the concentration of the active ingredient in the sample extract ($\mu g/mL$), V_{sample} is the volume of the sample extract (mL; 1 mL in the described method), d is the dilution factor, and V_{air} is the air volume (m^3).

3.3.4. Quality Assurance

1. Prepare blank samples (extract GFFs and PUF plugs; *see* **Subheadings 3.2.1.** and **3.2.2.**) and analyze them in SIM mode to check for blank values).
2. Check calibration by injecting a 1-µg/mL matrix-matched standard solution after every 10 samples (GFFs) and after every 5 samples (PUF plugs) (*see* **Note 12**).
3. Determine recoveries: Pipet 100 µL of the standard mix solution (concentration 10 µg/mL) onto the center of the GFF and into the PUF plugs (1 µg absolute), extract the filter (*see* **Subheading 3.2.1.**) and PUF plugs (*see* **Subheading 3.2.2.**), and analyze the ex-tracts as described in **Subheading 3.3.3.** (*see* **Note 13**). Use a minimum of three GFFs and PUF plugs for recovery studies (*see* **Note 14**).

4. Notes

1. The PUF plugs (30-kg/m³ density) are made by polymerization of toluene diisocyanate and polyoxypropylenetriol. For initial cleanup before use, extract the foam plugs in a 1-L Soxhlet apparatus (approx 30 extraction cycles each) successively with toluene (16 h),

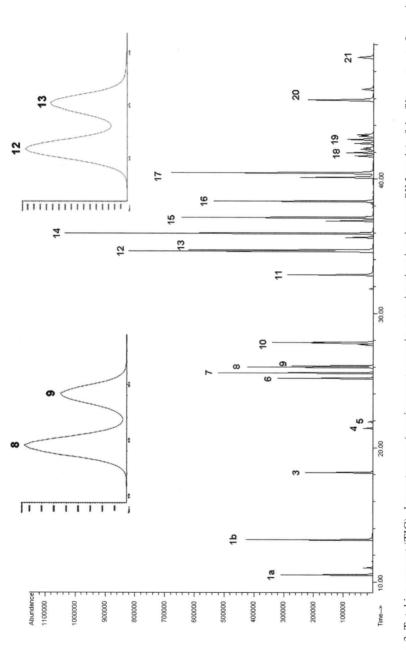

Fig. 3. Total ion current (TIC) chromatogram (y-axis: counts; x-axis: retention time in minutes; SIM mode) of the filter extract after an indoor spraying experiment: (1a) p-chlorobenzene-isocyanate; (1b) 2,6-difluorobenzene amide; (2) dichlorvos; (3) propoxur; (4) lindane; (5) diazinon; (6) fenitrothion; (7) malathion; (8) fenthion; (9) chlorpyrifos; (10) allethrin; (11) chlorodecone; (12) piperonyl butoxide; (13) resmethrin; (14) tetramethrin; (15) phenothrin; (16) cyhalothrin; (17) permethrin (two peak isomers); (18) cyfluthrin (four peak isomers); (19) cypermethrin (four peak isomers); (20) fenvalerate (two peak isomers); (21) deltamethrin.

185

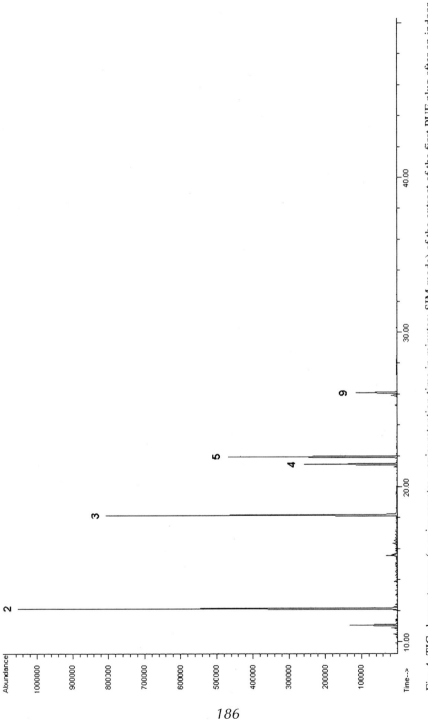

Fig. 4. TIC chromatogram (*y*-axis: counts; *x*-axis: retention time in minutes; SIM mode) of the extract of the first PUF plug after an indoor spraying experiment (peak assignment, *see* **Fig. 3**).

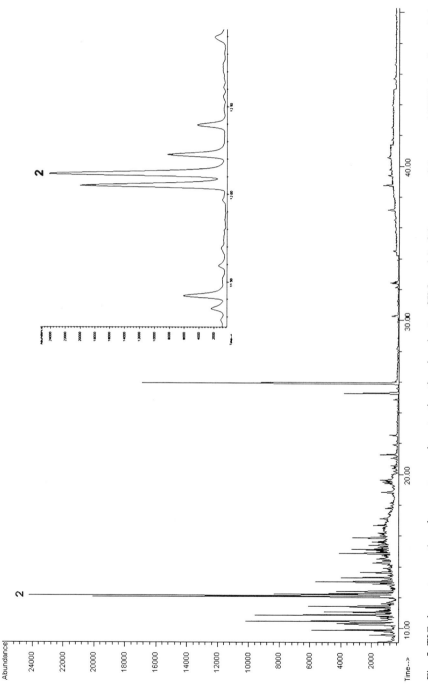

Fig. 5. TIC chromatogram (y-axis: counts; x-axis: retention time in minutes; SIM mode) of the extract of the second PUF plug after an indoor spraying experiment (peak assignment, see **Fig. 3**).

187

acetone (16 h), and ethyl acetate (16 h). After extraction, remove the bulk of the solvent by squeezing the plugs between the bottoms of two 600-mL glass beakers. Dry the plugs in an exsiccator using a small flow of nitrogen. Store the foams in 250-mL wide-opening amber bottles until use.

2. Standard solutions are stable at 4–7°C for more than 1 yr; matrix-matched calibration solutions must be prepared fresh because fenthion degrades in the presence of matrix. Matrix-matched calibration is necessary to obtain accurate results in sample measurements (matrix-induced signal enhancement) as described in detail in **ref. 20**.

3. Be careful during transport to avoid sample loss because of contact of the exposed side of the filter with the glass walls of the petri dish. Fold the exposed filter (exposed side against exposed side) and place into the petri dish.

4. Extract the filter and PUF plugs as soon as possible to avoid analyte loss caused by degradation or evaporation.

5. Use a Pasteur pipet, for instance, to blow nitrogen into the flask. The flask may be placed into a water bath (20–30°C) to facilitate solvent evaporation. Alternatively, a commercial sample concentration workstation with nitrogen-assisted solvent evaporation and sample heating block can be used.

6. Under the given conditions, diflubenzuron decomposes completely in the injection port, resulting in two thermal degradation products (i.e., para-chlorobenzene isocyanate and 2,6-difluorobenzene amide), which are monitored. Use 2,6-difluorobenzene amide (141/ 157) for quantification of diflubenzuron.

7. Use the underlined ion (m/z) for quantification (target ion) and the second ion (m/z) for the confirmation of a specific compound (qualifier ion). Identify every compound by retention time and target/qualifier ion response ratio.

8. Usually, the two most intense ion signals of each compound observed in the SCAN mode are chosen. The more characteristic ion, which usually is the one with the higher mass or the highest abundance, is used as target ion, the other one as qualifier ion. However for some compounds (piperonyl butoxide, cyfluthrin, deltamethrin), different ions are selected because of interferences with signals from ubiquitously present phthalates (m/z 149) and GC column bleeding substances (m/z 207, 253).

9. Inject matrix samples (five times) before starting a calibration procedure. This procedure allows the equilibration of the analytical system (injection port liner, column).

10. It was observed that quadratic regression functions are often better suited, especially for pyrethroids and organophosphorus compounds.

11. If the sample signal exceeds the range of the calibration curve, dilute the sample and note the dilution factor for calculations. Repeat the analysis once.

12. Recalibrate if deviation is 20% or more (GFF) or 25% or more (PUF).

13. Extract the GFFs immediately after evaporation of ethyl acetate. Otherwise, allethrin, phenothrin, resmethrin, and tetramethrin cannot be recovered quantitatively.

14. The recovery rates of the analytes for spiked filters (without air throughput) range from 87 to 118%, relative standard deviation (RSD) 3–10% (except 40% recovery for dichlorvos). The recovery rates for PUF plugs (without air throughput) are between 89 and 107%, RSD 5–9% (except 65% recovery for dichlorvos). The minimum method detection limits for most active ingredients in the air are 0.1–0.3 ng/m^3 (except 1 ng/m^3 for fenvalerate and 4–5 ng/m^3 for cyfluthrin, cypermethrin, and deltamethrin). The method validation is described in detail in **ref. 20**.

Acknowledgment

The financial support of the Federal Institute for Risk Assessment (BfR), formerly the Federal Institute for Health Protection of Consumers and Veterinary Medicine (BgVV) in Berlin, is gratefully acknowledged.

References

1. Class, T. J. (1991) Determination of pyrethroids and their degradation products in indoor air and on surfaces by HRGC–ECD and HRGC–MS(NCI). *J. High Res. Chromatogr.* **14,** 446–450.
2. Leidy, R. B. and Wright, C. G. (1991) Trapping efficiency of selected adsorbents for various airborne pesticides. *J. Environ. Sci. Health* **26,** 367–382.
3. Roinestad, K. S., Louis, J. B., and Rosen, J. D. (1993) Determination of pesticides in indoor air and dust. *J. AOAC Int.* **76,** 1121–1126.
4. Schenk, G., Rothweiler, H., and Schlatter, C. (1997) Human exposure to airborne pesticides in homes treated with wood preservatives. *Indoor Air* **7,** 135–142.
5. Riegner, K. and Schmitz, J. R. (1994) Production of a test atmosphere and separation of gaseous and particle-bound crop-protection agent residues from air in Tenax sampling tubes. *Planzenschutz-Nachrichten Bayer* **47,** 157–171.
6. Matoba, Y., Takimoto, Y., and Kato, T. (1998) Indoor behavior and risk assessment following space spraying of *d*-tetramethrin and *d*-resmethrin. *AIHA J.* **59,** 181–199.
7. Clément, M., Arzel, S., Le Bot, B., Seux, R., and Millet, M. (2000) Adsorption/thermal desorption–GC/MS for the analysis of pesticides in the atmosphere. *Chemosphere* **40,** 49–56.
8. Haraguchi, K., Kitamura, E., Yamashita, T., and Kido, A. (1994) Simultaneous determination of trace pesticides in urban air. *Atmospheric Environ.* **28,** 1319–1325.
9. Millet, M., Wortham, H., Sanusi, A., and Mirabel, P. (1996) A multiresidue method for determination of trace levels of pesticides in air and water. *Arch. Environ. Contam. Toxicol.* **31,** 542–556.
10. Murayama, H., Mukai, H., Mitobe, H., and Moriyama, N. (2000) Simple method for determining trace pesticides in air using extraction disks. *Anal. Sci.* **16,** 257–263.
11. Berger-Preiss, E., Preiss, A., Sielaff, K., Raabe, M., Ilgen, B., and Levsen, K. (1997) The behaviour of pyrethroids indoors. *Indoor Air* **7,** 248–261.
12. Turner, B. C. and Glotfelty, D. E. (1977) Field air sampling of pesticide vapors with polyurethane foam. *Anal. Chem.* **49,** 7–10.
13. Hsu, J. P., Wheeler, H. G., Jr., Camann, D. E., and Schattenberg, H. J. J. (1988) Analytical methods for detection of nonoccupational exposure to pesticides. *Chromatogr. Sci.* **26,** 181–189.
14. Foreman, W. T., Majewski, M. S., Goolsby, D. A., Wiebe, F. W., and Coupe, R. H. (2000) Pesticides in the atmosphere of the Mississippi River Valley, part II—air. *Sci. Total Environ.* **248,** 213–216.
15. Majewski, M. S., Foreman, W. T., Goolsby, D. A., and Nakagaki, N. (1998) Airborne pesticide residues along the Mississippi River. *Environ. Sci. Technol.* **32,** 3689–3698.
16. Ramesh, A. and Vijayalakshmi, A. (2001) Monitoring of allethrin, deltamethrin, esbiothrin, prallethrin and transfluthrin in air during the use of household mosquito repellents. *J. Environ. Monit.* **3,** 191–193
17. Van Dyk, L. P. and Visweswariah, K. (1975) Pesticides in air: sampling methods. *Residue Rev.* **55,** 91–134.

18. Kawata, K., Mukai, H., and Yasuhara, A. (1995) Monitoring of pesticides in air by gas chromatography–mass spectrometry and the use of quartz–fibre wool and activated carbon for sampling. *J. Chromatogr. A* **710,** 243–250.

19. Kawata, K. and Yasuhara, A. (1994) Determination of fenitrothion and fenthion in air. *Bull. Environ. Contam. Toxicol.* **52,** 419–424.

20. Elflein, L., Berger-Preiß, E., Levsen, K., and Wünsch, G. (2003) Development of a gas chromatography–mass spectrometry method for the determination of household insecticides in indoor air. *J. Chromatogr. A* **985,** 147–157.

15

Assessment of Dermal and Inhalatory Exposure of Agricultural Workers to Malathion Using Gas Chromatography–Tandem Mass Spectrometry

Francisco J. Egea González, Francisco J. Arrebola Liébanas, and A. Marín

Summary

The exposure of agricultural workers to malathion is assessed from different approaches. Methods for potential and actual exposure assessment are proposed and validated using gas chromatography with tandem mass spectrometry (GC–MS/MS) as the analytical technique. The metabolites α- and β-malathion monocarboxylic acids (α- and β-MMA) are determined after a derivatization process to obtain their hexafluoroisopropyl esters. Whole-body dosimetry is used for potential dermal exposure assessment. Inhalation exposure is estimated with active air sampling using personal air samplers and polyurethane foam (PUF) plugs as sorbents. The internal dose measurements are carried out by analyzing urine samples, which are extracted by applying solid-phase extraction (SPE) with C18. The recoveries of the analytes of the three matrices are between 90 and 102%. Quantification limits are lower than 0.24 ng/L. The proposed methods are applied to assess potential and actual exposure of applicators spraying malathion in greenhouses.

Key Words: Biomonitoring; dermal exposure; GC–MS/MS inhalation exposure; malathion; metabolites.

1. Introduction

Directive 91/414/EEC *(1)* is the legislative basis for the regulation of pesticides in the European Union. It states that members shall not allow a pesticide to be authorized unless it is scientifically shown that normal use has no risk of harmful effects on humans, establishing exposure data requirements. The approaches for assessing dermal exposure have been well described in several reviews *(2–4)* in which patch and whole-body dosimetry are discussed as sampling methods. It was concluded that there is a need for systematic research on sampling and analytical methods to choose the adequate sampling media and to establish performance parameters for analytical proce-

From: *Methods in Biotechnology, Vol. 19 Pesticide Protocols*
Edited by: J. L. Martínez Vidal and A. Garrido Frenich © Humana Press Inc., Totowa, NJ

dures. One advantage of whole-body dosimetry is its compatibility with biological monitoring. The analysis of the coverall worn by applicators to obtain potential dermal exposure and the use of biological monitoring to measure the internal dose appear the most sophisticated tools in exposure assessment.

The Organisation for Economic Cooperation and Development (OECD) *(5)* reports a guidance for the study of potential dermal occupational exposure, including a quality assurance and quality control procedure mainly to the sampling step. Nevertheless, from a practice point of view aspects that might affect the reliability of results should also be considered, such as the matrix effect in the quantification of pesticides or the influence of the retention characteristics of different clothes used for whole-body dosimetry, because they may affect the exposure assessment *(6,7)*. Relative to ambient air monitoring, most references address the active sampling using solid sorbents attached to personal pumped air samplers as the best methodology for determining pesticides in air. In this aspect, it is important to consider both the generation of pesticide standard vapors for validation purposes and the range of air concentrations in which the sorbent is suitable without saturation or breakthrough *(8,9)*.

Malathion, diethyl(dimethoxythiophosphorylthio) succinate, is a very widely used nonsystemic insecticide and acaricide with contact, stomach, and respiratory action. It is a cholinesterase inhibitor of low mammalian toxicity. Many studies have assessed the metabolic effects of malathion on insects *(10)*, rats, and humans *(11)*. The principal route of malathion metabolism in animals is via deesterification to the α- and β-malathion monocarboxylic acids (α- and β-MMA), followed by further metabolism to malathion dicarboxylic acid (MDA). This is a facile esterase-catalyzed detoxification route considered responsible for the low toxicity of malathion to vertebrates. Another metabolic pathway is the oxidative desulfuration that leads to the formation of malaoxon, an active acetylcholinesterase inhibitor.

The main metabolites detected in humans are α- and β-MMA. Appreciable amounts of MDA are also found, but malaoxon is a minor metabolite. Additional human metabolites identified were *O,O*-dimethyl phosphorodithioate, *O,O*-dimethyl phosphorothioate, dimethyl phosphate, and monomethyl phosphate. Malathion is metabolized and then excreted predominantly in the urine (85–89%) and feces (4–15%). Therefore, the best way to determine the actual exposure to malathion is by analyzing MMA and MDA in urine. The analysis of MDA presents difficulty in that it is not easily obtained in the pure state for use as an analytical standard. In addition, it is not very stable. Unlike MDA, MMA is stable in the pure state and can be commercially obtained.

Gas chromatography with tandem mass spectrometry (GC–MS/MS) has been successfully applied for determining other metabolites and pesticides in biological monitoring *(12–15)*. The combination of the above-mentioned techniques allows the analysis of ultratrace levels of target compounds in complex matrices such as biological fluids (urine or blood). Also, the confirmation of the results carried out by MS/MS spectra obtained in special experimental conditions reduces the possibility of false positives compared with other MS techniques (selected ion monitoring or full-scan modes) used as GC detectors.

It is evident that the likely consequences of exposure assessment data on human health and economics justify the need to use several approaches and powerful analytical tools. As presented in this chapter, the most complete information can be obtained by combining potential dermal, inhalation, and internal dose measurements supported in GC–MS/MS detection. Analytical methods for analyzing malathion in different types of sampling media, protective clothing, polyurethane foam (PUF), and urine (in which metabolites α- and β-MMA are also determined) have been validated, establishing the respective performance parameters and the quality control procedure. Finally, the exposure levels obtained for three pesticide applicators have been determined.

2. Materials

1. Pesticide-quality solvents: *n*-hexane, methanol, diethyl ether, and acetone.
2. Standards of the pesticides malathion, MMA, and chlorpyrifos-methyl used as internal standard (ISTDs) with purity higher than 99%.
3. Formulation of malathion 90 (malathion 90% w/v, EL, Lainco, SA).
4. Anhydrous sodium sulfate.
5. Potassium carbonate.
6. 1,1,1,3,3,3-Hexafluoroisopropanol (HFIP).
7. Diisopropylcarbodiimide (DIC).
8. Sep-Pak cartridge packed with 500 mg C18 (e.g., Waters, Milford, MA).
9. Latex gloves.
10. 65% cotton, 35% polyester disposable coveralls (Iturri, Sevilla, Spain).
11. Protective mask (3M model 4251).
12. SKC personal air samplers model PCEX3KB working at a sampling flow rate of 2 L/min.
13. PUF plugs 10 cm long, 2-cm diameter, and 0.022 g/cm^3 density.
14. Gas chromatograph (e.g., Varian 3800, Sunnyvale, CA) with a split/splitless programmed temperature injector (e.g., Varian model 1078) and an autosampler (e.g., Varian model 8200).
15. Ion trap mass spectrometer (e.g., Varian Saturn 2000).
16. Software for data acquisition and data analysis (e.g., Varian Saturn 2000).
17. GC column DB5-MS 30 m × 0.25 mm id × 0.25-μm film thickness (e.g., J&W Scientific, Folsom, CA).
18. Test tube shaker with a variable-speed controller (e.g., Ika-Works).
19. Overhead mixer (agitator) to hold containers (1-L capacity with lid) is used for cold extraction of contaminated clothes.
20. Soxhlet extractor used for the extraction of PUF plugs.

3. Methods

3.1. Sampling Procedure

Gloves and coveralls are the sampling media for the whole-body methodology. After applications, remove them carefully and dry in the shade (*see* **Note 1**).

Inhalation exposure during the applications is assessed for each operator with a personal sampler, operating at a 2-L/min flow rate (*see* **Note 2**), connected to a PUF plug, which is fitted downward, on the upper part of the chest, avoiding accidental contamination by dripping or contact with contaminated items. Replace the PUF plug each hour to avoid saturation and breakthrough. Store plugs in labeled bags and in a portable fridge.

Because usually the pharmacokinetics of compounds are unknown for human beings, for biomonitoring we recommend taking between 7 and 10 urine samples from each volunteer up to 24 h after the application of malathion *(see* **Note 3**). Urine samples are stored in sterilized polyethylene containers, frozen immediately, and kept at − 30°C until analysis.

A field quality control protocol *(see* **Note 4**) must be followed to ensure the integrity of samples and analytes during sampling, transport, storage, and analysis. Take an aliquot of the spray liquid from the gun 15 min after starting and at the end of each application to check its concentration and perform field spikes. Uncontaminated samples of each medium (coverall, cotton gloves, PUF, and urine) are taken from each operator before the applications to prepare field blanks and field spikes as field quality control samples. Three samples are labeled field blanks and stored in the same way as samples to check accidental contamination or degradation of sampling media. The rest of the blank samples are spiked as follows: Three cotton gloves and three pieces (30 × 30 cm^2) of protective coverall are spiked with 100 µL of the spray tank liquid (around 135 µg of malathion, depending on the spray tank concentration); three PUF plugs are spiked with 5 µL of the spray tank liquid (6.8 µg of malathion); and three aliquots of uncontaminated urine are also spiked with malathion and MMA standards at 40-µg/L concentration level. Field blanks, field spikes, and samples have to be stored, processed, and analyzed in the same batch. According to the acceptance criteria assumed *(5)*, recovery rates of field spikes should be between 70 and 120%, with a relative standard deviation (RSD) of less than 20%; field blanks should not evidence any contamination or sample decomposition; slopes of calibration curves should not differ more than 25% from those obtained in validation studies and should fit to straight lines with $r^2 > 0.95$.

3.2. Preparation of Stock Solutions

Prepare individual stock solutions of malathion and methyl-chlorpyrifos at 400 µg/ mL in acetone and store in a freezer (−30°C). The working solutions for biological monitoring determinations, obtained by appropriate dilution of the stock solution with the same solvent, should be stored in a refrigerator (4°C). Matrix-matched calibration solutions of malathion are prepared for whole-body analysis using acetone extracts from uncontaminated coveralls.

3.3. Extraction Procedures

3.3.1. Urine Extraction Procedure

1. Condition a C18 solid-phase extraction cartridge with 6 mL of methanol and 4 mL of distilled water in that order *(see* **Note 5**).
2. Pass a 3-mL aliquot of urine through the C18 cartridge previously conditioned.
3. To carry out a cleanup step, pass 4 mL of distilled water through the cartridge. The last drops of liquid from the cartridge are withdrawn with a vacuum pump.
4. Elute the analytes with 10 mL diethyl ether, which is passed through anhydrous sodium sulfate.
5. Add to the extract 100 µL of ISTD solution (500 ng/mL in acetone).
6. Remove the solvent under a soft stream of nitrogen without heating it.

7. Redissolve the residue in 1 mL *n*-hexane.
8. Derivatize, for metabolites analysis *(14,16)* (*see* **Note 6**), by adding to the extract 10 μL HFIP with gentle mixing while adding 15 μL DIC.
9. Shake for 3 min.
10. Wash the extract with 1 mL 5% aqueous potassium carbonate solution to neutralize the excess derivatizing agent.
11. Transfer the organic layer to a 2-mL autosampler vial for GC–MS/MS analysis.

3.3.2. Personal Protective Equipment Extraction Procedure

The extraction procedure is similar to that described in **ref. 6** based on the sectioning of coveralls in nine pieces and further extraction with different volumes of acetone.

1. Cut coverall in pieces as in **Fig. 1**, place each piece separately in 1-L bottles, and add the corresponding volume of acetone.
   a. Head and neck (250 mL).
   b. Left arm (250 mL).
   c. Right arm (250 mL).
   d. Chest (350 mL).
   e. Back (350 mL).
   f. Thighs/waist, front (350 mL).
   g. Thighs/waist, back (350 mL).
   h. Lower leg, left (250 mL).
   i. Lower leg, right (250 mL).
   j. Glove, left (150 mL).
   k. Glove, right (150 mL).
2. Agitate for 30 min in an overhead shaker at 18*g*.
3. Transfer an aliquot of this extract to a 10-mL volumetric flask containing 3.5 μg of ISTD to get ready for GC–MS/MS analysis (*see* **Note 7**).

3.3.3. PUF Plugs Extraction

1. Place PUF plugs in a Soxhlet extractor, siphoning at 20 min/cycle, with 100 mL of acetone for 8 h.
2. Evaporate the extract until almost dry, add ISTD (0.4 μg), and dilute the extract to 4 mL to get ready for GC–MS/MS analysis.

3.4. Instrumental Analysis

3.4.1. Chromatographic Conditions

1. Set the injector temperature from 90°C (hold 0.1 min at 90°C) to 280°C at the rate of 200°C/min and hold it at 280°C for 20 min.
2. Set the oven temperature varying from 60°C (hold for 1.75 min) to 270°C at the rate of 20°C/min (hold for 20 min).
3. Set the carrier gas (helium) flow rate at 1 mL/min.

3.4.2. Ion Trap Conditions

1. Set the ion trap mass spectrometer in the electron ionization (EI; 70 eV) and MS/MS modes.
2. Perform an MS/MS/EI library especially for the target analytes under the experimental conditions (*see* **Note 8**). This acts as a reference spectra library for identification purposes.

FRONT   BACK

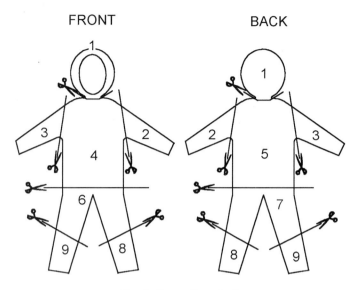

Fig. 1. Coverall sectioning.

3. Optimize the sensitivity for mass detection by filling the trap with target ions, switching on the automatic gain control (AGC) (*see* **Note 9**). For example, fix the AGC target at 2000 counts (higher values might cause electrostatic interactions among ions in the ion trap chamber).

4. Choose for analysis a parent ion for each analyte on the basis of its *m/z* and its relative abundance; both should be as high as possible to achieve greater sensitivity. In this case, a nonresonant waveform (second ionization) is selected for all the compounds.

5. Select the excitation storage level and excitation amplitude as appropriate to generate spectra with the parent ion as their molecular peaks (between 10 and 20% of relative abundance).

6. Calibrate the mass spectrometer weekly.

7. For the case study, the operating conditions are summarized in **Tables 1** and **2**. MS/MS is performed in a nonresonant mode for all compounds. The MS/MS spectra obtained in the selected experimental conditions are shown in **Fig. 2** (*see* **Note 10**).

3.5. Validation of the Analytical Methods

3.5.1. Potential Dermal Exposure Method

1. Prepare calibration curves in a wide concentration range (500 and 1500 µg/L) with pure standards and using blank extracts of uncontaminated coveralls to avoid matrix effects in the quantification. For that, cut a 30×30 cm^2 piece of coverall and extract it with 250 mL acetone following the method explained above. Use such extract for filling up to the volume of calibration solutions. The statistical data we obtained are summarized in **Table 3** (*see* **Note 11**).

2. Calculate limits of detection (LODs) and limits of quantification (LOQs) by analyzing 10 uncontaminated pieces of coveralls and following the IUPAC recommendations (*17*). The values obtained (**Table 3**) are in the ng/L level, a good indication of the sensitivity of MS/MS (*see* **Note 12**).

Table 1
Mass Spectrometer Operating Conditions

Ionization mode	EI
Multiplier voltage	1700 V
A/M amplitude voltage	4.0 V
Trap temperature	200°C
Manifold temperature	45°C
Transfer-line temperature	280°C
Emission current	80 mA
Automatic gain control (AGC)	On
AGC target	2000 counts

Table 2
MS/MS Conditions[a]

Compound	Activation time (min)	Range (*m/z*)	Parent ion (*m/z*)	Mass defect (mu/100 u)	Excitation amplitude (V)	Excitation storage level (*m/z*)
MMA	9.0–10.8	60–305	295	+74	37.5	80
m-clor (IS)	10.8–11.4	125–295	286	0	66.0	80
Malathion	11.4–18.0	90–185	173	0	70.0	89

[a]Excitation time = 40 µs; isolation window = 2 u; nonresonant waveform.

3. Calculate recovery rates and precision by spiking and analyzing 10 coverall pieces that are 30×30 cm^2 and 10 cotton gloves at low and high concentration levels considering the calibration curves (e.g., 720 and 1440 µg/L) (*see* **Note 13**)
4. Calculate intermediate precision by analyzing three pieces of coverall and gloves spiked as above every 2 wk for 4 mo (*see* **Note 14**).
5. State the stability of samples by spiking 36 pieces (30×30 cm^2) of each garment with 180 µg of malathion. Store them in darkness at 4°C for 4 mo and analyze a set of three samples of each sampling medium the first day and each week for 1 mo. The percentage recovery by storage time is calculated by comparing the amount recovered with the amount recovered on the first day after spiking (*see* **Note 15**).

3.5.2. Inhalation Exposure Method

1. Prepare calibration curves by diluting pure standard solution and using extract of uncontaminated PUF containing the ISTD (e.g., 100 µg/L) for filling to volume. Calculate linear ranges, LOD, and LOQ for the air analysis method using the same procedure as in the dermal exposure method *(17)*.
2. Select the appropriate sorbent by stating the trapping efficiency of sorbents such as Porapak R, Chromosorb 102, Supelpak, Amberlites XAD2 and XAD4, or PUF to sample malathion in greenhouse air.
3. Build up a system for generating pesticide standard atmospheres as described in previous work *(14,15)*; in our case, we used a chromatographic oven with a 0.5-cm diameter hollow glass column. This system allows study of the influence of such variables as sampling volume, sampling time, sampling flow rate (e.g., 1 or 2 L/min), air relative humidity (e.g., dry to 100% saturation), breakthrough volume (e.g., until 1 m^3), and concentration

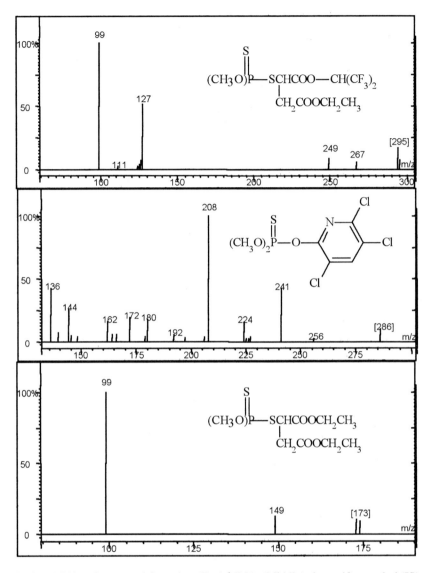

Fig. 2. MS/MS and structural formulas. (Top) MMA; (Middle) clorpyrifos-methyl (IS); and (Bottom) malathion.

of saturation (e.g., until 0.4 mg malathion), which have to be studied to validate the air-sampling method.

4. Optimize the optimum conditions for achieving the complete volatilization of compounds (*see* **Note 16**).

5. Obtain recovery rates by injecting in the device different volumes of a standard solution, sampling a known volume of air using PUF plugs, and analyzing them. Check the influence of air humidity by saturating the air with water (*see* **Note 17**).

Table 3
Limits of Detection (LOD) and Quantification (LOQ) and Calibration Data

Compound	LOD (ng l^{-1})	LOQ (ng l^{-1})	a	b	r^2
MMA	0.01	0.03	3.09	-0.08	0.9988
Malathion	0.07	0.24	0.67	-0.11	0.9831

a, intercept; *b*, slope; *r*^{2}, determination coefficient.

6. Check the breakthrough volume by connecting two PUF plugs in series and sampling different volumes of standard atmospheres (*see* **Note 18**). The high sorption capacity of PUF allows sampling of a wide range of pesticide concentrations in the air.
7. Study the stability of samples at least for a 1-mo period, for example, by storing spiked PUF plugs at ambient temperature 4 and −18°C) and checking recovery rates weekly (three replicates each) (*see* **Note 19**).

3.5.3. Actual Exposure (Biomonitoring) Method

1. Calibrate using a blank urine sample spiked with each analyte, such as the range 1–500 μg/L, which is an adequate calibration range for the expected analyte concentrations in urine samples.
2. Add the ISTD at 100 μg/L.
3. Obtain LODs and LOQs by analyzing 10 control urine samples without malathion contamination (**Table 3**). The best results are obtained for MMA-hexafluoroisopropyl derivatives (*see* **Note 20**). The LOD and LOQ are low enough for monitoring exposure of pest control operators to malathion.
4. Use solid-phase extraction C18 cartridges for sample preparation because these eluates are better suited for direct GC analysis, and they seldom require cleanup prior to GC analysis. Liquid–liquid extraction must involve a cleanup step to avoid reduction in column efficiency and contamination of injector and ion trap (*see* **Note 21**).
5. State the dependence of extraction efficiency on the initial concentration of analytes by spiking 10 urine samples with two different concentrations (e.g., 40 and 200 μg/L) and checking recoveries (*see* **Note 22**).

3.6. Application of Methods

3.6.1. Field Trial Design

The exposure is assessed during a total of three applications of malathion in greenhouses (*see* **Note 23**). Three flat-roof polyethylene greenhouses (200-μm thick, 15 × 40 × 2.50 m) located in Almería (Spain) were selected for the applications. The lateral windows remained closed during the experiment, and climatological conditions are registered. In the studied case, crops were green beans (cultivar Helda), tomato (cultivar Daniela), and cucumber (cultivar Almería). They were 2 m high with a 1-m interrow distance, allowing the applicators to walk between each row during the applications. For spraying, a semistationary high-volume application equipment with one circular nozzle operating at a flow rate of 4 L/min was used. The volume sprayed was 375 L in each case, corresponding approximately to a dose of 1 Kg/ha of malathion. For the three applications, the spray liquid was prepared by dispersing 600 mL of malathion 90 EL in a tank containing 400 L water.

Each of the treatments should take at least 90 min. Applicators spray following an application pattern, in this case walking between the rows and spraying one side of the crop and returning along the same row and spraying the other side of the row. The protection equipment of the three applicators consisted of a cotton t-shirt for the first applicator and a full cotton–polyester coverall for the other two. All wore a mask. In the case study, applicator 1 did not wear any protective equipment, so that potential dermal exposure was not assessed.

As a case study, inhalation, internal dose, and potential dermal exposure to malathion is determined for the applicators who wore personal protective equipment (operators 2 and 3), the latter expressed as milliliters of spray tank deposited on the garment per hour of application. The total volume of spray liquid deposited on the suit of the operators and its distribution on the body was very similar for both applicators (72.3 and 73.0 mL/h, respectively) (*see* **Note 24**).

Considering the distribution on the body, approx 75% of total potential exposure is found on the lower body (thighs and lower legs), with both lower legs the most contaminated sections (approx 19 mL/h each leg). The highest amount of pesticide on these regions was caused by the operators, who directed the spray gun to cut the flow downward and pointing to the legs when they passed from one row to another row. On the upper body (head, torso, back, and arms), the amount found on both applicators was approx 25% of the total, with the left arm (4.8 and 5.1 mL/h, applicators 1 and 2, respectively) and right arm (4.3 and 3.9 mL/h, respectively) the most exposed areas.

The concentrations of malathion in the breathing area during the application were 69.4 and 85.9 $\mu g/m^3$ for applicators 2 and 3, respectively.

Concerning biomonitoring results, the proposed method has been applied to the analysis of the urine of the three applicators. A number of samples ranging between 7 and 10 were collected from each worker in the 24 h after the applications. Malathion and MMA were not detected in the samples taken as blanks before the applications. **Figure 3** shows the results obtained in the postapplication samples. It can be seen that malathion was not detected, but its metabolite MMA was present in most of samples. In all cases, the highest concentration found in urine was 5–8 h after spraying malathion (except applicator 3, which was 20 h after application). The total amount of MMA excreted was calculated considering the concentration found in each sample and the volume collected. The total amount excreted ranged between 133.75 and 671.42 µg. The highest amount determined corresponds to the urine of the applicator who did not wear a cotton coverall during the experiment. This fact reveals the important use of protective equipment (coverall) to spray malathion.

4. Notes

1. Workers are placed in a clean location, where coveralls are removed carefully and held in the shade until dry, then the coveralls are extended on aluminum foil and rolled together, avoiding contact between different parts of the suit. Store them in appropriate labeled bags and store in a portable refrigerator until arrival at the laboratory.

2. The flow rate of the personal pump must be well calibrated, taking into account that the final result of the air concentration is referred to the volume of air sampled. Keep records of the certificates of calibration.

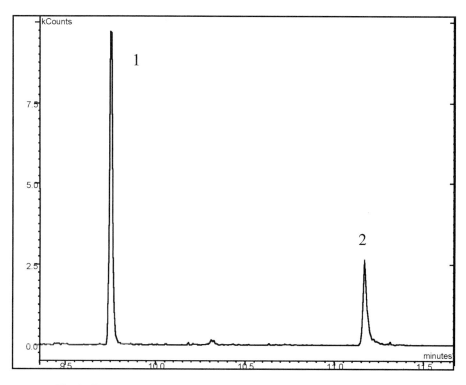

Fig. 3. Gas chromatogram of a urine sample containing (1) MMA and (2) IS.

3. Take the whole urine sample, indicating on the label the date and hour in which the sample is taken.

4. The field quality control checks if any accidental contamination occurs by analyzing blank samples of each sampling media, urine, PUF plugs, and coveralls. Prepare a clean location that is safe from contamination and is close to the greenhouse. Identical samples (three replicates) are spiked in the field to check recoveries of each sampling media. The sample of spray tank is collected (three replicates) directly from the nozzle about 15 min from the start of application to ensure the homogeneity of the tank. The concentration of the spray tank must be calculated to express the results as milliliters of spray tank per hour of application.

5. The cartridge should not become dry during conditioning.

6. A derivatization step necessary for the analysis of metabolites because of the low volatility and high polarity of MMA is applied. Several derivatizing agents (e.g., fresh diazomethane, Sigmasil A, and HFIP) could be used. In our case, the methyl esters obtained with diazomethane presented questionable stability and low sensitivity. In addition, it is a carcinogenic reactive. The trimethylsilyl esters obtained by a reaction of dried eluent with Sigmasil A are much more sensitive than methyl derivatives, but the rest of the derivatizing agent and its by-products yield several interfering peaks. So, we prefer HFIP derivatives, which are produced by rapid coupling of the eluent with HFIP in the presence of DIC reaction, are highly sensitive and do not produce interfering substances.

7. A specific concentration of the solvent is unnecessary because the amount of pesticide on the suit is high enough to be determined, and lower concentrations are not relevant for exposure assessment.

8. For confidence in results, an in-house library is necessary, prepared in the working conditions of the MS/MS, with standards. The fit is a parameter to quantify the match between the spectra of the library and the spectra of samples, with identical spectra having a fit value of 1000.

9. The AGC continuously adapts in an automatic mode the ionization time of the ion trap detector to avoid a reduction of the sensitivity. For the optimization of the sensitivity of the detector, the trap should be filled with the target ions. A higher AGC target would cause electrostatic interactions between the ions (space–charge effects), degrading performance.

10. The α- and β-MMA HFIP derivatives coelute in the selected GC conditions at the 9.79-min retention time and can be quantified as the sum of both isomers. It can be seen in the chromatogram (**Fig. 4**) that the retention times for the ISTD and malathion were 11.20 and 11.57 min, respectively.

11. The whole-body method shows a matrix effect in the quantification of the analyte because of the composition of the garment and the presence of products that are different from the active ingredient in the commercial formulation of malathion. This effect is shown when recovery rates of malathion from garments are determined in the method validation. Recovery rates obtained when spiked samples are quantified with a calibration curve prepared in solvent are lower than the ones obtained when a matrix-matching calibration is used (83.4 vs 90.2%, respectively). Thus, matrix-matching calibration was chosen for the quantification of malathion in personal protective equipment samples. Matrix-matching calibration is performed by preparing the working standard solutions with an extract of an uncontaminated sample containing all the matrix components. The use of matrix-matching calibration solutions is extended in the field of pesticide analysis. Matrix may affect the analytical signal, introducing constant error when the intercept is modified in relation with a calibration prepared in solvent or a proportional error when the slope is modified or both if either intercept or slopes are different.

12. Lower limits achieved with this technique are low enough for exposure assessment purposes without the need for concentrating the sample extract. A linear range in high concentrations is useful to avoid dilutions and reanalysis of samples with a high concentration.

13. In our case, recovery rates ranged between 90.2 and 103.2%, with the precision (expressed as RSD) lower than 9.1% in all cases (**Table 4**).

14. Typical values for RSD of these measures is less than 15%.

15. In our case, it was above 91% in all cases during the period studied.

16. Try different oven injector and detector temperatures and different caudal of air; for example, by setting the chromatographic oven at 150°C and the injector and detector at 200°C, we achieved the volatilization of a known amount of pesticides. Passing air for 20 min at a flow rate of 2 L/min is enough to achieve the transfer of the analyte to the sorbent.

17. The recovery rate in the case study for malathion was 93.2% (5.7% RSD). Using PUF, no influence of the air humidity on recovery rates obtained using dry or saturated air (93.2 and 94.1%, respectively) was observed.

18. A sampling method has to be versatile enough to sample air either with a high concentration of the target compound for a short period of time or with a low pesticide concentration for a longer period of time for application in different working conditions. For

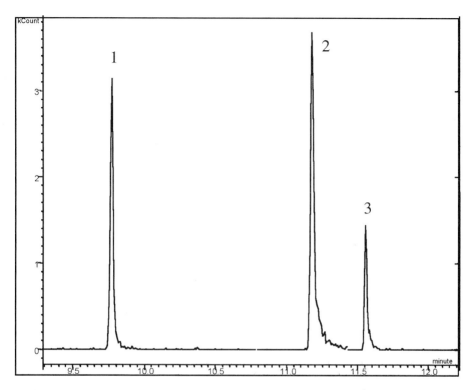

Fig. 4. Gas chromatogram of a clean urine fortified with 10 mg L^{-1}of each analyte: (1) MMA; (2) IS; and (3) malathion.

example, in indoor applications the air concentration is expected to be higher than in the outdoor environment when the method is applied for bystander exposure assessment. In the first case, a short sampling time is appropriate, but in the second a high volume of air should be sampled to determine trace levels. For example, testing should be done that breakthrough does not occur by sampling for 30 min a standard atmosphere contaminated with 10 mg/m^3 of pesticides and for 8 h a standard atmosphere containing 0.2 μg/m^3 of pesticides.

19. In our case, we observed that light affects PUF plugs, resulting in a decrease of recovery rates.
20. Worse results are found if the analyses are carried out in full-scan mode because of the presence of interfering substances, which coelute with the target analytes and are difficult to remove by background subtraction. The average spectral fit obtained when analyzing 10 clean urine samples spiked with 200 ng/mL of each compound are 531 (MMA) and 397 (malathion), using full-scan mode, and 882 (MMA) and 820 (malathion) using MS/MS mode. When MS/MS is used, if a coeluted interfering substance has the same identification ion as the analyte, it can be avoided using special experimental conditions for the collision-induced dissociation and quantifying with a specific ion from the analyte.
21. After testing several elution solvents (acetone, *n*-hexane, diethyl ether, methanol, ethyl acetate, and dichloromethane), the maximum elution is found with 10 mL of diethyl ether.

Table 4
Recoveries (R%) and Relative Standard Deviation (RSD%)

		Compound	
Sample	Fortification level	MMA	Malathion
Coverall	720 μg/L	—	90.2 (8.5)
	1440 μg/L	—	91.7 (7.6)
Air	0.2 μg/m³	—	93.2 (5.7)
	10 μg/m³	—	94.1 (6.0)
Cotton gloves	720 μg/L	—	103.2 (6.9)
	1440 μg/L	—	98.4 (9.1)
Urine	40 μg/L	107.2 (6.6)	114.3 (15.2)
	200 μg/L	102.1 (6.1)	109.4 (12.3)

$n = 10$.

22. In our case, the average recoveries obtained were between 102.1 and 114.3% in all cases. The repeatability expressed as RSD was smaller than 15.2%. The results are summarized in **Table 4**.

23. Keep a record of a plan of the greenhouse and plot and take pictures of the distribution of crops (including height and interrow distance) and the dimensions of the greenhouse. Obtain the flow rate of the gun by measuring the volume sprayed in a baker for 1 min by replicating (at least three times) in the same operating conditions as during the pesticide applications. Application times have to be well controlled to refer the results to the amount of pesticide per hour of work. Take pictures of the applicators during the treatments and describe carefully the application patterns. The applicators will fill an agreement form of cooperation and a declaration that they have been informed about the potential risk of their work.

24. We prefer the expression of results for potential dermal exposure in milliliters of spray liquid contaminating the suit per hour because it is independent of the concentration of spray liquid and gives the possibility of comparing with other experiments. To obtain such information, the spray liquid must be well analyzed because the raw data from the mass spectrometer provides the amount of pesticide as the milliliters of spray liquid obtained by dividing such raw data by the concentration of the spray. Other ways to express such exposure is as milligrams of active ingredient per square centimeter of body or milligrams per hour.

Acknowledgments

This research was supported by the European Union Project ALFA (DG-12-RSMT) and by the Comisión Interministerial de Ciencia y Tecnología (CICYT) Project AMB97-1194-CE.

References

1. Commission of the European Communities, Council Directive 91/414/EEC, Official Journal of the European Communities No. L230, Luxembourg, 19 August 1991.

2. Chester, G. (1995) Methods of pesticide exposure assessment. In: Curry, P. B., ed., Plenum Press, New York, 179–215.

3. Fenske, R. A. (1993) Dermal exposure assessment techniques. *Ann. Occup. Hyg.* **37**, 687–706.

4. Van Hemmen, J. J., Brouwer, D. H. (1995) Assessment of dermal exposure to chemicals. *Sci. Total Environ.* **168**, 131–141.

5. OECD, Guidance document for the conduct of studies of occupational exposure to pesticides during agricultural application, Series on Testing and Assessment No. 9, OECD/GD (1997).

6. Castro Cano, M. L., Martínez Vidal, J. L., Egea González, F. J., Martínez Galera, M., Cruz Márquez, M. (2000) Gas chromatographic method and whloe body dosimetry for assessing dermal exposure of greenhouse applicators to chlorpyrifos-methyl and fenitrothion. *Anal. Chim. Acta* **423**, 127–136.

7. Martínez Vidal, J. L., Glass, C. R., Egea González, F. J., Mathers, J. J., Castro Cano, M. L. (1998) Techniques for potential dermal exposure assessment in Southern Europe. Proceedings of 9th International Congress, Pesticide Chemistry, The Royal Society of Chemistry and the IUPAC, London.

8. Martínez Vidal, J. L., Egea González, F. J., Glass, C. R., Martínez Galera, M., Castro Cano, M. L. (1997) Analysis of lindane, α- and β-endosulfan and endosulfan sulfate in greenhouse air by gas chromatography. *J. Chromatogr. A* **765**, 99–108.

9. Egea González, F. J., Martínez Vidal, J. L., Castro Cano, M. L., Martínez Galera, M. (1998) Levels of methamidophos in air and vegetables after greenhouse applications by gas chromatography. *J. Chromatogr. A* **829**, 251–258.

10. Doichuanngam, K., Thornhill, R. A. (1992) Penetration, excretion and metabolism of ^{14}C malathion in susceptible and resistant strains of *Plutella xylostella Comp. Biochem. Physiol.* **101C**, 583–588.

11. PSD, Pesticide Safety Directorate, Evaluation (1995) **135**, UK MAFF.

12. Martínez Vidal, J. L., Arrebola, F. J., Fernández-Gutiérrez, A., Rams, M. A. (1998) Determination of endosulfan and its metabolites in human urine using gas chromatography–tandem mass spectrometry. *J. Chromatogr. B* **719**, 71–78.

13. Arrebola, F. J., Martínez Vidal, J. L., Fernández-Gutiérrez, A. (1999) Excretion study of endosulfan in urine of a pest control operator. *Toxicol. Letters* **107**, 15–20.

14. Arrebola, F. J., Martínez Vidal, J. L., Fernández-Gutiérrez, A., Akhtar, M. H. (1999) Monitoring of pyrethroid metabolites in human urine using solid-phase extraction followed by gas chromatography-tandem mass spectrometry *Anal. Chim. Acta*. **401**, 45–54.

15. Arrebola, F. J., Martínez Vidal, J. L., Fernández-Gutiérrez, A. (2001) Analysis of endosulfan and its metabolites in human serum using gas chromatography-tandem mass spectrometry (GC-MS/MS). *J. Chromatogr. Sci.* **39**, 177–182.

16. Glastrup, J. (1998) Diazomethane preparation for gas chromatographic analysis. *J. Chromatogr. A.* **827**, 133–136.

17. González Casado, A., Cuadros Rodríguez, L., Alonso Hernández, E., Vílchez, J. L. (1996) Estimate of gas chromatographic blanks Application to detection limits evaluation as recommended by IUPAC. *J. Chromatogr. A.* **726**, 133–139.

16

Pesticides in Human Fat and Serum Samples vs Total Effective Xenoestrogen Burden

Patricia Araque, Ana M. Soto, M. Fátima Olea-Serrano, Carlos Sonnenschein, and Nicolas Olea

Summary

Tests to screen for estrogenicity and appropriate biomarkers of human exposure are required for epidemiological studies of endocrine disruption. This requirement is addressed here by developing a protocol that investigates bioaccumulated xenoestrogens that are candidates for estrogenicity and assesses their combined estrogenic effect. It comprises two major parts. The first part consists of an exhaustive chemical determination of the compounds under study, including procedures for the extraction, purification, identification, and confirmation of the organochlorine pesticides and the quantification of lipids in the samples. This goal is achieved using a semipreparative high-performance liquid chromatographic (HPLC) separation of xenoestrogens from endogenous hormones and subsequent chromatographic analysis of the fractions. The second part corresponds to the testing of HPLC eluted fractions in the E-Screen test for estrogenicity, which yields the estimation of the total effective xenoestrogen burden of each sample.

Key Words: Biomarkers; endocrine disruption; estrogenicity; exposure assessment; xenoestrogens.

1. Introduction

Many studies of the association between environmental estrogens and adverse health effects are based on the identification and quantification of chemical residues in biological samples from human populations for the purpose of exposure assessment. Despite the merits of this approach, it is proposed that study of the combined effect of these compounds would be a better method for the assessment of exposure *(1)*. We have faced this challenge by developing a mixed methodology using semipreparative high-performance liquid chromatography (HPLC), followed by chemical identification, quantification, and application of an E-Screen assay. By this means, organohalogenated xenoestrogens are efficiently separated from ovarian estrogens, and the exposure is quantitatively assessed in terms of hormonal activity (total effec-

From: *Methods in Biotechnology, Vol. 19, Pesticide Protocols*
Edited by: J. L. Martínez Vidal and A. Garrido Frenich © Humana Press Inc., Totowa, NJ

tive xenoestrogen burden, TEXB) *(2–5)*. The distinction between the estrogenicity of the fractionated tissue samples is of maximum interest because preliminary studies have indicated that estrogenicity caused by organohalogenated chemicals is a risk factor for breast cancer *(6)*.

In this chapter, the methodology proposed opens up a new approach to xenoestrogen exposure assessment because it can focus on particular chemicals present in tissue fractions. In fact, the list of xenoestrogens has been expanded using this method *(7,8)*, and chlorinated bisphenols, brominated nonylphenols, and organochlorine pesticides have been identified as putative candidates responsible for the estrogenicity observed in the bioassay. The semipreparative HPLC fractions that have been associated with cancer risk contain organochlorine pesticides, among other organohalogenated chemicals. DDT (dichlordiphenyltrichlor) and derivatives are present in almost 100% of the samples assayed in populations from southern Spain *(9)*.

Natural endogenous estrogens together with phytoestrogens and nonhalogenated xenoestrogens elute in fractions collected later in the chromatographic run. Natural estrogens are produced and accumulate in adipose tissue *(10,11)*, but the combined role of natural estrogens and xenoestrogens in the estrogenicity of these fractions is not known. In fact, estradiol esters, together with bisphenols, polyphenols, phytoestrogens, mycoestrogens, and ovarian estrogens, are eligible candidates to explain the estrogenicity of these fractions.

Most human epidemiological studies of xenoestrogen exposure estimate the serum or adipose levels of one chemical or of a small number of chemicals, ignoring the impact of other chemicals and the cumulative effects of mixtures in the cell environment. The protocol presented here helps establish a relationship between the content of xenoestrogens in adipose tissue and the risk of hormonal diseases because it measures the combined effects of organohalogenated xenoestrogens separately from the hormonal activity of endogenous estrogens.

2. Materials

2.1. Equipment

1. HPLC (e.g., model 501, Waters, Milford, MA) equipped with two pumps and an injector with 500-µL load capacity with an ultraviolet/visible detector (e.g., Waters model 490) and Millennium Chromatography Manager software (Varian, Sunnyvale, CA). A column packed with Lichrospher Si-60 with 5-µm particle size is used (e.g., Merck, Darmstadt, Germany).
2. Gas chromatography with electron capture detector (GC/ECD; e.g., Varian 3350) with a ^{63}Ni electron capture detector and Chromatography Manager software (e.g., Millennium). A CP SIL8 CB column (30 m × 0.25 mm) is used.
3. GC with mass spectrometry (GC-MS; e.g., Saturn 2000, Varian) detector equipment and CP5860 WCOT fused silica column (30 m × 0.25 mm).
4. Pyrex glass column (6 mm id) for the chromatography.
5. Sep-Pak Cartridge part WATO51900 (e.g., Waters).
6. Vortex mixer.
7. Rotavapor.
8. Titertek multiscan apparatus.
9. 0.22-mm filter (e.g., Millipore, Bedford, MA).

2.2. Chemicals

1. Methanol.
2. Isopropanol.
3. Hexane.
4. Ethanol.
5. Chloroform.
6. Hydrochloric acid.
7. Alumina 90 (70–230 mesh) (e.g., 1097, Merck).
8. Aldrin.
9. Dieldrin.
10. Endrin.
11. Lindane.
12. Methoxychlor.
13. Endosulfan I.
14. Endosulfan II.
15. Mirex.
16. *p,p′*-DDT.
17. *o,p′*-DDT.
18. *o,p′*-DDD.
19. Hexachlorobenzene.
20. Vinclozolin.
21. *p,p′*-Dichlorobenzophenone.
22. *p,p′*-DDE.
23. Endosulfan diol.
24. Endosulfan sulfate.
25. Endosulfan lactone .
26. Endosulfan ether.
27. MCF-7 cell line (Tufts University, School of Medicine, Boston, MA).
28. Dulbecco's modification of Eagle's medium (DME) (e.g., Norit A, Sigma Chemical, Poole, UK).
29. Trypsin.
30. Charcoal.
31. Sulforhodamine-B (SRB).
32. 17β-Estradiol (E2).
33. 17β-Ethynylestradiol.
34. Progesterone.
35. Testosterone.
36. Nonylphenol.
37. Bisphenol A.
38. Genistein.
39. Dextran T-70.
40. Charcoal–dextran-treated human serum (CDHuS) (e.g., Irvine Scientific, Santa Ana, CA).
41. Phenol red-free DME (e.g., BioWittaker, Walkersville, MD).
42. Fetal bovine serum (FBS; e.g., BioWittaker).
43. Trichloracetic acid.
44. Acetic acid.
45. 24-Well plates.
46. 96-Well plates.

3. Methods

The methods described below outline the chemical determination of organochlorine compounds and the biological assay.

3.1. Chemical Analysis

The steps described under **Subheadings 3.1.1.–3.1.7.** include the procedures for the extraction, purification, identification, and confirmation of the organochlorine pesticides and the lipid determination. The samples are frozen at $-80°C$ until their use (*see* **Notes 1** and **2**).

3.1.1. Extraction of Xenoestrogens From Adipose Tissue

Biocumulative compounds are extracted from adipose tissue (reviewed in **ref. *12***, with slight modifications).

1. Extract 200 mg adipose tissue with 20 mL hexane and elute in a glass column filled with 200 mg alumina previously hydrated with 5% distilled water (*see* **Note 3**).
2. Concentrate the eluate obtained at reduced pressure and then under a stream of nitrogen to a volume of 1 mL or until total dryness (*see* **Note 4**).

The alumina (70–230 mesh) used in the extraction is previously dried at $600°C$ for 4 h.

3.1.2. Extraction of Xenoestrogens From Serum

1. Incubate a 4-mL aliquot of serum with 2 mL methanol in a glass tube and shake the mixture vigorously.
2. Add 5 mL ethylic ether and hexane (v/v) to the same tube, centrifuge for 10 min at 1800*g*, and collect the organic phase in another tube.
3. Repeat **step 2** three times.
4. Concentrate the organic phases under a stream of nitrogen to a volume of 1 mL.
5. Add to the residue 0.5 mL of H_2SO_4 and 1 mL hexane, centrifuge, and collect the organic phase.
6. Add 1 mL hexane twice, centrifuge, and collect the organic phases.
7. Concentrate the organic phases under a stream of nitrogen until total dryness.

3.1.3. Semipreparative HPLC

A preparative liquid chromatographic method is used to separate xenoestrogens from natural estrogens without destroying them. Extracts are eluted by a gradient program based on a previously described method (reviewed in **ref. *13***) with modifications: Two phases are used as the mobile phase: *n*-hexane (phase A) and *n*-hexane:methanol:2-isopropanol (40:45:15 v/v/v) (phase B) at a flow rate of 1.0 mL/min *(13)* (*see* **Note 5**). The normal phase column separates xenoestrogens according to their polarity, with the most lipophilic compounds eluting in the shortest times (*see* **Note 6**). This is described under **Subheadings 3.1.3.1.–3.1.3.2.** and includes the separation of the compounds in three fractions and an extensive separation in 32 fractions.

3.1.3.1. HPLC Separation of Xenoestrogens in Three Fractions

1. Resuspend the residue extracted in 1 mL hexane, mix, and inject 500 µL.

2. Collect the result of the first 11 min in a tube. This fraction is called the α-fraction.
3. Collect the fraction between minutes 11 and 13. This fraction is called the x fraction.
4. Collect the fraction between minutes 13 and 32 in another tube. This fraction is called the β-fraction.
5. Inject the other 500 μL and repeat **steps 2–4**.
6. Join together the α-, x, and β-fractions collected in both injections.
7. Concentrate until total dryness under a stream of nitrogen.

3.1.3.2. EXTENSIVE HPLC SEPARATION OF XENOESTROGENS

When a much more concentrated fraction is needed or when a more exhaustive method is required, an extensive fractionation method is proposed. The method is the same as described under **Subheading 3.1.3.1.** with some modifications.

1. After extracting three extracts of 500 mg each from homogenized fat samples, resuspend the concentrate in 3 mL hexane, mix, and inject 500 μL.
2. Pass through the HPLC column and collect fractions of 1 mL from min 1 to 32.
3. Inject an additional 500 μL and repeat **step 2**, collecting minute 1 with min 1 of **step 2**, and so on.
4. Repeat **step 3** four additional times and join together the fractions collected at the same times. Thus, each fat sample gives 32 fractions of 6 mL that correspond to 1500 mg of tissue.

3.1.4. Cleanup of Serum

The organic extracts are purified using silica Sep-Pak. The extract of serum is dried in a tube (tube A).

1. Treat the cartridge with 2 mL hexane to condition the column.
2. Eluate the organic extract by adding 10 mL hexane to tube A (*see* **Note 7**), mix, add to the cartridge, and collect the eluate in another glass tube (tube B).
3. Add 10 mL hexane:methanol:isopropanol (45:40:15 v/v/v) to tube A and repeat **step 2**.
4. Concentrate the eluate obtained at reduced pressure in a rotavapor and then under a stream of nitrogen to a volume of 1 mL or until total dryness (*see* **Note 4**).

3.1.5. Identification and Quantification of Organochlorine Compounds by GC

The α-fraction from HPLC and the serum eluated from Sep-Pak are identified using GC/ECD (*see* **Notes 8** and **9**).

1. The HPLC α-fraction and serum extracts are dissolved, respectively, in 100 μL and 1 mL of hexane labeled with *p,p'*-dichlorobenzophenone internal standard.
2. Mix vigorously.
3. Inject 1 μL of the solution into the chromatograph.
4. Quantitative analysis, internal standard method: *p,p'*-dichlorobenzophenone is selected as THE internal standard because it appears in the chromatogram near the components under study, is chemically similar to these components, and is expressed in comparable concentrations. The reproducibility of the internal standard is excellent, and the calibration curve has a fit of $R^2 = 0.988$ in a range of concentrations between 0.03 and 1.50 μg/mL. The concentrations of the pattern solutions for the tested products are between 0.02 and 0.0001 μg/mL. The detection limits of the compounds studied have been established following International Union of Pure and Applied Chemistry (IUPAC) norms, obtaining

values of 0.1 to 3.0 ng/mL. Once the response factor is calculated, the equations of the curves for each product are obtained, and the corresponding correlation coefficient is calculated from the ratio of the areas to the concentrations, with a value more than 0.922.

3.1.6. Confirmation by GC/MS

The presence of organochlorines in HPLC α-fraction and in serum eluate is confirmed by GC–MS (*see* **Note 10**).

1. Inject 2 µL of the solution prepared under **Subheading 3.1.5.** (*see* **Note 11**).
2. Dry until total dryness to store.

3.1.7. Lipid Determination

Total lipid content is quantified gravimetrically.

1. Weigh the flask in which the extraction is to be performed (empty flask weight).
2. Homogenize 100 mg of adipose tissue in 5 mL of chloroform:methanol:hydrochloric acid (20:10:0.1) in a glass potter and pass to a centrifuge tube.
3. Extract again with 5 mL of chloroform:methanol:hydrochloric acid (20:10:0.1) and pass to another centrifuge tube.
4. Add 5 mL of 0.1 N HCl to each tube.
5. Centrifuge at 1800g for 10 min.
6. Collect the organic phases in the previously weighed flask.
7. Extract the nonorganic phases again and add them to the first extraction products.
8. Weigh the flask (full flask weight).

Calculate percentage lipid according to the formula

$$\text{Lipid } \% = \frac{\text{Filled flask weight (g)} - \text{Empty flask weight (g)}}{\text{Adipose tissue weight (g)}} \times 100$$

The total lipid content is expressed in grams of lipid per grams of adipose tissue.

3.2. Bioassay for Measuring Estrogenicity

The steps described under **Subheadings 3.2.1.–3.2.5.** include the description of the human breast cancer cell line used, the procedures for the E-Screen assay, the standard curve of estradiol, and the description of the transformation of cell proliferation into estradiol equivalent units.

3.2.1. MCF-7 Cell Line

Cloned MCF-7 cancer cells are grown for routine maintenance in DME supplemented with 5% FBS in an atmosphere of 5% CO_2/95% air under saturating humidity at 37°C. The cells are subcultured at weekly intervals using a mixture of 0.05% trypsin and 0.01% EDTA.

3.2.2. Charcoal–Dextran Treatment of Serum

1. Add calcium chloride to outdated plasma to a final concentration of 30 mM to facilitate clot formation.
2. Remove sex steroids from serum by charcoal–dextran stripping.
3. Prepare a suspension of 5% charcoal with 0.5% dextran T-70.

4. Centrifuge at 1000g for 10-min aliquots of the charcoal–dextran suspension of similar volume to the serum aliquot to be processed.
5. Aspirate supernatants.
6. Mix serum aliquots with the charcoal pellets.
7. Maintain this charcoal–serum mixture in suspension by rolling at 6 cycles/min at 37°C for 1 h.
8. Centrifuge the suspension at 1000g for 20 min.
9. Filter the supernatants through a 0.22-μm filter.
10. Store the CDHuS at −20°C until needed.

3.2.3. Cell Proliferation Experiments

MCF-7 cells are used in the test of estrogenicity according to a technique slightly modified (reviewed in **ref. 14**) from that originally described (reviewed in **ref. 10**).

1. Trypsinize cells and plate in 24-well plates at initial concentrations of 20,000 cells per well in 5% FBS in DME.
2. Allow the cells to attach for 24 h.
3. Resuspend α-, x, and β-fractions obtained by preparative HPLC chromatography in 5 mL CDHuS-supplemented phenol red-free medium.
4. Shake vigorously and leave to rest for 30 min.
5. Filter through a 0.22-μm filter.
6. Replace the seeding medium with 10% CDHuS-supplemented phenol red-free DME.
7. Test the product of **step 5** in the assay for estrogenicity at dilutions from 1:1 to 1:10. (*see* **Note 12**).
8. Stop the assay after 144 h by removing medium from wells.
9. Fix the cells and stain them with SRB.
10. Treat the cells with cold 10% trichloroacetic acid and incubate at 4°C for 30 min.
11. Wash the cells five times with tap water and leave to dry.
12. Stain for 10 min trichloroacetic-fixed cells with 0.4% (w/v) SRB dissolved in 1% acetic acid.
13. Rinse the wells with 1% acetic acid and air dry.
14. Dissolve the bound dye with 10 mM Tris-HCl base, pH 10.7, in a shaker for 20 min.
15. Transfer aliquots to a 96-well plate.
16. Read in a Titertek Multiscan apparatus at 492 nm.

The linearity of the SRB assay with cell number is verified prior to the cell growth experiments.

3.2.4. Estradiol Dose–Response

The first step toward the estradiol dose–response is to define the dose-proliferative response curve for estradiol in MCF-7 cells as a reference curve (reviewed in **ref. 9**). At concentrations below 1 pM estradiol, equivalent to 1 fmol in 1 mL of culture medium, mean cell numbers do not significantly differ from those in the steroid-free control. Thus, 1 fmol estradiol/well is determined as the lowest detectable amount of estrogen in this assay.

3.2.5. Estimation of the TEXB

The 100% proliferative effect (PE) is calculated as the ratio between the highest cell yield obtained with 50 pM of estradiol and the proliferation of hormone-free control cells. The PE of α- and β-fractions was referred to the maximal PE obtained with estradiol and transformed into estradiol equivalent units (Eeq) by reading from the dose–response curve prepared using estradiol (concentration range 0.1 pM to 10 nM) (reviewed in **ref. *15***).

The TEXB is then expressed as the concentration of estradiol in picomolar units that results in a similar relative PE to that obtained with the α- and β-fractions and referred to the percentage of lipid in the sample.

4. Notes

1. Adipose tissue samples are taken from patients undergoing surgical treatments for breast cancer. Breast adipose tissues are obtained in the operating room during the course of surgery of primary lesions. Tissues are placed in a glass vial on ice, coded, and frozen to −80°C, always within 30 min of excision or aspiration.
2. Blood is obtained by venipuncture before surgery using glass tubes with no substance added. Serum is separated from cells by centrifugation, divided into 4-mL aliquots, coded, and stored at −80°C until analysis.
3. For successful extraction, it is most important to extract the adipose tissue gradually, adding the hexane to the potter in 3-mL amounts.
4. The extract obtained should be concentrated to total dryness if processing is not immediately performed.
5. The working conditions are as follows: gradient $t = 0$ min, 100% phase A; $t = 17$ min, 60% phase A; $t = 22$ min, 100% phase B; $t = 32$ min, 100% phase A.
6. The α-fraction contains organochlorine pesticides; the β-fraction contains natural endogenous estrogens together with phytoestrogens and nonhalogenated xenoestrogens.
7. The hexane and hexane:methanol:isopropanol should be added to tube A in 2-mL amounts to ensure better purification.
8. Working conditions are as follows: ECD at 300°C; injector at 250°C; for the program, initial $T = 130$°C (1 min), 20°C/min to 150°C, 10°C/min to 200°C, 20°C/min to 260°C (20 min). The carrier gas is nitrogen at a flow of 30 mL/min, and the auxiliary gas is nitrogen at a flow of 40 mL/min
9. The detection limits are between 0.1 and 1 ng/mL.
10. The working conditions are as follows: 250°C injector temperature; 50°C initial column temperature (2 min), 30°C/min to 185°C (5.5 min), 2°C/min to 250°C (32.5 min), and finally 30°C/min to 300°C (6.67 min). The carrier gas is helium with an injector flow of 1 mL/min. Manifold, transfer-line, and trap temperatures are 50, 230, and 200°C, respectively.
11. If the GC–MS is performed immediately after the injection into the GC–ECD, it is not necessary to dry or resuspend again in hexane.
12. Each sample is done in triplicate with a negative (vehicle) and a positive (10 pM estradiol) control in each plate.

References

1. Rice, C., Birnbaum, L. S., Cogliano, J., et al. (2003) Exposure assessment for endocrine disruptors: some considerations in the design of studies. *Environ. Health Perspect.* **111,** 1683–1690.
2. Damstra, T., Barlow, S., Bergman, A., Kavlock, R., and Van der Kraak, G. (eds.). (2002) *Global Assessment of the State-of-the-Science of Endocrine Disruptors.* WHO, Geneva.
3. Herbs, A. L., Ulfelder, H., and Poskanzer, D. C. (1971) Adenocarcinoma of the vagina. *N. Engl. J. Med.* **284,** 878–881.

4. Colborn, T., vom Saal, F. S., and Soto, A. M. (1993) Developmental effects of endocrine-disrupting chemicals in wildlife and humans. *Environ. Health Perspect.* **101**, 378–384.

5. Snedeker, S. M. (2001) Pesticides and breast cancer risk: a review of DDT, DDE, and dieldrin. *Environ. Health Perspect.* **109**(Suppl. 1), 35–47.

6. Sonnenschein, C. and Soto, A. M. (1998) An updated review of environmental estrogen and androgen mimics and antagonists. *J. Steroid Biochem. Mol. Biol.* **65**, 143–150.

7. Borgert, C. J., LaKind, J. S., and Witorsch, R. J. (2003) A critical review of methods for comparing estrogenic activity of endogenous and exogenous chemicals in human milk and infant formula. *Environ. Health Perspect.* **111**, 1020–1036.

8. Sonnenschein, C., Soto, A. M., Fernandez, M. F., Olea, N., Olea-Serrano, M. F., and Ruiz-Lopez, M. D. (1995) Development of a marker of estrogenic exposure in human serum. *Clin. Chem.* **41**(12 Pt. 2), 1888–1895.

9. Rivas, A., Fernandez, M. F., Cerrillo, I., et al. (2001) Human exposure to endocrine dis-rupters: standardisation of a marker of estrogenic exposure in adipose tissue. *APMIS* **109**, 185–197.

10. Soto, A., Lin, T. M., and Justicia, H. (1992) An "in culture" bioassay to assess the estroge-nicity of xenobiotics (E-SCRREEN), in *Chemically Induced Alterations in Sexual Devel-opment: The Wildlife/Human Connection* (Colborn, T. and Clement, C., eds.), Princeton Scientific, Princeton, NJ, pp. 295–309.

11. Nilsson, S. and Gustafsson, J. A. (2002) Estrogen receptor action. *Crit. Rev. Eukaryot. Gene Expr.* **12**, 237–257.

12. Okond'ahoka, O., Lavaur, E., Le Sech, J., Phu Lich, N., and Le Moan, G. (1984) Etude de l'impregnation humaine par les pesticids organo-halogenes au Zaire. *Ann. Fals Exp. Chim.* **77**, 531–538.

13. Medina, M. B. and Sherman, J. T. (1986) High performance liquid chromatographic sepa-ration of anabolic oestrogens and ultraviolet detection of 17 β-oestradiol, zeranol, diethylstilboestrol or zearalenone in avian muscle tissue extracts. *Food Addit. Contam.* **3**, 263–272.

14. Villalobos, M., Olea, N., Brotons, J. A., Olea-Serrano, M. F., Ruiz de Almodovar, J. M., and Pedraza, V. (1995) The E-screen assay: a comparison of different MCF7 cell stocks. *Environ. Health Perspec.* **103**, 844–850.

III

PESTICIDE ANALYSIS IN FOOD

17

Quality Criteria in Pesticide Analysis

Antonia Garrido Frenich, José L. Martínez Vidal, Francisco J. Egea
González, and Francisco J. Arrebola Liébanas

Summary

There is an increasing concern in pesticide residue analysis laboratories to ensure the quality of their analytical results. Internal quality control (IQC) measures are an essential element to ensure reliable results because they allow both the continuous monitoring of the process and measurements and the elimination of causes of unsatisfactory performance. IQC measures involve the use of blanks, certified reference materials (CRMs), quality control samples, calibrating standards, spiked samples, replicated samples, and blind samples. IQC measures are included in the analytical batch, but it is important not to forget that IQC criteria must be consistent with the cost of analyses, so that the number of IQC measures must not exceed 10% of the total number of samples. Finally, quality criteria in the identification and confirmation of pesticides are considered.

Key Words: Confirmation; identification; internal quality control; pesticides.

1. Introduction

Analytical laboratories, as entities that provide analytical information, have developed a quality system to ensure the quality of their results. There are three basic elements of that system: quality assurance (QA) (*1,2*), quality control, and quality assessment. Quality management is related to planification, assurance, control, and assessment of quality activities. QA is the essential organizational infrastructure that underlies all reliable analytical measurements. It is concerned with achieving, among others, appropriate levels in matters such as staff training and management; adequacy of the laboratory environment; safety; storage, integrity, and identity of samples; record keeping; maintenance and calibration of instruments; and use of technically validated and properly documented methods to obtain measurement traceability. These practices have been recognized as essential. However, the adequate introduction and management of these elements are not enough to guarantee the quality of analytical results obtained at a given time. For this aim, specific activities are also necessary that, in the framework of the quality system, are termed quality control activities.

From: *Methods in Biotechnology, Vol. 19, Pesticide Protocols*
Edited by: J. L. Martínez Vidal and A. Garrido Frenich © Humana Press Inc., Totowa, NJ

The International Standards Organization (ISO) 8402 *(3)* defines *quality control* as the techniques and activities of operative character used to fulfill the requirements of quality. It is a general definition that shows that quality control refers to operative-type activities; organizational and management activities would correspond to the QA concept.

In the context of analytical laboratories *(4)*, internal quality control (IQC) is defined as the set of procedures undertaken by laboratory staff for the continuous monitoring of an operation and the measurements to decide whether results are reliable enough to be released. IQC is characterized as the responsibility of the staff (analysts, technicians) who directly carry out the activities.

In the literature, IQC generally is also termed quality control, although it should not be confused with external quality control (EQC). This provides both evidence of the quality of the laboratory's performance and individual analyst proficiency. Typical EQC activities are proficiency testing or collaborative studies. Although important, participation in EQC activities is not a substitute for IQC measures or vice versa.

IQC includes measures that involve both monitoring the process and eliminating causes of unsatisfactory performance *(5)*. IQC activities must be sufficient to ensure that the measurement chemical processes (MCPs) are under statistical control *(6)*. This goal is achieved when the quality level is good enough to detect whether unexpected or unwanted changes have occurred during analysis of samples. The frequency of IQC measures will depend on the volume of work, the experience of the analysts, and costs involved in ensuring that the analytical system is under control. Some IQC operations must be carried out routinely (e.g., daily quality checks through control charting).

IQC is an important determinant of the quality of data, and as such is recognized by international organizations and accrediting agencies. IQC is based on a number of measures that an analyst can make to ensure the data obtained are fit for their intended purpose. So, IQC measures enable the analyst to accept a result or reject it and repeat the analysis. IQC influences the accuracy and the precision of the measurements.

IQC is undertaken by the inclusion of particular samples in the analytical sequence and by replicate analysis. Some measures of IQC are the analysis of blanks, quality control samples, certified reference material (CRM) or in-house reference material, blind samples, replicate samples, calibrating standards, and laboratory or field-spiked samples.

QA and quality control are essential in helping analytical chemists deliver reliable and meaningful answers to their customers. Analytical results delivered without any quality statements now should be considered counterproductive data.

This chapter describes IQC measures as well as other basic activities necessary to obtain quality measurements in pesticide analysis.

2. Basic Activities in Pesticide Analysis

1. The laboratory must be distributed in well-defined work areas to avoid contamination of standards, samples, and extracts. So, separate areas for sample reception, processing, and extraction and for the storage and preparation of analytical standards must be defined.
2. All equipment (measurement instruments, balances, flask, pipets) must be calibrated regularly (to establish the relation between the value indicated by the equipment and the value indicated by a standard) and records of calibration maintained. In addition, measurement

instruments should be maintained (to avoid breakdown and bad operation) and verified (to check that some requirements are fulfilled) according to an established program. These activities are essential to ensure the traceability of measurements.

3. Laboratory operations should be carried out according to the requirements of international standards, such as ISO 17025, ISO 9001:2000, or good laboratory practices.

4. Analytical methods used must be validated in agreement with intended purpose and documented as standard operating procedures (SOPs).

5. The laboratory should analyze appropriate CRMs, when available, and participate in interlaboratory studies when appropriate samples are available that may be used as indicators of the comparability of data generated by the laboratory.

6. Personnel must be qualified, trained, and motivated. The required qualification, apart from personnel experience, is obtained through job-oriented internal and external trainings. Staff must be trained in the appropriate laboratory skills, in the use of specialized analytical instrumentation, and in the evaluation and interpretation of the data they produce. Training and experience are documented in records that must be kept for all staff. Personnel must not carry out a sample analysis without demonstrating previously its ability to give good results.

7. Sampling, usually done by the customer or by the laboratory, requires highly specialized analytical expertise. The laboratory should state in the report that the samples were analyzed as received if it was not responsible for sampling. Samples must be transported to the laboratory in clean containers in no more than 1 d. When it is not possible to send samples of fresh products to the laboratory within a few hours, they must be frozen and transported in boxes containing dry ice to avoid thawing. Samples must be clearly identified by legible labels that cannot be detached easily.

8. Samples preferably must be processed within 1 or 2 d of sample receipt. Analysis of samples should be carried out immediately after processing unless it can be demonstrated that residues in extracts are not affected by longer storage (freezer, overnight storage at 1–5°C in the dark). Analysis of certain volatile or labile pesticides must be carried out quickly and without storage of the samples. The use of a smaller test portion may increase the uncertainty of the MCP.

9. The quality of reagents used must be appropriate for the test concerned. The grade of any reagent used should be stated in the SOP. If the quality of reagents is critical to a test, changes in supply batches should be verified.

10. Reference materials, CRMs, and pesticide standards must be unambiguously identified on receipt and accompanying certificates stored.

11. Reagents prepared in the laboratory should be labeled to identify compound, concentration, solvent, and date of preparation or expiry. Also, the analyst who prepares the reagent shall be identifiable either from the label or from records.

12. Individual primary calibration solutions, generally 100–1000 mg/L, for the majority of the pesticides are stable for several months if kept at low temperature (in a refrigerator or freezer) in the dark and sealed against evaporation. For its preparation, not less than about 10 mg of the pesticide standard must be weighed. In any case, the stability of the standard solutions must be checked and documented.

13. Methods developed in-house must be fully validated *(7–14)* using internationally accepted criteria *(15–17)*. The laboratory must verify its capability to obtain good results when using standard methods before any sample is analyzed. For that, the laboratory must recalculate the characteristic parameters of the standard method to decide if it is "fit to purpose." It may be possible to incorporate a new analyte to an existing method, or this method may be used to analyze a different matrix with appropriate additional validation.

14. The validation process of the method has to include the estimation of uncertainty *(18–21)* as an essential tool for ongoing IQC. The uncertainty of the analytical method is indispensable in establishing the comparability of the measurement.
15. The laboratory's quality system must attend to compliance with safety, chemical hygiene, and all applicable regulations. SOPs should include the safe disposal of all waste materials.
16. The laboratory environment should be clean enough to ensure that it has no influence on analytical measurements.
17. IQC must be applied after analyzing a number of samples. It is very important to establish the right set size, based on experimental studies, to demonstrate that the MCP does not suffer changes intraset. On the other hand, the application of IQC should be consistent with the cost of analysis, so the number of analyses because of IQC must not exceed 10% of the total number of samples. In routine analysis, an IQC level of 5% is reasonable; however, for complex MCP, levels between 20 and 50% are not unusual.

3. Internal Quality Control Measures

Quality control is a key element for ensuring reliable results. IQC measures *(22)* must be included in the analytical batches to enable analysts to decide whether the batches satisfy the preset quality criteria and that a set of the results can be accepted. IQC measures involve the use of blanks, CRMs, calibrating standards, spiked samples, replicate samples, and blind samples, among others (**Fig. 1**).

3.1. Blanks

Laboratory reagent blanks eliminate false positives by contamination in the extraction process, instruments, or chemicals used. A matrix blank has to be analyzed to detect interferences of sample matrix. If a matrix blank is not available, the use of a simulated home-made matrix is allowed *(23)*. Analyses of blanks with each set of samples ensures that the analytical signal is attributable just to the analyte *(24,25)*.

3.2. CRMs and Quality Control Samples

Reference materials can be used to study the variation between batches of samples, verifying the comparability of the measures. However, CRMs are expensive to use in daily IQC of a laboratory. In addition, few reference material producers are dedicated to pesticides in water or foods. This lack of reference materials may be overcome by the preparation and use of home-made reference materials (QC sample), spiking a blank sample matrix with 5–10% of the target pesticides. This sample, properly stored, must be checked for stability and homogeneity and then used as a quality control sample. This sample must be analyzed every day by applying the analytical method, providing additional information about instrument performance (instrument sensitivity, column performance, etc.).

The variation in the data obtained from the analysis of the quality control sample is normally monitored on a quality control chart, setting warning limits at $\pm 2s$ (s = standard deviation of at least 10 replicate analyses) and action limits at $\pm 3s$. To set realistic limits, the mean and standard deviation must include the effect of interday precision, considering variations of analyst, chemicals, standards, and so on. Some rules based on statistical criteria may be applied to the interpretation of data trends in quality control charts *(8,26,27)*.

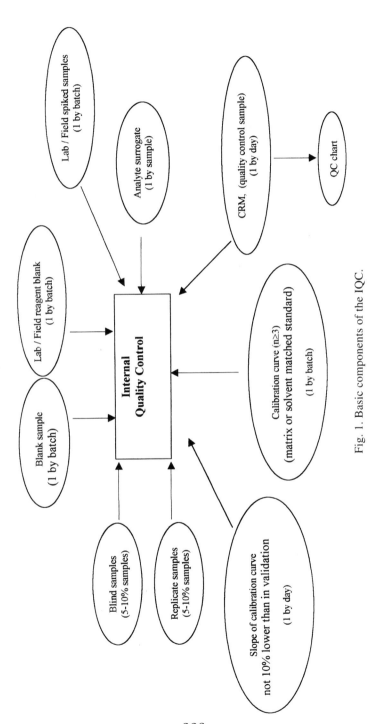

Fig. 1. Basic components of the IQC.

3.3. Calibrating Standards

Methodological calibration involves the use of chemical calibrating standards, which may be analyzed separately from samples (external standards) or as part of the samples (internal standards). A calibration curve must be carried out for every batch of samples. A minimum of three standard concentrations, in routine use, has to be used for the calibration of each pesticide. The first of them has to be equal or preferably lower than the maximum residue level (MRL) allowed for each pesticide in the target matrices. This point must be higher than or equal to the limit of quantitation (LOQ). The fit (linear, quadratic, etc.) of the calibration function must be inspected visually to ensure that it is satisfactory; also, the lack-of-fit test *(8)* may be used if replicate calibration points are available. Individual points of the calibration curve must not differ more than ±20% (±10% if the MRL is exceeded or approached). So, the difference between the concentration of analyte in each calibrating standard and the concentration calculated from the calibration curve must be lower than ±20% (±10% if the MRL is exceeded or approached). In contrast, a more appropriate fit must be used or the individual points must be repeated.

Calibration can also be performed by interpolation between two concentration levels. It is acceptable when the mean response factors (signal-to-concentration ratio), also called *sensitivity*, calculated from replicate analysis at each level indicate good linearity of the response. The higher response factor should not be more than 120% of the lower one (110% if the MRL is exceeded or approached).

Finally, single-level calibration is also used and may provide more accurate results than multilevel calibration if the detector response varies with the time. The sample response should be within ±50% (±10% if the MRL is exceeded) of the calibration response.

The slopes of the different calibration graphs must not differ significantly from the value found in the validation of the method, when the LOQs were established, because of the dependence of the quality of an analytical method for compounds at trace levels of its sensitivity. A decrease in the slope of the calibration curve may be critical when analyzing samples with analyte concentrations close to the LOQs. **Figure 2a** shows calibration curves with slopes that are 30, 20, 15, and 10% lower than the slope value found in the validation. These variations involve a decrease of the LOQ level according to its definition *(27,28)* of 45, 25, 18, and 12%, respectively (**Fig. 2b**). So, the slope of the daily calibration curve must not be lower than 10% of the slope value found in the validation step of the analytical method, mainly working close to the LOQs. In a consistent way, a good option may be not to establish the characteristic parameters of the analytical method (i.e., LOQs) using the best sensitivity of the MCP, but instead a more realistic option should be taken into account, considering that, after a long period of time, decreases in the capability of the method to carry out analytical measures occur because of replacement of equipment parts, variations in the sample matrix, changes in the environment, among others.

On the other hand, matrix influence *(8,29)* on the calibration step has to be considered at method validation. Unfortunately, the reliability of the above-described calibration approaches may not be absolute when the sample matrix is too complex and

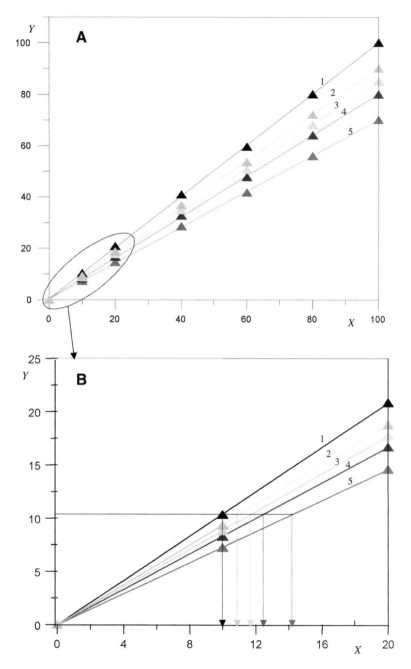

Fig. 2. Calibration plots with different slopes: (1) value found in the validation of the method and varying by (2) 10%, (3) 15%, (4) 20%, and (5) 30% considering (**A**) all the linear range and (**B**) the three first points of the linear range.

standards are prepared in solvent. If this is not taken into account in the developed method, then systematic errors can affect the result and cause bias in the final results. Matrix effect may affect both the slope and the intercept of the calibration line. In these cases, calibration using external standards prepared in the sample matrix *(6,30–34)* or standard addition procedure are then a recommended choice to avoid mistakes in the quantitation step. However, they also suffer some problems, such as that different commodities and different concentrations of matrix may give matrix effects of different magnitudes or the difficulties of obtaining a blank matrix for every sample type. Most of them would be avoided if a matrix, selected as representative matrix, would be used for calibration purposes for the same and similar matrices.

3.4. Spiked Samples

Recoveries measured from spiked matrix blanks are used to check the extraction efficiency in each batch of samples (because of the lack of reference materials for pesticide analysis). If the detection system allows, all recoveries may be combined in a single analysis by adding all analytes to a matrix blank at a concentration level about 30% above the LOQ. This level of addition may be varied to have information over a range of concentrations.

Mean recovery values are acceptable at the validation step within the range 70–110% *(16)*. Routine recovery values are acceptable in the range 60–140%, but recoveries for pesticides detected in the analyzed samples must be within the 70–110% range. If a routine recovery result is too low, the batch of samples should be reanalyzed. If the routine recovery is too high and no positive residues are detected, it is not necessary to reanalyze the samples, although high recovery should be investigated.

The above criteria about recoveries cannot be applied to solid-phase microextraction and head space analyses.

An analyte surrogate can be added to all the samples before starting the analysis to demonstrate that the MCP has been satisfactory for each sample and that mistakes have been avoided (i.e., to monitor method performance for each sample).

3.5. Replicated Samples

Replicated samples provide a less-formal means for checking for drift than quality control samples. Normally, the replicated sample is a conventional sample for which the analysis is repeated later within the batch of samples or perhaps in different batches. The results obtained from the analysis of these samples must be comparable, taking into account the uncertainty of the method. They are located in the batch every certain number of samples (i.e., 1 in every 10 samples). The analyst knows when replicated samples are included in the batch.

3.6. Blind Samples

Blind samples are replicate samples placed in the analytical batch without being known by the analyst. They may be sent by the laboratory management to check a particular system or by the customer to check the laboratory. Together with replicated samples, they are used to check inter- or intrabatch precision. In this way, blind samples are complementary to replicated samples, providing information about an analyst's proficiency.

Results of the replicated and blind samples have to be in agreement with the original analysis in the range established by the uncertainty of the method.

4. Quality Criteria in the Identification and Confirmation of Pesticide Residues

4.1. Identification

For the retention time windows (RTWs), reproducible retention times are important in helping to identify an analyte in chromatographic techniques. For that, it is recommended to use relative retention data expressed with respect to the retention time of a standard substance (e.g., internal standard). RTWs are calculated by injection ($N = 10$) of a calibrating standard of an analyte and are defined as retention time averages plus or minus three standard deviations of retention time. Identification criteria set a RTW based on the reference retention time within which the analyte peak for a real sample must occur *(35)*.

4.2 Approaches to Confirmation

Confirmation of results is an essential step. In the case of negative results, the results are considered confirmed if the recovery result is acceptable for the batch in which the samples are analyzed. This is also of application in the case of positive results, but they require additional qualitative and quantitative confirmation. Several alternatives may be used in chromatographic techniques, depending on the detection method: liquid chromatography (LC) with diode array detection if the ultraviolet spectrum is characteristic; LC with ultraviolet/visible investigation or fluorescence in combination with other techniques; gas chromatography with electron capture detection, nitrogenous phosphorus detection, or flame phosphorus detection applying stationary or mobile phases of different selectivity or other techniques *(36)*. However, mass spectrometry (MS) is often used for the confirmation of organic microcontaminants because it can provide potentially unequivocal confirmation.

4.2.1. Confirmation by MS

MS detection is carried out by recording full-scan mass spectra or selected ion monitoring (SIM), as well as MS-MS^n techniques such as selected reaction monitoring or other suitable MS or tandem MS (MS/MS) techniques in combination with appropriate ionization modes such as electron impact (EI) or chemical ionization.

4.2.1.1. FULL SCAN

The confirmation of the results must be performed by comparing the spectrum of the previously identified compound with that generated using the instrument employed for analysis of the samples. It must be considered that distortion of ion ratios is observed when analyte overloads the detector, and background spectra are not carefully subtracted to obtain representative spectra of the chromatographic peaks.

The target spectra must contain at least four measured diagnostic ions (the molecular ion, characteristic adducts of the molecular ion, characteristic fragment, and isotope ions) with a relative intensity of more than 10% in the reference spectrum of the calibration standard. The spectrum may be accepted as sufficient evidence of identity when ions unrelated to the analyte in a full-scan spectrum (i.e., from m/z 50 to 50 mass units greater than the molecular ion) present relative intensities that do not exceed 25% (for EI) or 10% (for other ionization methods). When computer-aided library searching is used, the comparison of mass spectral data has to exceed a critical match factor (fit threshold) that shall be determined during the validation process for every analyte. Such a factor must be calculated by considering variability caused by the sample matrix and the detector performance. The fit thresholds are usually scaled to 1000 for the best match (identical spectra). The critical match factor shall be set by the average fit of the spectra from 10 injections of the midlevel standard of calibration and subtracting a factor that considers the variability above described (usually 250 units).

4.2.1.2. Selected Ion Monitoring and MS/MS

The molecular ion shall preferably be one of the selected diagnostic ions, which should not exclusively originate from the same part of the molecule and must have a signal-to-noise ratio of 3 or greater. Two ions of $m/z > 200$ or three ions of $m/z > 100$ are the minimum data requirement when SIM mode is used. The most abundant ion that shows no evidence of chromatographic interference and the best signal-to-noise ratio should normally be used for quantification. When full-scan and SIM modes are employed, the relative intensity of the detected ions shall correspond to those of the calibration standard at comparable concentrations and measured under the same experimental conditions and be within a tolerance of ±20%.

The acquisition of spectra performed by MS/EI or MS/MS/EI may provide good evidence of identity and quantity in many cases. However, when mass spectra are produced by other process such as chemical ionization or atmospheric pressure ionization, further evidence may be required because of the simpleness of the obtained mass spectra. The additional evidence can be sought using (1) a different chromatographic separation system; (2) a different ionization technique; (3) MS/MS; (4) medium/high-resolution MS; or (5) altering fragmentation by changing the "cone voltage" in LC–MS.

When MS/MS is employed, the following can be selected for confirmatory purposes: (1) one precursor and two daughter ions; (2) two precursor ions, each with one daughter. When third order MS (MS/MS/MS) is used, the confirmation can be done by comparing one precursor, one daughter, and two granddaughter ions. Finally, in high-resolution mass spectrometry, the resolution shall typically be greater than 10,000 for the entire mass range at 10% valley and shall require three ions for confirmation.

References

1. Funk, W., Dammann, V., and Donnevert, G. (1995) *Quality Assurance in Analytical Chemistry*, VCH, Weinheim.
2. Günzler, H. (ed.) (1994) *Accreditation and Quality Assurance in Analytical Chemistry*, Springer-Verlag, Berlin.
3. International Organisation for Standardisation. (1986) *Quality Vocabulary (ISO/IEC Standard 8402)*, International Organisation for Standardisation, Geneva.
4. Inczedy, J., Lengyel, T., and Ure, A.M. (1997) *Compendium of Analytical Nomenclature. Definitive Rules*, 3rd ed., Blackwell Science, Oxford, UK.

5. Thompson, M., and Wood, R. (1995) Harmonized guidelines for internal quality control in analytical chemistry laboratories (technical report). *Pure Appl. Chem.* **67**, 649-666.

6. Analytical Methods Committee. (1989) Principles of data quality control in chemical analysis. *Analyst* **114**, 1497–1503.

7. Huber, L. (1998) Validation of analytical methods: review and strategy. *LC–GC Int.* **11**, 96–105.

8. Massart, D. L., Vandeginste, B. G. M., Buydens, L. M. C., de Jong, S., Lewi, P. J., and Smeyers-Verbeke, J. (1997) *Handbook of Chemometrics and Qualimetrics: Part A*, Elsevier, Amsterdam.

9. Swartz, M. and Krull, I. S. (1997) *Analytical Method Development and Validation*, Dekker, New York.

10. Feinberg, M. (1996) *La Validation des Méthodes D'analyse*, Masson, Paris.

11. Green, J. M. (1996) A practical guide to analytical method validation. *Anal. Chem.* **68**, 305A–309A.

12. Jenke, D. R. (1996) Chromatographic method validation: a review of current practices and procedures. I. General concepts and guidelines. *J. Liq. Chromatogr. Rel. Technol.* **19**, 719–736.

13. Jenke, D. R. (1996) Chromatographic method validation: a review of current practices and procedures. II. Guidelines for primary validation parameters. *J. Liq. Chromatogr. Rel. Technol.* **19**, 737–757.

14. Jenke, D. R. (1996) Chromatographic method validation: a review of current practices and procedures. III. Ruggedness, revalidation and system suitability. *J. Liq. Chromatogr. Rel. Technol.* **19**, 1873–1891.

15. http://www.eurachem.ul.pt/guides/CITAC%20EURACHEM%20GUIDE.pdf

16. S. Reynolds. *Quality Control Procedures for Pesticide Residues Analysis. Guidelines for Residues Monitoring in the European Union*, 3rd ed., (2003) Document no. SANCO/10476/2003. European Commission. York, United Kingdom.

17. Eurachem Guide: The Fitness for Purpose of Analytical Methods. A Laboratory Guide to Method Validation and Related Topics, (1998). LGC, Teddington, United Kingdom. Also available from the EURACHEM Secretariat and Web site. 18. Cuadros-Rodríguez, L., Hernández Torres, M. E., Almansa López, E., Egea González, F. J., Arrebola Liébanas, F. J., and Martínez Vidal, J. L. (2002) Assessment of uncertainty in pesticide multiresidue analytical methods: main sources and estimation. *Anal. Chim. Acta* **454**, 297–214.

19. Thompson, M., Ellison, S.L.R., and Wood, R. (2002) Harmonized Guidelines for single-laboratory validation of methods of analysis (IUPAC technical report). *Pure Appl. Chem.* **74**, 835-855.

20. Williams, A., Ellison, S. L. R., and Roesslein, M. (Eds.). *EURACHEM Guide, Quantifying Uncertainty in Analytical Measurement*, 2nd ed.,(2000) available at: http://www.eurachem.ul.pt/guides/QUAM2000-1.pdf

21. International Organisation for Standardisation. (1993) *Guide to the Expression of Uncertainty in Measurement*, International Organisation for Standardisation, Geneva.

22. Martínez Vidal, J. L., Garrido Frenich, A., and Egea González, F. J. (2003) Internal quality control criteria for environmental monitoring of organic micro-contaminants in water. *Trends Anal. Chem.* **22**, 34–40.

23. Keith, L., Libby, R., Crummer, W., Taylor, J., Deegan, J., and Wentler, G., Jr. (1983) Principles of environmental analysis. *Anal. Chem.* **55**, 2210–2218.

24. Pérez-Bendito, D. and Rubio, S. (1999) *Environmental Analytical Chemistry*, Elsevier, Amsterdam.

25. Quevauviller, P. (ed.). (1995) *Quality Assurance in Environmental Monitoring—Sampling and Sample Pre-treatment*, VCH, Weinheim.
26. International Organisation for Standardisation. (1995) *ISO Standards Handbook, Statistical Methods for Quality Control, Vol. 2 Measurement Methods and Results. Interpretation of Statistical Data. Process Control*, 4th ed., International Organisation for Standardisation, Geneva.
27. Long, E. L. and Winefordner, J. D. (1983) Limit of detection: a closer look at the IUPAC definition. *Anal. Chem.* **55,** 712–724.
28. Miller, J. N. (1991) Basic statistical methods for analytical chemistry, Part 2: calibration and regression methods, *Analyst* **116,** 3–14.
29. Kellner, R., Mermett, M., Otto, M., Valcárcel, M., and Widmer, H. M. (eds.). (2004) *Analytical Chemistry*, 2nd ed., Wiley-VCH, Weinheim, chapter 6, p 69–89
30. Quevauviller, P. (2002). *Quality Assurance for Water Analysis*, Wiley/European Commission, Brussels.
31. Taylor, M. J., Hunter, K., Hunter, K., Lindsay, D., and Le Bouhellec, S. (2002) Multiresidue method for rapid screening and confirmation of pesticides in crude extracts of fruits and vegetables using isocratic liquid chromatography with electrospray tandem mass spectrometry. *J. Chromatogr. A* **982,** 225–236.
32. Zrostlíkivá, J., Hajslová, J., Poustka, J., and Begany, P. (2002) Alternative calibration approaches to compensate the effect of co-extracted matrix components in liquid chromatography–electrospray ionisation tandem mass spectrometry analysis of pesticide residues in plant materials. *J. Chromatogr. A* **973,** 13–26.
33. Egea González, F. J., Hernández Torres, M. E., Almansa López, E., Cuadros-Rodríguez, L., and Martínez Vidal, J. L. (2002) Matriz effects of vegetable commodities in electrón-capture detection applied to pesticide multiresidue analysis. *J. Chromatogr. A* **966,** 155–165.
34. Martínez Vidal, J. L., Arrebola, F. J., Garrido Frenich, A., Martínez Fernández, J., and Mateu-Sanchez, M. (2004) Validation of a gas chromatographic–tandem mass spectrometric method for analysis of pesticide residues in six food commodities. Selection of a reference matrix for calibration. Chomatographia **59,** 321—327.
35. Hernández Torres, M. E., Egea González, F. J., Cuadros-Rodríguez, L., Almansa López, E., and Martínez Vidal, J. L. (2003) Assessment of matrix effects in gas chromatography electron capture pesticide residue analysis. *Chromatographia* **57,** 657–664.
36. http://www.iupac.org/symposia/conferences/method_validation_4nov99/report.html

18

Immunoassay Methods for Measuring Atrazine and 3,5,6-Trichloro-2-Pyridinol in Foods

Jeanette M. Van Emon and Jane C. Chuang

Summary

This chapter describes the use of enzyme-linked immunosorbent assay (ELISA) methods for the analysis of two potential environmental contaminants in food sample media: atrazine and 3,5,6-trichloro-2-pyridinol (3,5,6-TCP). Two different immunoassay formats are employed: a magnetic particle immunoassay testing kit and a 96-microwell plate immunoassay. Diluted, filtered, nonfat baby foods were analyzed directly by a commercial magnetic particle immunoassay testing kit for the determination of atrazine. Fatty baby foods were extracted with water for the magnetic particle immunoassay analysis of atrazine. For 3,5,6-TCP analysis, the food samples from a duplicate diet exposure study were analyzed by a laboratory-based 96-microwell plate immunoassay. Acidic methanol (72% methanol, 26% water, and 2% acetic acid) was the solvent for the food samples.

Key Words: Atrazine; baby food; duplicate-diet food; enzyme-linked immunosorbent assay (ELISA); magnetic particle immunoassay testing kit; 96-microwell plate immunoassay; 3,5,6-trichloro-2-pyridinol (3,5,6-TCP).

1. Introduction

Children can be exposed to pesticides by inhaling contaminated air, ingesting tainted food as well as nondietary substances (i.e., dust or soil), or absorption through the skin from contaminated media. With the passage of the U.S. Food Quality and Protection Act of 1996 *(1)*, new, more stringent standards for pesticide residues in foods were established to provide increased emphasis on health protection for infants and children. Pilot-scale field studies suggested that dietary ingestion was a major route of exposure to some pollutants and that the doses received by children could exceed those of adults living in the same household *(2–5)*.

Determination of pesticides in food is often complicated by the presence of fats. Samples typically require extensive cleanup steps before final analysis, greatly increasing analytical costs. Simple, rapid, cost-effective methods are also needed for

From: *Methods in Biotechnology, Vol. 19, Pesticide Protocols*
Edited by: J. L. Martínez Vidal and A. Garrido Frenich © Humana Press Inc., Totowa, NJ

environmental and biological media to determine aggregate pesticide exposures. Enzyme-linked immunosorbent assay (ELISA) methods are generally sensitive, specific, and cost-effective. They can facilitate a high sample throughput and can be used as a quantitative tool for monitoring pesticides in various matrices. Several ELISA methods have been developed for the detection of environmental pollutants in various sample media *(6–16)*. Performance data have been reported for real-world samples using ELISA testing kits *(12–16)* and laboratory-based 96-microwell plate immunoassays *(9,10)*. This chapter describes analytical methods utilizing two immunoassay method formats (magnetic particle and 96-microwell plate immunoassays) for the determination of atrazine in nonfatty and fatty baby foods and 3,5,6-TCP in composite diet food samples.

2. Materials

The materials listed next are grouped by the two immunoassay method formats: (1) magnetic particle immunoassay and (2) 96-microwell plate immunoassay. Within each immunoassay format, materials are listed in two subsections: one for preparing samples for immunoassay and one for performing the immunoassay.

2.1. Magnetic Particle Immunoassay for Atrazine in Baby Foods

2.1.1. Preparing Baby Food Samples

1. Vortex mixer.
2. Stainless steel spatulas.
3. Kim wipes.
4. Clean glass wool.
5. Silanized glass vials with Teflon-lined screw caps.
6. Distilled, deionized water.
7. Glass rods.
8. Pressurized solvent extraction system (ISCO SFX-200 or Dionex ASE 200 or equivalent).
9. Glass funnels.
10. Silanized glass wool.
11. Solid-phase extraction manifold and reservoir.
12. Graduated cylinders.
13. Disposable Pasteur glass pipets.
14. Sand (reagent grade).
15. Latex or nitrile gloves.
16. Methanol (distilled-in-glass or reagent grade).
17. Quartz fiber filters.
18. Acrodisc polytetrafluoroethylene (PTFE) 25-mm, 0.45-μm syringe filters.
19. Nitrogen evaporator.
20. Volumetric flasks.
21. Eppendorf pipets (or equivalent).

2.1.2. Magnetic Particle Immunoassay

1. Atrazine assay testing kit (e.g., Strategic Diagnostics, Newark, DE).
2. Magnetic separator rack (e.g., Strategic Diagnostics).
3. Atrazine diluent (e.g., Strategic Diagnostics).

4. Adjustable volume pipet.
5. RPA-1 RaPID analyzer (e.g., Strategic Diagnostics).
6. Vortex mixer.
7. Digital timer.
8. Kim wipes.
9. Pipet tips.
10. Waste container.
11. Latex or nitrile gloves.
12. Glass vials with Teflon-lined screw caps.

2.2. 96-Microwell Plate Immunoassay for 3,5,6-TCP in Duplicate-Diet Foods

2.2.1. Preparing Food Samples

1. Vortex mixer.
2. Latex or nitrile gloves.
3. Methanol (distilled-in-glass or reagent grade).
4. Distilled, deionized water.
5. Centrifuge.
6. Nitrogen evaporator.
7. Volumetric flasks.
8. Eppendorf pipets (or equivalent).
9. Quartz fiber filters.
10. Acrodisc PTFE 25-mm, 0.45-μm syringe filters.
11. Glass funnels.
12. Graduated cylinders.
13. Glass vials with Teflon-lined screw caps.

2.2.2. 96-Microwell Plate Immunoassay

1. High-binding, flat-bottom 96-well microtiter plates (Nunc Maxsorb I or equivalent).
2. Plate washer (Skatron 300 or equivalent).
3. Calibrated, single-channel, adjustable-volume pipets (1000, 250, and 10 μL).
4. Calibrated, multichannel, adjustable-volume pipet (250 μL).
5. Serological pipets (5, 10, 25 mL).
6. Presterilized pipet tips (1000, 250, and 10 μL).
7. Microfuge tubes.
8. Glass scintillation vials.
9. Spectrophotometer (microplate reader).
10. Vortex mixer.
11. Digital timer.
12. Latex or nitrile gloves.
13. Distilled, deionized water.
14. Goat antirabbit immunoglobulin G alkaline phosphatase conjugate.
15. Microwell substrate system: 5X diethanolamine substrate solution and 5-mg *p*-nitrophenyl phosphate tablets.
16. Phosphate-buffered saline with 0.05% Tween-20 and 0.02% sodium azide (PBST buffer) at pH 7.4.
17. 3,5,6-TCP monoclonal antibody *(17)*.
18. Coating antigen *(17)*.

3. Methods

The methods described include (1) magnetic particle immunoassay for atrazine in baby foods and (2) 96-microwell plate immunoassay for 3,5,6-TCP in composite duplicate-diet foods.

3.1. Magnetic Particle Immunoassay for Atrazine in Baby Foods

The method for the determination of atrazine in nonfatty baby foods consists of (1) diluting the baby food with water and (2) analyzing the sample extract with the magnetic particle immunoassay testing kit. The method for the determination of atrazine in fatty baby foods consists of (1) extracting the baby food with water, (2) mixing the extract with sand, and (3) analyzing the sample extract with the magnetic particle immunoassay.

3.1.1. Sample Preparation for Atrazine Magnetic Particle Immunoassay

1. For nonfatty food samples, dilute an aliquot (1 g) of baby food with 5 mL deionized or reagent water. For the matrix-spiked sample, add a known amount of atrazine into the food sample prior to the dilution process. Filter the diluted baby food through silanized glass wool into a 60-mL reservoir attached to a solid-phase extraction manifold (*see* **Note 1**). The aqueous extract is now ready for the immunoassay. The immunoassay analysis procedure used is described in **Subheading 3.1.2.**
2. For fatty food samples, remove an aliquot (nominal 4 g) of baby food after thoroughly mixing the contents of the glass jar with a glass stirring rod. Mix the baby food with 4 g sand. For the matrix-spiked sample, add a known amount of atrazine into the baby food-and-sand mixture. Transfer the mixture to an extraction cell (the baby food-and-sand mixture is sandwiched between two plugs of silanized glass wool and sand) for pressurized liquid extraction. Deionized water is used as the extraction solvent. The instrumental pressurized liquid extraction conditions are: (1) 2000-psi extraction pressure, (2) 150°C extraction temperature, and (3) 20 min static and 20 min dynamic extraction times. The resulting aqueous extract is then analyzed by the magnetic particle immunoassay for atrazine (*see* **Note 2**).

3.1.2. Atrazine Magnetic Particle Immunoassay Procedures

1. Carefully place 200-µL aliquots of the calibration solution (0, 0.1, 1.0, and 5.0 ng/mL), control solution (3 ng/mL), sample extracts, and quality control samples (duplicates, matrix spikes, and method blanks) into the bottom of individually labeled test tubes secured onto a magnetic rack (*see* **Note 3**).
2. Add a 250-µL aliquot of the atrazine enzyme conjugate and a 500-µL aliquot of the atrazine antibody-coupled paramagnetic particles to the inside wall of each tube and allow to flow down the side (*see* **Note 4**).
3. Mix the above solution on a vortex mixer and then incubate at room temperature for 15 min.
4. Affix the magnetic separation rack to the magnetic base and let the samples sit for 2 min to allow the magnetic particles to separate and adhere to the wall of the tube.
5. Invert the rack assembly over a waste container to decant the solution containing any unbound reagents.
6. Blot the rim of each test tube on several layers of clean paper towels.
7. Add a 1-mL aliquot of the atrazine washing solution down the inside wall of each test tube and allow the solution to stand for 2 min.

8. Decant the solution into the waste container as before and blot the rim of each test tube on paper towels. Repeat this washing step.
9. Remove the magnetic rack from the magnetic base.
10. Add a 500-µL aliquot of the color reagent down the inside wall of each tube and mix this solution on the vortex mixer.
11. Allow the solution to incubate for 20 min at room temperature.
12. At the end of the incubation period, add a 500-µL aliquot of the stopping solution down the wall of each tube to stop color development.
13. Analyze each test tube on the RPA-I RaPID photometric analyzer at 450 nm within 15 min of the addition of stopping solution (*see* **Note 5**).

3.2. 96-Microwell Plate Immunoassay for 3,5,6-TCP in Duplicate-Diet Food

The method for the determination of 3,5,6-TCP in food consists of (1) extracting the food with acidic methanol (72% methanol, 26% water, and 2% acetic acid), (2) diluting the sample extract with PBST, and (3) analyzing the sample extract with the 96-microwell plate immunoassay.

3.2.1. Sample Preparation Procedures for Duplicate-Diet Foods

Duplicate-diet sampling methodology has been used for determining exposures to potential pesticide residues in food in the context of aggregated exposure measurements via dietary ingestion route (*2,4,5,18–20*). Duplicate-diet samples are the duplicate portions of each food and beverage that the study participants consumed. Depending on study designs of the exposure field studies, composite duplicate-diet solid food samples may represent the entire solid food intake of the study participants over several days. Therefore, the composite duplicate-diet solid food samples are extremely complex in food composition, with mixtures of the four basic food groups present in any sample.

1. Prior to the immunoassay, mix the composite solid food samples with dry ice in equal proportions.
2. Homogenize the mixture with a commercial mechanical food processor.
3. Mix thoroughly individual aliquots of the food sample homogenates with Celite 545 and sonicate with acidic methanol (72% methanol, 26% water, and 2% acetic acid) for 30 min.
4. Centrifuge the mixture at 1260*g* (2500 rpm) at 4°C for 20 min.
5. Remove a 200-µL aliquot of the supernatant, located beneath the fat layer, and dilute with 800 µL of PBST for subsequent ELISA analysis.

3.2.2. Procedures for 96-Microwell Plate Immunoassay

1. Perform the ELISA analyses for food samples using a laboratory-based 96-microwell format (*10,16*).
2. Coat a 96-microwell plate with 100 µL of coating antigen (125 ng/mL) and incubate overnight at 4°C.
3. Wash each plate three times with PBST in a plate washer programmed for a three-cycle wash. Rotate the plates 180° between wash cycles to remove unbound antigens effectively.
4. Dry the plates further by tapping on absorbent paper, seal each plate with an acetate cover, and store in a refrigerator until needed.
5. Prepare the calibration standard solutions in clean borosilicate vials by serial dilutions with 10% methanol in PBST from a stock solution of 3,5,6-TCP to final concentrations of 25, 12.5, 6.25, 3.13, 1.56, 0.78, 0.39, 0.198, and 0.1 ng/mL.

6. Also prepare a fresh aliquot of PBST solution containing none of the 3,5,6-TCP (designated as 0.0-ng/mL standard solution) for each assay.
7. Add individual 100-µL aliquots of the standard solutions, sample extracts, and blanks to appropriate microwells in the antigen-coated plate.
8. Add a 100-µL aliquot of a 3,5,6-TCP monoclonal antibody (1:4000) in PBST to all microwells except those used for instrument blanks.
9. Incubate the plates for 2 h at room temperature on an orbital shaker at low speed.
10. Remove the excess reagent, not bound to the plate, by washing with PBST as described above.
11. Dry the plate further by tapping on absorbent paper.
12. Add a 100-µL aliquot of goat antimouse immunoglobulin G alkaline phosphatase conjugate at a 1:1000 dilution in PBST to each microwell.
13. Incubate the plates again for 2 h at room temperature on the orbital shaker at low speed.
14. Remove the excess conjugate by washing with PBST and add 100 µL of para-nitrophenyl phosphate at 1 ng/mL in diethanolamine buffer to each microwell.
15. After a 30-min room temperature incubation, read each microwell using a Molecular Devices Spectra Max Plus microplate spectrophotometer (or equivalent).
16. Determine the absorbance of the microwells at 405 nm and normalize to a 1-cm path length.
17. Perform data processing with SOFTmaxPro software version 2.1.1 or equivalent interfaced to a personal computer using a four-parameter curve fit.

4. Notes

1. Quality control samples for each sample set should include a method blank, matrix spike, and duplicate aliquots of samples. In a previous study *(15)*, no atrazine was detected in any of the nonspiked baby food samples. Quantitative recoveries of spiked baby foods (pear, carrot, green bean, chicken noodle, broccoli/chicken/cheese) were achieved using the methods described in **Subheading 3.1.** Overall method accuracy was $100 \pm 20\%$, and method precision was within $\pm 20\%$.
2. If there are particles present in the extract, the sample extract should be filtered again with an Acrodisc PTFE 25-mm, 0.45-µm syringe filter. The filtrate will then be ready for the magnetic particle immunoassay. A solvent spike sample should also be prepared for the Acrodisc filtration to assess any loss occurring during the filtration step.
3. It is important to warm the reagents to room temperature before conducting the magnetic particle immunoassay. If the reagents are not at the proper temperature, the calibration curve may not meet the specified requirements provided by the testing kit. It is also a good practice to bring reagents to room temperature prior to measuring for the 96-microwell plate immunoassay protocol.
4. The commercially available magnetic particle immunoassay testing kits typically have a small dynamic optical density (od) range (i.e., 1.0–0.35 od) and small changes in od correlate to large changes in derived concentrations. Note that the differences between absorbance values from duplicate assays are generally small and are well within the acceptance requirement (%CV < 10%) for the calibration standard solutions. However, the percentage difference of the derived concentrations of the standard solution from duplicate assays sometimes exceeds 30%. The greater percentage difference values obtained for some of the measured concentrations for the standards and samples may be caused by a small volume of standard or sample retained in the pipet tip during the transfer step. Extreme care should be taken when transferring each aliquot of standard or sample to the test tube. A trace amount of aliquot not delivered to the test tube may result in a large variation in the data from duplicate assays.

5. For quantitative analysis, if the result of a sample extract is above the assay calibration range (>5 ng/mL), the sample extract should be diluted accordingly with diluent and reanalyzed by the magnetic particle immunoassay.

Acknowledgment

The US Environmental Protection Agency (EPA), through its Office of Research and Development, funded and collaborated in the research described here under Contract 68-D-99-011 to Battelle. It has been subjected to agency review and approved for publication. Mention of trade names or commercial products does not constitute endorsement or recommendation by the EPA for use. We gratefully acknowledge J. J. Manclus and A. Montoya, Laboratorio Integrado de Biogenieria, Universidad Politecnica de Valencia, Spain, for the generous gift of TCP antibody and antigen.

References

1. Food Quality Protection Act of 1996. Available at: http://www.epa.gov/opppspsl/fqpa/fqpa-iss.htm.
2. Wilson, N. K., Chuang, J. C., Lyu, C., and Morgan, M. K. (2003) Aggregate exposures of nine preschool children to persistent organic pollutants at day care and at home. *J. Expo. Anal. Environ. Epidemiol.* **13,** 187–202.
3. MacIntosh, D. L., Kabiru, C., Echols, S. L., and Ryan, P. B. (2001) Dietary exposure to chlorpyrifos and levels of 3,5,6-trichloro-2-pyridinol in urine. *J. Expos. Anal. Environ. Epidemiol.* **11,** 279–285.
4. Chuang, J. C., Callahan, P. J., Lyu, C. W., and Wilson, N. K. (1999) Polycyclic aromatic hydrocarbon exposures of children in low-income families. *J. Expo. Anal. Environ. Epidemiol.* **2,** 85–98.
5. Chuang, J. C., Lyu, C., Chou, Y.-L., et al. (1998) *Evaluation and Application of Methods for Estimating Children's Exposure to Persistent Organic Pollutants in Multiple Media,* EPA 600/R-98/164a, 164b, and 164c. US EPA, Research Triangle Park, NC.
6. Van Emon, J. M. and Lopez-Avila, V. (1992) Immunochemical methods for environmental analysis. *Anal. Chem.* **64,** 79A–88A.
7. Van Emon, J. M., Brumley, W. C., and Reed, A. W. (2001) Elegant environmental immunoassays paper presented at the 221st ACS National Meeting, American Chemical Society, Division of Environmental Chemistry, April 1–5, 2001.
8. Lopez-Avila, V., Charan, C., and Van Emon, J. M. (1996) Quick determination of pesticides in foods by SFE-ELISA. *Food Test. Anal.* **2,** 28–37.
9. Chuang, J. C., Miller, L. S., Davis, D. B., Peven, C. S., Johnson, J. C., and Van Emon, J. M. (1998) Analysis of soil and dust samples for polychlorinated biphenyls by enzyme-linked immunosorbent assay (ELISA). *Anal. Chim. Acta* **376,** 67–75.
10. Van Emon, J. M., Reed, A. W, Chuang, J. C., Montoya, A., and Manclus, J. J. (2000) Determination of 3,5,6-Trichloro-2-pyridinol (TCP) by ELISA. Paper presented at Immunochemistry Summit 7, 219th National ACS Meeting, San Francisco, CA, March 26–31.
11. Lucas, A. D., Goodrow, M. H., Seiber J. N., and Hammock, B. D., (1995) Development of an ELISA for the *N*-dealkylated *s*-triazines: application to environmental and biological samples. *Food Agric. Immunol.* **7,** 227–241.
12. Mackenzie, B. A., Striley, C. A. F., Biagini, R. E., Stettler, L. E., and Hines, C. J. (2001) Improved rapid analytical method for the urinary determination of 3,5,6-trichloro-2-pyridinol, a metabolite of chlorpyrifos. *Bull. Environ. Contam. Toxicol.* **65,** 1–7.

13. Shackelford, D. D., Young, D. L., Mihaliak, C. A., Shurdut, B., A., and Itak, J. A. (1999) Practical immunochemical method for determination of 3,5,6-trichloro-2-pyridinol in human urine: applications and considerations for exposure assessment. *J. Agric. Food Chem.* **47,** 177–182.
14. Chuang, J. C., Pollard, M. A., Chou, Y.-L., and Menton, R. G. (1998) Evaluation of enzyme-linked immunosorbent assay for the determination of polycyclic aromatic hydrocarbons in house dust and residential soil. *Sci. Total Environ.* **224,** 189, 199.
15. Chuang, J. C., Pollard, M. A., Misita, M., and Van Emon, J. M. (1999) Evaluation of analytical methods for determining pesticides in baby food. *Anal. Chim. Acta* **399,** 135–142.
16. Chuang, J. C., Van Emon, J. M., Reed, A. W., and Junod, N. (2004) Comparison of immunoassay and gas chromatography/mass spectrometry methods for measuring 3,5,6-trichloro-2-pyridinol in multiple sample media. Anal. Chim. Acta **517,** 177–185.
17. Manclús, J. J. and Montoya, A. (1996) Development of an enzyme-linked immunosorbent assay for 3,5,6-trichloro-2-pyridinol. 1. Production and characterization of monoclonal antibodies. *J. Agric. Food Chem.* **44,** 3703–3709.
18. Wilson, N. K., Chuang, J. C., Iachan, R., et al. (2004) Design and sampling methodology for a large study of preschool children's aggregate exposures to persistent organic pollutants in their everyday environments. *J. Expo. Anal. Environ. Epidemiol.* **14,** 260–274.
19. Thomas, K., Sheldon, L. S., Pellizzari, E., Handy, R., Roberts, J., and Berry, M. (1997) Testing duplicate diet sample collection methods for measuring personal dietary exposures to chemical contaminants. *J. Expo. Anal. Environ. Epidemiol.* **7,** 17–36.
20. Scanlon, K., Macintosh, D., Hammerstrorm, K., and Ryam, P. (1999) A longitudinal investigation of solid-food based dietary exposure to selected elements. *J. Expo. Anal. Environ. Epidemiol.* **9,** 485–493.

19

Quick, Easy, Cheap, Effective, Rugged, and Safe Approach for Determining Pesticide Residues

Steven J. Lehotay

Summary

This chapter describes a simple, fast, and inexpensive method for the determination of pesticides in foods and potentially other matrices. The method, known as the quick, easy, cheap, effective, rugged, and safe (QuEChERS) method for pesticide residues involves the extraction of the sample with acetonitrile (MeCN) containing 1% acetic acid (HAc) and simultaneous liquid–liquid partitioning formed by adding anhydrous magnesium sulfate ($MgSO_4$) plus sodium acetate (NaAc), followed by a simple cleanup step known as dispersive solid-phase extraction (dispersive-SPE). The QuEChERS method is carried out by shaking a fluoroethylenepropylene (FEP) centrifuge tube that contains 1 mL 1% HAc in MeCN plus 0.4 g anhydrous $MgSO_4$ and 0.1 g anhydrous NaAc per gram wet sample. The tube is then centrifuged, and a portion of the extract is transferred to a tube containing 50 mg primary secondary amine (PSA) and 50 mg C18 sorbents plus 150 mg anhydrous $MgSO_4$ per milliliter extract (the dispersive-SPE cleanup step). Then, the extract is centrifuged and transferred to autosampler vials for concurrent analysis by gas chromatography–mass spectrometry (GC–MS) and liquid chromatography–tandem mass spectrometry (LC–MS/MS). Different options in the protocol are possible depending on alternate analytical instrumentation available, desired limit of quantitation (LOQ), scope of targeted pesticides, and matrices tested.

Key Words: Food; fruits; gas chromatography; liquid chromatography; mass spectrometry; pesticide residue analysis; sample preparation; vegetables.

1. Introduction

Multiresidue analysis of pesticides in fruits, vegetables, and other foods is a primary function of many regulatory, industrial, and contract laboratories throughout the world. It is estimated that more than 200,000 food samples are analyzed worldwide each year for pesticide residues to meet a variety of purposes. Once analytical quality requirements (trueness, precision, sensitivity, selectivity, and analytical scope) have been met to suit the need for any particular analysis, all purposes for analysis favor

From: *Methods in Biotechnology, Vol. 19, Pesticide Protocols*
Edited by: J. L. Martínez Vidal and A. Garrido Frenich © Humana Press Inc., Totowa, NJ

practical benefits (high sample throughput, ruggedness, ease of use, low cost and labor, minimal solvent usage and waste generation, occupational and environmental friendliness, small space requirements, and few material and glassware needs).

A number of analytical methods designed to determine multiple pesticide residues have been developed in the time since this type of analysis became important *(1–10)*. However, few if any of these methods can simultaneously achieve high-quality results for a wide range of pesticides *and* the practical benefits desired by all laboratories. In 2003, the QuEChERS (quick, easy, cheap, effective, rugged, and safe) method for pesticide residue analysis was introduced *(11)*; it provides high-quality results in a fast, easy, inexpensive approach. Follow-up studies have further validated the method for more than 200 pesticides *(12)*, improved results for the remaining few problematic analytes *(13)*, and tested it in fat-containing matrices *(14)*.

The QuEChERS method has several advantages over most traditional methods of analysis in the following ways: (1) high recoveries (>85%) are achieved for a wide polarity and volatility range of pesticides, including notoriously difficult analytes; (2) very accurate (true and precise) results are achieved because an internal standard (ISTD) is used to correct for commodity-to-commodity water content differences and volume fluctuations; (3) high sample throughput of about 10–20 preweighed samples in about 30–40 min is possible; (4) solvent usage and waste are very small, and no chlorinated solvents are used; (5) a single person can perform the method without much training or technical skill; (6) very little labware is used; (7) the method is quite rugged because extract cleanup is done to remove organic acids; (8) very little bench space is needed, thus the method can be done in a small mobile laboratory if needed; (9) the acetonitrile (MeCN) is added by dispenser to an unbreakable vessel that is immediately sealed, thus minimizing solvent exposure to the worker; (10) the reagent costs in the method are very inexpensive; and (11) few devices are needed to carry out sample preparation.

This chapter provides the protocol for the QuEChERS method that is currently undergoing an extensive interlaboratory trial for evaluation and validation by pesticide-monitoring programs in several countries. In brief, the method uses a single-step buffered MeCN extraction while salting out water from the sample using anhydrous magnesium sulfate ($MgSO_4$) to induce liquid–liquid partitioning. For cleanup, a simple, inexpensive, and rapid technique called dispersive solid-phase extraction (dispersive-SPE) is conducted using a combination of primary secondary amine (PSA) and C18 sorbents to remove fatty acids among other components and anhydrous $MgSO_4$ to reduce the remaining water in the extract. Then, the extracts are concurrently analyzed by liquid chromatography (LC) and gas chromatography (GC) combined with mass spectrometry (MS) to determine a wide range of pesticide residues.

The final extract concentration of the method in MeCN is 1 g/mL. To achieve <10-ng/g limit of quantitation (LOQ) in modern GC–MS, a large volume injection (LVI) of 8 µL is typically needed, or the final extract can be concentrated and solvent exchanged to toluene (4 g/mL), in which case 2-µL splitless injection provides the anticipated degree of sensitivity. If MS instruments are not available in the laboratory, other options are also possible to analyze the samples using LC and GC coupled to element-selective detectors. These aspects are discussed in more detail in **Subheadings 1.2., 1.3.**, and **3.**

Table 1 lists the many pesticides that have been successfully evaluated with the QuEChERS method. Many other untested pesticides in the same classes can be analyzed by the method, and the final choice of analytes for this protocol depends on the analyst's particular needs. The only pesticides that have failed to be validated successfully in studies thus far include asulam, daminozide, dicofol, captan, folpet, pyridatc, and thiram. The method does not work for carboxylic acids, such as daminozide, because of their strong retention on PSA during the cleanup step: dicofol degrades rapidly to dichlorobenzophenone in samples, which is why it was not found in the extracts; asulam, pyridate, and thiram are exceptionally difficult and are not currently analyzed in multiclass, multiresidue methods; and the problems with captan and folpet are not likely to be because of the QuEChERS sample preparation method, but related to their GC–MS analysis, which is especially difficult because of their degradation on active sites in the GC system. (Unfortunately, these pesticides cannot be readily analyzed by LC–tandem mass spectrometry [MS/MS] either.)

1.1. Calibration in Pesticide Methods

In any quantitative method, the accuracy of the result cannot be better than the accuracy of the calibration. Pesticide residue analysis using chromatographic methods nearly always utilizes external calibration, in which analyte solutions of known concentrations are injected contemporaneously (in the same sequence) as the sample extracts, and the intensity of the analyte peaks in the standards is compared with those from the samples to determine the pesticide concentrations. The number of calibration standards needed in the determination and their concentrations depend on the quality assurance (QA) requirements for the analysis or laboratory, but generally four calibration standards (plus the matrix blank, or zero standard) dispersed from the LOQ to the highest expected analyte concentration are accepted practice *(15,16)*. Reported results should not list concentrations outside the concentration range covered by the calibration standards.

Furthermore, QA guidelines generally dictate that analytical methods be evaluated to determine the effects of matrix components in the extracts on the quantitative results *(15,16)*. If it is demonstrated that no differences are observed between analyte peak intensities in matrix extracts vs those in solvent only over the entire concentration range, then calibration standards may be prepared in solvent-only solutions. Each pesticide–matrix pair must be evaluated in this case, which can be a great deal of work. Otherwise, the matrix effects must be overcome empirically because the determined results may not be altered with a "fudge factor" in most pesticide analysis applications.

In general, GC methods for organochlorine insecticides are not affected by matrix, and LC using non-MS detection techniques dos not encounter matrix effects (unless there are chemical interferences in the signals). However, LC–MS techniques, particularly using electrospray ionization (ESI), are susceptible to ion suppression effects from coeluting components in the chromatogram, even though direct interference in the MS spectrum is seldom observed *(17)*. This indirect matrix effect in LC–MS tends to yield falsely low results in the samples when compared to standards that do not contain matrix components. In the case of GC, matrix components tend to fill active sites in the system (mainly in the injector liner and capillary column). This reduces the

Table 1
Possible Pesticide Analytes That Have Been Shown to Yield >90% Recoveries Using the QuEChERS Method

acephate,[a] acetamiprid, Acrinathrin, aldicarb, aldicarb sulfone, aldicarb sulfoxide, Aldrin, azaconazole, azamethiphos, azinphos-methyl, azoxystrobin, Bifenthrin, bitertanol, Bromopropylate, bromuconazole, Bupirimate, buprofezin, butocarboxim, butocarboxim sulfone, butocarboxim sulfoxide, Cadusafos, carbaryl, carbendazim, carbofuran, 3-hydroxy-carbofuran, chlorbromuron, (α-, γ-)Chlordane, (α-, β-)Chlorfenvinphos, Chlorpropham, Chlorpyrifos, Chlorpyrifos-methyl, Chlorthaldimethyl, Chlorothalonil,[a] Chlozolinate, clofentezine, Coumaphos, cycloxydim,[a] (λ-)Cyhalothrin, cymoxanil, Cypermethrin, cyproconazole, cyprodinil, (2,4'-, 4,4'-)DDE, (2,4'-, 4,4'-)DDT, Deltamethrin, demeton, demeton-*O*-sulfoxide, demeton-*S*-methyl, demeton-*S*-methyl sulfone, desmedipham, Diazinon, dichlofluanid,[a] Dichlorobenzophenone, dichlorvos, diclobutrazole, Dicloran, dicrotophos, Dieldrin, Diethofencarb, difenoconazole, Diflufenican, dimethoate, dimethomorph, diniconazole, Diphenyl, Diphenylamine, disulfoton, disulfoton sulfone, diuron, dmsa, dmst, dodemorph, (α-, β-)Endosulfan, Endosulfan sulfate, EPN, epoxiconazole, Esfenvalerate, etaconazole, ethiofencarb sulfone, ethiofencarb sulfoxide, Ethion, ethirimol, Ethoprophos, etofenprox, Etridiazole, Famoxadone, fenamiphos, fenamiphos sulfone, Fenarimol, Fenazaquin, fenbuconazole, fenhexamid,[a] Fenithrothion, fenoxycarb, Fenpiclonil, Fenpropathrin, Fenpropidine, fenpropimorph, fenpyroximate, Fenthion, fenthion sulfoxide, Fenvalerate, florasulam,[a] Flucythrinate I and II, Fludioxonil, flufenacet, Flufenconazole, flusilazole, Flutolanil, Fluvalinate, Fonophos, fosthiazate, Furalaxyl, furathiocarb, furmecyclox, Heptachlor, Heptachlor epoxide, Heptenophos, Hexachlorobenzene, hexaconazole, hexythiazox, imazalil, imidacloprid, Iprodione, iprovalicarb, isoprothiolane, isoxathion, kresoxim-methyl, Lindane, linuron, Malathion, malathion oxon, Mecarbam, mephosfolan, Mepronil, Metalaxyl, metconazole, methamidophos,[a] Methidathion, methiocarb, methiocarb sulfone,[a] methiocarb sulfoxide, methomyl, methomyl-oxime, metobromuron, metoxuron, Mepanipyrim, Mevinphos, monocrotophos, monolinuron, myclobutanil, nuarimol, Ofurace, omethoate, oxadixyl, oxamyl, oxamyl-oxime, oxydemeton-methyl, paclobutrazole, Parathion, Parathion-methyl, penconazole, pencycuron, (*cis*-, *trans*-)Permethrin, phenmedipham, *o*-Phenylphenol, Phorate, phorate sulfone, Phosalone, Phosmet, Phosmet-oxon, phosphamidon, Phthalimide, picoxystrobin, Piperonyl butoxide, pirimicarb, pirimicarb-desmethyl, Pirimiphos-methyl, prochloraz, Procymidone, profenofos, Prometryn, Propargite, Propham, propiconazole, propoxur, Propyzamide, Prothiofos, pymetrozine,[a] Pyrazophos, pyridaben, pyridaphenthion, pyrifenox, pyrimethanil, Pyriproxyfen, Quinalphos, Quinoxyfen, Quintozene, sethoxydim,[a] spinosad, spiroxamine, tebuconazole, tebufenozide, Tebufenpyrad, tetraconazole, Tetradifon, Tetrahydrophthalimide, Terbufos, Terbufos sulfone, thiabendazole, thiacloprid, thiamethoxam, thiodicarb, thiofanox, thiofanox sulfone, thiofanox sulfoxide, thiometon, thiometon sulfone, thiometon sulfoxide, thiophanate-methyl, Tolclofos-methyl, tolylfluanid,[a] triadimefon, triadimenol, Triazophos, trichlorfon, tricyclazole, tridemorph, trifloxystrobin, trifluminazole, Trifluralin, Triphenylphosphate, vamidothion, vamidothion sulfone, vamidothion sulfoxide, Vinclozolin

GC-amenable pesticides are capitalized; those preferentially analyzed by LC–MS/MS are not capitalized; those that can be analyzed by either technique are underlined.

[a]Or >70%.

number of active sites exposed to those analytes that also tend to adsorb on the sites. Therefore, the common effect of matrix in GC is to cause greater response of the susceptible analytes in the sample extracts than in solvent only because more of the analytes are lost to active sites in calibration standards in solvent-only solutions *(18–25)*. Those pesticides most strongly affected in GC tend to contain hydroxy, amino, phosphate, and other relatively polar groups *(21)*.

Several approaches have been devised in an attempt to overcome matrix effects in LC–MS *(26–28)* and GC *(11,19–25)*, but in both instrumental methods, the most common approach is the use of matrix-matched calibration standards *(20–23)*. Matrix matching has been shown to work better than most other approaches, but it is not ideal because it requires many blank matrices (which may be hard to find), entails extra extractions, and reduces ruggedness by introducing more matrix to the analytical instrument in a sequence than would be injected otherwise.

In the future, enough evidence may accumulate for two promising approaches to replace matrix-matching calibration: (1) the echo technique in LC–MS and (2) analyte protectants in GC. The echo technique involves injection of a calibration standard in solvent only just prior to (or immediately after) the sample extract when the mobile phase gradient has just started. This leads to two peaks adjacent to each other per analyte; one is the standard, and the other is from the sample. If ion suppression effects are the same for both peaks, then this will lead to accurate results *(26–28)*. In GC, the use of analyte protectants takes advantage of the increased response provided by the matrix-induced enhancement effect to equalize the signals of susceptible analytes in sample extracts and calibration standards alike *(11,19,24,25)*. This is done by adding a high concentration of components (analyte protectants) with multiple hydroxy groups to sample extracts and calibration standards in solvent. The analyte protectants have been shown to work well in providing accurate results, better peak shapes, and lower LOQ, and they surprisingly increase ruggedness of the analysis by continuing to work even in a very dirty GC system *(11,24,25)*.

Although the two alternate approaches may become the standard methods in the near future, it is too early to make this assertion now. Also, the careful choice of analytes quantified by LC–MS/MS and GC–MS may bypass matrix effects altogether. In the meantime, instructions in this protocol (*see* **Subheading 3.3.**) are given for the use of four matrix-matched calibration standards (plus the zero standard) to cover the concentration range of the pesticides that need to be detected in the samples.

1.2. Analysis of GC-Amenable Pesticide Residues

Traditionally, selective detectors in GC have been used to detect individual classes of GC-amenable pesticides, such as organochlorines, organophosphates, and organonitrogens *(1–6)*. Either multiple injections were necessary or split flows would be made to multiple detectors. GC–MS has become the primary approach to analyze all classes of GC-amenable pesticides in the same chromatogram *(7–10)*. Traditionally, GC–MS was mainly used for confirmation of analytes previously detected by selective detectors, but modern GC–MS instruments are sensitive, easy to use, reliable, and affordable for most laboratories. GC–MS has become a standard laboratory instrument and can provide qualitative and quantitative information for essentially any GC-amenable analyte in a single injection. Especially when fitted with LVI, GC–MS can provide comparable sensitivity of selective detectors even in complicated extracts.

Several MS techniques are available, the most common of which use a quadrupole design that is very rugged and practical. Ion trap MS instruments provide the advantages of lower LOQ in full-scan operation and the option for conducting MS^n of targeted analytes. Time-of-flight (TOF) instruments are more expensive, but may provide greater speed or higher mass resolution in the analysis. Magnetic sector is a fourth MS instrument option, but they are very large and expensive and generally reserved for special applications. Any of these MS techniques may be coupled with GC for pesticide residue analysis and should produce equal high-quality results *(26,27)*. Any difference in analytical accuracy between these types of MS systems is most likely a function of the injection process and not related to detection *(10)*.

Each MS approach also has multiple modes of operation. The most common ionization approach for GC–MS analysis of pesticides is electron impact (EI) ionization, which often yields many mass fragments to aid analyte identification. EI at 70 eV is the standard used for generating spectra with commercial instruments, and mass spectral libraries are available that contain full-scan spectra for as many as 300,000 compounds at these conditions. Another facet in MS analysis involves whether selected ion monitoring (SIM) or MS^n should be employed to provide lower LOQ and greater selectivity in the analysis of *targeted* pesticides *(8,9)*, or whether full-scan MS should be conducted to potentially identify *any* GC-amenable chemical in the chromatogram *(7)*. The targeted approach limits the number of analytes to about 60 that can be detected in a typical 30- to 40-min GC chromatogram, but full-scan operation permits a nearly unlimited number of analytes in a single injection. The analyst should refer to the literature if needed for further discussion *(29–31)*.

1.3. Analysis of LC-Type Pesticide Residues

Since the development of robust atmospheric pressure ionization (API) ion source designs, which consist of ESI and atmospheric pressure chemical ionization (APCI), very powerful and reliable LC–MS instruments have been introduced commercially. Depending on the source design, APCI works equally well or better as ESI for many pesticides, but APCI heats the analytes more than ESI, which potentially leads to problems for thermolabile pesticides. Thus, ESI has greater analytical scope and has become the primary ionization technique in LC–MS, but if all of the analytes in a method are compatible with APCI, then APCI may provide benefits of fewer ion suppression effects and a higher flow rate.

Because of the soft ionization nature of API, high background of LC mobile phases, and relatively low separation efficiency of LC, tandem MS (or high resolution) is often required to determine pesticide residues in complex extracts. Just as quadrupole, ion trap, TOF, and magnetic sector instruments may be coupled to GC, they may also be used in LC with the same advantages and disadvantages *(30)*. Moreover, just as trueness and precision in the analytical result are generally influenced by injection in

GC–MS more than detection, the performance of the ion source is typically the limiting factor in LC–MS techniques.

LC–MS/MS is rapidly becoming an indispensible analytical tool in analytical chemistry, and most pesticide-monitoring laboratories in developed countries have access to LC–MS/MS instruments. Many modern pesticides are not GC amenable, and if they do not fluoresce or contain a strong chromophore for ultraviolet/visible absorption, then LC–MS/MS is the only way to detect the chemical in its underivatized form. Derivatization of these types of analytes followed by GC analysis was often done in the past, but such methods are usually problematic to develop and implement in practice, and they do not lend themselves to multiclass, multiresidue applications *(1,7)*. Despite the great capital expense of the instruments, the powerful attributes of LC–MS/MS provide exceptional analytical performance, save time in method development, and can be used robustly in a variety of routine or special projects *(12,17,26–28,32)*.

The quality of LC–MS/MS analyses and instruments has reached the point that LC–MS/MS provides superior results than GC–MS even for many GC-amenable pesticides. This is indicated in **Table 1**; 90% of the underlined pesticides are not capitalized, which means that LC–MS/MS provided better sensitivity, greater trueness, or more precision than GC–MS for that pesticide *(12)*. The broad peaks in LC separations allow plenty of time in the MS/MS data collection process to monitor many other coeluting peaks without affecting quality of the results. Thus, hundreds of pesticide analytes can be monitored by LC–MS/MS in a single chromatogram *(12,26–28)*, which is not possible in GC–MS using SIM or MS^n techniques. Alternate methods for LC analysis using selective detectors rely on the LC separation to resolve the difference analytes from each other and matrix interferences. This is acceptable in a few multiresidue applications, such as *N*-methyl carbamate insecticides *(1,6–8)*, but traditional LC methods cannot meet multiclass, multiresidue analytical needs.

Indeed, the concurrent use of LC–MS/MS and (LVI) GC–MS for nearly any pesticide constitutes the state-of-the-art approach to multiclass, multiresidue analysis of pesticides in a variety of matrices. The QuEChERS method is an effective sample preparation procedure that very efficiently produces sample extracts suitable for both of these powerful analytical tools. This approach can be improved further in the near future by integrating other advanced techniques, such as direct sample introduction *(33–35)* and fast-GC–MS separations *(30,36–38)*, which may someday become the ultimate approach to pesticide residue analysis. The following protocol is an important step to meeting that challenge.

2. Materials

2.1. Sample Comminution

1. Food chopper (e.g., Stephan or Robotcoupe vertical cutters).
2. Probe blender (e.g., Ultraturrax) or Polytron homogenizers.
3. Container jars.
4. Blank sample verified to contain no detectable analytes.
5. Samples to be analyzed.
6. Freezer.

Optional items are:

1. Dry ice or liquid nitrogen.
2. Cryogenic chopper.

2.2. QuEChERS Sample Preparation

1. Analytical-grade MeCN.
2. High-performance liquid chromatographic (HPLC)-grade glacial acetic acid (HAc).
3. 1% HAc in MeCN (v/v) (e.g., 10 mL glacial HAc in 1 L MeCN solution).
4. Reagent-grade anhydrous sodium acetate (NaAc) (*see* **Note 1**).
5. Powder form anhydrous $MgSO_4 > 98\%$ pure (*see* **Note 2**).
6. PSA sorbent with 40-μm particle size (e.g., Varian, Harbor City, CA) (*see* **Note 3**).
7. Analytical-grade toluene.
8. Pesticide reference standards, typically above 99% purity (e.g., Chemservice, Accustandard, Dr. Ehrenstorfer).
9. Pesticide stock solutions (10 mg/mL): add 5 mL toluene to each 50 mg pesticide reference standard in 8-mL dark glass vials with Teflon-lined caps and store at 20°C or below (*see* **Note 4**).
10. ISTD stock solution (2 mg/mL): add 5 mL toluene to 10 mg d_{10}-parathion (e.g., C/D/N Isotopes or Cambridge Isotope Laboratories) in 8-mL dark glass vial with Teflon-lined cap and store at 20°C or below (*see* **Note 5**).
11. Triphenylphosphate (TPP) stock solution (2 mg/mL): add 5 mL toluene to 10 mg TPP in 8-mL dark glass vial with Teflon-lined cap and store at 20°C or below.
12. Working standard pesticides solution (40 ng/μL): add 400 μL of each pesticide stock solution at room temperature (RT) to a 100-mL volumetric flask containing 10 mL 1% HAc in MeCN and dilute with MeCN to the mark. Transfer four roughly equal portions of the solution to 40-mL dark glass vials with Teflon-lined caps and store at 20°C or below (*see* **Note 6**).
13. ISTD working solution (20 ng/μL): add 250 μL of the ISTD stock solution at RT to a 25-mL volumetric flask and dilute with MeCN to the mark. Transfer the solution to a 40-mL dark glass vial with Teflon-lined cap and store at 20°C or below.
14. TPP working solution (2 ng/μL): add 25 μL of the TPP stock solution at RT to a 25-mL volumetric flask and dilute with 1% HAc in MeCN to the mark. Transfer the solution to a 40-mL dark glass vial with Teflon-lined cap and store at 20°C or below (*see* **Note 7**).
15. Calibration standard spiking solutions w, x, y, and z (for w, x, y, and z standards): add 50 μL of ISTD stock solution, 2.5 mL of 1% HAc in MeCN solution, and 12.5•(w, x, y, and z) μL of the 40-ng/μL working standard pesticides solution at RT per (w, x, y, and z) ng/g desired equivalent calibration standard concentration into a 25-mL volumetric flask and fill to the mark with MeCN. For example, if the w standard is to be 10 ng/g, then add 125 μL of the 40-ng/μL working standard pesticides solution to the flask. Transfer the solutions to four 8-mL dark glass vials with Teflon-lined caps and store at 20°C or below.
16. 50-mL Fluorocthylenepropylene (FEP) centrifuge tubes (e.g., Nalgene 3114-0050 or equivalent) (or 250-mL FEP centrifuge bottles for 16- to 75-g samples).
17. Top-loading balance.
18. Solvent dispenser (15 mL for 15-g sample) and 1- to 4-L bottle.
19. Centrifuges.
20. Vials containing anhydrous NaAc plus anhydrous $MgSO_4$: add 1.5 g anhydrous NaAc plus 6 g anhydrous $MgSO_4$ to each vial for use with 15-g sample size (*see* **Note 8**).

21. Sealable centrifuge tubes (2–15 mL) containing powders for dispersive SPE: add 50 mg PSA sorbent plus 150 mg anhydrous $MgSO_4$ per 1 mL of extract to undergo cleanup (*see* **Note 8**).

Optional items:

1. Mechanical shaker, probe blender, or sonication device.
2. C18 sorbent with 40-µm particle size (*see* **Note 9**).
3. Graphitized carbon black (GCB; e.g., Supelco or Restek) (*see* **Note 10**).
4. Vortex mixer.
5. Minicentrifuge.
6. Evaporator (e.g., Turbovap or N-Evap).
7. Graduated centrifuge tubes (10–15 mL) for use in evaporator.
8. Calibration standard spiking solutions w, x, y, and z in toluene (for w, x, y, and z standards): add 50 µL of ISTD stock solution and 12.5•(w, x, y, and z) µL of the 40-ng/µL working standard pesticides solution at RT per (w, x, y, and z) ng/g desired equivalent calibration standard concentration into a 25-mL volumetric flask and fill to the mark with toluene. For example, if the w standard is to be 10 ng/g, then add 125 µL of the 40-ng/µL working standard pesticides solution to the flask. Transfer the solutions to four 8-mL dark glass vials with Teflon-lined caps and store at 20°C or below.

2.3. Analysis of GC-Amenable Pesticides

1. GC–MS system.
2. Programmable temperature vaporizer for LVI.
3. Autosampler.
4. A 30-m analytical capillary column with 0.25 mm id, 0.25 µm of (5% phenyl)-methylpolysiloxane low-bleed stationary phase (e.g., DB-5ms or equivalent).
5. Retention gap such as a 1- to 5-m, 0.25 mm id deactivated capillary column.
6. Helium at 99.999% purity.

Alternatives:

1. GC system(s) coupled with selective detector(s) such as pulsed flame photometric detector, flame photometric detector, halogen-specific detector, electron capture detector, electrolytic conductivity detector, atomic emission detector, nitrogen–phosphorus detector.
2. Split/splitless injector.

2.4. Analysis of LC-Type Pesticides

1. LC–MS/MS system.
2. ESI ion source.
3. Automated divert valve placed between analytical column and ion source.
4. Syringe pump for direct infusion of solutions into ion source.
5. Autosampler.
6. HPLC-grade methanol (MeOH).
7. HPLC-grade water.
8. Double-distilled, 88% formic acid.
9. 5 mM Formic acid in MeOH: add 214 µL formic acid to MeOH in 1 L solution.
10. 5 mM Formic acid in water: add 214 µL formic acid to water in 1 L solution.
11. 6.7 mM Formic acid in water: add 72 µL formic acid to water in 250 mL solution.
12. 15 cm long, 3.0 mm id, 3-µm particle size C18 analytical column.
13. 4 cm long, 3.0 mm id C18 guard column.

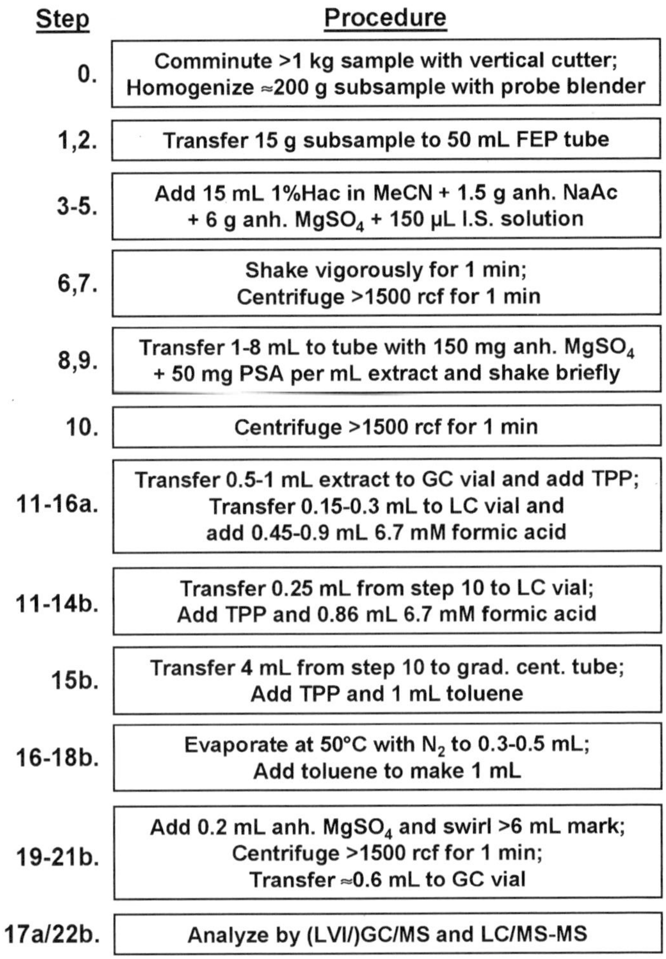

Step	Procedure
0.	Comminute >1 kg sample with vertical cutter; Homogenize ≈200 g subsample with probe blender
1,2.	Transfer 15 g subsample to 50 mL FEP tube
3-5.	Add 15 mL 1%Hac in MeCN + 1.5 g anh. NaAc + 6 g anh. $MgSO_4$ + 150 μL I.S. solution
6,7.	Shake vigorously for 1 min; Centrifuge >1500 rcf for 1 min
8,9.	Transfer 1-8 mL to tube with 150 mg anh. $MgSO_4$ + 50 mg PSA per mL extract and shake briefly
10.	Centrifuge >1500 rcf for 1 min
11-16a.	Transfer 0.5-1 mL extract to GC vial and add TPP; Transfer 0.15-0.3 mL to LC vial and add 0.45-0.9 mL 6.7 mM formic acid
11-14b.	Transfer 0.25 mL from step 10 to LC vial; Add TPP and 0.86 mL 6.7 mM formic acid
15b.	Transfer 4 mL from step 10 to grad. cent. tube; Add TPP and 1 mL toluene
16-18b.	Evaporate at 50°C with N_2 to 0.3-0.5 mL; Add toluene to make 1 mL
19-21b.	Add 0.2 mL anh. $MgSO_4$ and swirl >6 mL mark; Centrifuge >1500 rcf for 1 min; Transfer ≈0.6 mL to GC vial
17a/22b.	Analyze by (LVI/)GC/MS and LC/MS-MS

Fig. 1. Outline of the protocol in the QuEChERS method.

Alternatives:

1. LC system(s) coupled with selective detector(s) (e.g., fluorescence, diode array detector, ultraviolet/visible absorbance).
2. Postcolumn derivatization system and reagents.

3. Methods

Figure 1 shows a flowchart of the overall protocol of the approach, including the QuEChERS sample preparation method and its two main options that essentially depend on the desired LOQ in GC–MS. **Option A** relies on LVI to achieve the low LOQ if needed, and **Option B** entails solvent evaporation and exchange to toluene to increase the amount of equivalent sample injected in splitless mode. Once all the materials are ready and the 15-g homogenized subsamples have been weighed into the 50-mL tubes, a single analyst can prepare 10–20 extracts with the QuEChERS method in approx 30–40 min in **Option A**. The solvent exchange and evaporation step in **Option B** approximately doubles the time needed for the analyst to complete the method.

3.1. Sample Comminution

For food samples, an appropriate chopper (e.g., vertical cutter) must be used to comminute large, representative sample portions up to 9 kg *(1)*. Blend the sample until it gives a consistent texture. Transfer approx 200 g to a sealable container for freezer storage after further comminution with a probe blender. Blend the subsample with the mixer until it is homogeneous. A second subsample (e.g., 15 g) is taken for extraction immediately, and the container is then sealed and stored in the freezer in case reanalysis is necessary (*see* **Notes 11** and **12**).

3.2. QuEChERS Sample Preparation

The QuEChERS method may be scaled appropriately to any subsample size shown to be adequately representative of the original sample. If LVI is not used for GC–MS, then 12 g or more must be extracted to typically detect <10 ng/g of the pesticides in food. The method is designed for samples with >75% moisture. If needed, add water to hydrate drier samples so that moisture becomes approx 80% and pores in the sample are more accessible to the extraction solvent. The following instructions are scaled for 15-g samples (after hydration, if needed) extracted in 50-mL FEP centrifuge tubes. *Safety note*: Work with pesticides and solvents in a hood and wear appropriate laboratory safety glasses, coat, and gloves; ensure that the centrifuge is balanced and do not exceed the safety limits of the tubes or rotors used.

3.2.1. Extraction and Cleanup

1. Weigh 15 g sample into each tube (use 13 mL water for a reagent blank).
2. Weigh 15 g blank(s) to attain enough extract for five matrix-matched calibration standards as described in **Subheadings 3.2.2.** and **3.2.3.** Add 75 µL working standard pesticides solution to an additional matrix blank (this will yield 200 ng/g) as a quality control (QC) spike for evaluating recoveries.
3. Add 15 mL 1% HAc in MeCN into each tube using the solvent dispenser.
4. Add 150 µL of ISTD solution (this will yield 200 ng/g) to samples, reagent blank, and QC spike, but not to blank(s) used for matrix-matched calibration standards (*see* **Note 13**).
5. Add 6 g anhydrous $MgSO_4$ plus 1.5 g anhydrous NaAc (or 2.5 g NaAc•$3H_2O$) to all tubes (the extract will reach 40–45°C) and seal the tubes well (ensure that powder does not get into the screw threads or rim of the tube).
6. Shake the tubes vigorously by hand for 1 min (using a motion from the arms more than the wrist) with 3–5 tubes at once in each hand, ensuring that the solvent interacts well with the entire sample and that crystalline agglomerates are broken up sufficiently during shaking. (*see* **Note 14**).
7. Centrifuge the tubes at more than 3000*g*. The greater the force, the better for forming a solid sample plug and providing potentially cleaner extracts.
8. Transfer needed amount (1–8 mL) of the MeCN extract (upper layer) at RT to the dispersive-SPE tubes containing 50 mg PSA (and C18 for fatty samples) plus 150 mg anhy-

drous $MgSO_4$ per milliliter extract. For matrix blanks to be used for the five matrix-matched calibration standards, first combine the blank extracts (if multiple blanks were extracted), then either transfer the needed amounts (1–8 mL) into separate dispersive SPE tubes as with the sample extracts or proportionately scale up the dispersive SPE step to obtain the extract volume needed for the standards after cleanup (*see* **Subheadings 3.2.2.** and **3.2.3.** for further explanation).

9. Seal the tubes well and mix by hand (or use a vortex mixer) for approx 30 s.
10. Centrifuge the dispersive SPE tubes at more than 3000*g*.

3.2.2. Options for Handling Extracts for Analysis

Depending on the LOQ needed, the chosen pesticide analytes, and analytical instruments and techniques used, 1–8 mL of the extract will be taken for dispersive-SPE cleanup. This cleanup technique loses half of the extract volume to the powder reagents, and the extraction method yields 1-g/mL equivalent sample concentrations. For GC–MS, approx 8 mg should be injected to generally achieve an LOQ below 10 ng/g, assuming that matrix interferences are not the limiting source of noise. If this degree of sensitivity is needed, then either LVI (e.g., 8-µL injection) must be used or the extracts must be concentrated. LVI is the simpler option, but if such a device is not available on the GC instrument (or it does not provide acceptable results for certain pesticide analytes), then splitless injection of the concentrated extract is the remaining option. When performing the MeCN evaporation step in this option, it is convenient to exchange solvent to toluene, which acts as a good keeper for the pesticides and has benefits in traditional GC analysis (e.g., smaller vaporization expansion volume). Further details, including a comparison of GC injection solvents, are provided elsewhere for this application *(13,39)*.

In **Option A**, if the desired LOQ can be achieved in GC with injection of the MeCN extract (using LVI or not), then a 1-mL aliquot is taken to minimize reagent costs (or a larger volume is taken, and the procedure is scaled up appropriately at slightly greater materials cost). In **Option B**, if direct injection of the MeCN extract in GC cannot achieve the necessary LOQ using the available instrumentation, then 8 mL is taken for dispersive SPE cleanup, and an extract concentration and solvent exchange step is performed prior to GC analysis (LC injection volume can be more easily increased, thus extract concentration is less of an issue in that case). Each of these options is described as follows:

Option A. Use 1 mL extract in **step 8**, and then after **step 10**:

11a. Transfer 500 µL of the final extracts from the dispersive SPE tubes (or five 500-µL aliquots of the combined matrix blank extract after dispersive SPE) to autosampler vials for (LVI) GC–MS.
12a. Add 50 µL of the 2-ng/µL TPP working solution at RT to all extracts (to make 200-ng/g equivalent concentration and 0.09% HAc, which improves stability of certain pesticides).
13a. Add 25 µL of MeCN to all sample extracts, the QC spike, the reagent blank, and the zero standard (to compensate for the volume to be added to the calibration standards in the next step).
14a. Follow procedures described in **Subheading 3.2.3., Option A**, for the four matrix blank extracts to be used for matrix-matched calibration standards (*w, x, y,* and *z* standards).

15a. Cap and shake the vials to mix solutions, then uncap them.

16a. Transfer 150 μL of the extracts from each vial to a counterpart LC autosampler vial into which 0.45 mL of 6.7 m*M* formic acid solution has been added (this is done to match the organic solvent and formic acid contents in the initial LC mobile phase of 5 m*M* formic acid in 25% MeOH).

17a. Cap all vials and conduct (LVI) GC–MS and LC–MS/MS analytical sequences according to **Subheadings 3.3.** and **3.4.**

Option B. Use 8 mL extract in **step 8**, and then after **step 10**:

11b. Transfer 250 μL of the MeCN extracts from the dispersive SPE tubes (or five 250-μL aliquots of the combined matrix blank extract after dispersive SPE) to autosampler vials for LC–MS/MS.

12b. Add 25 μL of the 2-ng/μL TPP working solution at RT to all vials and 12.5 μL of MeCN to all sample extracts, the QC spike, the reagent blank, and the zero standard.

13b. Follow procedures described in **Subheading 3.2.3.**, **Option B**, for the four matrix blank extracts to be used for the *w*, *x*, *y*, and *z* standards.

14b. Add 860 μL of 6.7 m*M* formic acid solution to achieve the acid concentration and organic solvent content at the initial LC mobile phase and cap all vials.

For evaporation and solvent exchange to toluene for GC–MS (without LVI):

15b. Transfer 4 mL of each extract (or five 4-mL aliquots of the combined matrix blank extract after dispersive SPE) to 10- to 15-mL graduated centrifuge tubes containing 1 mL of toluene and 400 μL of the 2-ng/μL TPP working solution added at RT.

16b. Evaporate the extracts at 50°C and sufficient N_2 gas flow until volume is 0.3–0.5 mL.

17b. Follow procedures described in **Subheading 3.2.3.**, **Option B**, for the four matrix blank extracts to be used for the *w*, *x*, *y*, and *z* standards.

18b. Add toluene to take each extract up to the 1-mL mark

19b. Add anhydrous $MgSO_4$ to reach the 0.2-mL mark on the tube and swirl to rinse above the 6-mL mark.

20b. Centrifuge the tubes at more than 600*g*.

21b. Transfer ≈0.6 mL of the final extract to the GC autosampler vials, and cap all vials.

22b. Conduct (LVI/)GC/MS and LC/MS-MS analytical sequences according to **Subheadings 3.3.** and **3.4.**

3.2.3. Preparation of Matrix-Matched Calibration Standards

The concentration range of the matrix-matched calibration standards is to be decided by the analyst, and these concentrations are listed as *w*, *x*, *y*, and *z* (given as nanograms-per-gram equivalent concentrations with respect to the original sample). As an example, if the LOQ of the method is 10 ng/g, then the four suggested concentrations of the standards are 10, 50, 250, and 1250 ng/g. In continuation of the procedures above, the instructions for the preparation of the matrix-matched calibration standards are as follows:

Option A. If 1- to 2-mL aliquots of the extracts are taken for dispersive SPE in **step 8**, then only a single 15-g matrix blank is typically needed to provide enough extract for the zero, *w*, *x*, *y*, and *z* standards. For the 0.5-g equivalent extracts described in **step 14a**, add 25 μL of the respective calibration standard spiking solution (*w*, *x*, *y*, and *z*) at RT to the appropriate four matrix blank extracts (*w*, *x*, *y*, and *z* standards). Similarly, if

2-mL aliquots are taken in **step 8,** then 1-mL extracts are to be transferred in **step 11a,** in which case add 50 µL of the respective calibration standard spiking solution (*w*, *x*, *y*, and *z*) to the appropriate four matrix blank extracts (*w*, *x*, *y*, and *z* standards) in **step 14a.**

Option B. At least 22 mL of matrix blank extract is needed after dispersive SPE cleanup (or ≥44 mL of initial extract) to prepare the zero, *w*, *x*, *y*, and *z* standards. Depending on the matrix and water content, a 15-g sample will typically yield 11 mL MeCN extract after centrifugation, thus four (but maybe five) 15-g blank samples need to be extracted. For the *w*, *x*, *y*, and *z* standards in LC–MS/MS described in **step 13b,** add 12.5 µL of the respective calibration standard spiking solutions *w*, *x*, *y*, and *z* at RT. For the *w*, *x*, *y*, and *z* standards in toluene for GC–MS as described in **step 17b,** add 200 µL of the respective calibration standard spiking solution (*w*, *x*, *y*, and *z*) at RT. The calibration standard spiking solutions for GC in this case should preferably be in toluene. If the spiking solution is in MeCN, then 200 µL MeCN should also be added to the other extracts in **step 18b.** Be aware that the presence of 20% MeCN may lead to poor chromatography, and MeCN should not be added if an N-sensitive GC detector (e.g., nitrogen–phosphorus detector) is used without a detector bypass vent.

3.3. Analysis of GC-Amenable Pesticides

Generic conditions are given next and in **Table 2** for the GC–MS analysis of selected pesticides from the list in **Table 1.** The analyst may use many different sets of conditions that offer equally valid results in the separation and detection of pesticides of their particular interest. In fact, the analyst should optimize the given conditions to yield the lowest LOQ for their chosen analytes in the shortest amount of time. The selected ions for quantitation and identification should be made to maximize S/N ratios of the analytes while avoiding matrix interferences. Information about the expected retention times (t_R) and intense ions in the mass spectra for hundreds of pesticides are listed elsewhere *(1,7,8).* Commercial mass spectral libraries (e.g., National Institute of Standards and Technology [NIST] and Wiley) also contain the EI spectra of hundreds of pesticides, which can help determine their t_R and choose quantitation masses when optimizing the GC conditions.

Otherwise, the way to determine the t_R and mass spectrum for a pesticide is to inject > 1 ng and look for the peak(s). The presence of the molecular ion (M^+) in the spectrum helps ensure that the pesticide does not degrade during injection, and if no library spectrum is available, it should be verified that the spectrum makes sense relative to the structure of the pesticide. In general, the analyst should choose the ion(s) for quantitation with the highest intensity at higher mass, but all selections should be verified to meet LOQ requirements in the matrix(es) of interest. Proper choice of quantitation ions can often substantially reduce LOQ, especially in complex backgrounds.

For extracts in MeCN, inject only as much as needed to achieve the LOQ desired in the analysis. Split mode (e.g., 10:1 split ratio for a 1-µL injection) may be all that is needed for applications designed to detect pesticides >1 µg/g in the samples, but LVI is required for maximal sensitivity. In most applications, 8 mg equivalent sample injected onto the column should be sufficient to achieve <10 ng/g LOQ for most pesticides. This would necessitate 8-µL injection in LVI of the 1-g/mL MeCN extracts from **Subheading 3.2.2., Option A.** In this case, it is suggested to program the pro-

Table 2
Conditions for the GC–MS Analysis of Selected Analytes, TPP, and the ISTD Using the Generic GC Method Described in the Text

Analyte	t_R (min)	Quant. ion(s) (m/z)	Analyte	t_R (min)	Quant. ion(s) (m/z)
Methamidophos	5.95	94,95, 141	Dichlorobenzophenone	13.85	139, 250
Dichlorvos	6.02	109,185	Cyprodinil	14.39	224
Propoxur	8.96	110,152	Penconazole	14.57	213,248
Ethoprophos	9.22	158,200	Tolylfluanid	14.68	137,181,238
Hexachlorobenzene	10.15	282,284,286	Heptachlor epoxide	14.80	237,353
Lindane	10.79	181,183,219	cis-Chlordane	15.80	237,272,375
Diazinon	10.89	179,276,304	p,p′-DDE	16.47	246,316,318
Chlorothalonil	11.26	264, 266,268	Dieldrin	16.63	263,277
Chlorpyrifos-methyl	12.17	199,286,288	Endosulfan sulfate	18.27	237,272,387
Carbaryl	12.49	115,116,144	TPP (QC)	18.50	326
Dichlofluanid	13.20	123, 224	cis-Permethrin	20.54	183
Chlorpyrifos	13.44	199,314,316	trans-Permethrin	20.66	183
d_{10}-Parathion (ISTD)	13.61	301	Coumaphos	20.73	226,334,362

Quant., quantitation.

grammable temperature vaporizer to start at 75°C for 3 min followed by a 200°C/min temperature ramp to 275°C. A solid plug of Carbofrit or inert sorbent is typically needed to contain the liquid solvent during the vaporization process in LVI *(12,40,41)*. For the 4-g/mL extracts in toluene, 2-µL splitless injection at 250°C should be satisfactory to achieve <10 ng/g LOQ. It is recommended to use a 1–5-m phenylmethyldeactivated guard column as a retention gap to minimize solvent condensation effects and better protect the analytical column. This also serves to reduce the effect of GC maintenance on the t_R of the analytes because the retention gap is shortened, not the analytical column.

For GC, set helium head pressure on the column at 10 psi or flow to 1 mL/min with systems capable of electronic pressure/flow control. After an appropriate time for solvent delay (e.g., approx 1.5 min in splitless and approx 4 min in LVI), a generic oven temperature program for MeCN extracts is 75°C initial temperature ramped to 175°C at 25°C/min, then to 225°C at 5°C/min, followed by a 25°C/min ramp to 290°C, at which it is held for 10 min. In the case of toluene injections, the initial oven temperature should be increased to 100°C and everything else kept the same. Of course, other temperature programs may be used, but in any case, peak shapes should be Gaussian, and peak widths at half-heights should be less than 5 s.

For the MS, the analyst should follow the instructions provided by the instrument manufacturer to optimize the system for detection of the pesticide analytes. Prior to injection of the sequence, a system suitability test should be made, such as autotuning of the MS, to help ensure that analytical quality is acceptable *(15,16,42)*. It is suggested that the calibration standards be dispersed throughout the sequence to demonstrate adequate instrument performance over the entire time-frame that the samples are injected.

3.4. Analysis of LC-Type Pesticides

In the case of LC–MS/MS, the analyst should follow the instrument manual guidelines to set the ion source temperature, gas flows, voltage potentials, and other general parameters for the particular instrument and analytical needs. Based on the LC conditions described below, the MS/MS detection parameters for each analyte should be optimized by using a syringe pump to infuse approx 1 ng/µL of the pesticide in 5 mM formic acid in 1/1 MeOH/water solution at 0.3 mL/min into the source. Most of the pesticides will ionize well in ESI-positive (+) mode, thus the M^{+1} mass spectral peak should first be optimized, and then conditions for collision-induced dissociation to maximize the S/N ratio of the first MS/MS transition should be determined. Most instruments have automated programs to optimize the parameters with little analyst intervention. It is not difficult also to test the signals in negative mode for comparison purposes, but nearly all of the LC-type analytes listed in **Table 1** ionize sufficiently well in ESI⁺. In some cases, Na⁺ adducts with the ionized pesticides form in the ion source. This is not necessarily problematic if all of the analyte generates the adduct and if the result is quantitatively reproducible.

Each instrument will give somewhat different optimized settings, even for the same model, but sensitivity will not typically be a problem. With the latest instruments,

Table 3
Conditions for the LC–MS/MS Analysis of Selected Pesticides in a Triple Quadrupole Instrument Using ESI$^+$ Mode at the LC Conditions Given in the Text

Analyte	t_R (min)	Precursor ion (*m/z*)	Product ion (*m/z*)	Collision energy (V)
Methamidophos	9.6	141.8	112.0	17
Pymetrozine	10.0	217.9	105.0	27
Acephate	11.1	138.8	143.0	19
Carbendazim	11.9	191.8	160.0	25
Thiabendazole	13.2	201.8	174.9	37
Imidacloprid	15.9	255.9	209.0	21
Imazalil	16.3	296.8	159.0	31
Thiophanate-methyl	18.8	342.8	151.0	29
Dichlorvos	19.0	220.7	127.0	23
Carbaryl	19.3	202.2	145.0	13
Dichlofluanid	20.9	332.7	223.8	17
Ethoprophos	21.2	242.8	173.0	21
Cyprodinil	21.2	225.9	108.0	35
Tolylfluanid	21.3	346.7	237.9	15
Penconazole	21.5	283.8	159.0	39

LOQ < 10 ng/g can typically be achieved with a 10-µL injection volume for the extracts of 0.25-g/mL equivalent sample concentrations. The water and acid contents of the extract closely match the initial LC mobile phase; thus, a larger volume may be injected without seriously affecting peak shapes. Therefore, a lower LOQ can usually be achieved by injecting a larger volume of the extract if needed.

As an example, generic LC conditions for the analysis of multiple pesticide residues are as follows: 0.3-mL/min flow rate; reservoirs to contain 5 m*M* formic acid in (1) water and (2) MeOH; gradient program of 25% solution B ramped to 100% linearly over 15 min, then held for an additional 15 min. After 30 min, the flow can be increased to 0.5 mL/min and the mobile phase returned to 75% solution A over the course of 2 min and allowed to equilibrate for 6 min. A divert valve should be placed between the column outlet and ion source to eliminate the introduction of salts and early eluting matrix components into the MS instrument before the t_R of the first analyte and any coextracted matrix components that may elute after the last pesticide of interest. Some pesticides may give broad or dual peaks if the mobile phase pH is not acidic enough. In this case, the formic acid content may be increased in an attempt to provide better chromatographic peak shapes. **Table 3** lists sample conditions for 15 selected pesticides in LC–MS/MS.

Just as with GC–MS, QA protocols should be followed and system suitability tests conducted prior to analyzing a sequence of samples. Regular preventive maintenance must be done to ensure adequate operation of the instruments. Inject the matrix blank to determine if a significant interferant is present at the t_R of the analytes. No evidence of carryover should be present in the reagent blank, which should be injected after the most highly concentrated standard in the sequence.

3.5. Data Analysis

Quantitation is based on least-linear-squared calibration of analyte peak areas divided by the ISTD peak areas plotted vs analyte concentration. The ratio of analyte peak area to ISTD peak area becomes the signal S. The analyte concentrations in the matrix-matched standards on a per-sample basis (nanogram per gram) can be determined by multiplying the volume (microliters) added to the extract by the analyte concentrations in the added solutions (nanograms per microliters) and dividing by the equivalent amount (grams) of sample in the extract. The concentrations (in nanograms per gram) C of the pesticide analytes in the samples and QC matrix spike are determined from the equation

$$C = (S - y \text{ intercept})/\text{Slope}$$

If there are no interferences, the y-intercept should be nearly zero, and the correlation coefficient of the slope should be >0.99. In some circumstances, a nonlinear relationship occurs in the calibration plot and a quadratic best-fit curve may provide better correlation and results. The TPP is a QC measure to isolate the variability of the analytical step from the sample preparation method. The volumes of the final extracts and each preceding step are carefully controlled in the sample preparation protocol, and ideally the ISTD would not need to be used to achieve equally accurate results. Pipets, syringes, and balances should be periodically calibrated to ensure accuracy. However, random and systematic errors in volumetric transfers are inherent in analytical methods, and the ISTD should improve the accuracy of the results. The recoveries of the ISTD can be assessed by comparing the peak areas of the ISTD in the samples with those from the calibration standards. The TPP/ISTD peak area ratio should remain consistent ($<10\%$ relative standard deviation) in the method, and if any extract gives a substantially different ratio from the others, then the results from that extract should be questioned. Furthermore, if the QC spike yields recoveries $<70\%$ or $>120\%$, then the results from all the samples should be questioned. If all pesticide recoveries outside the acceptable range are the same, then a systematic bias is indicated. If variable recoveries are obtained, then a systematic bias is not likely to be the source of the problem. Many pesticides can be analyzed by both LC–MS/MS and GC–MS, and the comparison of their results from both distinct methods can be invaluable in isolating any problems that may occur with one of the instruments or techniques.

4. Notes

1. NaAc•3H$_2$O may be substituted for anhydrous NaAc, but 1.7 g per gram sample must be used rather than 1 g anhydrous NaAc per gram sample.
2. Heat bulk quantities of anhydrous MgSO$_4$ to 500°C for longer than 5 h to remove phthalates and any residual water prior to its use in the laboratory, but this is not critical.
3. Aminopropyl SPE sorbent can be substituted for PSA, but 75 mg per milliliter of extract should be used.
4. Toluene is the most suitable solvent for long-term storage (>10 yr) of pesticide stock solutions in general because of its slower evaporation rate, miscibility with MeCN, and the higher solubility and stability of pesticides in toluene compared to other solvents. However, not all pesticides can be dissolved at 10 mg/mL in toluene. In these cases,

MeCN, acetone, MeOH, or ethyl acetate should be used, but long-term stability may become an issue, and old solutions should be replaced more often.

5. The choice of ISTD is very important because it must not already be present in the sample, but be completely recovered in the method for detection in both GC–MS and LC–MS/MS (otherwise, a pair of ISTDs may be used). A relatively inexpensive deuterated pesticide (d_{10}-parathion) was chosen as the ISTD in this protocol. Deuterated chlorpyrifos or chlorpyrifos-methyl are more suitable for a greater range of selective GC detectors, but they are more expensive. It is also possible to use an uncommon compound as the ISTD, but its suitability would have to be determined.

6. In this protocol, 250 is the maximum number of pesticides that can be added to make this solution, which will consist of 100% toluene if the stock standards are all in toluene. This is not ideal for spiking of the sample or preparation of calibration standard spiking solutions. The stock standards can be prepared in MeCN, but some pesticides will have reduced stability, which is significantly improved in 0.1% HAc solution, but degradation is not eliminated *(39)*. An alternative approach is to prepare mixtures of pesticides in stock solutions by dissolving multiple pesticide reference standards in the same vial.

7. The TPP working standard is prepared in MeCN with 1% HAc because when it is added to the MeCN extracts in **steps 12a** or **12b** and **15b**, the extract will contain 0.09% HAc for improved stability of certain pesticides (e.g., chlorothalonil, captan, folpet, tolylfluanid, dichlofluanid, carbaryl).

8. To speed the process greatly, the density of the powders can be determined and scoops made of the appropriate volume, but weighing should still be done to check consistency (reagent weights ± 5% deviation from the stated amount are acceptable). The containers should be sealed during storage and can be refilled and reused without cleaning between uses. A commercial product (#CUMPSC2CT) containing 50 mg of PSA + 150 mg anh. MgSO$_4$ is available from United Chemical Technologies.

9. If samples contain ≥ 1% fat, add the same amount of C18 sorbent as PSA (in addition to PSA) in centrifuge tubes for dispersive SPE. As the fat content increases, a third phase (lipids) will form in the extraction tube, and the recoveries of nonpolar pesticides will decrease as they partially partition into the lipid layer. The most nonpolar pesticides (e.g., hexachlorobenzene, DDE) will give < 70% recovery at approx 5% fat content, but relatively polar GC-amenable and LC-type pesticides are completely recovered at >15% fat *(14)*. A commercial product (#CUMPSC18CT) containing 50 mg each of PSA and C18 + 150 mg anh. MgSO$_4$ is available from United Chemical Technologies.

10. If none of the analytes have planar structures, then GCB can be used to provide additional cleanup, especially for removal of chlorophyll, sterols, and planar matrix coextractives. In this case, add the same amount of GCB as PSA and C18 (50 mg each per 1 mL of extract) in centrifuge tubes for dispersive-SPE. Planar pesticides include terbufos, thiabendazole, hexachlorobenzene, and quintozene, among many others *(11)*.

11. The advantages of this approach include (1) the extracted portion is highly representative of the initial sample; (2) the sample is well comminuted to improve extraction by shaking rather than blending; (3) less time is spent on the overall homogenization process than trying to provide equivalent homogenization of the large initial sample using the chopper alone; and (4) a frozen subsample is available for reanalysis if needed. The sample homogenization step is a critical component in the overall sample preparation process; unfortunately, many analysts do not pay adequate attention to this important step. If the sample is not homogenized properly, then the analytical results will not be as accurate as they could be, independent of the performance of the sample preparation and analytical steps.

12. To provide the most homogeneous comminuted samples, frozen conditions, sufficient chopping time, and appropriate sample size to chopper volume should be used. Use of frozen samples also minimizes degradative and volatilization losses of certain pesticides (e.g., dichlorvos, chlorothalonil, dichlofluanid). If best results of susceptible pesticides are needed, then cut the food sample into 2- to 5-cm³ portions with a knife and store the sample in the freezer prior to processing. Cryogenic blending devices, liquid nitrogen, or dry ice may also be used (but make sure all dry ice has sublimed before weighing samples and ensure that water condensation is minimal, especially in a humid environment). For further information about sample processing in pesticide residue analysis of foods, the analyst should refer to several publications on the topic *(43–48)*.

13. An uncommon or deuterated pesticide standard may be spiked into the sample during homogenization to determine the effectiveness of the procedure through the measurement of recovery and reproducibility using the technique and specific devices. For typical applications, the recovery should be > 70%, with relative standard deviation < 20% for a 100- to 500-ng/g fortification level.

14. Alternately, do not seal the tubes and use a probe blender for extraction, taking care not to overheat the extract. Another option is to extract using sonication. These stronger measures may be needed to ensure that any bound residues are extracted. Fruits, vegetables, and other high-moisture samples do not typically interact strongly with the residues, and shaking alone is usually acceptable for extraction of nearly all pesticides. However, dry or porous/sorptive sample types, such as grains and soils, require blending, higher temperature, more acidic or basic conditions, or more time to completely extract those residues prone to strong matrix interactions.

Acknowledgment

Mention of brand or firm name does not constitute an endorsement by the US Department of Agriculture above others of a similar nature not mentioned.

References

1. Food and Drug Administration. (1999) *Pesticide Analytical Manual Volume I: Multiresidue Methods*, 3rd ed., US Department of Health and Human Services, Washington, DC. Available at: http://www.cfsan.fda.gov/~frf/pami3.html
2. Luke, M. A., Froberg, J.E., and Masumoto, H. T. (1975) Extraction and cleanup of organochlorine, organophosphate, organonitrogen, and hydrocarbon pesticides in produce for determination by gas–liquid chromatography. *J. Assoc. Off. Anal. Chem.* **58**, 1020–1026.
3. Specht, W. and Tilkes, M. (1980) Gas chromatographische bestimmung von rückständen an pflanzenbehandlungsmitteln nach clean-up über gel-chromatographie und mini-kieselgel-säulen-chromatographie. *Fresenius J. Anal. Chem.* **301**, 300–307.
4. Lee, S. M., Papathakis, M. L., Hsiao-Ming, C. F., and Carr, J. E. (1991) Multipesticide residue method for fruits and vegetables: California Department of Food and Agriculture. *Fresenius J. Anal. Chem.* **339**, 376–383.
5. Andersson, A. and Pålsheden, H. (1991) Comparison of the efficiency of different GLC multi-residue methods on crops containing pesticide residues. *Fresenius J. Anal. Chem.* **339**, 365–367.
6. Cook, J., Beckett, M. P., Reliford, B., Hammock, W., and Engel, M. (1999) Multiresidue analysis of pesticides in fresh fruits and vegetables using procedures developed by the Florida Department of Agriculture and Consumer Services. *J. AOAC Int.* **82**, 1419–1435.

7. General Inspectorate for Health Protection. (1996) *Analytical Methods for Pesticide Residues in Foodstuffs*, 6th ed., Ministry of Health Welfare and Sport, The Netherlands.
8. Fillion, J., Sauvé, F., and Selwyn, J. (2000) Multiresidue method for the determination of residues of 251 pesticides in fruits and vegetables by gas chromatography/mass spectrometry and liquid chromatography with fluorescence detection. *J. AOAC Int.* **83,** 698–713.
9. Sheridan, R. S. and Meola, J. R. (1999) Analysis of pesticide residues in fruits, vegetables, and milk by gas chromatography/tandem mass spectrometry. *J. AOAC Int.* **82,** 982–990.
10. Lehotay, S. J. (2000) Determination of pesticide residues in nonfatty foods by supercritical fluid extraction and gas chromatography/mass spectrometry: collaborative study. *J. AOAC Int.* **83,** 680–697.
11. Anastassiades, M., Lehotay, S. J., Stajnbaher, D., and Schenck, F. J. (2003) Fast and easy multiresidue method employing acetonitrile extraction/partitioning and "dispersive solid-phase extraction" for the determination of pesticide residues in produce. *J. AOAC Int.* **86,** 412–431.
12. Lehotay, S. J., Hiemstra, M., van Bodegraven, P., and de Kok, A. (2005) Validation of a fast and easy method for the determination of more than 200 pesticide residues in fruits and vegetables using gas and liquid chromatography and mass spectrometric detection. *J. AOAC Int.* **88,** 595–614.
13. Lehotay, S. J., Mastovská, K., and Lightfield, A. R. (2005) Use of buffering to improve results of problematic pesticides in a fast and easy method for residue analysis of fruits and vegetables. *J. AOAC Int.* **88,** 615–629.
14. Lehotay, S. J., Mastovská, K., and Yun, S.-J. (2005) Evaluation of two fast and easy methods for pesticide residue analysis in fatty food matrices. *J. AOAC Int.* **88,** 630–638.
15. Fajgelj, A. and Ambrus, Á. (eds.) (2000) *Principles and Practices of Method Validation*, Royal Society of Chemistry, Cambridge, UK, pp. 179–295.
16. Hill, A. R. C. and Reynolds, S. L. (1999) Guidelines for in-house validation of analytical methods for pesticide residues in food and animal feed. *Analyst* **124,** 953–958.
17. Stry, J. J., Amoo, J. S., George, S. W., Hamilton-Johnson, T., and Stetser, E. (2000) Coupling of size-exclusion chromatography to liquid chromatography/mass spectrometry for determination of trace levels of thifensulfuron-methyl and tribenuron-methyl in cottonseed and cotton gin trash. *J. AOAC Int.* **83,** 651–659.
18. Erney, D. R., Gillespie, A. M., Gilvydis, D. M., and Poole, C. F. (1993) Explanation of the matrix-induced chromatographic enhancement of organophosphorus pesticides during open tubular column gas chromatography with splitless or hot on-column injection and flame photometric detection. *J. Chromatogr.* **638,** 57–63.
19. Erney, D. R. and Poole, C. F. (1993) A study of single compound additives to minimize the matrix induced chromatographic response enhancement observed in the gas chromatography of pesticide residues. *J. High Resolut. Chromatogr.* **16,** 501–503.
20. Erney, D. R., Pawlowski, T. M., and Poole, C. F. (1997) Matrix-induced peak enhancement of pesticides in gas chromatography: is there a solution? *J. High Resolut. Chromatogr.* **20,** 375–378.
21. Schenck, F. J. and Lehotay, S. J. (2000) Does further clean-up reduce the matrix enhancement effect in gas chromatographic analysis of pesticide residues in food? *J. Chromatogr. A* **868,** 51–61.
22. Hajslová, J., Holadová, K., Kocourek, V., et al. (1998) Matrix-induced effects: a critical point in gas chromatographic analysis of pesticide residues. *J. Chromatogr. A* **800,** 283–295.

23. Hajslová, J. and Zrostlíková, J. (2003) Matrix effects in (ultra)trace analysis of pesticide residues in food and biotic matrices. *J Chromatogr A*. **1000**, 181–197.

24. Anastassiades, M., Mastovská, K., and Lehotay, S. J. (2003) Evaluation of analyte protectants to improve gas chromatographic analysis of pesticides. *J. Chromatogr. A* **1015**, 163–184.

25. Mastovská, K. and Lehotay, S.J. (submitted) Optimization and evaluation of analyte protectants in gas chromatographic analysis. *Anal. Chem.*

26. Mol, H. G., van Dam, R. C., and Steijger, O. M. (2003) Determination of polar organophosphorus pesticides in vegetables and fruits using liquid chromatography with tandem mass spectrometry: selection of extraction solvent. *J. Chromatogr. A* **1015**, 119–127.

27. Klein, J. and Alder, L. (2003) Applicability of gradient liquid chromatography with tandem mass spectrometry to the simultaneous screening for about 100 pesticides in crops. *J. AOAC Int.* **86**, 1015–1037.

28. Zrostlíková, J., Hajslová, J., Poustka, J., and Begany, P. (2002) Alternative calibration approaches to compensate the effect of co-extracted matrix components in liquid chromatography–electrospray ionisation tandem mass spectrometry analysis of pesticide residues in plant materials. *J. Chromatogr. A* **973**, 13–26.

29. Niessen, W. M. A. (ed.) (2001) *Current Practice of Gas Chromatography–Mass Spectrometry*, Dekker, New York.

30. Mastovská, K. and Lehotay, S. J. (2003) Practical approaches to fast gas chromatography–mass spectrometry. *J. Chromatogr. A* **1000**, 153–180.

31. Cochran, J. W. (2002) Fast gas chromatography–time-of-flight mass spectrometry of polychlorinated biphenyls and other environmental contaminants. *J. Chromatogr. Sci.* **40**, 254–268.

32. Niessen, W. M. (2003) Progress in liquid chromatography–mass spectrometry instrumentation and its impact on high-throughput screening. *J Chromatogr A* **1000**, 413–436.

33. Amirav, A. and Dagan, S. (1997) A direct sample introduction device for mass spectrometry studies and gas chromatography mass spectrometry analyses. *Eur. Mass Spectrom.* **3**, 105–111.

34. Lehotay, S. J. (2000) Analysis of pesticide residues in mixed fruit and vegetable extracts by direct sample introduction/gas chromatography/tandem mass spectrometry. *J. AOAC Int.* **83**, 680–697.

35. Patel, K., Fussell, R. J., Goodall, D. M., and Keely, B. J. (2003) Analysis of pesticide residues in lettuce by large volume–difficult matrix introduction–gas chromatography–time of flight–mass spectrometry (LV–DMI–GC–TOF–MS). *Analyst* **128**, 1228–1231.

36. Matisová, E. and Domotorová, M. (2002) Fast gas chromatography and its use in trace analysis. *J. Chromatogr A* **1000**, 199–221.

37. Amirav, A., Gordin, A., and Tzanani, N. (2001) Supersonic gas chromatography/mass spectrometry. *Rapid Commun. Mass Spectrom.* **15**, 811–820.

38. Mastovská, K., Lehotay, S. J., and Hajslová, J. (2001) Optimization and evaluation of low-pressure gas chromatography–mass spectrometry for the fast analysis of multiple pesticide residues in a food commodity. *J. Chromatogr. A* **926**, 291–308.

39. Mastovská, K. and Lehotay, S. J. (submitted) Evaluation of common organic solvents for gas chromatographic analysis and stability of multiclass pesticide residues. *J. Chromatogr. A.*

40. Martinez Vidal, J. L., Arrebola, F. J., and Mateu-Sanchez, M. (2002) Application to routine analysis of a method to determine multiclass pesticide residues in fresh vegetables by gas chromatography/tandem mass spectrometry. *Rapid Commun. Mass Spectrom.* **16**, 1106–1115.

41. Rosenblum, L., Hieber, T., and Morgan, J. (2001) Determination of pesticides in composite dietary samples by gas chromatography/mass spectrometry in the selected ion monitoring mode by using a temperature-programmable large volume injector with preseparation column. *J. AOAC Int.* **84,** 891–900.

42. Soboleva, E. and Ambrus, Á. (2004). Application of a system suitability test for quality assurance and performance optimisation of a gas chromatographic system for pesticide residue analysis. *J. Chromatogr. A* **1027,** 55–65.

43. Young, S. J., Parfitt, C. H., Jr., Newell, R. F., and Spittler, T. D. (1996) Homogeneity of fruits and vegetables comminuted in a vertical cutter mixer. *J. AOAC Int.* **79,** 976–980.

44. Lyn, J. A., Ramsey, M. H., Fussell, R. J., and Wood, R. (2003) Measurement uncertainty from physical sample preparation: estimation including systematic error. *Analyst* **128,** 1391–1398.

45. Hill, A. R. C., Harris, C. A., and Warburton, A. G. (2000) Effects of sample processing on pesticide residues in fruits and vegetables, in *Principles and Practices of Method Validation* (Fajgelj, A. and Ambrus, Á., eds.), Royal Society of Chemistry, Cambridge, UK, pp. 41–48.

46. Maestroni, B., Ghods, A., El-Bidaoui, M., et al. (2000) Testing the efficiency and uncertainty of sample processing, in *Principles and Practices of Method Validation* (Fajgelj, A. and Ambrus, Á., eds.), Royal Society of Chemistry, Cambridge, UK, pp. 49–88.

47. Lehotay, S. J., Aharonson, N., Pfeil, E., and Ibrahim, M. A. (1995) Development of a sample preparation technique for supercritical fluid extraction in the multiresidue analysis of pesticides in produce. *J. AOAC Int.* **78,** 831–840.

48. Fussell, R. J., Jackson Addic, K., Reynolds, S. L., and Wilson, M. F., (2002) Assessment of the stability of pesticides during cryogenic sample processing. 1. Apples. *J. Agric. Food Chem.* **50,** 441–448.

20

Determination of Organophosphorus Pesticide Residues in Vegetable Oils by Single-Step Multicartridge Extraction and Cleanup and by Gas Chromatography With Flame Photometric Detector

Alfonso Di Muccio, Anna M. Cicero, Antonella Ausili, and Stefano Di Muccio

Summary

The method presented is applicable to the determination of organophosphorus (OP) pesticide residues in vegetable oils. The method performs in a single step an on-column extraction and cleanup of OP pesticide residues by means of a system of three cartridges. A solution of 1 g oil in *n*-hexane is loaded into an Extrelut-NT3 cartridge (large-pore diatomaceous material). The OP pesticide residues are extracted by eluting the cartridge with 20 mL acetonitrile, which is cleaned up by passing through a silica and a C18 cartridge connected on-line to the Extrelut NT-3 cartridge. A few milligrams of lipid is carried over into the eluate, which after concentration and solvent exchange is directly amenable to determination by gas chromatography (GC) with flame photometric detector with optical filter for phosphorus compounds (FPD-P). Recovery values for 45 OP pesticide residues are reported at two spiking levels. In the lower concentration range tested (0.09–0.60 mg/kg), satisfactory results (74–86%) were obtained for 39 OP pesticides; exceptions were formothion (5%), disulfoton (32%), phosalone (54%), demeton-*S*-methyl sulfone (60%), fenthion (62%), and borderline phosphamidone (68%). In the higher concentration range tested (0.38–2.35 mg/kg), satisfactory results (82–109%) were obtained for 43 OP pesticides; exceptions were formothion (48%) and disulfoton (53%). Compared to conventional techniques, such as liquid–liquid partition and size-exclusion chromatography (SEC), the method described is faster and simpler, requires minimal solvents and essentially only disposable items, and does not require skilled operators or maintenance of costly apparatus.

Key Words: C18 cleanup; extraction; flame photometric detector; gas chromatography; multicartridge cleanup; organophosphorus pesticide residues; partition cleanup; solid–matrix partition; vegetable oil.

From: *Methods in Biotechnology, Vol. 19, Pesticide Protocols*
Edited by: J. L. Martínez Vidal and A. Garrido Frenich © Humana Press Inc., Totowa, NJ

1. Introduction

Organophosphorus pesticides (OP) are a class of compounds that includes derivatives of phosphoric, phosphorous, thiono-phosphoric, and thion-thiolo phosporic acids esterified with methyl or ethyl and different alcohol groups. Depending on the alcoholic moiety, the polarity of OP pesticides ranges from water-soluble compounds such as dimethoate to lipophilic compounds such as bromophos-ethyl. The majority of OP pesticides are insecticides with anticholinesterase activity. OP pesticides are used in agriculture, for the control of eso- and endoparasites of animals, and for civil uses.

Depending mainly on the chemical stability and the dose, time, and mode of application, pesticide residues may occur in crops and derived products. Regarding vegetable oils, residues of OP pesticides are likely to occur essentially in "virgin" olive oil because it is produced without chemical treatments. In contrast, seed oils usually undergo chemical refining, including alkaline treatment, which cleaves the ester bond of OP pesticides.

Extraction of OP pesticide residues from oils is usually carried out by (1) liquid–liquid partition in a separatory funnel between *n*-hexane and acetonitrile *(1,2)*, in which the OP pesticides partition into the acetonitrile phase, and the major part of the oil remains in the apolar phase (*n*-hexane), or (2) size-exclusion chromatography *(3–7)*, in which the major part of the lipid matrix is eluted in the excluded volume and is separated from the OP pesticides, which have relatively lower molecular masses than triglycerides.

The method described here is based on a separation of OP pesticide residues from lipid material by a liquid–liquid partition carried out on a disposable, ready-to-use cartridge filled with large-pore diatomaceous material (solid–matrix partition). A silica gel cartridge and a C18 cartridge, connected in series downstream of the partition cartridge, provide further removal of polar and apolar components. The identification and determination are carried out by gas chromatography (GC) with flame photometric detector with optical filter for phosphorus compounds (FPD-P).

The solid–matrix partition has been able to extract different classes of pesticide residues from fatty matrices *(8–12)*. Compared to conventional methods, it is simple and fast, requires a minimum of solvents and essentially only disposable items, and does not require skilled operators or maintenance of costly apparatus.

2. Materials

1. Extrelut NT-3 cartridge (e.g., E. Merck, cat. no. 1.15095.0001, Darmstadt, Germany).
2. Sep-Pak silica cartridge (e.g., Waters, part no. 51900, Milford, MA).
3. Sep-Pak C18 cartridge (e.g., Waters, part no. 51910).
4. Solvents: *n*-hexane, acetonitrile, methanol, acetone, pesticide residues grade, or analytical reagent grade redistilled from an all-glass apparatus.
5. Solvent mixtures: acetonitrile saturated with *n*-hexane, used for solid–matrix partition (in a separatory funnel with polytetrafluoroethylene stopcock, shake acetonitrile with some *n*-hexane and let phases separate; prepare daily some time before use to allow for phase separation); iso-octane plus acetone (80:20 v/v), used to dissolve the sample extracts for the GC analysis.
6. Triphenylphosphate (TPP) internal standard.

7. Gas chromatograph with twin injectors, twin columns, twin detectors, and typical instrument setup and operating conditions:

- autosampler/autoinjector
- split/splitless injector, operated in pulsed (25 psi for 2 min) splitless mode with a 1-min purge-off time, with a dual-tapered deactivated glass liner and temperature set at 240°C
- capillary, fused-silica columns, either SPB 1 or SPB 5, and SPB-1701, 30 m × 0.25 mm id × 0.25-µm film thickness, with an uncoated deactivated retention gap, 2.5 m × 0.25 mm id
- temperature program of the column oven: 60°C (2 min), 10°C/min to 160°C, then 3°C/min to 260°C, finally at 260°C (20 min); total time 65 min
- helium carrier gas supplied through electronic pressure control module at 1.5 mL/min (set at room temperature) in constant-flow mode
- FPD-P; temperature of detector base set at 250°C; nitrogen as auxiliary gas; hydrogen and airflow rates to the detector set according to the manufacturer's directions

3. Methods

3.1. Preparation of the Analytical Sample and the Reduced Analytical Sample

If the laboratory sample has been kept in a refrigerator, allow it to warm to room temperature. Shake the bottle or can so that the material inside can be considered homogeneous (*analytical sample*). If the laboratory sample is made up of several bottles or cans, take from each vessel an amount of material proportional to the content size. Combine the aliquots and mix thoroughly (*reduced analytical sample*).

3.2. Analytical Procedure

1. Weigh in a 10-mL test tube an amount of the analytical sample (or reduced analytical sample) close to 1 g (±0.01 g) (*test sample*). Add 1 mL *n*-hexane and mix well. Using a Pasteur pipet, transfer the solution into an Extrelut NT-3 cartridge. Apply the solution as close as possible to the upper surface of the Extrelut NT-3 cartridge. Avoid touching the inner wall of the cartridge. Let the solution drain into the filling material. Wash the test tube three times with 0.5 mL *n*-hexane and transfer the washings into the Extrelut NT-3 cartridge. Let the washings drain into the filling material. Wait 10 min to obtain an even distribution (*see* **Note 1**).
2. Connect the short end of a Sep-Pak silica cartridge to the Luer tip of the Extrelut NT-3 cartridge. Cut away 0.5 cm of the longer end of the silica cartridge. Then, connect a Sep-Pak C18 cartridge to the Sep-Pak silica cartridge using a short length of glass tube, typically 10 × 4 mm ed × 2 mm id (*see* **Fig. 1**). Cut the end of the C18 cartridge in a fluted mouth shape. Arrange for collection of the eluate in a preweighed (±0.0001 g) 50-mL Erlenmeyer flask. Elution is carried out under gravity alone (*see* **Note 2**).
3. Wash the test tube with five 1-mL portions of acetonitrile saturated with *n*-hexane and transfer the washings into the Extrelut NT-3 cartridge. Elute the system of the three combined cartridges with three additional 5-mL portions of acetonitrile saturated with *n*-hexane. For both washings and eluting solvent, apply each portion after the previous one has just disappeared into the filling material. If elution does not start spontaneously, it is possible to force the flow with slight pressure (*see* **Note 3**).

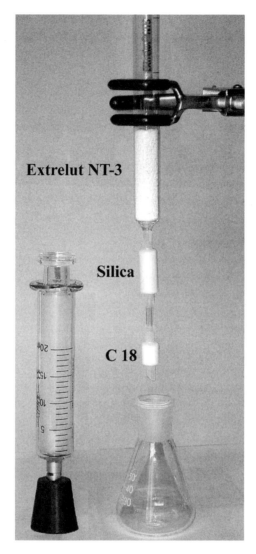

Fig. 1. Assembly of the three cartridges for extraction and cleanup and syringe assembly to force the flow temporarily.

4. Add 4 mL methanol to the combined eluates and concentrate cautiously to dryness by rotatory evaporator (55°C bath temperature, reduced pressure) (*see* **Note 4**).
5. To check the efficiency of the lipid removal, weigh the Erlenmeyer flask and calculate by difference the mass of lipid residue. For olive oil, it is generally a few milligrams and is compatible with the GC injection systems (*see* **Note 5**).
6. Dissolve the residue in 1 mL iso-octane plus acetone (80:20 v/v) and add the internal standard solution.

3.3. Identification

Analyze by GC/FPD-P by injecting into two GC columns of different polarity. The minimum identification criterion is based on matching the retention times of the analytes (both absolute and relative to the internal standard) in the sample run with those obtained in the calibrant solution run in the same sequence of analyses. Whenever possible, a confirmation by GC–mass spectrometry (GC–MS) or liquid chromatography–mass spectrometry (LC–MS) should be carried out.

3.4. Determination and Calculation

On both GC columns, the quantitation of each compound is carried out by comparing its peak area in the sample run to that in the standard solution run. The working standard solution used for quantitation should contain an amount of sample residue comparable to that in the extract of the sample to be analyzed (matrix-matched calibrant) (*see* **Note 6**).

Quantitation can be carried out according to a calibration curve, by interpolation between the responses of two calibrant solutions that encompass the response in the sample run, or by proportion to the response of a calibrant solution comparable within ±50% to that in the sample run (±10% if the level of the determination is close to the maximum residue limit). The concentration of the pesticide in the sample is given by the formula

$$mg/kg = \mu g/g = [(Ra,i/IS)_s \times (R\mu g,i/IS)_c/(Ra,i/IS)_c] \times [(\mu g_{IS})_s/g]$$

where $(Ra,i/IS)_s$ and $(Ra,i/IS)_c$ are the ratios of the peak area of the *i*th compound to the area of ISTD in the sample and in the calibrant solution, respectively; $(R\mu g,i/IS)_c$ is the ratio of the amount of the *i*th compound to the amount of ISTD in the calibrant solution; $(\mu g_{IS})_s$ is the amount in μg of ISTD in the sample solution; and g is the amount in grams of the test sample in the sample solution. Usually, the result is not corrected for the recovery value.

3.5. Checking the Recovery of the Method

Prepare *spiked samples* as follows: Analyze the sample of oil to be spiked and check the absence of interferences with the compounds to be determined. Samples spiked at different levels can be obtained by adding to a test tube 1 g oil and 1 mL standard solution in *n*-hexane at appropriate concentration and then processing the spiked sample as described in the method; otherwise, in a 10-mL test tube, add 1 g oil and the following volumes of standard solution:

Standard solution		Compound	Spiking level
(µL)	(µg/mL)	(µg)	(mg/kg)
5	10	0.05	0.05
10	10	0.1	0.1
20	10	0.2	0.2
50	10	0.5	0.5

Table 1
Mean (N = 6) Recovery Values of 45 Organophosphorus Pesticides From 1 g Olive Oil Spiked at Two Different Levels (ISO Names Used)

Pesticide	p-value[a]	Spiking level (mg/kg)		Recovery (%), mean (N = 6) ± SD	
		L[b]	H[c]	L[b]	H[c]
Azinphos-Et	<0.01	0.20	0.77	77.2 ± 5.7	98.2 ± 4.7
Azinphos-Me	<0.01	0.49	1.92	77.5 ± 5.1	91.8 ± 3.7
Bromophos	0.26	0.12	0.60	95.5 ± 3.5	94.1 ± 3.7
Bromophos-Et	0.36	0.15	0.58	79.5 ± 5.2	96.2 ± 4.3
Cadusafos	0.30	0.21	1.03	99.9 ± 4.1	93.3 ± 4.2
Carbophenothion	0.15	0.15	0.60	98.1 ± 6.2	95.6 ± 4.8
Chlorfenvinphos	0.05	0.21	0.79	81.6 ± 3.9	98.4 ± 5.4
Chlorpyrifos-Et	0.25	0.10	0.46	97.0 ± 6.2	103.0 ± 5.7
Chlorpyrifos-Me	0.13	0.15	0.76	98.2 ± 4.7	101.1 ± 5.0
Coumaphos	0.01	0.11	0.58	92.7 ± 4.8	93.1 ± 4.0
		0.60	2.35	80.5 ± 4.3	100.0 ± 5.2
DDVP	0.04	0.24	1.17	97.0 ± 3.4	87.4 ± 4.7
Demeton-S-methyl sulfone	<0.01	0.41	1.60	59.7 ± 4.8	81.6 ± 4.8
Diazinon	0.24	0.11	0.53	97.9 ± 2.8	97.8 ± 3.8
Diazoxon	0.06	0.10	0.38	74.4 ± 4.7	95.7 ± 5.3
Dichlofenthion	0.26	0.10	0.39	81.7 ± 5.0	98.6 ± 4.5
Dimethoate	<0.01	0.10	0.40	94.6 ± 4.6	90.1 ± 4.4
Dioxathion	<0.01	0.16	0.80	93.4 ± 3.5	95.2 ± 3.7
Disulfoton	0.15	0.16	0.62	31.6 ± 4.1	53.0 ± 3.7
Ethion	0.11	0.09	0.45	96.0 ± 6.1	98.5 ± 5.6
Ethoprophos	0.17	0.15	0.76	98.6 ± 2.8	95.3 ± 3.9
Etrimfos	0.16	0.19	0.95	97.9 ± 5.8	96.3 ± 4.6
Fenamiphos	0.03	0.34	1.71	84.7 ± 5.5	99.3 ± 5.1
Fenchlorphos	0.19	0.19	0.73	81.1 ± 5.9	96.8 ± 3.7
Fenitrothion	0.02	0.24	0.93	80.2 ± 6.0	94.0 ± 3.5
Fenthion	0.06	0.12	0.47	62.3 ± 5.5	83.9 ± 4.7
Fenthoate	0.06	0.15	0.76	94.6 ± 5.2	93.3 ± 4.6
Fonofos	0.16	0.10	0.38	81.3 ± 3.9	96.6 ± 4.9
Formothion	<0.01	0.28	1.36	5.2 ± 4.7	47.7 ± 5.1
Heptenophos	0.04	0.10	0.38	85.9 ± 4.8	96.1 ± 4.2
Isofenphos	0.08	0.11	0.55	109.5 ± 4.6	97.8 ± 4.4
Malathion	0.03	0.19	0.93	97.8 ± 3.9	94.7 ± 4.1
Methacrifos	0.04	0.12	0.60	82.3 ± 6.2	94.7 ± 5.6
Methidathion	<0.01	0.26	0.99	84.6 ± 5.7	96.4 ± 5.1
Parathion-Et	0.03	0.18	0.69	80.8 ± 4.2	92.7 ± 4.7
Parathion-Me	<0.01	0.10	0.38	81.5 ± 3.2	92.9 ± 4.8
Phosalone	0.02	0.31	1.52	53.7 ± 3.4	91.3 ± 4.8
Phosphamidon (E + Z)	<0.01	0.39	1.54	68.2 ± 3.8	91.5 ± 4.8
Pirimiphos-Me	0.16	0.10	0.38	81.9 ± 3.4	96.7 ± 6.0
Pyrazophos	0.04	0.31	1.52	92.3 ± 4.2	98.9 ± 4.4
Pyridafenthion	0.01	0.33	1.67	88.1 ± 6.3	99.4 ± 5.8
Sulfotep	0.08	0.09	0.44	96.5 ± 3.3	94.4 ± 4.5

(continued)

Table 1 *(continued)*
Mean (N = 6) Recovery Values of 45 Organophosphorus Pesticides From 1 g Olive Oil Spiked at Two Different Levels (ISO Names Used)

| Pesticide | p-value[a] | Spiking level (mg/kg) | | Recovery (%), mean (N = 6) ± SD | |
		L[b]	H[c]	L[b]	H[c]
Tetrachlorvinphos	0.03	0.27	1.33	95.7 ± 3.8	96.1 ± 3.6
Tolclofos-Me	0.12	0.17	0.83	97.4 ± 3.6	98.2 ± 4.2
Triazophos	<0.01	0.32	1.29	75.5 ± 4.8	101.6 ± 4.6
Trithion	0.17	0.15	0.76	106.0 ± 4.9	95.5 ± 4.7

[a] p value between n-hexane and acetonitrile by GC/FPD-P after a single distribution (p-value is the fraction of solute partitioning into the nonpolar phase of an equivolume two-phase system).
[b] L is the lower spiking level, range 0.09–0.60 mg/kg.
[c] H is the higher spiking level, range 0.38–2.35 mg/kg.
SD, standard deviation.

The sample spiked at about 0.1 mg/kg is used as the quality control sample for routine recovery check with each batch of analyses. The quantitation of the spiked samples is carried out using a matrix-matched calibrant solution.

3.6. Performance of the Method

Typical recovery values and standard deviation for 45 OP pesticides are reported in **Table 1** (*see* **Note 7**) in the lower (0.09–0.60 mg/kg) and higher (0.38–2.35 mg/kg) ranges of concentration.

In the lower concentration range tested, satisfactory (70–110%) results (74–86%) were obtained for 39 OP pesticides, but not for formothion (5%), disulfoton (32%) phosalone (54%), demeton-S-methyl sulfone (60%), fenthion (62%), and phosphamidone (which was borderline at 68%). In the higher concentration range tested, satisfactory results (82–103%) are obtained for all OP pesticides except formothion (48%) and disulfoton (53%).

4. Notes

1. Because the nominal volume of the Extrelut NT-3 cartridge is 3 mL, no solvent is expected to drain from the cartridge at this stage.
2. The flow rate is dropwise and is self-regulated by the pressure drop through the cartridge system.
3. Sometimes, the air bubble at the end of the Extrelut NT-3 cartridge or in the glass connector between the silica and C18 cartridges can block the elution. To start the elution, temporarily apply, as far it is needed, slight pressure to the top of the Extrelut NT-3 cartridge by a syringe connected to an "inverted" stopcock (*see* **Fig. 1**). When removing the syringe, slide the stopcock off to avoid sucking back the eluting solvent. Although the flow rate is not critical regarding extraction efficiency, avoid increasing the flow rate too much.
4. Because traces of pesticides can be lost during concentration of the solution, the distillation temperature is lowered by adding methanol prior to concentration to form an azeotrope with acetonitrile. Avoid bringing the solution to complete dryness with the rotatory

evaporator. Instead, stop the concentration when a few drops of solvent are left. Then, remove the flask from the evaporator and bring to dryness by manually rotating the flask kept in an inclined position. The minute amount of lipid left after the cleanup acts as "keeper," that is, it helps reduce losses by volatilization.

5. Regarding the efficiency of the cleanup, the amount of the lipid matrix carried over into the acetonitrile eluate mainly depends on the composition (triglycerides and minor components) of the fat analyzed. Solid fats such as pork fat are almost completely removed, but highly unsaturated fish oils are less efficiently removed, and a further cleanup, for instance, by high-performance size-exclusion chromatography *(3–7)*, should be given.

6. *Matrix effect* in analytical chemistry is defined as the combined effect of all components of the sample other than the analyte on the measurement of the quantity. If a specific component can be identified as causing an effect, then it is referred to as *interference*. Interferences can most likely produce effects at the detector level by enhancing or quenching the response for the analyte. For instance, matrix components not removed by the cleanup step and eluting from the GC column at the same time as an OP pesticide may influence its quantitation by FPD-P. Indeed, combustion in the detector flame produce CO_2, which absorbs part of the radiation emitted by the OP pesticide, quenching the response of the detector. Instead, the bulk of matrix components can produce a matrix effect at the injector level. Depending on the type, geometry, and operating conditions of the injector, labile compounds, including certain OP pesticides, can undergo partial degradation in the injector. The coextractives injected generally have a protective action on the analyte, so the proportion of the injected amount of the analyte transferred to the GC column (and hence detected) is influenced by the amount of the coextractives injected together with the analyte. Consequently, the matrix effect results in enhancement of the signal in a sample extract compared to the signal for a solvent-based calibration standard of the same concentration. In these cases, the calculated concentration of an affected pesticide in the sample is higher than the true concentration. The entity of this matrix effect depends on the pesticide, concentration, matrix components, and condition of the GC system. The most common approach to counteract this matrix enhancement effect is the use of matrix-matched standards, that is, calibration standards in blank sample extracts *(13,14)*. To prepare a matrix-matched calibrant solution, process a blank sample with the described method to obtain the final blank extract. Redissolve with the working standard solution and use as calibrant solution. The operator should check under the instrument and operating conditions if a relevant matrix effect occurs with the analyte(s) of interest.

7. The ability of the method to recover compounds of different polarity through the on-column partition step can be predicted by calculating the *p* value *(15,16)*, defined as the fraction of solute partitioning into the nonpolar phase of an equivolume two-phase system. The lower the *p* value, the more efficiently the polar compounds are partitioned into the acetonitrile eluate. The 45 OP pesticides tested have *p* values between *n*-hexane and acetonitrile in the range <0.01–0.36, indicating the ability of the method to recover compounds with a wide range of polarity.

References

1. McMahon, B. M., and Hardin, N. F. (eds.). (1994) *Pesticide Analytical Manual*, US Department of Health and Human Services, Food and Drug Administration, Washington, DC, Volume I, Section 304, pp. 15–17.

2. Parfitt, C. H. (ed.). (2000) Pesticide and industrial chemical residues, in *Official Methods of Analysis of AOAC International*, 7th ed. (Horwitz, W., ed.), AOAC International, Gaithersburg, MD, pp. 6–7.
3. Blaha, J. J. and Jackson, P. (1985) Multiresidue method for quantitative determination of organophosphorus pesticides in foods. *J Assoc. Off. Anal. Chem.* **68,** 1095–1099.
4. Spccht, W. and Tilkes, M. (1985) Gas-chromatographische Bestimmung von Rückständen an Pflanzebehandlungsmitteln nach Clean-up über Gel-Chromatographie und Mini-Kiesel-Säulen-Chromatografie. *Fresenius Z. Anal. Chem.* **322,** 443–455.
5. Roos, A. H., van Munsteren, A. J., Nab, F. M., and Tuinstra, L. G. M. Th. (1987) Universal extraction/clean-up procedure for screening of pesticides by extraction with ethyl acetate and size exclusion chromatography. *Anal. Chim. Acta* **196,** 95–102.
6. Seymour, M. P., Jefferies, T. M., and Notarianni, L. J. (1986) Large-scale separation of lipids from organochlorine pesticides and polychlorinated biphenyls using a polimeric high-performance liquid chromatographic column. *Analyst* **111,** 1203–1205.
7. McMahon, B. M. and Hardin, N. F. (eds.). (1994) *Pesticide Analytical Manual*, 3rd ed., US Department of Health and Human Services, Food and Drug Administration, Washington, DC, Vol. 1, Section 304, pp. 21–24.
8. Di Muccio, A., Pelosi, P., Attard Barbini, D., et al. (1999) Determination of pyrethroid pesticide residues in fatty materials by solid-matrix dispersion partition followed by mini-column size-exclusion chromatography. *J. Chromatogr. A* **833,** 19–34.
9. Di Muccio, A., Pelosi, P., Attard Barbini, D., Generali, T., Ausili, A., and Vergori, F. (1997) Selective extraction of pyrethroid pesticide residues from milk by solid-matrix dispersion. *J. Chromatogr. A* **765,** 51–60.
10. Di Muccio, A., Ausili, A., Vergori, L., Camoni, I., Dommarco, R., and Gambetti, L. (1990) Single-step multi-cartridge clean up for organophosphate pesticide residue determination in vegetable oil extracts by gas chromatography. *Analyst* **115,** 1167–1169.
11. Di Muccio, A., Generali, T., Attard Barbini, D., et al. (1997) Single-step separation of organochlorinated pesticide residues from fatty material by combined use of solid-matrix partition and C18 cartridges. *J. Chromatogr. A* **765,** 61–68.
12. Di Muccio, A., Pelosi, P., Attard Barbini, D., et al. (2002) Application of single step, solid matrix partition-C18 cleanup to the analysis of *N*-methylcarbamate residues in fats and oils, in Abstract Book Fourth European Pesticide Residues Workshop (EPRW 2002), Rome, Italy, May 28–31, Poster 143.
13. Schenck, F. J. and Lehotay, S. J. (2000) Does further clean-up reduce the matrix enhancement effect in gas chromatographic analysis of pesticide residues in food? *J. Chromatogr. A* **868,** 51–61.
14. Hajslová, J. and Zrostlíková, J. (2003) Matrix effects in (ultra)trace analysis of pesticide residues in food and biotic matrices. J. Chromatogr. A **1000,** 181–197.
15. Bowman, M. C. and Beroza, M. (1965) Extraction *p*-values of pesticides and related compounds in six binary solvent systems. *J. Assoc. Off. Anal. Chem.* **48,** 943–965.
16. Beroza, M., Inscoe, M. N., and Bowman, M. C. (1969) Distribution of pesticides in immiscible binary solvent systems for cleanup and identification and its application in the extraction of pesticides from milk. *Residue Rev.* **30,** 1–61.

21

Multiclass Pesticide Analysis in Vegetables Using Low-Pressure Gas Chromatography Linked to Tandem Mass Spectrometry

Francisco J. Arrebola Liébanas, Francisco J. Egea González, and Manuel J. González Rodríguez

Summary

A method for determining pesticide residues in vegetables by low-pressure gas chromatography–tandem mass spectrometry (LPGC–MS/MS) is proposed after a fast and simple extraction of the vegetable with dichloromethane and without cleanup steps. The technique reduces the total time required to determine 72 pesticides to less than 31 min, increasing the capability of a monitoring routine laboratory with respect to conventional capillary GC. The use of a guard column and a carbofrit plug into the glass liner in combination with LPGC avoids previous cleanup steps, simplifying the manipulation of the sample and reducing time, cost, and uncertainty of the final result. The method presents limits of quantitation (LOQs) low enough to determine the pesticide residues at concentrations below the maximum residue levels stated by legislation.

Key Words: Fast gas chromatography; low-pressure gas chromatography; pesticides; tandem MS; vegetables.

1. Introduction

The monitoring of pesticide residues in fruits and vegetables became a priority problem because of the large list of active ingredients and large amounts applied throughout the world. Several factors favor the application of pesticides to minimize the incidence of pests and diseases caused by climatic conditions, crop cycles, or agronomic aspects such as high plants density or excessive use of fertilizers.

Pesticide levels are lower in agricultural commodities now because of the application of good farming practices. However, one must be completely certain that the residues are below the maximum residue levels (MRLs) established by governments or regulatory bodies to minimize their consumer intake.

From: *Methods in Biotechnology, Vol. 19, Pesticide Protocols*
Edited by: J. L. Martínez Vidal and A. Garrido Frenich © Humana Press Inc., Totowa, NJ

The need is evident for programs for monitoring pesticide residues in fresh products to ensure compliance with MRLs *(1)* and to estimate better the actual exposure of consumers to pesticide residues. Because of the trace detection levels required and the complex matrices in which the target analytes have to be determined, efficient sample preparation, detection at very low levels, and unambiguous identification are essential steps in analytical methodology development. Probably two of the most important features of an analytical method are the degree of selectivity and the detection limit.

Gas chromatography (GC)–tandem mass spectrometry (MS/MS) using benchtop ion trap systems has been a relevant approach in pesticide residue analysis, allowing both increased selectivity and decreased detection limits *(2,3)*. However, a main goal in the development of new GC–MS methods is the increase of analysis speed to reduce the analysis time because the multiresidue methods currently used in pesticide residue laboratories are relatively slow, restricting the number of analyses per day.

An approach to perform very fast separations, as Gidding first demonstrated *(4)*, is to conduct GC at subambient pressure conditions, that is, by creating a high vacuum in the analytical column, what is called low-pressure gas chromatography (LPGC) *(4–8)*. This is achieved by coupling a wide-bore capillary column (0.53 mm id) to a restriction capillary that is positioned at the injector part. As a consequence, the optimal linear gas velocity for capillary columns operated under vacuum can be increased by a factor of 10. These high velocities result in very short analysis times compared to conventional GC. The application of LPGC provides large sample capacity and originates chromatographic signals with similar peak widths as in traditional GC methods. The last feature allows the use of a common scanning mass spectrometer (ion trap or quadrupole) to obtain enough points for building the signals. In addition, LPGC provides a high signal-to-noise ratio because of the increased peak heights obtained at the high gas velocities, improving detection limits. Another benefit is the low column temperatures, which provide the possibility of analyzing thermolabile compounds at trace levels as bleed will also be low owing to elution temperatures. A drawback of this approach is a loss only in separation efficiency because of the use of shorter columns (10 m). The use of MS detection does not need complete chromatographic resolution between all compounds for analysis *(9,10)*.

The present method is based on simple and fast solvent extraction of the vegetables without postextraction cleanup steps before analysis. After that, an LPGC–MS/MS determination is proposed for analyzing 72 multiclass pesticides in fresh vegetables in an instrumental step of less than 31 min (a reduction of more than 50% with respect to an alternative method based on conventional capillary GC).

2. Materials

2.1. Reagents

1. Pesticide standards with purity certified higher than 99% (e.g., Riedel-de-Haën, Seelze-Hannover, Germany).
2. Caffeine used as internal standard (ISTD) with purity certified (e.g., Riedel-de-Haën).
3. Pesticide-quality solvents *n*-hexane, dichloromethane, methanol, and acetone.
4. Residue analysis-quality anhydrous sodium sulfate.

2.2. Stock and Working Standard Solutions

1. Stock standard solutions (between 75 and 550 μg/mL must be prepared by exact weighing and dissolution in acetone and stored in a freezer at −30°C without exposure to light (*see* **Note 1**).
2. Working standard solutions (5 μg/mL) are prepared by appropriate dilution of the stock standard solutions with cyclohexane and stored in a refrigerator (4°C) (*see* **Note 2**).

2.3. Apparatus

1. Gas chromatograph with an ion trap tandem mass spectrometry detector (e.g., Varian Instruments, Sunnyvale, CA). It must be fitted with electronic flow control, an autosampler with a 100-μL syringe and a temperature-programmable injector.
2. Glass liner equipped with a carbofrit plug (e.g., Resteck Corp., Bellefonte, PA) (*see* **Note 3**).
3. Guard column consisting of a fused silica-untreated capillary column 2 m × 0.25 mm id (e.g., Supelco, Bellefonte, PA) (*see* **Note 4**).
4. Wall-coated, open tubular, fused-silica column reference CP-Sil 8 CB with low bleed of 10 m × 0.53 mm id × 0.25-μm film thickness (e.g., Varian Instruments).
5. Helium (at least 99.999% pure).
6. Chopper.
7. Homogenizer (e.g., Polytron PT2100; Kinematica A.G., Littan/Luzern, Switzerland).
8. Rotary evaporator.

3. Methods

The method described outlines (1) the pesticide extraction procedure from vegetables, (2) the instrumental determination, and (3) the identification and quantification of the pesticide residues.

3.1. Pesticide Extraction From Vegetables

1. Chop and homogenize 2 kg vegetable sample (*see* **Note 5**).
2. Weight exactly a 15-g aliquot into a glass and mix with 50 mL of the dichloromethane in the Polytron for 2 min.
3. Add 50 g anhydrous sodium sulfate and mix with the Polytron for 1 min.
4. Allow the mixture to rest for 2 min and then filter through a 9-cm Büchner funnel and filter again through a paper filter with anhydrous sodium sulfate.
5. Evaporate the solvents to dryness in a rotary evaporator (35–40°C).
6. Redissolve the dried residue with 5 mL cyclohexane.
7. Add 1 mL of this solution to a 2-mL volumetric flask containing 50 μL of ISTD solution of 20 mg/L and reach the final 2-ml volume with cyclohexane.

3.2. Instrumental Determination

1. Inject aliquots of 10 μL of sample into the gas chromatograph operating at a syringe injection flow rate of 10 μL/s.
2. Hold the initial injector temperature at 70°C for 0.5 min and then increase at 100°C/min to 310°C, which is held for 10 min (*see* **Note 6**). Hold the column temperature at 70°C for 3.5 min, increase at 50°C/min to 150°C, and then at 3°C/min to 235°C, and finally increase to 300°C at 50°C/min and hold for 3 min (*see* **Notes 7** and **8**).
3. Operate the ion trap mass spectrometer in MS/MS with electron impact (EI) (*see* **Note 9**).
4. Set the transfer line, manifold, and trap temperatures at 280, 50, and 200°C, respectively.

5. Perform the analysis with a filament-multiplier delay of 4.75 min to prevent instrument damage.
6. Activate the automatic gain control (AGC) with an AGC target of 5000 counts (*see* **Note 10**).
7. Set the emission current for the ionization filament at 80 μA, generating electrons with 70-eV energy, and set the axial modulation amplitude 4.0 V.
8. Carry out the MS/MS process by collision-induced dissociation (CID) with a nonresonant excitation for all the compounds studied.
9. Set the electron multiplier voltage at 1700 V (+200 V offset above the autotuning process).
10. Scan rate and mass range scanned depend on the number of pesticides analyzed simultaneously (*see* **Note 11**). Typical parameters found using the described experimental conditions are shown in **Table 1**.

3.3. Identification and Quantification

1. Calculate the retention time window (RTW) for each pesticide like the retention time average plus or minus three standard deviations of retention time when 10 blank samples spiked at the second calibration level of each compound are analyzed.
2. Compare the retention time obtained in the analyzed sample with the RTW as identification criterion.
3. Confirm a previously identified compound by comparing the MS/MS spectrum obtained in the sample with another stored as a reference spectrum in the same experimental conditions.
4. Obtain the reference spectra daily by injecting a blank cucumber sample spiked at the concentration of the second calibration point (*see* **Note 12**). **Figure 1** shows a fragment of a gas chromatogram obtained in the selected experimental conditions monitoring the quantification ion of each pesticide.
5. Build a calibration line, using at least three calibration points prepared into blank matrix extract (*see* **Notes 13** and **14**), by plotting peak areas related to the ISTD against concentration of pesticide in milligrams per kilogram.
6. Calculate the amount of pesticide in the sample in milligrams per kilogram from the calibration curve.
7. Use as quality control criteria the analyses of a blank extract, a calibration set, and a blank extract spiked with a known amount of the pesticides (*see* **Note 15**).

4. Notes

1. Some pesticides are light sensitive. They can be stored for at least 1 yr in the established conditions.
2. They can be stored for at least 2 mo.
3. The size of the carbofrit plug must be adapted to the inner diameter of the glass liner. For that, over a clean surface, the carbofrit must be softly rolled by pushing with the finger. The reduction of the carbofrit size must be until just the internal glass liner diameter to avoid plug movement during instrumental operation.
4. The use of a guard column together with the carbofrit plug reduces considerably the frequency of maintenance operations over the analytical column.
5. No degradation of the pesticides was detected in samples stored at 4°C for 24 h.
6. Several injection techniques are available with LPGC because, despite the fact that the analytical column has to be kept under low-pressure conditions, the injector works at conventional column head pressures. As a consequence, typical injection volumes can be used, and the sample capacity is not limited. The amount injected (10 μL) allows the determination of pesticide residues at concentrations below or equal to the MRLs with a

good peak shape. Typical LOQs obtained for the studied compounds using similar instrumentation ranged from 0.2 to 22 µg/kg. The injector temperature is increased from 70 to 310°C to minimize breakdown of the most thermolabile pesticides.

7. Electronic flow control of the carrier gas is recommended to optimize separation of the target analytes when a GC analysis with LPGC columns is performed as well as with ovens with temperatures that can be programmed to provide fast heating rates and high-frequency digital output detection.

8. In general, LPGC columns show lower separation efficiency than conventional capillary columns in the analysis of complex compound mixtures. The lack of selectivity in the analysis of complex pesticide mixture is compensated by using the MS/MS detector, which allows essentially error-free identification of the compounds if the optimization of the MS/MS parameters is properly adjusted.

9. Ion trap detectors that carry out the ionization inside the trap chamber can quickly switch between chemical ionization and EI modes along the same run. This allows optimization of the ionization step for each pesticide. However, because of the high number of pesticides frequently analyzed, sometimes it is necessary compromise when various chemical ionization and EI pesticides coelute.

10. The AGC continuously adapts in an automatic mode the ionization time of the ion trap detector to avoid a reduction of the sensitivity. For the optimization of detector sensitivity, the trap should be filled with the target ions. A higher AGC target would cause electrostatic interactions between the ions (space–charge effects), providing degraded performance.

11. MS/MS detection mode can apply different CID conditions that switch quickly. Each CID condition is specific for each coeluted analyte. After acquisition, the plotting of the experimental data obtained separately for each pesticide offers a selective chromatogram for each compound. Nevertheless, the capability of the detector to analyze a large number of coeluting analytes is limited. With a constant mass spectrometer scanning speed (for the Saturn 2000, it is 5600 thomson/s), the scan time for a MS/MS experiment depends, among other things, on the m/z range scanned and the number of microscans acquired to obtain an averaged spectrum of adequate signal-to-noise ratio. The lack of optimization of these parameters may generate chromatographic peaks defined by too few points or by enough points obtained with an elevated uncertainty; that is, it can affect the precision of the chromatographic data. For this reason, the parameters mentioned above have been extensively studied for the pesticides that coelute to obtain good-quality spectra and peak shapes. The m/z range scanned is adjusted for each pesticide to cover only the product ions obtained under the MS/MS conditions selected. In the case of MS/MS spectra with several not-very-intense qualifier ions, some of these ions are discarded and not scanned to shorten in the scan time. **Figure 2** represents the influence of m/z range scanned on the total scan time. The number of microscans averaged to obtain a spectrum is relevant in the peak shape and spectral fit. Typical chromatographic peak shapes obtained are shown in **Fig. 3**. Acceptable peak shapes are obtained for two and three microscan cases, but appreciably worse results are generated when one microscan is selected (some intensity points can decrease or increase abnormally). This shows that the process of spectral averaging improves the precision of the total intensity of the spectrum, and that at least two microscans should be obtained to average the final spectrum.

12. For some pesticides, small differences have been observed between spectra obtained with and without matrix influence. The comparison results (fit) are scaled to 1000 for the best match (identical spectra). To set the fit threshold for each pesticide, the average fits of the

Table 1
GC–MS/MS Conditions (*see* Note 16)

Pesticide	Parent ion (*m/z*)	CID Amplitude (V)	CID Rf (*m/z*)	Quantification ion (*m/z*)	Range (*m/z*)
Dichlorvos	185	78	81	109 + 131	80–190
Acephate	136	37	47	107 + 119	80–190
Heptenophos	124	37	47	89	80–160
Propoxur	152	41	66	110	80–160
Ethoprophos	158	27	47	94 + 114 + 130	70–230
Dimethoate	125	55	60	79	70–230
Lindane	219	70	100	180 + 183	140–220
Pyremethanil	198	100	81	98	90–300
Chlorthalonil	266	90	85	133	90–300
Disulfoton	186	60	71	97	90–300
Etrimphos	292	45	70	181	90–300
Pirimicarb	166	49	53	83	80–200
Caffeine	194	56	60	120	80–200
Formothion	170	38	70	107	80–200
Ethiofencarb	168	39	63	107	80–200
Chlorpirifos-m	286	72	85	208	100–300
Vinclozoline	285	34	100	241 + 213	100–300
Parathion-m	263	48	80	136 + 216	100–300
Metalaxyl	206	54	75	132 + 162	100–300
Pirimiphos-m	290	64	85	151	90–300
Fenitrotion	260	65	71	122 + 138 + 170	90–300
Malathion	173	51	75	99	90–320
Chlorpyrifos	314	100	170	258	90–320
Fenthion	278	92	112	135	90–320
Triadimefon	208	62	75	144	90–345
Tetraconazole	336	96	108	218	90–345
Dicofol	250	49	90	215	90–345
Pendimethalin	252	62	95	208 + 191 + 162	90–345
Penconazole	248	77	89	192 + 157	90–300
Chlozolinate	331	88	145	259	90–300
Isonfenphos	213	52	93	185	90–300
Pyrifenox	263	90	100	192 + 228	90–300
Chlorfenvinphos	267	82	100	159	90–300
Procymidone	283	57	80	253:257	90–300
Quinometionathe	234	60	83	196	69–250
Endosulfan I	241	84	80	170 + 172	69–250
Fenaminphos	303	56	95	195	120–305
Fludioxinil	248	84	89	152 + 154 + 127	120–305
Buprofezin	249	50	80	191:195	120–305
Hexaconazole	231	100	100	159	120–275
Bupimirate	273	77	120	193	120–275
Endosulfan II	241	84	80	170 + 172	120–275

(continued)

Table 1 (Continued)
GC–MS/MS Conditions (see Note 16)

Pesticide	Parent ion (m/z)	CID Amplitude (V)	CID Rf (m/z)	Quantification ion (m/z)	Range (m/z)
Oxadixyl	163	46	71	132	100–235
Ethion	231	63	100	175 + 203	100–235
Benalaxyl	148	46	50	91	90–345
Carbofenothion	342	64	131	199 + 157	90–345
Endosulfan III	272	64	80	235 + 238	90–345
Propiconazole	259	78	114	191 + 173	100–260
Nuarimol	235	56	75	139	100–260
Tebuconazole	250	63	75	125	100–260
Propargite	173	56	66	117 + 145	100–260
Iprodione	314	88	125	245 + 271	140–345
Bromopropylate	341	45	70	181:187	140–345
Bifenthrin	181	40	50	165	140–275
Fenpropathrin	265	72	95	210	140–275
Tetradifon	229	95	100	197:203	140–275
Furathiocarb	325	77	140	194	100–330
Phosalone	182	70	80	111 + 138	100–330
Piriproxifen	136	57	59	96	70–140
Cyhalothrin	181	90	80	152	120–290
Amitraz	162	50	71	132 + 147	120–290
Pirazofos	265	87	120	210	120–290
Acrinathrin	181	87	80	152	70–200
Permethrin	183	74	70	152	70–200
Pyridaben	147	53	64	111 + 105	70–200
Cyfluthrin	206	96	86	149:152	100–325
Cypermethrin	163	53	70	127	100–325
Flucythrinate	157	69	79	107	100–325
Esfenvalerate	225	51	70	119	100–325
Difenoconazole	323	87	122	265	100–325
Deltramethrin	253	57	90	172 + 174	120–350
Azoxistrobina	345	92	115	329	120–350

MS/MS spectra from 10 injections of the midlevel standard are determined, and 250 units are subtracted from the average. It often can be automatically carried out by the instrument software if the signal-to-noise ratio of the chromatographic peak is above 3. Typical confirmation threshold fits are obtained between 500 and 725.

13. More calibration points would increase the fit of the calibration function to the experimental data but reduce the total number of real samples analyzable in routine laboratories. It has been observed that three calibration levels can be considered appropriate for quantifying all the pesticides included in the method. The concentration level for each calibration point should be stated considering the MRL regulated for each pesticide and food commodity. For all compounds, the first point of the calibration curve is set at a concentration between LOQ and the smallest MRL found for the vegetables studied at

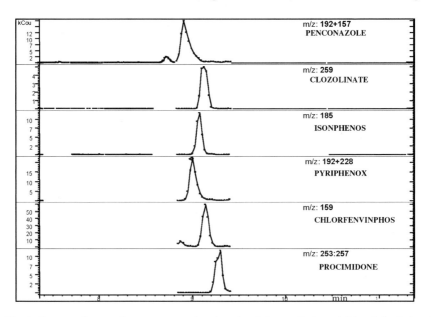

Fig. 1. Extract of a gas chromatogram showing six of the studied pesticides. It includes the product ion monitored for quantitation.

different legislations. In cases such as for acephate, when the LOQ is not much lower than the minimum MRL, the first calibration point is the MRL. On the other hand, for all pesticides, the highest calibration concentration is set to 15 times the first calibration concentration. Therefore, the range of calibration must include the MRL stated. Typical correlation coefficients between 0.98 and 0.99 can be obtained using instrumentation similar to the above described using linear calibration functions.

14. The origin can be added as a calibration point, but the calibration function should not be forced through the origin. The fit of the calibration function must be plotted and inspected visually, avoiding reliance on correlation coefficients, to ensure that the fit is satisfactory in the region relevant to the residues detected. Extracts containing high-level residues may be diluted to bring them within the calibrated range, adjusting accordingly the concentration of matrix extract. It is recommended to evaluate potential matrix effects at method validation. Food commodities should be classified according to certain reference matrices for calibration purposes. Such matrices must correspond at least to the next crop groups: (1) cereals and other dry crops, (2) commodities with high water content, (3) commodities with high fat content, and (4) fruits with high acid content. The validation process must evidence the representativeness of such reference matrices used for calibration purposes. Analysis of spiked blank samples with the pesticides originates typical recovery data between 71 and 108% with relative standard deviations always lower than 18%.

15. A proposal for quality control criteria is as follows:

> For blank analysis, no signals should be detected in the RTWs of the target analytes. Spiked sample should be spiked at the second calibration level, and the pesticides should present routine recovery rates between 60 and 140% (70–110% when residue levels higher than the MRLs have been determined).

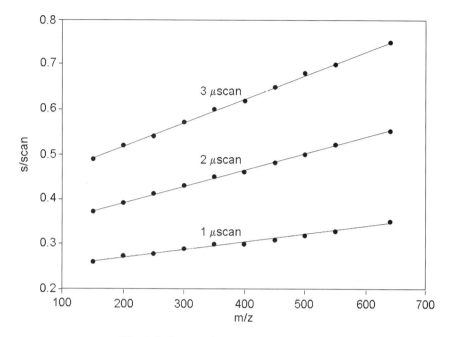

Fig. 2. Influence of *m/z* range on the scan time.

The calibration curve should present a coefficient of determination R^2 higher than 0.95, and the individual points must not exceed more than ±20% (±10% when the MRL is approached or exceeded) from the calibration curve in the relevant region. Otherwise, a more satisfactory calibration function must be used, or the determinations must be repeated.

16. For the dissociation conditions, in general, the precursor ion isolated is selected with the aim of achieving a compromise between both selectivity (the highest *m/z* ion) and sensitivity (the highest intensity ion). To achieve this, precursor ions are fragmented (CID) with the goal of generating spectra with multiple fragment ions of relatively high intensity that permitted accurate quantification while preserving a certain proportion (10 and 20%) of the precursor ions. This objective can be achieved by appropriate selection of the nonresonant excitation amplitude and excitation storage level. The excitation storage level, which is related to the trapping field that stabilizes the parent ion, is selected as the minimum value that allows the dissociation of the parent ion. If a higher excitation storage level is applied, the ions are held in the trapping field more strongly, allowing the application of a higher excitation voltage, which is able to produce more product ions. Nevertheless, if the excitation storage level is too high, the lower mass product ions will not be trapped and will not be observed in the spectrum. The MS/MS spectra obtained under the final experimental conditions are stored in a home-made MS/MS library.

Acknowledgments

We are grateful to Instituto Nacional de Investigación y Tecnología Agraria y Alimentaria (INIA) (Project CAL03-055) and FIAPA (2000/01), Spain for financial support.

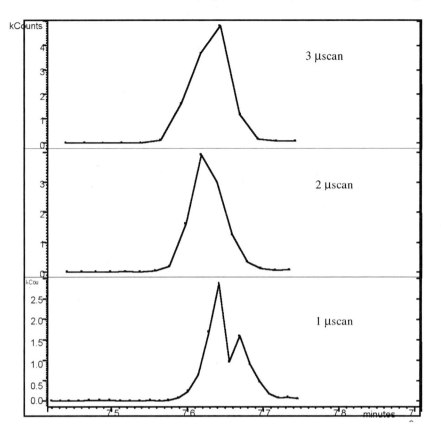

Fig. 3. Influence of number of microscans used to obtain averaged spectra on GC peak shape for pirimiphos methyl.

References

1. *Community Directive 93/58 EEC*, Off. Eur. Commun. L 211/6, European Community, Brussels, 1993.
2. Martínez Vidal, J. L., Arrebola, F. J., and Mateu-Sánchez, M. (2002) Application to routine analysis of a method to determine multiclass pesticide residues in fresh vegetables by gas chromatography/tandem mass spectrometry. *Rapid Commun. Mass Spectrom.* **16,** 1106–1115.
3. Martínez Vidal, J. L., Arrebola, F. J., and Mateu-Sánchez, M. (2002) Application of gas chromatography–tandem mass spectrometry to the analysis of pesticides in fruits and vegetables. *J. Chromatogr. A* **959**, 203–213.
4. Giddings, J. C. (1962) Theory of minimum time operation in gas chromatography. *Anal. Chem.* **34,** 314–319.
5. Van Deursen, M. M., Janssen, H. G., Beens, J., et al. (2000) Fast gas chromatography using vacuum outlet conditions. *J. Microcol. Sep.* **12,** 613–622.

6. Amirav, A., Tzanami, N., Wainhaus, S. B., and Dagan, S. (1998) Megabore vs microbore as the optimal column for fast gas chromatography/mass spectrometry. *Eur. Mass Spectrom.* **4,** 7–13.
7. De Zeeuw, J., Peene, J., Janssen, H.-G., and Lou, X. (2000) A simple way to speed up separations by GC–MS using short 0.53 mm columns and vacuum outlet conditions. *J. High Resolut. Chromatogr.* **23,** 677–680.
8. Mastovská, K., Lehotay, S. J., and Hajslová, J. (2001) Optimization and evaluation of low-pressure gas chromatography–mass spectrometry for the fast analysis of multiple pesticide residues in a food commodity. *J. Chromatogr. A* **926,** 291–308.
9. González Rodríguez, M. J., Garrido Frenich, A., Arrebola Liébanas, F. J., and Martínez Vidal, J. L. (2002) Evaluation of low-pressure gas chromatography linked to ion-trap tandem mass spectrometry for the fast trace analysis of multiclass pesticide residues. *Rapid Commun. Mass Spectrom.* **16,** 1216–1224.
10. Arrebola, F. J., Martínez Vidal, J. L., González-Rodríguez, M. J., Garrido-Frenich, A., and Sánchez Morito, N, (2003) Reduction of analysis time in gas chromatography— application of low-pressure gas chromatography–tandem mass spectrometry to the determination of pesticide residues in vegetables. *J. Chromatogr. A* **1005,** 131–141.

22

Use of Matrix Solid-Phase Dispersion for Determining Pesticides in Fish and Foods

Steven A. Barker

Summary

We describe here a method using matrix solid-phase dispersion (MSPD) for the analysis of atrazine, metribuzin, cyanazine, and simazine in fish muscle tissue with a minimal amount of time and solvent consumed. A tissue sample is placed in a glass mortar containing a bonded-phase solid support material. The solid support and sample are manually blended together using a glass pestle. The blended material is packed into a column containing activated Florisil and eluted with acetonitrile:methanol (9:1). The eluate is concentrated and analyzed by high-performance liquid chromatography with ultraviolet detection. The linearity of response ranges from 0.35 to 8.0 ppm, and the average percentage recoveries are between 75 ± 3 and $106 \pm 4\%$, for a spiking level of 0.35 ppm. The intraassay variability is <3%, and the interassay variability is <5%.

Key Words: Atrazine; cyanazine; fish; high performance liquid chromatography; matrix solid-phase dispersion; metribuzin; MSPD; simazine; triazines.

1. Introduction

The most difficult biological samples to extract and subsequently analyze for pesticides are solids, semisolids, and highly viscous samples. A method first introduced in 1989 has greatly simplified this task and found extensive application for drug and pesticide analysis of biological samples. In the years since its introduction, this method has become a proven protocol for the isolation of a wide range of pesticides from a variety of biological samples (1–52). The method is known as matrix solid-phase dispersion or MSPD (52).

The utility of MSPD in this regard has been based on the fact that the MSPD process simultaneously and efficiently accomplishes several steps in the more classical approach to sample preparation and solid-phase extraction (SPE)/fractionation, is somewhat generic, and eliminates many of the confounding properties of more classical extraction procedures. The application of MSPD to difficult analytical problems has shown that it can greatly reduce analyst time, increase sample throughput and

From: *Methods in Biotechnology, Vol. 19, Pesticide Protocols*
Edited by: J. L. Martínez Vidal and A. Garrido Frenich © Humana Press Inc., Totowa, NJ

shorten turnaround time, reduce solvent use and the attendant expense of solvent purchase and disposal, and eliminate emulsion formation seen in classical extractions while providing analytical results equal to or better than classical or "official" methods.

Thus, a sample (tissue, fruit, etc.) is placed in a glass mortar containing a bonded-phase solid support material, such as octadecylsilyl-derivatized silica (C18). The solid support and sample are manually blended using a glass pestle. In this manner, the irregularly shaped silica-based solid support serves as an abrasive that promotes disruption of the sample's general architecture and acts as a bound solvent that appears to further disrupt the sample by inducing lysis of cells and other cellular structures. Studies have shown that the sample is dispersed over the surface of the solid support material, greatly enhancing sample surface area available to extraction *(24,25,47,52)*.

The blended material is packed into a column (or a co-column containing other column packings) suitable for conducting sequential elution with solvents. The blended sample components and their distribution in the bonded phase and support provide a new phase that exhibits a unique character for performing sample fractionation.

We describe here the details of an MSPD method *(23)*, having selected members of the family of triazines, specifically atrazine [1912-24-9], metribuzin [21097-64-9], cyanazine [21725-46-2], and simazine [122-34-9], to serve as models to define a complete process for the rapid extraction, detection, and analysis of compounds of concern in the incidence of fish kills. Triazines are highly effective for pre- and postemergent weed control but are often associated with unintentional fish kills from overspraying, runoff, and improper disposal.

The multiresidue method detailed here is designed to detect lethal concentrations of triazine pesticides in fish (usually parts per million) with a minimal amount of time and solvent consumed and is adaptable to analysis of a variety of different classes of pesticides.

2. Materials

1. Liquid chromatographic (LC)-grade solvents from commercial sources: acetonitrile (ACN), hexane, dichloromethane, water, and methanol.
2. 0.01 *M* Phosphate buffer, pH 6.0 (NaOH).
3. 0.45-µm Membrane filter (e.g., Metricel®, Gelman Sciences, Inc., Ann Arbor, MI).
4. Atrazine [1912-24-9], metribuzin [21097-64-9], cyanazine [21725-46-2], and simazine [122-34-9] reference standards (99% purity; e.g., Crescent Chemical Co., Inc., Haupauge, NY).
5. Preparative-grade (bulk) C18 SPE material, 40-µm particle size, 18% carbon load, end capped (e.g., Analytichem Bondesil™, Varian Associates, Harbor City, CA; *see* **Note 1**). The C18 is cleaned prior to use by washing with two volumes each of hexane, dichloromethane, and methanol. The material is then vacuum aspirated overnight to remove excess solvent. This material is stored at room temperature in a closed plastic container until used (*see* **Note 2**).
6. Pesticide residue (PR)-grade Florisil (e. g. Alltech Assoc., Inc., Deerfield, ILUSA), 60–100 mesh [1343-88-0]. Activate by heating to 130°C for at least 1 wk before use. Activated Florisil is stored at 130°C in a solvent-washed, aluminum-foil-covered beaker.
7. Paper filters (1.5 cm, Whatman no. 1, Whatman, Florham Park, NJ).
8. 10-mL Disposable plastic syringe barrels and plungers (e.g., Becton Dickinson Co., Franklin Lakes, NJ) washed with hot soapy water, triple rinsed with tap water, double-distilled water, and methanol. The syringe barrels and plungers are air dried prior to use.

9. 1-mL Disposable plastic syringe barrels and plungers (e.g., Becton Dickinson Co.) are washed with hot soapy water, triple rinsed with tap water, double distilled water, and methanol. The syringe barrels and plungers are air dried prior to use.
10. Plastic 250-µL pipet tips.
11. Glass mortars and pestles (*see* **Note 3**).
12. 10-mL Disposable glass tubes.
13. Rubber bulb.
14. 1.5-mL Polypropylene test tube (e.g., MicroFuge, Sarstedt, Inc., Newton, NC).
15. 0.45-µm Syringe filters (e.g., Nalgene R, Rochester, NY).
16. LC autosampler vial (e.g., Hewlett-Packard, Wilmington, DE).
17. Narrow-bore column packed with octadecysilyl (ODS)-derivatized silica (200×2.1 mm, ODS Hypersil, 5 µm; e.g., Hewlett-Packard, Wilmington, DE).
18. ODS guard column cartridge (20×2.1 µm, 5 µm; e.g., Hewlett-Packard).
19. 0.45-µm Membrane filter (e.g., FP Vericel™, Gelman Sciences, Inc.).
20. Centrifuge (e.g., Centra-M, International Equipment Co., Needham Heights, MA).
21. Software for LC control, data acquisition, and peak integration (e.g., Hewlett-Packard ChemStation software).

3. Methods

The MSPD process is illustrated in **Fig. 1A,B**. The major steps involve column preparation, sample dispersion onto a solid support material, analyte isolation, and extract concentration, followed by analytical determination.

3.1. Preparation of Sample Extraction Columns

1. Tissue extraction columns are fashioned from 10-mL disposable plastic syringe barrels washed with hot soapy water, triple rinsed with tap water, double-distilled water, and methanol.
2. The syringe barrels and plungers are air dried prior to use.
3. Paper filters (two) are placed inside the base of the columns to serve as a frit.
4. Place 2 g of activated Florisil in the column.
5. A plastic pipet tip is secured to the outlet of the 10-mL barrels.

3.2. Extraction Procedure

1. In glass mortars containing 2.0 g of C18, place 0.5-g catfish fillet samples (*see* **Note 4**).
2. The samples are allowed to reach ambient temperature and, for standard curves, are injected with 40 µL of each triazine stock solution (concentrations of 4.37, 8.75, 25, 50, and 100 µg/mL) to achieve tissue concentrations of 0, 0.35, 0.7, 2, 4, and 8 µg/g, respectively. Blanks (no tissue) and negative controls (tissue without analyte spiking) receive solvent only.
3. The samples are allowed to stand 5 min and then are blended using glass pestles (*see* **Note 5**).
4. The resulting mixture is transferred to a 10-mL column fitted with two 1.5-cm disk paper filters and containing 2.0 g activated Florisil.
5. Another paper filter is placed on top of the transferred material and compressed to a volume of approx 7.5 mL using a syringe plunger (*see* **Note 6**).
6. The column is placed in a rack over a clean 10-mL glass test tube.
7. Add 8 mL of a 9:1 ratio of ACN and MeOH (*see* **Note 7**).
8. The triazines are eluted over a 15- to 25-min period using gravity flow. Slight positive pressure is applied to assist in initiating flow via a rubber bulb. The resulting eluent is approx 5 mL (*see* **Note 8**).

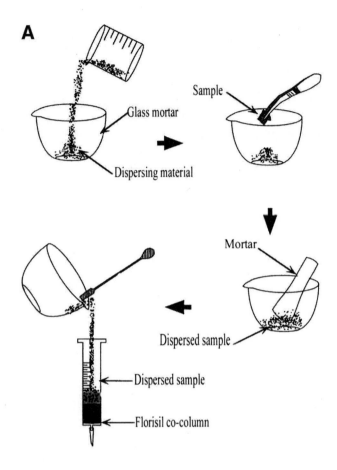

9. The extract is placed in a water bath (35°C) and evaporated under nitrogen.
10. The eluent residue is reconstituted by the addition of 500 μL of mobile phase and vortexed.
11. The sample extracts are placed in a 1.5-mL polypropylene test tube and centrifuged for 5 min at 15,600*g*.
12. The supernatant is placed in a previously cleaned 1-mL syringe fitted with a 0.45-μm syringe filter, and the supernatant is passed into an LC autosampler vial.

Total extraction time is under 40 min.

3.3. LC Analysis

All high-performance liquid chromatographic (HPLC) analyses are performed using a liquid chromatograph with an autosampler. Data are acquired with a photodiode array detector set at 220 nm with a 4-nm bandwidth and a reference signal at 450 nm with an 80-nm bandwidth. Detector sensitivity is set at 20 mAU full scale.

A reversed-phase, narrow-bore, packed-silica column fitted with an ODS guard column cartridge maintained at 40°C is used for all analyses.

B

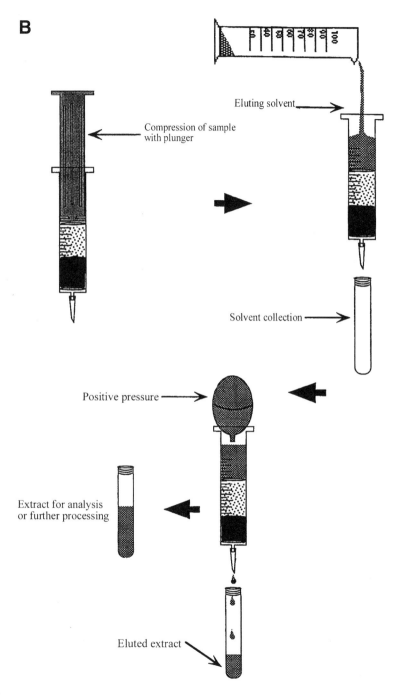

Eluting solvent

Compression of sample
with plunger

Solvent collection

Positive pressure

Extract for analysis
or further processing

Eluted extract

Fig. 1. *(opposite page)* (**A**) Preparation of the sample by MSPD; (**B**) preparation and elution of the MSPD column.

The mobile phase composition is a 7:3 ratio of 0.01 M aqueous NaH_2PO_4 (pH 6.0) (phosphate buffer) and ACN, respectively, delivered at a flow rate of 0.4 mL/min. Mobile phase solvents are filtered over a 0.45-μm membrane filter. All solvents are thoroughly purged with helium before use. A 10-μL aliquot of muscle extract is injected for LC analysis (*see* **Note 9**).

3.4. Identification and Quantification

Individual triazines are identified by their retention time relative (±1%) to reference standards and their ultraviolet diode array spectrum. Peak purity measurements are also conducted to ensure the absence of coeluting interferences. Quantification of the individual triazines is conducted using the relative peak height (or area) of the detected peak to those observed from fortified samples plotted as a standard curve, fitted to the linear regression line and equation for the line so derived (*see* **Notes 9–14**).

3.5. Data Analysis

The statistical parameters used to validate this method for triazine analysis include linearity of response, precision, and accuracy. Specificity and sensitivity are also evaluated.

1. Linearity of response: A standard curve covering the range of concentrations from 0.35 to 8 ppm for standards and extracts is made using the mean of the area of the replicate samples (*see* **Note 10**).
2. Precision is evaluated by examining intraassay variability and interassay variability. The values for the intraassay variation are derived from the coefficient of variation (CV) of means plus or minus standard deviation (SD) of peak areas of three repetitions of analysis of the same sample for each concentration. Overall intraassay variability of the method is determined by averaging the CVs for the five concentrations (0.35–8 ppm) examined. The values for the interassay variability are derived from the CV of peak areas of three distinct samples extracted and assayed on three different days. Overall interassay variation (±SD) of the method is obtained by averaging the CV of each individual concentration (0.35–8 ppm; *see* **Note 11**).
3. Recovery is evaluated by dividing the average peak area of each concentration obtained after extraction by the average peak area of the respective standard concentration obtained without extraction. The average percentage recovery P is calculated by summing the recovery percentage for each sample concentration on each separate day and dividing by three. The relative standard deviation of P is calculated by determining the standard deviation of P and dividing by P (*see* **Note 12**).
4. Specificity is calculated as the percentage of true negative samples of the number of known negatives tested (*see* **Note 13**).
5. Sensitivity is calculated as the percentage of correctly identified positive samples (true positives) of the total number of known positives tested (*see* **Note 14**) within the limits of the assay.

4. Notes

1. To date, silica-based support materials have been the most frequently used for MSPD. The use and effect of synthetic polymer-based solid supports is a subject for further study, particularly supports that possess unique surface and pore chemistries, such as hydrophobic interaction supports. For silica-based materials, however, studies have shown that the pore size is of minor importance in MSPD. This effect could vary with the sample and

should be considered, however. The effect of average particle size diameter has also been examined. As may well be expected, the use of very small particle sizes (3–10 μm) leads to extended solvent elution times for a MSPD column, requiring excessive pressures to obtain adequate flow. However, 40-μm particle size materials (60-Å average pore diameter) have been used most frequently and quite successfully. It has been reported that a blend of silicas possessing a range of particle sizes (40–100 μm) also works quite well, and such materials also tend to be less expensive. Depending on the application, non-end-capped materials or materials having a range of carbon loading (8–18%) may also be used. It is a simple matter to examine these variables for a given application and should be considered for obtaining the best extraction efficiency and the cleanest sample. The bonded phase will, of course, play a pivotal role. Depending on the polarity of the phase chosen, rather dramatic effects on the results may be observed, and a range of available phases should be examined for each application. It has been reported that, in applications requiring a lipophilic bonded phase, C18 and C8 can be used interchangeably. In general, it has been observed that the isolation of more polar analytes from biological samples is assisted by the use of polar phases (e.g., cyanopropyl) and less-polar analytes by less-polar phases. This would be expected based on retention characteristics of compounds discerned from classical SPE *(24,25,47,52)*.

2. Preconditioning of the materials used for MSPD greatly enhances analyte recovery, as has been established with SPE. However, in MSPD it also appears to speed the process of sample blending and dispersal by breaking the surface tension differences that may exist between the sample and bonded-phase solid support. As with SPE, washing or rinsing the solid support materials prior to use also eliminates contaminants from the final eluates obtained for analysis.

3. Glass or agate mortars and pestles are preferred. Porcelain mortars and pestles have been observed to cause loss of analyte and cross-contamination to subsequent samples, perhaps because of the more porous nature of this material.

4. It is important that the sample used is truly representative. Homogeneity issues can be reconciled by previous homogenization of a larger sample. Taking 0.5-g subsamples of the homogenate for MSPD analysis is simply a smaller scale version of the more classical approach to the problem. The best ratio of sample-to-bonded-phase solid support material is 1 to 4. Most applications have employed lipophilic bonded phase (C18) materials, blending 2.0 g solid support with 0.5 g sample. The best ratio, of course, is dependent on the application and should be examined as a variable during method development. Ratios of bonded phase solid support to sample less than 4:1 have been used successfully, and samples have been scaled up to 2 g from the typical 0.5 g used in most MSPD procedures, blended with a proportionately greater amount of solid support *(24,25,47,52)*.

5. The process of blending requires only moderate pressure to produce sample dispersion onto the solid support surface. It is not necessary, and is perhaps deleterious, to use excessive force, "grinding" the sample and solid support into the mortar surface. The blending process does not require vigorous effort to accomplish but is dependent on the degree of connective tissue or other more rigid biopolymer content of the sample. Different analysts may apply differing amounts of pressure or attain different degrees of dispersion of the sample. This may be reflected in different levels of recovery in analyst-to-analyst comparisons during method validation studies. However, excellent agreement is still usually attained in the final result, with good overall accuracy and precision and low variability. It should be kept in mind that in MSPD the sample itself is dispersed throughout the column. In contrast, much of the sample is retained only in the first few millimeters of the column bed in SPE. In MSPD, the sample matrix components cover much of the bonded-

phase solid support surface, creating a new phase that is dependent on the interactions with the solid support and bonded phase, interactions that give the MSPD column its unique character. This new phase, in association with the analyte's distribution and own interactions with it, are perhaps the most important controlling factors in elution vs retention *(24,25,47,52)*.

6. The column should be loosely packed, keeping in mind the basic principles of chromatography, avoiding voids or channels. Tightly compressing the column causes the elution flow rate to be greatly diminished.

7. The correct choice of elution solvents and the sequence of their application to a column are of utmost importance to the success of MSPD or SPE fractionation of samples. Elution solvent sequence and composition can be varied to obtain the best analytical results, attempting to isolate the analyte or further clean the column of interfering substances with each solvent step. The use of co-columns, such as the Florisil used here, is also available to enhance sample cleanup. The nature of MSPD columns and the enhanced degree of interaction with matrix components permit isolation of different polarity analytes or entire chemical classes of compounds in a single solvent or in differing polarity solvents passed through the column. This characteristic makes MSPD amenable to conducting multiresidue isolation and analysis on a single sample. In this regard, true gradient elution of a MSPD column has not been reported to date but should nonetheless prove applicable to the complete fractionation of samples for a range of pesticides. It has been observed that, in an 8-mL elution of a 2-g C18 column blended with 0.5 g sample, that most of the target analytes elute in the first 4 mL or in approx 1 column volume. This will of course vary with each application and with appropriate solvent selection but should be examined to reduce the use of solvent and the unintended elution of other potentially interfering components.

8. The solvent elution rate optimum has not been established. However, the general principles of chromatography apply; the solvent should not flow through the column so fast that it precludes adequate distribution from the solid phase into the mobile phase or flow so slowly that analysis time is unnecessarily long. Although gravity flow is used here, use of a vacuum box is recommended, giving greater control over flow rates and permitting greater ease of handling larger numbers of samples *(24,25,47,52)*.

9. HPLC was selected for analysis of the triazines. After reviewing the literature on HPLC of nonvolatile organic compounds, such as the triazines, in environmental samples, it was decided to use $0.1M$ ammonium acetate (NH_4Ac) as the aqueous portion of the mobile phase. However, spectroscopic analysis of this solution showed absorbance in the 220- to 240-nm range that interfered with maximum triazine absorbance. A ($0.01M$) phosphate buffer ($NaH_2PO_4·H_2O$) that offers no interference with triazine absorption is satisfactorily substituted in this method. Initially, five wavelengths are monitored from 200 to 260 nm, but 220 nm consistently gives the best absorbance maximum for all the triazines in this study. Comparison of the ultraviolet spectra of extracts to standards showed excellent peak purity. This method has 100% specificity for the four triazines examined individually at the concentrations tested. However, the LC method does not resolve all four compounds simultaneously because cyanazine and simazine have approximately the same retention times *(23)*. Adjustments in the mobile phase could resolve this problem as would the use of LC–mass spectrometry or LC–tandem mass spectrometry.

10. A linear response is observed for daily five-point calibration curves for standards and MSPD extracts of muscle using mean peak area ratio values of triplicate injections. Correlation coefficients of 0.999 or more were obtained for all calibration curves using standard solutions. Daily standard curves for MSPD-extracted muscle samples have

correlation coefficients ranging from 0.998 to 1.000, with average daily values of 0.999 ± 0.001 for each triazine *(23)*.

11. This method measured the amount of variability of the test results from the repeated application of the procedure. The intraassay variability or repeatability of the method as determined by the CV of triplicate injections is below 3% for the triazines at concentrations of 0.35–8 ppm for extracts and all of the standards *(23)*. The interassay variability or reproducibility of the method varies from approx 5% for metribuzin at concentrations of 0.35 µg/g muscle to 0.39% for muscle containing 2 µg/g using the mean peak area ratio values for triplicate injections. The overall daily variation for muscle extracts containing 0.35 to 8.0 ppm ranged from 0.89 to 2.43% *(23)*.

12. Average percentage recoveries range from 75 ± 3% for 0.35-ppm metribuzin to 106 ± 4% for 0.35-ppm atrazine ($N = 9$ for extracted pesticides). Data indicate that the MSPD methodology allows successful extraction and determination of all four triazines tested in catfish muscle *(23)*.

13. This method correctly identifies negative samples because neither method blanks nor control blanks produces interfering peaks. Therefore, this method has 100% specificity for the four triazines when examined under the conditions detailed here *(23)*.

14. The sensitivity of this test is evaluated by its ability to correctly identify the samples known to be positive for triazines in fortified tissues ranging from 0.35 to 8 ppm. This method has 100% sensitivity for the four triazines examined individually at the concentrations tested. The levels tested are not meant to reflect the limit of detection of the method but rather the limit of quantitation as determined from the concentrations examined. This method has a proven limit of quantitation of 350 ppb for each triazine and represents a starting point in quantitating toxic levels in fish muscle, which are significantly higher *(23)*. Indeed, the sensitivity of the present method could be immediately enhanced by injection of 50–100 µL of reconstituted extract vs 10 µL as used here.

References

1. Albero, B., Sanchez-Brunete, C., and Tadeo, J.L. (2003) Determination of organophosphorus pesticides in fruit juices by matrix solid-phase dispersion and gas chromatography. *J. Agric. Food Chem.* **51**, 6915–6921.

2. Giza, I. and Sztwiertnia, U. (2003) Gas chromatographic determination of azoxystrobin and trifloxystrobin residues in apples. *Acta Chromatogr.* **13**, 226–229.

3. Husain, S. W., Kiarostami, V., Morrovati, M., and Tagebakhsh, M. R. (2003) Multiresidue determination of diazinon and ethion in pistachio nuts by use of matrix solid phase dispersion with a lanthanum silicate co-column and gas chromatography. *Acta Chromatogr.* **13**, 208–214.

4. Kastelan-Macan, M. and Babic, S. (2003) Pesticides, in *Handbook of Thin-Layer Chromatography*, 3rd ed. Chromatographic Science Series, Vol. 89, pp. 767–805.

5. Tadeo, J. L. and Sanchez-Brunete, C. (2003) Analysis of pesticide residues in fruit juices by matrix solid-phase dispersion and gas chromatographic determination. *Chromatographia* **57**, 793–798.

6. Li, Z.-Y., Zhang, Z.-C,, Zhou, Q.-L., Wang, Q.-M., Gao, R.-Y., and Wang, Q.-S. (2003) Stereo- and enantioselective determination of pesticides in soil by using achiral and chiral liquid chromatography in combination with matrix solid-phase dispersion. *J. AOAC Int.* **86**, 521–528.

7. Morzycka, B. (2002) Determination of organophosphorus pesticides in fruits and vegetables by matrix solid-phase dispersion method. *J. Plant Protect. Res.* **42**, 17–22.

8. Fernandez, M., Pico, Y., and Manes, J. (2002) Rapid screening of organophosphorus pesticides in honey and bees by liquid chromatography–mass spectrometry. *Chromatographia* **56,** 577–583.

9. Morzycka, B. (2002) Simple method for the determination of trace levels of pesticides in honeybees using matrix solid-phase dispersion and gas chromatography. *J. Chromatogr. A* **982,** 267–273.

10. Michel, M. and Buszewski, B. (2002) HPLC determination of pesticide residue isolated from food matrices. *J. Liq. Chromatogr. Relat. Technol.* **25,** 2293–2306.

11. Blasco, C., Font, G., and Pico, Y. (2002) Comparison of microextraction procedures to determine pesticides in oranges by liquid chromatography-mass spectrometry. *J. Chromatogr. A* **970,** 201–212.

12. Bordajandi, L. R., Merino, R., and Jimenez, B. (2001) Organochlorine exposure in peregrine falcon (*Falco peregrinus*) eggs and its avian prey (*Columbia livia*). *Organohalogen Comp.* **52** (Dioxin 2001), 139–142.

13. Perret, D., Gentili, A., Marchese, S., Sergi, M., and D'Ascenzo, G. (2002) Validation of a method for the determination of multiclass pesticide residues in fruit juices by liquid chromatography/tandem mass spectrometry after extraction by matrix solid-phase dispersion. *J. AOAC Int.* **85,** 724–730.

14. Blasco, C., Pico, Y., and Font, G. (2002) Monitoring of five postharvest fungicides in fruit and vegetables by matrix solid-phase dispersion and liquid chromatography/mass spectrometry. *J. AOAC Int.* **85,** 704–711.

15. Sanchez-Brunete, C., Albero, B., Miguel, E., and Tadeo, J. L. (2002) Determination of insecticides in honey by matrix solid-phase dispersion and gas chromatography with nitrogen-phosphorus detection and mass spectrometric confirmation. *J. AOAC Int.* **85,** 128–133.

16. Ahmed, F. E. (2001) Analyses of pesticides and their metabolites in foods and drinks. *Trends Anal. Chem.* **20,** 649–661.

17. Yague, C., Bayarri, S., Lazaro, R., Conchello, P., Arino, A., and Herrera, A. (2001) Multiresidue determination of organochlorine pesticides and polychlorinated biphenyls in milk by gas chromatography with electron-capture detection after extraction by matrix solid-phase dispersion. *J. AOAC Int.* **84,** 1561–1568.

18. Pous, X., Ruiz, M. J., Pico, Y., and Font, G. (2001) Determination of imidacloprid, metalaxyl, myclobutanil, propham, and thiabendazole in fruits and vegetables by liquid chromatography–atmospheric pressure chemical ionization-mass spectrometry. *Fresenius J. Anal. Chem.* **371,** 182–189.

19. Albero, B., Sanchez-Brunete, C., and Tadeo, J. L. (2001) Multiresidue determination of pesticides in honey by matrix solid-phase dispersion and gas chromatography with electron-capture detection. *J. AOAC Int.,* **84,** 1165–1171.

20. Valenzuela, A. I., Pico, Y., and Font, G. (2001) Determination of five pesticide residues in oranges by matrix solid-phase dispersion and liquid chromatography to estimate daily intake of consumers. *J. AOAC Int.* **84,** 901–909.

21. Kristenson, E. M., Haverkate, E. G. J., Slooten, C. J., Ramos, L., Vreuls, R. J. J., and Brinkman, U. A. T. (2001) Miniaturized automated matrix solid-phase dispersion extraction of pesticides in fruit followed by gas chromatographic–mass spectrometric analysis. *J. Chromatogr. A* **917,** 277–286.

22. Curren, M. S. S. and King, J. W. (2001) Ethanol-modified subcritical water extraction combined with solid-phase microextraction for determining atrazine in beef kidney. *J. Agric. Food Chem.* **49,** 2175–2180.

23. Gaunt, P. and Barker, S. A. (2000) Matrix solid phase dispersion extraction of triazines from catfish tissues; examination of the effects of temperature and dissolved oxygen on the toxicity of atrazine. *Int. J. Environ. Pollut.* **13,** 284–312.

24. Barker, S. A. (2000) Matrix solid-phase dispersion. *J. Chromatogr. A* **885,** 115–127.
25. Barker, S. A. (2000) Applications of matrix solid-phase dispersion in food analysis. *J. Chromatogr. A* **880,** 63–68.
26. Valenzuela, A. I., Pico, Y., and Font, G. (2000) Liquid chromatography/atmospheric pressure chemical ionization–mass spectrometric analysis of benzoylurea insecticides in citrus fruits. *Rapid Commun. Mass Spectrom.* **14,** 572–577.
27. Valenzuela, A. I., Redondo, M. J., Pico, Y., and Font, G. (2000) Determination of abamectin in citrus fruits by liquid chromatography–electrospray ionization mass spectrometry. *J. Chromatogr. A* **871,** 57–65.
28. Fernandez, M., Pico, Y., and Manes, J. (2000) Determination of carbamate residues in fruits and vegetables by matrix solid-phase dispersion and liquid chromatography-mass spectrometry. *J. Chromatogr. A* **871,** 43–56.
29. Valenzuela, A. I., Lorenzini, R., Redondo, M. J., and Font, G. (1999) Matrix solid-phase dispersion microextraction and determination by high-performance liquid chromatography with UV detection of pesticide residues in citrus fruit. *J. Chromatogr. A* **839,** 101–107.
30. Dorea, H. S. and Lancas, F. M. (1999) Matrix solid-phase dispersion extraction of organophosphorus and synthetic pyrethroid pesticides in cashew nut and passion fruit. *J. Microcolumn Sep.* **11,** 367–375.
31. Torres, C. M., Pico, Y., Marin, R., and Manes, J. (1997) Evaluation of organophosphorus pesticide residues in citrus fruits from the Valencian Community (Spain). *J. AOAC Int.* **80,** 1122–1128.
32. Torres, C. M., Pico, Y., and Manes, J. (1997) Comparison of octadecylsilica and graphitized carbon black as materials for solid-phase extraction of fungicide and insecticide residues from fruit and vegetables. *J. Chromatogr. A* **778,** 127–137.
33. Crouch, M. D. and Barker, S. A. (1997) Analysis of toxic wastes in tissues from aquatic species. Applications of matrix solid-phase dispersion. *J. Chromatogr. A* **774,** 287–309.
34. Yago, L. S. (1996) Matrix solid-phase dispersion: the next step in solid phase extraction for food samples? An overview of the technology. *Sem. Food Anal.* **1,** 45–54.
35. Viana, E., Molt, J. C., and Font, G. (1996) Optimization of a matrix solid-phase dispersion method for the analysis of pesticide residues in vegetables. *J. Chromatogr. A* **754,** 437–444.
36. Tekel, J. and Hatrik, S. (1996) Pesticide residue analyses in plant material by chromatographic methods: clean-up procedures and selective detectors. *J. Chromatogr. A* **754,** 397–410.
37. Obana, H. and Hori, S. (1996) Latest analytical methods for the residual pesticides in foods. *Jpn. J. Toxicol. Environ. Health* **42,** 1–16.
38. Torres, C. M., Pico, Y., and Manes, J. (1995) Analysis of pesticide residues in fruit and vegetables by matrix solid-phase dispersion (MSPD) and different gas chromatography element-selective detectors. *Chromatographia* **41,** 685–692.
39. Torres, C. M., Pico, Y., Redondo, M. J., and Manes, J. (1996) Matrix solid-phase dispersion extraction procedure for multiresidue pesticide analysis in oranges. *J. Chromatogr. A* **719,** 95–103.
40. Schenck, F. J. and Wagner, R. (1995) Screening procedure for organochlorine and organophosphorus pesticide residues in milk using matrix solid phase dispersion (MSPD) extraction and gas chromatographic determination. *Food Addit. Contam.* **12,** 535–541.
41. Ling, Y.-C. and Huang, I.-P. (1995) Multiresidue-matrix solid-phase dispersion method for determining 16 organochlorine pesticides and polychlorinated biphenyls in fish. *Chromatographia* **40,** 259–266.
42. Ling Y. C. and Huang, I. P. (1995) Multi-residue matrix solid-phase dispersion method for the determination of six synthetic pyrethroids in vegetables followed by gas chromatography with electron capture detection. *J. Chromatogr. A,* **695,** 75–82.

43. Pico, Y., Molto, J. C., Manes, J., and Font, G. (1994) Solid phase techniques in the extraction of pesticides and related compounds from foods and soils. *J. Microcolumn Sep.* **6,** 331–359.

44. Lott, H. M. and Barker, S. A. (1993) Comparison of a matrix solid phase dispersion and a classical extraction method for the determination of chlorinated pesticides in fish muscle. *Environ. Monit. Assess.* **28,** 109–116.

45. Walker, C. C., Lott, H. M., and Barker, S. A. (1993) Matrix solid-phase dispersion extraction and the analysis of drugs and environmental pollutants in aquatic species. *J. Chromatogr.* **642,** 225–242.

46. Lott, H. M. and Barker, S. A. (1993) Extraction and gas chromatographic screening of 14 chlorinated pesticides in crayfish (*Procambarus clarkii*) hepatopancreas. *J. AOAC Int.* **76,** 663–668.

47. Barker, S. A., Long, A. R., and Hines, M. E., II. (1993) Disruption and fractionation of biological materials by matrix solid-phase dispersion. *J. Chromatogr.* **629,** 23 34.

48. Lott, H. M. and Barker, S. A. (1993) Matrix solid-phase dispersion extraction and gas chromatographic screening of 14 chlorinated pesticides in oysters (*Crassostrea virginica*). *J. AOAC Int.* **76,** 67–72.

49. Stafford, S. C. and Lin, W. (1992) Determination of oxamyl and methomyl by high-performance liquid chromatography using a single-stage postcolumn derivatization reaction and fluorescence detection. *J. Agric. Food Chem.* **40,** 1026–1029.

50. Long, A. R., Crouch, M. D., and Barker, S. A. (1991) Multiresidue matrix solid-phase dispersion (MSPD) extraction and gas chromatographic screening of nine chlorinated pesticides in catfish (*Ictalurus punctatus*) muscle tissue. *J. AOAC Chem.* **74,** 667–670.

51. Long, A. R., Soliman, M. M., and Barker, S. A. (1991) Matrix solid phase dispersion (MSPD) extraction and gas chromatographic screening of nine chlorinated pesticides in beef fat. *J. AOAC Chem.* **74,** 493–496.

52. Barker, S. A., Long, A. R., and Short, C. R. (1989) Isolation of drug residues from tissues by solid phase dispersion. *J. Chromatogr.* **475,** 353–361.

23

Analysis of Fungicides in Fruits and Vegetables by Capillary Electrophoresis–Mass Spectrometry

Yolanda Picó

Summary

A method based on solid-phase extraction (SPE) and capillary electrophoresis–mass spectrometry (CE–MS) for determining fungicides in fruits and vegetables is described. The sample, previously chopped and homogenized, is extracted with a methanol–water mixture by sonication, filtrated, and adjusted to an appropriate volume with water. The target compounds are isolated by solid-phase extraction (SPE) on C8, eluted with dichloromethane, evaporated to dryness, and reconstituted in buffer. Fungicide residues present in the sample are preconcentrated by stacking with buffer removal. Separation is achieved by capillary zone electrophoresis (CZE) using a buffer of formic acid–ammonium formate at pH 3.5 with 2% methanol and an untreated 150 cm × 75 µm fused-silica capillary placed in a special external detector adapter (EDA) cartridge. The CE device is coupled to an electrospray interface by a commercial sheath flow adapter, in which the CE eluate is mixed with a suitable sheath liquid at the probe tip and then nebulized using nitrogen gas. The detection of fungicides is carried out on a single quadrupole using positive ionization and selected ion monitoring modes.

Key Words: Capillary electrophoresis–mass spectrometry; food analysis; fruits; fungicides; solid-phase extraction; vegetables.

1. Introduction

Capillary electrophoresis (CE) is an alternative technique for the rapid and highly efficient separation of pesticide residues; it resolves the different analytes with an efficiency similar to gas chromatography (GC) and determines thermally degradable or nonvolatile molecules like liquid chromatography (LC) *(1–6)*. However, CE suffers from poor concentration sensitivity because of the small injection volumes (typically from 1 to 10 nL), which represent <1% capillary length *(1,4,5,7,8)*. This limitation constitutes a significant obstacle for routine analysis of levels of pesticides below milligrams per kilogram in real samples by this technique *(9,10)*. Alternative detector formats, such as mass spectrometry (MS), offer greater sensitivity and selectivity than

From: *Methods in Biotechnology, Vol. 19, Pesticide Protocols*
Edited by: J. L. Martínez Vidal and A. Garrido Frenich © Humana Press Inc., Totowa, NJ

ultraviolet absorbance, increasing the applications of this technique to wider types of analytes and matrices *(11–18)*. Coupling of CE and MS has been described, and the most common technique for hyphenating capillary separation techniques to the MS has been electrospray ionization (ESI) *(16,19–22)*.

The routine application of CE–MS still suffers from "electrodrilling" inside the capillary, large capillary length, and low throughput *(15–17)*. This is reflected by the fact that almost all CE–MS interfaces described in the literature are custom made *(7,8, 12,14,15,18,23)*. The combination of off-line preconcentration by solid-phase extraction (SPE), which enriches analytes and performs sample cleanup prior to analysis, and on-line preconcentration or "stacking" represents an effective and versatile way to enhance concentration sensitivity in CE *(4,5,7,8,23)*. The determination of two fungicides, thiabendazole and procymidone, in fruits and vegetables illustrates how these approaches are combined to detect their residues at the levels below parts per million, levels at which they are usually present in real samples.

2. Materials

1. P/ACE MDQ CE system (e.g., Beckman, Fullerton, CA) with an external detector adapter (EDA; e.g., Beckman).
2. CE–MS/ESI (e.g., Agilent Sprayer Kit, G1607A, Waldbronn, Germany).
3. Mass spectrometer with atmospheric pressure ionization (API)/electrospray source (e.g., Hewlett-Packard HP1100, Palo Alto, CA).
4. Pump to add the sheath liquid (e.g. Hewlett-Packard HP1100 series).
5. Procymidone and thiabendazole standards, 99.99% pure.
6. Gradient-grade methanol for LC.
7. Dichloromethane for organic trace analysis.
8. Deionized water.
9. Ammonium formate.
10. Formic acid.
11. Octylsilica sorbent (particle diameter in the range 45–55 µm).
12. 100 mm × 9 mm id glass column fitted with a coarse frit (no. 3) with a standardized hollow key (14/23) and standard ground cone (29/32) that can be combined with several laboratory glassware items, such as funnels, Kitasato flasks, vacuum adapters, and the like. Prepare the glass column by transferring 1 g C8 into it and covering with a plug of silanized glass wool.
13. An untreated 150 cm × 75 µm id fused-silica capillary.
14. For stock standard solution of procymidone, add 50 mg thiabendazole to 50-mL volumetric flask. Dissolve and bring to volume with methanol. Store in stained glass bottles. Stable at 4°C for up to 1 mo.
15. For standard mixtures of fungicides, add appropriate volumes of the standard solutions of procymidone in methanol to 5 mL volumetric flask. Bring to volume with sample buffer. Make fresh mixtures as required.
16. Capillary conditioning solution 1: 100 m*M* NaOH.
17. Capillary conditioning solution 2: deionized water.
18. Separation buffer: 12 m*M* ammonium formate, 20 m*M* formic acid, pH 3.5, 2% methanol.
19. Sheath liquid: 12 m*M* ammonium formate, 20 m*M* formic acid, pH 3.5, 2% methanol.
20. Sample buffer for stacking of 1.5 m*M* ammonium formate, 2.5 m*M* formic acid, at pH 3.5 can be prepared by diluting the separation buffer eight times.

21. Fruit and vegetable samples are taken at various places distributed throughout the lot; they should weigh at least 1 kg and consist of at least 10 individual pieces. The sample is collected on a net bag and transported to the laboratory within 24 h of pickup. Samples are stored at 4°C until the analysis.
22. Filter standard solutions, samples, and buffers through a 0.22-μm membrane filter.

3. Methods

The methods detailed below summarize (1) the extraction of fungicides from fruits using an isolation-and-preconcentration method based on SPE, (2) the coupling of CE to MS, (3) the capillary electrophoretic separation of both fungicides using a simple stacking procedure, and (4) their identification and quantification by MS.

3.1. Sample Extraction

Frequently, the fungicides of interest are present at levels too low for detection. Sample preparation can concentrate the components to adequate amounts for measurement (*see* **Note 1**). The extraction, isolation, and concentration of thiabendazole and procymidone from fruits and vegetables are explained in **Subheadings 3.1.1.– 3.1.4.** This involves (1) the description of sample pretreatment, (2) the extraction of fungicides from the solid matrix, (3) the isolation of fungicides using SPE, and (4) their elution and concentration.

3.1.1. Sample Pretreatment

1. Samples are to be analyzed within 3 d of sample receipt or within 4 d of sample pickup. Samples are kept at 4°C prior to analysis to prevent disappearance or degradation of fungicide residues.
2. Analyze the samples unwashed, with the peel intact and, if applicable, stoned.
3. Chop the samples into small pieces using a conventional kitchen knife.
4. Homogenize a representative portion of the sample (200 g whole fruit or vegetable) in a food chopper (*see* **Note 2**).

3.1.2. Pesticide Extraction From the Solid Matrix

1. Weigh a 5-g portion of fruit or vegetable sample and place it into an Erlenmeyer flask. Add process control or spiking solutions at this step and let the samples stand at room temperature for 3 h to achieve the solvent evaporation and the fungicide distribution in the fruit or vegetable.
2. Add 5 mL methanol and 5 mL water and homogenize by sonication for 15 min.
3. Filter the resulting suspension through a Whatman 40-μm filter and wash the filter cake twice with 5 mL deionized water.
4. Adjust the filtrate to a volume of 100 mL with deionized water.

3.1.3. Solid-Phase Extraction

The SPE is carried out using conventional C8-bonded silica and a glass column produced according to the design specification described in the **Subheading 2.** (*see* **Note 3**).

1. Connect the glass column containing the solid phase to a Kitasato flask.
2. Condition or activate the solid phase using 10 mL methanol and 10 mL water to solvate the bonded phase. Do not allow the solid phase to dry out to prevent reversion to its original state.

3. Attach the conditioned glass column to a 200-mL glass funnel.
4. Pump samples through the conditioned solid phase with the aid of vacuum at a flow rate between 10 and 25 mL/min.
5. Dry the solid phase with air to avoid the coelution of the remaining water with the elution solvent.

3.1.4. Elution and Concentration

1. Elute fungicides by gravity into a 15-mL conical glass tube with 10 mL of dichloromethane.
2. Apply vacuum to ensure that all dichloromethane is removed from the solid phase.
3. Evaporate the eluate to dryness using a gentle stream of nitrogen at 40°C.
4. Reconstitute the dry residue in 500 µL of 4 mM sample stacking buffer, thoroughly mixing in an ultrasonic bath for 5 min.
5. Transfer to polymerase chain reaction vial for the EC–MS analysis. These vials allow small sample volumes to be handled. Conventional vials for electrophoresis require at least 2 mL of sample.

This extraction procedure yields a mean recovery of 65% for thiabendazole and 75% for procymidone. In addition, it provides at least a 10-fold concentration factor, and the extracts obtained are quite clean, suitable for further determination without any additional cleanup.

3.2. Coupling CE to the Mass Spectrometer

The connection of a P/ACE System MDQ to a mass selective detector (MSD) is carried out using a liquid sheath electrospray (ESI) source (*see* **Note 4** and **Fig. 1**), as described in **Subheadings 3.2.1.–3.2.4**. This includes (1) the connection of the system to deliver the sheath liquid, (2) the preparation of the capillary, (3) the capillary electrophoretic system, and (4) the mass spectrometer. The combination of both instruments requires careful optimization to prevent some coupling problems encountered in practice (*see* **Note 5**). Most frequent difficulties are siphoning, longer migration times, and lower peak resolution. To avoid siphoning the CE system, put the inlet vial at the same height as the sprayer tip of MSD (*see* **Note 6**). To minimize the second effect, place both instruments close to each other to allow a short capillary length (*see* **Note 7**).

3.2.1. Preparing the System to Deliver the Sheath Liquid

A syringe pump, which can deliver a precise flow rate of 1–25 µL/min, or a conventional LC pump with a flow splitter are the alternatives to deliver the sheath liquid into the ESI source. In this case, the sheath flow splitter (**Fig. 1**) (included in the Agilent CE–MS sprayer kit) has been connected to the pump outlet.

1. Connect the tubing labeled "Pump" to the pump outlet. The recycle or waste tubing may be drawn back into the sheath liquid bottle to reuse the solvent.
2. Leave the tubing labeled "Out" disconnected from the sprayer but place its end in a beaker. It will be connected later.
3. Fill the sheath liquid in solvent the bottle A of the 1100 pumping system. Open the purge valve and flush the pump at 2 mL/min (100% A) for 5 min. Reduce the flow rate to 1.3 mL/min and close the purge valve. The splitter splits the sheath liquid in the ratio 1:100, so the sheath liquid flow rate will be 13 µL/min. The backpressure at 13 mL/min is approx 130 bar.

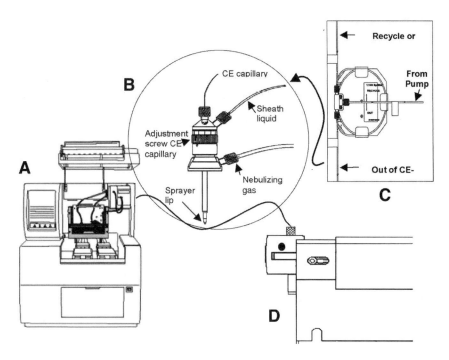

Fig. 1. Scheme of the instrumental setup of CE coupled to the MS system: (**A**) capillary electrophoresis; (**B**) CE–MS/ESI sprayer; (**C**) sheath flow splitter; and (**D**) mass spectrometer.

4. The sheath liquid must be degassed prior to use, which can be done by sonication or by a vacuum degasser. The vacuum degasser can be directly connected to the pump; in this case, the flush time should be longer.

3.2.2. Preparing the P/ACE MDQ System

The P/ACE MDQ system works with the EDA, which is designed for connection to external detectors such as mass spectrometers. The EDA consists of instrument modifications, software changes, and hardware to carry the capillary outside the MDQ. This modification is not user installable.

The instrument is used with a special EDA cartridge, in which the capillary exits the instrument and an external ground electrode completes the high-voltage circuits (**Fig. 2**). Basic difference with a conventional cartridge is that the inlet side of the cartridge is blocked (only maintains the cooling liquid circuit). Using an EDA cartridge, the MDQ detector is not available, and the length of capillary from the point sample injection to the outside of the MDQ is substantially reduced.

The experiments are carried out in the anionic mode (cathode at the inlet and anode at the outlet) of the MDQ system. When the EDA is installed, the function of the polarity indicator lights on the front of the MDQ is reversed (the reverse indicator light will light in normal polarity and vice versa).

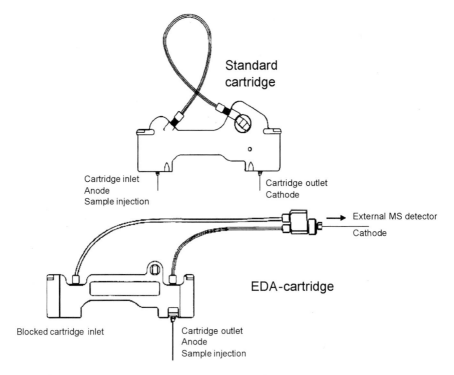

Fig. 2. EDA-only vs conventional cartridge.

1. Insert the capillary through the outlet side of the cartridge. Push the capillary through the cartridge and the outlet side coolant tube until it emerges from the capillary adapter body.
2. Trim the capillary to length using the MDQ trimming template before installing it in the instrument.
3. Insert the capillary adapter body into the slot in the right side of the MDQ instrument. Secure it with the locking ring.

3.2.3. Inserting and Adjusting the CE Capillary in the CE–MS/ESI Sprayer

1. Cut the outlet end of the capillary. Make sure that the cut is flat and the capillary is not shattered at the end because the quality of the cut determines the precision of the spray and avoids the presence of adsorptive sites for analytes.
2. Remove 2 to 3 mm of the polyamide coating at the tip of the CE capillary by burning it off and cutting the capillary to make the MS signal more stable.
3. Turn the adjustment screw counterclockwise (+ direction) until its mechanical stop. Then, turn two complete turns clockwise (– direction).
4. Fix the capillary so that it still can be moved up or down. The capillary should be aligned flat with the sprayer tip.
5. Tighten the fitting screw tight enough to keep the capillary in place.
6. Turn the adjustment screw a quarter turn counterclockwise (2 mark in + direction). Finally, the capillary should protrude approx 0.1 mm out of the sprayer tip to obtain a stable MS signal.

3.2.4. Preparing the MSD

1. Connect the nebulizing gas and the sheath liquid to the CE sprayer.
2. Carefully insert the CE sprayer into the electrospray chamber of the MSD. Do not touch the electrospray chamber with the sprayer tip because it can easily damage the sprayer tip. Do not hold the sprayer at the adjustment screw while inserting it into the ion source. This can misalign the sprayer.

3.3. CE Separation

The next step is the separation of the thiabendazole and procymidone using a simple stacking procedure to enhance sensitivity (*see* **Note 8**). CE instrumentation is controlled from a different computer with a System Gold software selecting the EDA mode so that the adapter functions correctly. It is important to remember that, in the EDA configuration, the capillary inlet is in the outlet side of the instrument. This affects programming for both voltage and pressure. The parameters giving in **Subheadings 3.3.1.** and **3.3.2.** are typical for determining the selected analytes in fruit and vegetable samples. However, they might have to be adapted when other analytes or other samples are analyzed.

3.3.1. Capillary Equilibration

Prior to first use, a new capillary should be properly conditioned. The CE capillary should also be reconditioned daily prior to experimental use to obtain a stable baseline in the detection system.

Equilibrate the capillary, executing in order the following high-pressure (20-psi) forward rinse steps:

> $0.1M$ sodium hydroxide for 20 min
> Deionized water for 10 min
> Separation buffer for 80 min

When the sheath liquid is running, there is no need to remove the capillary from the system.

3.3.2. Stacking and Separation of the Compounds

1. Perform a rinse flushing the capillary with the running buffer, 2 min at 20 psi, from a vial that contains separation buffer at a temperature of 16°C because application of the lowest thermostated temperature improves the resolution of the compounds.
2. Inject the sample hydrodynamically by pressure (0.5 psi) for 15 s.
3. Apply a negative voltage of –30 kV to the inlet end of the capillary.
4. When the current is 95% of the original running buffer, turn off the electrodes (*see* **Note 8**).
5. Carry out the separation with a positive power supply of 30 kV (average current 5.5 µA) at a temperature of 16°C.

Using the "stacking" technique, the analytes are concentrated fivefold compared with the standard injection (hydrodynamic injection, 0.5 s, 0.5 psi) (*see* **Fig. 3**). The sensitivity is not dramatically improved by the stacking, but the detection limits are better, and the peak shape is maintained.

The separation technique provides peak efficiencies higher than 4,520,000 (plates number) and resolution factor *(R's)* between both compounds up to 0.6.

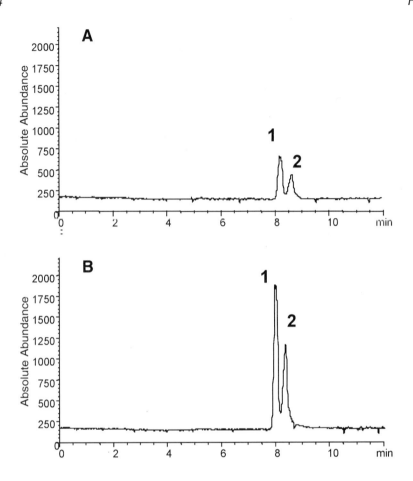

Fig. 3. CE–MS/ESI electropherograms in the selected ion monitoring mode of thiabenda-zole and procymidone standards (0.1 and 1 μg/mL) obtained by (**A**) standard injection and (**B**) stacking with matrix removal.

3.4. MS Identification and Quantification

Described below are the steps that can be utilized in the determination of thia-bendazole and procymidone by MS. The MSD, data acquisition, and handling are con-trolling by an independent computer by the LC ChemStation software (Agilent).

1. Set the spray chamber and mass spectrometer parameter (*see* **Table 1** and **Note 9**).
2. Analysis start/stop signals are provided by relay contacts on the CE instruments that are closed only during separation.

The peaks corresponding to thiabendazole and procymidone can be seen in the chro-matograms at the retention times of 7.8 and 8.5 min, respectively (*see* **Fig. 4**). The detection limits obtained are 0.03 and 0.003 μg/mL for procymidone and thiabenda-

Table 1
Parameters of the Spray Chamber and Mass Spectrometer

Parameter	Value
General information	
Use MSD	Enable
Ionization mode	API/ES
Polarity	Positive
Stop time	12 min
SIM parameters	
Ions selected	m/z 202 for thiabendazole
	m/z 284 for procymidone
Gain	3
Dwell time (ms)	349
Spray chamber parameters	
Drying gas flow	11 L/min
Drying gas temperature	150°C
Nebulizing gas pressure	40 psi
Voltage of the capillary	4000 V
Fragmentor voltage	100 V

zole, respectively (*see* **Fig. 4A**). The electropherograms displayed after extraction of fruit or vegetables samples that did not contain the studied fungicides show a stable baseline without interfering peaks caused by endogenous compounds (*see* **Fig. 4B**). The analysis of spiked sample extract shows no matrix interferences. The method provides quantification limits of 0.05 mg/kg for procymidone and 0.005 mg/kg for thiabendazole, making the determination reliable in real matrices (*see* **Fig. 4C**).

4. Notes

1. Multiple methods can be utilized to extract, isolate, and concentrate the fungicides present in fruit and vegetables. Methods such as solid–liquid extraction with organic solvents followed by liquid–liquid partitioning or solid-phase cleanup have been widely reported in the literature, and all can be combined with CE to analyze fungicides (*1–3,6,23*). However, the detection limits obtained should meet the criterion of being lower (at least twice) than the maximum residue limits established by the European Union (EU). Because of this, it is interesting that the selected method achieves a concentration factor of at least fivefold. The SPE procedure proposed easily provides concentration factors of 10-fold or more.
2. Good sample preparation is essential for accurate results with only a few sample measurements. Sample homogenization is the most neglected step in the analysis of fungicides in fruit and vegetables. If too large samples are taken, the time required for the homogenization is much too long, and the matrix is not homogeneous enough. On the contrary, small sample sizes can be little representative of the entire sample.
3. The SPE system reported here has been utilized very effectively to analyze different pesticides from water and aqueous extracts of fruit and vegetables (*10,24–26*). Other SPE systems have also proven to be very effective in isolating and concentrating pesticides prior to CE. Some of these include different formats (cartridges, syringes, and disks) and

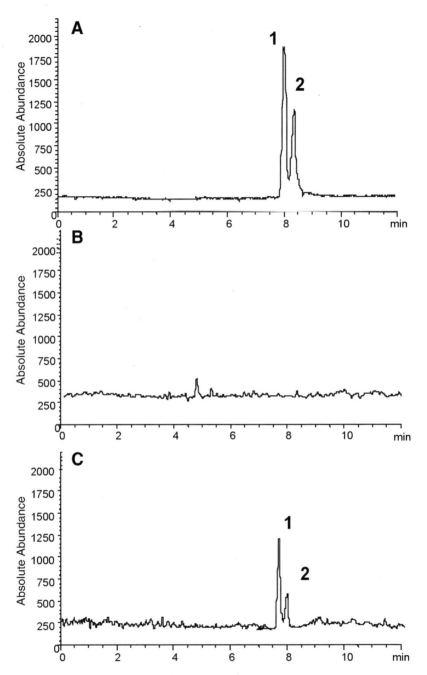

Fig 4. CE–MS/ESI electropherograms in the selected ion monitoring mode of (**A**) thiabenda-zole and procymidone standards (0.1 and 1 μg/mL), (**B**) untreated control orange sample, and (**C**) control orange sample spiked at 0.01 and 0.1 mg/kg, respectively. Reprinted from **ref. 25** with permission from Elsevier Science.

several sorbents [in addition to the traditional bonded silicas, C18 especially designed for polar compounds, several styrene–divinylbenzene copolymers, graphitized carbon blacks, and poly(divinylbenzene-*co-N*-vinylpyrrolidone) copolymers) *(4–6,23)*.

4. CE is most often interfaced to MS with the aid of an ESI source. The liquid sheath ESI source, first introduced by Smith et al. *(27)*, has become the most popular choice because it can most easily accommodate the high-concentration buffer systems typically used as eluents in CE and can tolerate both high and low electrosmotic flow rates as well. In a typical liquid sheath ESI configuration, the CE capillary is inserted in a stainless steel needle with an inner diameter that closely matches the outer diameter of the fused-silica capillary. The ESI voltage is applied to this metal needle. A liquid sheath, usually a mixture of $CH_3OH/H_2O/CH_3COOH$, is supplied through the needle around the last section of the CE capillary. The liquid sheath serves two main purposes: (1) it provides electrical contact at the CE capillary terminus, and (2) it dilutes the CE background electrolyte, making it electrosprayable. The composition of sheath liquid affects ESI sensitivity and CE resolution. The best results are mostly obtained when the composition of sheath liquid is similar to the running buffer.

5. When a liquid sheath ESI source is used to couple CE to MS, variations in migration time and loss of separation efficiency are sometimes experienced. Increased migration times (compared to ultraviolet detection) can simply be a result of the extra column length necessary to allow interfacing of the two techniques. However, there are several other factors that may interfere with and alter the separation efficiency and migration times, such as moving ionic boundaries; buffer depletion (toward the end of the CE capillary in a hot ESI source); penetration of the liquid sheath by diffusion into the CE capillary; electrode reaction at the ESI needle, which may result in pH modification; suction effect induced by the sheath gas flow used in addition to the liquid sheath flow; and alteration of **electrosmotic flow** by the radial electric field that penetrates across the fused silica capillary wall. In addition, the separation efficiency may also be compromised by the data acquisition speed, the response time of the mass spectrometer, the liquid sheath flow rate, and the relative positioning of the CE capillary and the ESI needle.

6. Siphoning of the sample is caused by a pressure difference between the inlet and the outlet. The solution is to keep both capillary ends at approximately the same height, which is ± 0.5 cm around the sprayer's levels, using a height-adjustable table placed under the CE instrument. However, typically this position is reached by placing both instruments in the same bench without more adjustment.

7. A standalone CE system maintains the capillary thermostated at a fixed temperature and handles separation capillary length from 20 to 50 cm because the detector is internal and close to the capillary. The coupling to an external mass spectrometer requires considerably extended capillaries with an overall length of 60 cm or more. As the mobility of an ion is proportional to the applied field strength, the latter is inversely proportional to the capillary length, which means that migration time is increased. In addition, only a small portion of the capillary is thermostated, and a large part of it (that between the CE outlet and the MS/ESI) is at room temperature. As the resolution between analytes is inversely proportional to the Joule effect, the peak resolution diminishes.

8. The term *stacking* was coined to describe any on-capillary mode of concentrating analytes based on changes of electrophoretic mobility because of the electric field across the concentration boundaries. The injected samples are prepared in a lower ionic strength buffer than the separation one. When the separation voltage is applied across the capillary, the analytes move quickly through the sample until they reach the buffer zone, where they slow because the high ionic strength of the separation buffer achieving analytes' concen-

tration. The optimum sample plug length that can be injected into the capillary without loss of separation efficiency is the limitation of the stacking. The sample volume injected can be increased if the sample buffer is removed after stacking by reversing the polarity. The stacking efficiency depends on the injection time, injection plug length, and ratio of conductivity between the separation sample and buffers.

9. The parameters of the mass spectrometer should be checked and optimized in full-scan mode prior to setting the instrument in selected ion monitoring mode because some variations and particular characteristics can be observed depending on the type and model of mass spectrometer, the individual tuning of the instrument, or the trademark of the reactives used as buffer or sheath liquid. For example, procymidone tends to form cluster ions with NH_4^+ that can be the observed ion instead of the protonated molecule.

Acknowledgments

This research was supported by the Ministry of Science and Technology, Spain together with the European Regional Developments Funds (ERDF) (project AGL2003-01407).

References

1. Picó, Y., Rodríguez, R., and Mañes, J. (2003) Capillary electrophoresis for the determination of pesticide residues. *Trends Anal. Chem.* **22,** 133–151.
2. Karcher, A. and El Rassi, Z. (1999) Capillary electrophoresis and electrochromatography of pesticides and metabolites. *Electrophoresis* **20,** 3280–3296.
3. Malik, A. K. and Faubel, W. (2001) A review of analysis of pesticides using capillary electrophoresis. *Crit. Rev. Anal. Chem.* **31,** 223–279.
4. Martínez, D., Borrull, F., and Calull, M. (1999) Strategies for determining environmental pollutants at trace levels in aqueous samples by capillary electrophoresis. *Trends Anal. Chem.* **18,** 282–291.
5. Martínez, D., Cugat, M. J., Borrull, F., and Calull, M. (2004) Solid-phase extraction coupling to capillary electrophoresis with emphasis on environmental analysis. *J. Chromatogr. A* **902,** 65–89.
6. Menzinger, F., Schmitt-Kopplin, P., Freitag, D., and Kettrup, A. (2000) Analysis of agrochemicals by capillary electrophoresis. *J. Chromatogr. A* **891,** 45–67.
7. Shihabi, Z. K. (2000) Stacking in capillary zone electrophoresis. *J. Chromatogr. A* **902,** 107–117.
8. Osbourn, D. M., Weiss, D. J., and Lunte, C. E. (2000) On-line preconcentration methods for capillary electrophoresis. *Electrophoresis* **21,** 2768–2779.
9. Rodríguez, R., Picó, Y., Font, G., and Mañes, J. (2001) Determination of urea-derived pesticides in fruit and vegetables by solid-phase preconcentration and capillary electrophoresis. *Electrophoresis* **22,** 2010–2016.
10. Rodríguez, R., Boyer, I., Font, G., and Picó, Y. (2001) Capillary electrophoresis for the determination of thiabendazole, prochloraz and procymidone in grapes. *Analyst* **126,** 2138.
11. Song, X. and Budde, W. L. (1998) Determination of chlorinated acid herbicides and related compounds in water by capillary electrophoresis-electrospray negative ion mass spectrometry. *J. Chromatogr. A* **829,** 327–340.
12. Otsuka, K., Smith, C. J., Grainger, J., et al. (1998) Stereoselective separation and detection of phenoxy acid herbicide enantiomers by cyclodextrin-modified capillary zone electrophoresis–electrospray ionization mass spectrometry. *J. Chromatogr. A* **817,** 75–81.

13. Tsai, C. Y., Chen, Y. R., and Her, G. R. (1998) Analysis of triazines by reversed electroosmotic flow capillary electrophoresis–electrospray mass spectrometry. *J. Chromatogr. A* **813,** 379–386.
14. Lazar, I. M. and Lee, M. L. (1999) Capillary electrophoresis time-of-flight mass spectrometry of paraquat and diquat herbicides. *J. Microcolumn Sep.* **11,** 117–123.
15. Menzinger, F., Schmitt-Kopplin, P., Frommberger, M., Freigtag, D., and Kettrup, A. (2001) Partial-filling micellar electrokinetic chromatography and non-aqueous capillary electrophores for the analysis of selected agochemicals. *J. Chromatogr. A* **371,** 25–34.
16. Hau, J. and Roberts, M. (1999) Advantages of pressurization in capillary electrophoresis/electrospray ionization mass spectrometry. *Anal. Chem.* **71,** 3977–3984.
17. Goodwin, L., Startin, J. R., Keely, B. J., and Goodall, D. M. (2003) Analysis of glyphosate and glufosinate by capillary electrophoresis–mass spectrometry utilising a sheetless microelectrospray interface. *J. Chromatogr. A* **1004,** 107–119.
18. Núñez, O., Moyano, E., and Galceran, M. T. (2002) Capillary electrophoresis–mass spectrometry for the analysis of quaternary ammonium herbicides. *J. Chromatogr. A* **974,** 243–255.
19. Niessen, W. M. A., Tjaden, U. R., and van der Greef, J. (1993) Capillary electrophoresis-mass spectrometry. *J. Chromatogr. A* **636,** 3–19.
20. Ñielen, M. W. F. (1995) Industrial applications of capillary zone electrophoresis-mass spectrometry. *J. Chromatogr. A* **712,** 269–284.
21. Tanaka, Y., Otsuka, K., and Terabe, S. (2003) Evaluation of an atmospheric pressure chemical ionization interface for capillary electrophoresis–mass spectrometry. *J. Pharm. Biomed. Anal.* **30,** 1889–1895.
22. Schmitt-Kopplin, P. and Frommberger, M. (2003) Capillary electrophoresis–mass spectrometry. Fifteen years of developments and applications. *Electrophoresis* **24,** 3837–3867.
23. Da Silva, C. L., de Lima, E. C., and Tavares, M. F. M. (2003) Investigation of preconcentration strategies for the trace analysis of multi-residue pesticides in real samples by capillary electrophoresis. *J. Chromatogr. A* **1014,** 109–116.
24. Rodríguez, R., Mañes, J., and Picó, Y. (2003) Off-line solid-phase microextraction and capillary mass spectrometry to determine acidic pesticides in fruits. *Anal. Chem.* **75,** 452–459.
25. Rodríguez, R., Picó, Y., Font, G., and Mañes, J. (2002) Analysis of thiabendazole and procymidone in fruits and vegetables by capillary-electrophoresis-electrospray mass spectrometry. *J. Chromatogr. A* **949,** 359–366.
26. Rodríguez, R., Picó, Y., Font, G., and Mañes, J. (2001) Analysis of post-harvest fungicides by micellar electrokinetic chromatography. *J. Chromatogr. A* **924,** 387–396.
27. Smith, R. D., Barinaga, C. J., and Udseth, H. R. (1998) Improved electrospray ionization for capillary zone electrophoresis–mass spectrometry. *Anal. Chem.* **60,** 1948–1952.

24

Application of Supercritical Fluid Extraction for the Analysis of Organophosphorus Pesticide Residues in Grain and Dried Foodstuffs

Kevin N. T. Norman and Sean H. W. Panton

Summary

A supercritical fluid extraction (SFE) method using carbon dioxide (CO_2) and cleanup by solid-phase extraction (SPE) with graphitized carbon black is presented. Gas chromatography using flame photometric detection (GC/FPD) and gas chromatography–mass spectrometric detection (GC–MSD) allow the quantitative determination of organophosphorus (OP) pesticide residues in grains such as wheat, maize, and rice as well as other dried foodstuffs such as cornflakes and kidney beans. The method is robust and sensitive, allowing residues to be determined down to 0.05 mg/kg or lower using GC–MSD. Extraction, cleanup, and determination are automated, allowing higher sample throughput compared to methods using solvent extraction and gel permeation chromatography. Solvent consumption is minimal using the SFE method, and hazardous waste is reduced. The SFE and cleanup methods used are likely to be suitable for the determination of OP pesticide residues in other dried foodstuffs without the need to alter the SFE or SPE conditions.

Key Words: Cereals; dried foods; flame photometric detection; gas chromatography; grains; mass spectrometric detection; organophosphorus pesticides; supercritical fluid extraction.

1. Introduction

Current methodologies to determine pesticide residues usually use liquid extraction *(1,2)*, often incorporating a partitioning stage, concentration of extract, as well as a cleanup stage prior to determination using chromatography. Such manual methods are labor intensive, need many hours of analyst time, and use large volumes of solvent that require disposal. An attractive alternative is to use supercritical fluids, which are substances above their critical temperature and pressure. Supercritical fluids have densities similar to liquids, but lower viscosities and higher diffusion coefficients. This combination of properties results in a fluid that is more penetrative, has a higher solvating power, and will extract solutes faster than liquids.

From: *Methods in Biotechnology, Vol. 19, Pesticide Protocols*
Edited by: J. L. Martínez Vidal and A. Garrido Frenich © Humana Press Inc., Totowa, NJ

A number of supercritical fluid extraction (SFE) studies suggested recoveries of pesticides are higher using SFE than those obtained using liquid extraction *(3–6)*, and extracts are potentially cleaner *(7,8)*. Supercritical carbon dioxide (CO_2) is a suitable fluid for the extraction of pesticide residues and has the additional advantages of a low critical point (31.05°C, 74.8 bar) *(9)*, ready availability, and relatively low toxicity. Commercial SFE instruments control supercritical conditions and automatically extract samples and collect analytes on sorbent traps, cryogenic traps, or in solvent. Extracts may then be analyzed directly, concentrated, or cleaned up prior to determination by chromatography using an appropriate detector. Once conditions are established, routine extractions are straightforward, and labor costs are reduced compared to conventional liquid extraction.

In this chapter, an automated method is described for the SFE of organophosphorus (OP) pesticides from grains and dried foodstuffs. Pesticides are collected on an octadecylsilane (ODS) sorbent trap and desorbed by acetone prior to clean up using automated solid-phase extraction (SPE) with ENVI-Carb cartridges (e.g., Supelco, Poole, UK) *(10)*. This method allows the analyst to prepare eight extracts in 2.5 h compared to 9 h using conventional liquid extraction and gel permeation chromatography *(11)*. The approach is robust, avoids the need for cleanup using gel permeation chromatography, and is suitable for many low-moisture food commodities, including cereals such as wheat, maize, white rice, and brown rice, as well as other dried foodstuffs, such as kidney beans and cornflakes. Conditions for determination using gas chromatography with flame photometric detection (GC/FPD) and mass spectrometric (MS) detection are also provided.

2. Materials

1. Supercritical fluid extractor (e.g., Agilent 7680T, Palo Alto, CA,); 7-mL extraction thimbles; end caps (e.g., Crawford Scientific, Strathaven, UK).
2. Glen Creston (Stanmore, UK) Retsch sample divider.
3. Glen Creston type 11-500 Laboratory Disc Mill (*see* **Note 1**) using a setting of 10.
4. Screen that allows 1 mm or smaller particles to pass through when sieving (e.g., Endecotts Ltd., London, UK).
5. Organic solvents (acetone, acetonitrile, toluene) suitable for pesticide residue analysis.
6. Pesticide standards as required.
7. 50- to 100-mesh acid washed sand, general purpose reagent (GPR; *see* **Note 2**).
8. Whatman GF/F 15-mm diameter glass microfiber filter disk.
9. ODS 50-μm particles (IST, Hengoed, UK; *see* **Note 3**) for use in the SFE sorbent trap.
10. SFE-grade CO_2 (99.995% pure), industrial Grade cryogenic CO_2 with dip tube (99.95% pure; *see* **Note 4**).
11. 2-mL Autosampler vials sealed with a polytetrafluoroethylene-faced septum for collecting SFE extracts and GC autosampler vials.
12. Acetonitrile:toluene (3:1 v/v) SPE elution solvent.
13. Automated Gilson ASPEC XL (e.g., Villiers-Le-Bel, France) with 6-mL cartridges containing ENVI-Carb (500 mg).
14. Techne DB 3-A, Dri-block, and Techne sample concentrator for blowing down of extracts (Techne, Princeton, NJ).
15. Corning Pyrex disposable borosilicate culture tube, 85 mm × 15 mm (New York, NY).

16. Gas chromatograph (e.g., Agilent model 5890 Series II, Waldborn, Germany) fitted with an FPD; 30 m × 0.25 mm DB-5, with a film thickness of 0.25 μm (e.g., J&W Scientific, Folsom, CA); model 7673 autosampler; computer data acquisition system using ChemStation software, version A 08.03 (Agilent).

17. Gas chromatograph (e.g., Agilent model 5890 Series II, Waldborn, Germany) fitted with an electron impact ionization (70-eV) mass spectrometric detector (e.g., Agilent model 5971, Palo Alto, CA); 30 m × 0.25 mm DB-5 MS, with a film thickness of 0.25 μm (e.g., J&W Scientific); computer data acquisition system using ChemStation software, G1034.C, version C 03.00.

3. Methods

3.1. Preparation of Solutions for Calibration

1. Prepare individual pesticide standard stock solutions of 1000 μg/mL in acetone and a standard mix containing 40 μg/mL of each of the pesticides in acetone from the individual standard stocks.

2. Use the standard mix to prepare solutions for spiking samples and matrix-matched standards appropriately diluted (*see* **Notes 5** and **6**).

3.2. Sample Storage and Homogenization

1. Store samples in a freezer (–18°C or lower).

2. Homogenize laboratory samples (2 kg) using the sample divider and recombining fractions from the sample divider.

3. Take a portion of the recombined fractions (500 g) and mill to a coarse powder/particles (*see* **Note 7**).

4. Store milled samples at –18°C or lower.

3.3. SFE Unit Preparations

1. Prepare the SFE trap by removing the old packing material and frit, taking care not to damage the screw thread of the trap.

2. Insert new frit and add fresh ODS using the supplied Agilent funnel.

3. Pack down using gentle tapping on the bench; do not pack down using glass rod or similar (*see* **Note 8**).

4. Install the SFE trap in the SFE unit and perform a system check using an empty extraction thimble and end caps prior to an extraction sequence (*see* **Note 9**). When the system check is successful, prepare samples for SFE sequence.

3.4. Preparation of Samples and Spiked Samples for SFE Sequence (see Fig. 1)

1. Place a glass microfiber filter disk in the bottom of an SFE extraction thimble and pack with milled sample (3.1–3.8 g) using a funnel (*see* **Note 10**).

2. Fill any dead volume with sand, tap down (adding more sand if required), and top with a glass microfiber filter disk before sealing with an SFE thimble end cap. A typical extraction sequence may contain eight samples or fewer if quality control samples are required.

3. Prepare spiked samples by adding 90 μL of acetone containing OP pesticides (2 or 20 μg/mL) using a microliter syringe to milled sample (3.1–3.8 g) known not to contain pesticide residues (*see* **Note 6**). This corresponds to a residue of approx 0.05 or 0.5 μg/g.

4. Allow the solvent to evaporate for 10 min (*see* **Note 11**).

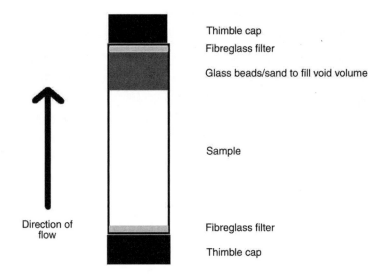

Fig. 1. Packing of the SFE extraction thimble.

3.5. Sample Extraction

1. Load extraction thimbles into SFE unit and insert appropriate number of capped autosampler vials in the SFE collection carousel.
2. Enter the SFE parameters into the SFE unit. Suggested extraction conditions are 5 min static extraction period at 245 bar and 70°C (0.73 g/mL density), followed by a 35-min dynamic extraction period at a flow rate of 1.0 mL/min.
3. Collect analytes following decompression of CO_2 on an ODS trap at 10°C with a nozzle temperature of 50°C.
4. Use acetone (1.5 mL) to desorb the ODS trap at 30°C with a flow rate of 2 mL/min into a 2-mL vial sealed with a polytetrafluoroethylene-faced septa.
5. Rinse the trap with 5 mL acetone at 30°C to waste to reduce the possibility of analyte carryover.
6. Remove SFE extracts from SFE unit, quantitatively transfer extracts to 2-mL volumetric flasks, and bring to volume with acetone (*see* **Note 12**).

3.6. SPE Cleanup of SFE Extract for GC Analysis

1. Transfer contents of 2-mL volumetric flask to an 85 × 15 mm borosilicate culture tube and seal with parafilm (*see* **Note 13**).
2. Enter the SPE parameters into the SPE unit.
3. Condition the cartridges with 5 mL SPE elution solvent at a flow rate of 5 mL/min.
4. Load 1 mL SFE extract onto the cartridge (5 mL/min) and elute with 5 mL elution solvent (1 mL/min).
5. Collect in a Corning Pyrex disposable borosilicate culture tube.
6. Reduce the extract to near dryness (0.2 mL) under nitrogen using a Techne dry block (45°C; *see* **Note 14**), and transfer the residuum to a 2-mL volumetric flask using three 0.5-mL aliquots of acetone, and bring to volume.
7. Transfer aliquots to GC autosampler vial ready for analysis.

3.7. Analysis of Extracts Using GC/FPD

1. Inject 1 μL of cleaned SFE extract into a split/splitless injector containing a 4-mm id deactivated single-taper liner at 250°C with the split vent opening after 0.75 min (*see* **Note 15**).
2. Use helium as carrier gas with electronic pressure control to maintain a linear velocity of about 30 cm/s.
3. Hold the column temperature at 100°C for 1 min, program at 20°C/min to 170°C for 15 min, 2°C/min to 180°C for 2 min, and at 30°C/min to 290°C and hold for 1.33 min.
4. Operate the detector in the phosphorous mode at 250°C.

3.8. Analysis of Extracts Using GC–Mass Spectrometric Detection

1. Inject 1 μL of cleaned SFE extract into a split/splitless injector containing a 4-mm id deactivated single taper liner at 250°C with the split vent opening after 0.75 min.
2. Use helium as carrier gas with electronic pressure control to maintain a linear velocity of about 27.5 cm/s.
3. Hold the column temperature at 100°C for 1 min, program at 25°C/min to 180°C for 2.1 min, 1°C/min to 190°C for 2.5 min, and at 25°C/min to 310°C and hold for 2.4 min.
4. Set the GC detector interface at 300°C and acquire data using selective ion monitoring to collect from 3 to 12 ions in a timed window (*see* **Note 15**).

3.9. Identification and Calibration

Pesticides can be identified by analyzing a single pesticide in acetone (1 μg/mL or higher) by GC/FPD or by GC–mass spectrometric detection (MSD) in full-scan mode using a single pesticide in acetone (10 μg/mL or higher; *see* **Note 16**) and recording retention times. The following calibration approach is suitable for the analysis of extracts using GC/FPD and GC–MSD. Use external standard calibration by bracketing blank, spiked sample, and samples with four calibrant levels. Prepare the lowest calibrant level equivalent to the required limit of quantitation. Choose a range of calibrants that covers the range of residues to be determined. Typical range for an initial run is 0.05–0.40 mg/kg (*see* **Note 17**).

4. Notes

1. A setting of 10 gives coarse powder/particles of which 99% passed through a 1-mm screen for dry commodities with moisture contents less than 20%.
2. Extract sand using the SFE method prior to use to ensure there are no contaminants present. If necessary, acid wash sand and recheck.
3. Any high-performance liquid chromatographic sorbent with particle size greater than 15 μm is suitable for use in the SFE sorbent trap.
4. SFE-grade CO_2 is used to extract samples. Cheaper, industrial grade CO_2 is suitable for cryogenic cooling of the SFE trap. Store a full cylinder of each type of gas in the laboratory so they are at ambient temperature and ready for immediate use when replacing empty cylinders.
5. Matrix-matched standards are prepared by adding, using microliter syringes, known amounts of pesticides to sample extracts that contain the same matrix concentration as samples. Calibrants prepared in this way protect analytes from degradation during GC injection, thus ensuring consistent responses are obtained for calibrants, spiked recover-

ies, and samples. Sufficient blank extract for matrix-matched standards should be prepared in advance by extracting samples known not to contain pesticide residues.

6. Any commodities needed as blank material used for matrix-matched standards or spiking must be checked before use by SFE and GC analysis. Organically grown commodities are usually suitable for commodity blanks but must also be checked before use.

7. Clean the sample divider and mill between samples by brushing and then processing a sample known not to contain pesticide residues and follow with further brushing.

8. It is not necessary to use a new trap for each sequence of eight sample extractions. Traps should only be repacked if it is suspected that extracts contain excessive amounts of coextractives.

9. Use the system check, which is a pressure drop test that automatically monitors leak rates. The user is notified of any leaks exceeding acceptable ranges; then, the user should leak test the SFE instrument and associated gas lines. Check gas lines and fittings for large CO_2 leaks, which will be seen as frosty deposits, and test for smaller leaks using 1% soap solution. The use of a separate extraction thimble and end caps, which are not used for samples, is recommended for the system check.

10. The mass of sample that can be loaded into an extraction thimble will vary depending on the sample density and particle size. Care should be taken to avoid contamination of extraction thimble/cap threads with sample or associated sample dust to ensure an effective seal is made.

11. Calculation of spiked recoveries gives an indication of how well the SFE unit is extracting and is a useful quality control tool. Do not use more than 100 µL solvent as this may lead to leaching of analytes through the bottom of the extraction thimble. It is important to allow the solvent to evaporate as any remaining solvent may act as a modifier, altering the polarity and extraction efficiency of the supercritical CO_2. A blank (i.e., a sample known not to contain pesticide residues) can be included in an extraction sequence following the spiked blank to show there is no carryover of pesticide residues to subsequent extractions if desired.

12. Avoid dependence on volumes of solvent dispensed by the SFE instrument trap as these may not be accurate. Collect solutions and bring to a fixed volume.

13. Parafilm used to cover the borosilicate culture tube reduces solvent evaporation while extracts are waiting SPE cleanup. It is not necessary to cover the collection vessels with parafilm because of the lower volatility of the SPE eluant solvents.

14. Care is required to prevent extract reaching complete dryness, which is likely to lead to the loss of more volatile pesticides such as dichlorvos and methacrifos.

15. Deterioration of chromatography, observed as peak broadening and change in detector response during the analysis of a batch of samples, may be observed because of the accumulation of coextractives or fats in the injector liner or the first 30 cm of analytical column. It is advisable to change the injection port liner and to remove the first 30 cm of analytical column following analytical runs of about 40 injections. Such a maintained system is capable of detecting pesticides at levels of 0.05 mg/kg with a minimum signal-to-noise ratio equal to 4 for the following OP pesticides: dichlorvos, methacrifos, diazinon, etrimfos, phosphamidon, chlorpyrifos methyl, fenitrothion, pirimiphos methyl, malathion, and chlorpyrifos ethyl. The extraction and cleanup technique is also suitable for additional OP pesticides and pesticides of other classes, such as organochlorine and synthetic pyrethroids.

16. For GC–MSD analysis, take a spectra of the pesticide peak and note its retention time. Select the most appropriate ion to be used for quantitation and select two qualifying ions

for each pesticide. Higher mass ions are most desirable, but also consider the abundance of the ions to ensure the limit of quantitation is achieved. Selecting a higher mass ion will limit any interference from commodity coextractives, some of which may coelute with pesticides.

17. It is convenient to prepare calibrants using levels that increase by a factor of two; this also facilitates interpretation of data. However, it may be more practical to use higher calibrant levels if anticipating samples that may contain high residue levels or samples known to contain residues at a higher level. The first and second set of calibrant response should not deviate by more than 20% from the mean response of the sets of calibrants to ensure accurate quantitation. If the response change is greater than 20%, it may be necessary to change the injection port liner and remove the first 30 cm of column or to calibrate more frequently.

References

1. Luke, M. A., Froberg, J. E., Doose, G. M., and Masumoto, H. T. (1981) Improved multiresidue gas chromatographic determination of organophosphorus, organonitrogen, and organohalogen pesticides in produce, using flame photometric and electrolytic conductivity detectors. *J. Assoc. Off. Anal. Chem.* **64,** 1187–1195.
2. Bottomley, P. and Baker, P. G. (1984) Multiresidue determination of organochlorine, organophosphorus and synthetic pyrethroid pesticides in grain by gas–liquid and high performance chromatography. *Analyst,* **109,** 85–90.
3. Norman, K. N. T. and Panton, S. W. (2001) Supercritical fluid extraction and quantitative determination of organophosphorus pesticide residues in wheat and maize using gas chromatography with flame photometric and mass spectrometric detection. *J. Chromatogr.* **907,** 247–255.
4. Valverde-García, A., Fernandez-Alba, A. R., Contreras, M., and Agüera, A. (1996) Supercritical fluid extraction of pesticides from vegetables using anhydrous magnesium sulfate for sample preparation. *J. Agric. Food Chem.* **44,** 1780–1784.
5. Faugeron, J., Tourte, J., Gros P., Charabel S., and Cooper J. F. (1997) Extraction par fluide supercritique de résidus de pesticides. Développement et validation d'une méthode appliquée aux ceréales et à leurs dérivés. *Analusis* **25,** 192–196.
6. Scopec, Z. V., Clark, R., Harvey, P. M. A., and Wells, R. J. (1993) Analysis of organophosphorus pesticides in rice by supercritical fluid extraction and quantitation using an atomic emission detector. *J. Chromatogr. Sci.* **31,** 445–449.
7. Kim, D. H., Heo, G. S., and Lee, D. W. (1998) Determination of organophosphorus pesticides in wheat flour by supercritical fluid extraction and gas chromatography with nitrogen-phosphorus detection. *J. Chromatogr. A,* **824,** 63–70.
8. Wang, J. H., Xu, Q., and Jiao, K. (1998) Supercritical fluid extraction and off-line cleanup for the analysis of organochlorine pesticide residues in garlic. *J. Chromatogr. A,* **818,** 138–143.
9. Anonymous. (1976) *International Thermodynamic Tables of the Fluid State, Carbon Dioxide*. Pergamon Press, Oxford, UK.
10. Fillion, J., Sauve, F., and Selwyn, J. (2000) Multiresidue method for the determination of 251 pesticides in fruits and vegetables by gas chromatography/mass spectrometry and liquid chromatography with fluorescence detection. *J. Assoc. Off. Anal. Chem.* **83,** 698–713.
11. Chamberlain, S. J. (1990) Determination of multi-pesticide residues in cereals, cereal products and animal feeds using gel-permeation chromatography. *Analyst,* **115,** 1161–1165.

25

Application of Microwave-Assisted Extraction for the Analysis of Dithiocarbamates in Food Matrices

Euphemia Papadopoulou-Mourkidou, Emmanuil Nikolaos Papadakis, and Zisis Vryzas

Summary

Microwave-assisted extraction (MAE) is a simple, fast, and accurate method developed for the analysis of N,N-dimethyldithiocarbamate (DMDTC) and ethylenebis(dithiobamate) (EBDTC) fungicides in fruits and vegetables. Residues are extracted from the plant matrices and hydrolyzed to CS_2 in a single step in the presence of 1.5% $SnCl_2$ in $5N$ hydrochloric acid using a laboratory microwave oven operated in the closed-vessel mode. The evolved CS_2, trapped in a layer of isooctane overlaying the reaction mixture, is analyzed by gas chromatography/flame photometric detection (GC/FPD). Sets of 12 samples are processed simultaneously. Quantification is based on external standard calibration curves made with either standard solutions of individual DMDTCs (thiram, ziram) or EBDTCs (maneb, zineb, mancozeb), which are processed as the field samples, or using standard solutions of CS_2 made in isooctane. Calibration curves are better fitted with quadratic equations with correlation coefficients >0.999; however, good linear correlation coefficients can also be obtained in narrower calibration ranges. Limits of detection (LODs) and limits of quantitation (LOQs) are in the range 0.005–0.1 mg/kg. Recoveries are >80%, with respective relative standard deviation values <20%.

Key Words: CS_2; Dithiocarbamates; fruits and vegetables; fungicides; gas chromatographic analysis; microwave-assisted extraction (MAE); residues.

1. Introduction

Dithiocarbamate (DTC) fungicides are very important nonsystemic fungicides used worldwide for the control of fungal and bacterial pathogens on tomatoes, grapes, apples, peaches, tobacco, lettuce, and other fruit and vegetable crops. DTCs used as plant protection products are divided into two subgroups: the N,N-dimethyldithiocarbamates (DMDTCs; thiram and ziram) and ethylenebis(dithiocarbamates) (EBDTCs; maneb, zineb, and mancozeb).

From: *Methods in Biotechnology, Vol. 19, Pesticide Protocols*
Edited by: J. L. Martínez Vidal and A. Garrido Frenich © Humana Press Inc., Totowa, NJ

Several methods have been published for the determination of DTC residues in fruits and vegetables. Most of these methods are based on an initial acid hydrolysis step of the DTCs in the presence of stannous chloride, as proposed by Keppel *(1,2)*, and subsequent analysis of the evolved CS_2 by spectrophotometry *(1,3)* or gas chromatography *(4–8)*. Methods employing a separation technique, such as high-performance liquid chromatography, prior to a spectophotometric-based technique for the final qualitative and quantitative analysis of residues of individual members of the DTCs have been also reported *(9–12)*. However, it should be emphasized here that maximum residue limits of DTCs in food are set collectively for the entire group of fungicides expressed as carbon disulfide; thus, irrespective of the analytical approach employed for their determination, residues are expressed in milligrams of carbon disulfide equivalents per kilogram of food.

Microwave-assisted extraction (MAE) techniques have been incorporated into many residue methods developed for the analysis of pesticides in a variety of substrates *(8,13–21)* and offer advantages such as improved accuracy and precision, reduced extraction time, low solvent consumption, and high level of automation compared to conventional extraction techniques *(22–24)*. Such an MAE-based method for the analysis of DTCs in tobacco and peaches has been published *(8)*, and this method as adapted for the analysis of DTCs in a variety of fresh fruits and vegetables is presented here.

Briefly, DTCs are subjected to acid hydrolysis in the presence of stannous chloride in an MAE system. The evolved CS_2 trapped in an overlaying isooctane layer is subjected to gas chromatographic analysis for qualitative and quantitative determinations. Sets of 12 samples are processed simultaneously. Two DTC fungicides were selected for method validation: thiram (a DMDTC) and mancozeb (an EBDTC). The selected commodities for method validation were apples, green beans, and peaches for thiram and potatoes and tomatoes and green peppers for mancozeb. Two additional commodities representing low-moisture processed foods containing fruits and vegetables were analyzed for mancozeb (baby fruit cream powder and dehydrated vegetable cubes).

2. Materials

1. Professional analysis-grade isooctane.
2. Hydrolysis solution of 1.5% stannous (II) chloride in $5M$ HCl, freshly prepared daily.
3. Mancozeb diluent of 0.2 M ethylenediaminetetraacetic acid and 0.4 M NaOH.
4. Microwave extraction system (e.g., MSP 1000, CEM, Matthews, NC) equipped with polytetrafluoroethylene-lined extraction vessels (*see* **Note 1**).
5. Gas chromatograph equipped with an FPD detector operated in the sulfur mode, split/splitless injector, and an autosampler (optional).
6. Chromatographic column 10 m × 0.53 mm (20 µm film thickness) with a 2.5-m particle trap (e.g., PORAPLOT Q column, Varian Analytical Instruments, The Netherlands).
7. Thiram stock and calibration solutions: Prepare a stock solution of 20 µg/mL in acetone (stable at −18°C for 5 days); dilute appropriately to produce calibration standards of 10, 5, 1, and 0.5 µg/mL.
8. Mancozeb stock and calibration solutions: Prepare a stock solution of 20 µg/mL in mancozeb diluent (use only freshly prepared solutions); dilute appropriately to produce calibration standards of 10, 5, 1, and 0.5 µg/mL.

Table 1
Operational Parameters of Microwave Oven

Magnetron power (W)	800
Maximum temperature (°C)	100
Maximum pressure (psi)	120
Time the system remains at above conditions (min)	25
Weight of sample (g)	10 for wet tissues
	2 for dry tissues
Number of vessels	12

9. CS$_2$ stock and calibration solutions: Prepare a stock solution of 1 mg/mL in isooctane; dilute appropriately to produce calibration standards of 0.05, 0.1, 0.5, 1, 5, and 10 µg/mL in isooctane (store at –18°C).

3. Methods

The methods presented deal with (1) safety and sample pretreatment, (2) the preparation of samples and calibration standards, (3) the chromatographic determination, and (4) the results.

3.1. Initial Considerations

3.1.1. Safety Precautions

Because of the handling of strong acid solutions and the fact that some of the DTCs are suspected carcinogens, all operations should be performed in a well-ventilated fume hood with the analyst wearing protective gloves.

3.1.2. Sample Pretreatment

1. Take a representative portion of the commodity to be analyzed.
2. If the sample is to be analyzed on the same day, homogenize it in a food processor at low speed. Immediately transfer the appropriate sample weight into the extraction vessels.
3. If the sample is to be analyzed on another occasion, keep the *intact* sample in a freezer (–18°C) (*see* **Note 2**).

3.2. Preparation of Samples and Calibration Standards

3.2.1. Sample Preparation

1. Weigh 10-g portions of sample (2 g in the case of low-moisture products) and transfer them into the sample vessels.
2. Add 10 mL isooctane and 30 mL hydrolysis solution.
3. Seal the vessels and shake them vigorously for approx 30 s (*see* **Note 3**).
4. Place the vessels into the MAE system.
5. Program the MAE system by entering the values depicted in **Table 1** (*see* **Note 4**).
6. Start the extraction.
7. When the extraction period is over, carefully remove the hot vessels from the MAE system, shake them for another 30-s period, and place them in an ice bath.
8. Monitor the temperature of the vessels at regular intervals by inserting the temperature probe into the control vessel.

Table 2
Preparation of Calibration Solutions to Construct a 10-Point Calibration Curve for Either DTC

Calibration level	Amount (μg) of pesticide added to each vessel	Volume (mL) of calibration solution transferred to each vessel	Calibration solutions (μg/mL)
1	0.1	0.2	0.5
2	0.25	0.5	0.5
3	0.5	1	0.5
4	1	1	1
5	2.5	0.5	5
6	5	1	5
7	10	1	10
8	20	1	20
9	40	2	20
10	50	2.5	20

9. Shake the vessels for 30 s when the temperature is in the range 50–60°C
10. When the temperature of the control vessel has reached 35°C, the sample vessels can be removed from the ice bath.
11. Carefully open the vessels and with the help of a glass Pasteur pipet transfer an aliquot of the isooctane phase into an autosampler vial (or other appropriate container). Seal the vial tightly to avoid carbon disulfide loss.

3.2.2. Preparation of Calibration Standards

Calibration curves are prepared by processing calibration standard solutions of either fungicide in the same manner as for the samples. To construct a 10-point calibration curve, pipet the appropriate volume of each calibration solution, as specified in **Table 2**, into each MAE vessel. Follow **steps 2–10** of the **Subheading 3.2.1.** (*see* **Note 5**).

3.2.3. Preparation of Fortified Samples

Fortified samples are prepared by spiking DTC–free portions of each commodity with appropriate volumes of thiram or mancozeb calibration solutions. After spiking, handle samples as described in **steps 2–10** of **Subheading 3.2.1.**

3.3. Gas Chromatography

3.3.1. Operational Conditions

Gas chromatographic analysis of carbon disulfide is carried out on a styrene–divinylbenzene copolymer porous column (*see* **Note 6**).

1. Use helium (99.999% pure) as a carrier gas at a constant linear velocity of 35 cm/s and nitrogen as makeup gas for the FPD at a flow rate of 30 mL/min. Also, set the air and hydrogen flow at 150 and 100 mL/min, respectively.

Table 3
Temperature Zones of Gas Chromatograph

Injector temperature	120°C
Detector temperature	200°C
Oven program	40–130°C at 8°C/min
	130–160°C at 10°C/min
	Hold at 160°C for 3 min

2. Program chromatograph temperature zones according to **Table 3**.
3. Set the detector in square root output mode (if applicable) (*see* **Note 7**).
4. Operate the split/splitless injector in the splitless mode.
5. Set the split valve to close 12 s prior to run start and to open 45 s after run start.
6. Use septum purge flow of about 0.5 mL/min and surge pressure of 2 psi from 4.8 s to 48 s after run start to ensure quick entry of carbon disulfide into the column.
7. Set the injection volume to 6 µL.
8. Start injection sequence.

3.3.2. Routine Sample Analysis

With every batch of samples, a blank run, three injections of a 10-µg/mL carbon disulfide standard solution and an injection of isooctane, should be performed to ensure acceptable retention time repeatability and sensitivity. A new calibration curve must be prepared each time a significant change in the sensitivity of the system is observed.

3.4. Results

3.4.1. Calibration Curves and Calculations

1. Construct calibration curves for each fungicide by plotting peak areas against the respective amount (micrograms) of fungicide added to each vessel.
2. Calculate the amount found for each pesticide, in micrograms, by referring to the appropriate calibration curve.
3. Report residue value by applying the following formula:

$$\text{Residue } (\mu g/g) = \mu g \text{ from calibration curve/Sample amount in g}$$

Sample calibration curves for carbon disulfide, thiram, and mancozeb are shown in **Figs. 1–3**. Calibration curves are better described by quadratic equations with correlation coefficients above 0.999. Good linear correlation coefficients ($r^2 = 0.999$) between concentrations and peak areas are also found at narrower concentration ranges (*see* **Note 8**).

Alternatively, results can be reported in micrograms CS_2 per gram of commodity. Calibration solutions of carbon disulfide are directly injected into the chromatographic system, and a calibration curve is constructed by plotting peak areas against respective concentrations (micrograms per microliter) of the carbon disulfide standard solutions injected. Results are calculated by the following formula:

$$CS_2 (\mu g/g) = 10 \text{ mL} \times (\mu g/mL \ CS_2 \text{ from calibration curve})/\text{Sample amount (g)}$$

Carbon disulfide

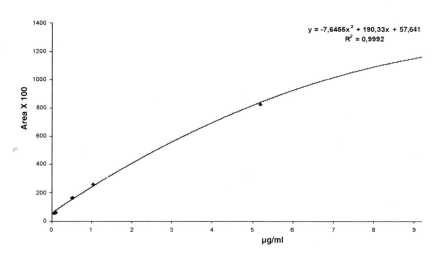

Fig. 1. Sample calibration curve for carbon disulfide.

Thiram

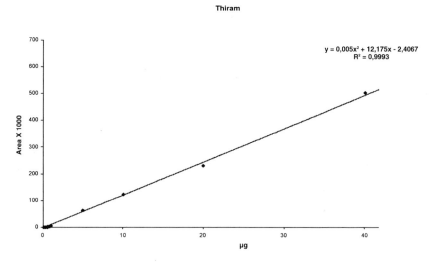

Fig. 2. Sample calibration curve for thiram.

3.4.2. Method Validation

The efficiency of the method is evaluated at regular intervals. This is performed by analyzing fortified samples of each commodity. When a new commodity is analyzed for the first time, recovery studies are performed to determine percentage recoveries in at least three fortification levels and establish respective LOD and LOQ values. In routine analysis, usually a single fortification level is adequate (*see* **Note 9**).

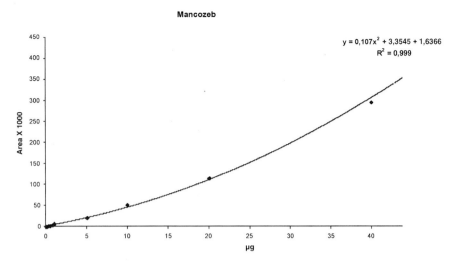

Fig. 3. Sample calibration curve for mancozeb.

Typical recovery values for thiram and mancozeb at three fortification levels, respectively, and for different commodities are given in **Table 4**. For all commodities, limits of detection (LODs) values are set at 0.005 µg/g for thiram and 0.01 µg/g for mancozeb. For thiram, limits of quantitation (LOQ) values of 0.01, 0.05, and 0.05 are determined for peach, green bean, and apple, respectively, whereas for mancozeb, LOQ values of 0.01 (tomato), 0.05 (potato, fruit cream powder, dehydrated vegetable cube), and 0.1 (green pepper) are obtained. Sample chromatograms are shown in **Figs. 4** and **5** (*see* **Note 10**).

4. Notes

1. Because the solvents used are mutually immiscible, sample particles tend to accumulate in the interface of the two layers; thus, contamination buildup is observed inside the vessel, especially when analyzing dry tissues. When this happens, it is possible to observe electric sparks inside the vessels, which could in extreme cases destroy the vessels. To avoid this, all vessels must be scrupulously cleaned with hot detergent solution and scrubbing after each extraction. In case of excess contamination buildup, the vessels can be cleaned with concentrated nitric acid after consulting the manufacturer.
2. Although it has been reported that DTC breakdown begins when residues are exposed to tissue juice, this seems to be a problem in methods in which the analysis of intact fungicide molecules is required *(7,9)*. However, it is advisable to proceed with sample preparation without undue delay. For tissues with high water content, they could be placed in the freezer and homogenized when partially frozen.
3. The vigorous shaking of the vessels facilitates tissue moistening, especially in the case of dry tissues; thus, sample accumulation in the interface is minimized.
4. The microwave system must be programmed to heat samples providing the maximum allowable energy at all times until the predetermined settings have been reached. This feature is called standard control in the specific instrument.

Table 4
Method Validation Data: Recoveries and Respective (Relative Standard Deviation) Values of the Selected DTCs in Fruits and Vegetables

Fortification level (µg/g)	Thiram					Mancozeb		
	Peach	Green bean	Apple	Tomato	Green pepper	Potato	Fruit cream powder	Dehydrated vegetable cube
0.5	90 (11)	103 (7)	92 (3)	110 (3)	94 (1)	85 (3)	81 (5)	83 (8)
1	100 (2)	98 (5)	105 (3)	113 (2)	98 (2)	93 (5)	95 (4)	82 (9)
5	85 (6)	100 (1)	95 (2)	87 (4)	95 (2)	97 (1)	94 (1)	93 (3)

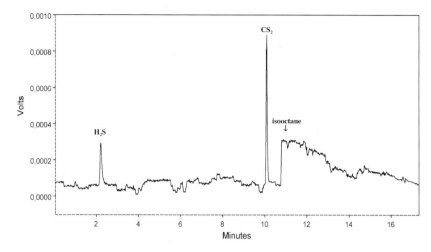

Fig. 4. Sample chromatogram from the analysis of a potato sample fortified with mancozeb at the 0.05-µg/g level.

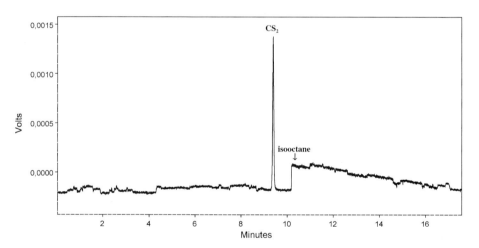

Fig. 5. Sample chromatogram from the analysis of an apple sample fortified with thiram at the 0.05-µg/g level.

5. Although visualization between the two phases is easy in the case of samples, it can be quite difficult to do when constructing a calibration curve; thus, there is a danger of taking up acidic reaction mixture along with isooctane. Also, avoid purging with air when transferring the sample with the Pasteur pipet into the autosample vials, or carbon disulfide loss may occur.

6. Porous layer open tubular (PLOT) columns are recommended for the analysis of solvents and volatile compounds such as CS_2, although traditional packed columns may also be used if carbon disulfide can be resolved from isooctane. The new type of bonded PLOT

columns promises many advantages compared to nonbonded predecessors, but they must be tested for the specific use. PLOT columns must be handled with extreme caution during installation and operation to avoid bed channeling.

7. When the FPD is operated in the sulfur mode, sulfur is detected as S_2, and the response is proportional to the square of the concentration of the sulfur-containing compound. When the square root output mode is selected, the detector's electrometer output is modified to correct for the nonlinear relationship between the detector output current and sulfur concentration. This feature is not applicable to all instruments.

8. When analyzing samples with residues near the LOQ levels, more accurate results can be obtained by constructing a calibration curve of only calibration points near the residue level.

9. The LOD is determined as the lowest concentration of the fungicide giving a signal-to-noise ratio of 3. The LOQ is determined as the lowest concentration of the fungicide giving a response that can be quantified with a relative standard deviation below 20%.

10. Except from the CS_2 and isooctane peaks, in some commodities peaks corresponding to other sulfur-containing compounds (mercaptans, hydrogen sulfide) appear in the chromatogram. In some instances, a compound with a retention time similar to carbon disulfide elutes from the column, and although it can be adequately resolved later, loss of column efficiency because of contamination from sample compounds can result in quantification errors. Also, loss of resolution between CS_2 and isooctane peaks can be observed because of column contamination.

References

1. Keppel, G. E. (1969) Modification of the carbon disulfide evolution method for dithiocarbamate residues. *J. AOAC* **52,** 162–167.
2. Keppel, G. E. (1971) Collaborative study of the determination of dithiocarbamate residues by a modified carbon disulfide evolution method. *J. AOAC* **54,** 528–532.
3. Kesari, R. and Gupta, V. K. (1998) A sensitive spectrophotometric method for the determination of dithiocarbamate fungicide and its application in environmental samples. *Talanta* **45,** 1097–1102.
4. Ahmad, N., Guo, L., Mandarakas, P., and Appleby, S. (1995) Determination of dithiocarbamate and its breakdown product ethylenethiourea in fruits and vegetables. *J. AOAC Int.* **78,** 1238–1243.
5. Woodrow, J. E. and Seiber, J. N. (1995) Analytical method for the dithiocarbamate fungicides ziram and mancozeb in air: Preliminary field results. *J. Agric. Food Chem.* **43,** 1524–1529.
6. Ahmad, N., Guo, L., Mandarakas, P., Farah, V., Appleby, S., and Gibson, T. (1996) Headspace gas–liquid chromatographic determination of dithiocarbamate residues in fruits and vegetables with confirmation by conversion to ethylenethiourea. *J. AOAC Int.* **79,** 1417–1422.
7. Royer, A., Ménand, M., Grimault, A., and Communal, P. Y. (2001) Development of automated headspace gas chromatography determination of dithiocarbamates in plant matrixes. *J. Agric. Food Chem.* **49,** 2152–2158.
8. Vryzas, Z., Papadakis, E. N., and Papadopoulou–Mourkidou, E. (2002) Microwave–Assisted Extraction (MAE)–acid hydrolysis of dithiocarbamates for trace analysis in tobacco and peaches. *J. Agric. Food Chem.* **50,** 2220–2226.
9. Gustafsson, K. H. and Fahlgren, C. (1983) Determination of dithiocarbamate fungicides in vegetable foodstuffs by high–performance liquid chromatography. *J. Agric. Food Chem.* **31,** 461–463.

10. Ekroth, S. B., Ohlin, B., and Österdahl, B.-G. (1998) Rapid and simple method for determination of thiram in fruits and vegetables with high-performance liquid chromatography with ultraviolet detection. *J. Agric. Food Chem.* **46**, 5302–5304.

11. Lo, C.-C., Ho, M.-H., and Hung, M.-D. (1996) Use of high–performance liquid chromatography and atomic absorption methods to distinguish propineb, zineb, maneb, and mancozeb fungicides. *J. Agric. Food Chem.* **44**, 2720–2723.

12. Weissmahr, K. W., Houghton, C. L., and Sediak, D. L. (1998) Analysis of the dithiocarbamate fungicides ziram, maneb, and zineb and the flotation agent ethylxanthogenate by ion–pair reversed–phase HPLC. *Anal. Chem.* **70**, 4800–4804.

13. Steinheimer, T. R. (1993) HPLC determination of atrazine and principal degradates in agricultural soils and associated surface and ground water. *J. Agric. Food Chem.* **41**, 588–595.

14. Molins, C., Hogendoorn, E.A., Heusinkveld, H. A. G., van Harten, D. C., van Zoonen, P., and Baumann, R.A. (1996) Microwave assisted solvent extraction for the efficient determination of triazines in soil samples with aged residues. *Chromatographia* **43**, 527–532.

15. Hoogerbrugge, R., Molins, C., and Baumann, R. A. (1997) Effects of parameters on microwave assisted extraction of triazines from soil: evaluation of an optimization trajectory. *Anal. Chim. Acta* **348**, 247–253.

16. Xiong, G., Liang, J., Zou, S., and Zhang, Z. (1998) Microwave–assisted extraction of atrazine from soil followed by rapid detection using commercial ELISA kit. *Anal. Chim. Acta* **371**, 97–103.

17. Papadakis, E. N. and Papadopoulou–Mourkidou, E. (2002) Determination of metribuzin and major conversion products in soils by microwave-assisted water extraction followed by liquid chromatographic analysis of extracts. *J. Chromatogr. A* **962**, 9–20.

18. Vryzas, Z. and Papadopoulou-Mourkidou, E. (2002) Determination of triazine and chloroacetanilide herbicides in soils by microwave-assisted extraction (MAE) coupled to gas chromatographic analysis with either GC-NPD or GC-MS. *J. Agric. Food Chem.* **50**, 5026–5033.

19. Molins, C., Hogendoorn, E. A., Dijkman, E., Heusinkveld, H. A. G., and Baumann, R. A. (2000) Determination of linuron and related compounds in soil by microwave–assisted solvent extraction and reversed–phase liquid chromatography with UV detection. *J. Chromatogr. A* **869**, 487–496.

20. Hogendoorn, E. A., Huls, R., Dijkman, E., and Hoogerbrugge, R. (2001) Microwave assisted solvent extraction and coupled–column reversed–phase liquid chromatography with UV detection–Use of an analytical restricted–access–medium column for the efficient multi–residue analysis of acidic pesticides in soils. *J. Chromatogr. A* **938**, 23–33.

21. Patsias, J., Papadakis, E. N., and Papadopoulou–Mourkidou, E. (2002) Analysis of phenoxyalkanoic acid herbicides and their phenolic conversion products in soil by microwave assisted solvent extraction and subsequent analysis of extracts by on–line solid–phase extraction–liquid chromatography. *J. Chromatogr. A* **959**, 153–161.

22. Eskilsson, C. S. and Björklund, E. (2000) Analytical–scale microwave–assisted extraction. *J. Chromatogr. A* **902**, 227–250.

23. Buldini, P. L., Ricci, L., and Sharma, J. L. (2002) Recent applications of sample preparation techniques in food analysis. *J. Chromatogr. A* **975**, 47–70.

24. Camel, V. (2000) Microwave–assisted solvent extraction of environmental samples. *Trends Anal. Chem.* **19**, 229–248.

26

Enantioselective Determination of α-Hexachlorocyclohexane in Food Samples by GC–MS

Chia-Swee Hong and Shaogang Chu

Summary

A method for the determination of individual α-hexachlorocyclohexane (α-HCH) enantiomers is described. The method consists of ultrasonic extraction, sulfuric acid-modified silica gel cleanup, fractionation on a polar silica solid-phase extraction (SPE) column, and final determination by gas chromatography/electron capture negative ion mass spectrometry (GC–MS/ECNI) on a Chirasil-Dex column. The MS is operated in selected ion monitoring mode. Racemic α-HCH is used as the calibration standard, and ε-HCH is used as the surrogate internal standard. Each α-HCH enantiomer is quantitated by the relative response factor method using the m/z 255 ion. The enantiomeric ratio (ER) is obtained as the area ratio of the (+)/(−) peaks eluting from the cyclodextrin column. ER determinations are important because enantiomers differ considerably in how they accumulate in organisms and decompose in the environment. This method provides a simple and useful tool to monitor the enantioselectivity of α-HCH in food samples.

Key Words: Chiral; food; GC; α-HCH; MS; SPE.

1. Introduction

Hexachlorocyclohexane (HCH) has been used as a pesticide since 1942, when the insecticidal properties of the γ-isomer were discovered. Technical HCH has an approximate composition of 60–70% α-HCH, 5–12% β-HCH, 10–15% γ-HCH, and minor percentages of other isomers; it is a widely used organochlorine compound and is found all over the world (*1–4*). Although γ-HCH is the isomer with the dominant insecticidal properties, the other isomers also have toxic properties (*4*). The technical mixture was banned in the United States and Canada in 1978 and 1971, respectively, but is still used in some developing countries.

α-HCH is easily transported via the atmosphere, and it has contaminated all environmental matrices, even those distant from sources (*4*). α-HCH is the only chiral HCH isomer and exists as two structurally distinct, mirror images called *enantiomers* (**Fig. 1**) on the basis of a lack of axial symmetry. Enantiomers have identical physical

From: *Methods in Biotechnology, Vol. 19, Pesticide Protocols*
Edited by: J. L. Martínez Vidal and A. Garrido Frenich © Humana Press Inc., Totowa, NJ

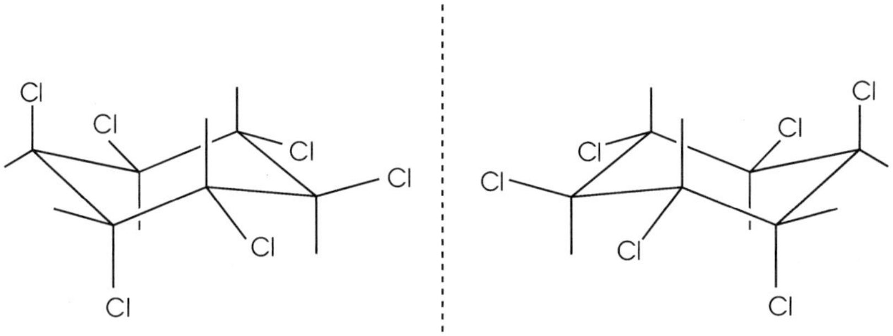

Fig. 1. Enantiomers of α-HCH.

properties and are designated as either (+) or (−) corresponding to their ability to rotate a plane of polarized light. The chiral compound is released into the environment as a racemic mixture *(5)*; however, the enantiomers may exhibit enantiomer-specific biological and toxicological characteristics *(6,7)*. Thus, the determination of enantiomeric ratios (ERs) or enantiomeric fractions (EFs) of chiral compounds can be used to assess the potential stereoselectivity of biotransformation pathways and enantiomer-specific biological activity because these processes may affect the relative accumulation of chiral compounds in biota. The ER (+/−) of α-HCH has been measured in many environmental samples, and different ratios have been observed in samples taken from air, water, fish, birds, and sea mammals *(6,8–11)*.

The determination of ERs of organochlorines by high-resolution gas chromatography became possible with the introduction of suitable chiral stationary phases (CSPs) a decade ago. Since then, a number of CSPs with modified cyclodextrins have been applied for the enantiomeric separation of chiral organochlorines *(12–14)*.

This chapter describes a method for determining the α-HCH enantiomers in food samples by gas chromatography/electron capture negative ion mass spectrometry (GC–MS/ECNI) *(12,15,16)* using a commercially available, modified cyclodextrin GC column.

2. Materials

1. (a) α-HCH standard; (b) a surrogate internal standard (ε-HCH; Ultra Scientific, Kingstown, RI; *see* **Note 1**).
2. Organic solvents (hexane, acetone, dichloromethane, methanol, isooctane) certified for ultratrace analyses or pesticide residue trace analyses.
3. Concentrated H_2SO_4 analytical reagent.
4. A food processor/blender or tissuemizer.
5. Ultrahigh-purity helium and methane (minimum purity 99.999%) for GC–MS/ECNI analysis; prepurified nitrogen (minimum purity 99.995%) for solvent evaporation after extraction, before fractionation, and before GC analysis.
6. Analytical balances (Ohaus, Florham, NJ; *see* **Note 2**).
7. Anhydrous sodium sulfate (*see* **Note 3**).
8. Sorvall RC-3 automatic refrigerated centrifuge, HL-8 rotor with 100-mL bucket.
9. Empty solid-phase extraction (SPE) reservoirs (20 mL) with frits (20-μm pore size) (Varian Inc., Walnut Creek, CA).

10. Silica gel (Aldrich; 70–230 mesh).
11. Polar silica SPE columns (Bond Elut straight-barrel columns, 40-μm particle size), 500 mg sorbent mass, 3-mL volume (Varian).
12. Kuderna-Danish (K-D) evaporator with 125-mL capacity fitted with Snyder column and 12-mL volumetric receiving tube, Kontes Glass Co.
13. Ultrasonic processor with titanium probe.
14. Agilent 6890 GC equipped with a programmed temperature vaporizing (PTV) injector and 5973 quadrupole mass spectrometer operated in ECNI mode (*see* **Note 4**); Chirasil-Dex column (Chrompack, Raritan, NJ; 25 m × 0.25 mm id and 0.25-μm film thickness; *see* **Note 5**) with immobilized permethyl 2,3,6-tri-*O*-methyl β-cyclodextrin on a polysiloxane backbone as the stationary phase.
15. A 3-m long retention gap with 0.25-mm id (Agilent Technologies, Palo Alto, CA; *see* **Note 6**).
16. 2-mL GC vials equipped with plastic screw caps and Teflon-lined septa.

3. Methods

3.1. Preparation of Standard Solutions (see Note 7)

1. Prepare stock solution A of standard α-HCH in isooctane (1 mg/mL) and a separate solution B of ε-HCH (1 mg/mL) in isooctane to be used as a surrogate internal standard.
2. Make a 1:10 dilution with isooctane to obtain the working solution.
3. Prepare the standard solution by adding 100 μL each of A and B work solutions to a 100-mL volumetric flask. Bring to volume with isooctane.

3.2. Sample Extraction

1. Homogenize the food sample (meat, fish fillet, egg) in a food processor/blender or with a tissuemizer.
2. Weigh 2–5 g of homogenized food sample (biological tissue) directly into a 100-mL centrifuge bottle on a top-loading balance.
3. Add 20 g of anhydrous sodium sulfate in a 100-mL centrifuge bottle, mix well with spatula. For egg samples, weigh 2–5 g of homogenized sample directly into a 100-mL centrifuge bottle, add 3 mL methanol, shake well, and let sit for 20 min before adding 20 g of anhydrous sodium sulfate.
4. Spike the sample with 100 μL surrogate internal standard (100 μg/mL).
5. Extract with 30 mL of 1:1 hexane:acetone by sonication for 5 min and then centrifuge at 1699*g* for 10 min (*see* **Note 8**).
6. Repeat **step 5** twice more, combining extracts.

3.3. Lipid Determination

1. Accurately measure the extract volume and remove 10 mL of extract to a preweighed and labeled test tube.
2. Evaporate the solvent by a gentle stream of nitrogen and then completely dry in oven at 105°C overnight.
3. Cool to ambient temperature and weigh the residue accurately on an analytical balance.

3.4. Evaporation

1. Transfer the remaining extract into a 125-mL K-D flask with 12-mL receiving tube.
2. Add small boiling chip to sample and concentrate the extracts to 6–8 mL using a three-ball Snyder column on a steam bath.

3. Rinse Snyder column twice with hexane; when dry, remove Snyder column and flask. Put Vigreaux column onto 12-mL receiving tube. Concentrate to approx 4 mL, rinse Vigreaux column twice with hexane, remove column, and continue concentrating to 0.5 mL with a gentle stream of nitrogen. Solvent exchange into hexane (*see* **Note 9**).

3.5. Lipid Removal

3.5.1. Preparation of 50% Sulfuric Acid on Silica

1. Rinse silica gel with consecutive methanol and dichloromethane.
2. Air-dry the dichloromethane-saturated material in the fume hood and then in an oven at 180°C for 3 h.
3. Cool and place in an appropriate size glass bottle.
4. Add sufficient concentrated sulfuric acid directly to the silica to yield an acid concentration of 50% based on total weight.
5. Shake manually until no clumping can be observed and store in a desiccator over phosphorous pentoxide until use. *Caution*: This reagent retains all of the properties of concentrated sulfuric acid and should be handled accordingly.

3.5.2. Acid-Modified Silica Cleanup (see **Note 10**)

1. Place 8 g acid-modified silica gel into an empty SPE column with a frit on the bottom. Settle the silica gel by tapping.
2. Quantitatively transfer the sample extract to the column (add the rinses).
3. Elute with 30 mL hexane and collect the eluate.
4. Concentrate to 0.5 mL as described in **Subheading 3.4**.

3.6. SPE Cleanup and Fractionation (see **Note 11**)

1. Preelute the polar silica SPE column (prepacked SPE tubes) with 5 mL hexane. Discard the eluate. Just before exposing the top frit to air, stop the flow.
2. Quantitatively transfer the concentrated extract to the column.
3. Elute with 10 mL hexane, discard the first 3 mL (contains most of the polychlorinated biphenyls [PCBs] and some pesticides) and collect 5 mL of the eluate, which contains the HCH isomers.
4. Reduce the eluate to near dryness with a gentle stream of nitrogen and reconstitute in 100 µL isooctane.
5. Transfer to a 150-µL glass insert in GC vial for the following GC–MS determination.

3.7. Gas Chromatography–Mass Spectrometry

1. Inject 25 µL (multiple injection, 5 × 5 µL) of sample with a large-volume PTV injector (*see* **Note 12**). The parameters for the injection are as follows: 60°C initial temperature, hold for 1 min, then increase to 300°C at 600°C/min. The vent time is 0.9 min, and the purge time is 2.5 min. The column temperature is held at 60°C for 3 min, ramped to 140°C at 10°C/min, ramped to a final temperature of 190°C at 5°C/min, and held at 190°C for 45 min, for a total analysis time of 66 min (*see* **Note 13**). The MS ion source is held at 250°C, and the quadrupoles are held at 150°C. Methane is used as the reagent gas and perfluoro-5,8-dimethyl-3,6,9-trioxadodecane is used as the calibrant.
2. In selected ion monitoring (SIM) mode, record the ions at m/z 253 and m/z 255 and use ion at m/z 255 for quantitation and ion at m/z 253 for qualification. In addition to retention time, use relative intensities of the monitored ions for identification (require to be within 10% of standard values).

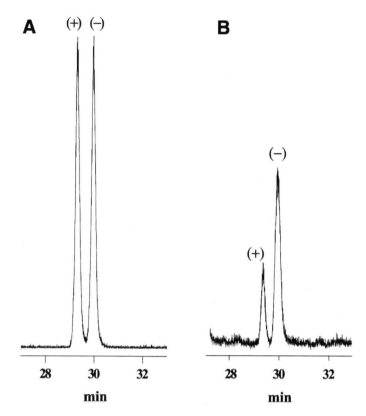

Fig. 2. ECNI–selected ion monitoring chromatograms (*m/z* at 255) showing elution of α-HCH on the Chirasil-Dex column in (**A**) a standard and (**B**) a chicken egg sample.

3. Run a hexane solvent blank prior to analyzing standards (*see* **Note 14**) and samples (**Fig. 2**).
4. Calculate the ER (*see* **Note 15**) as the area ratio of the (+)/(−) peaks eluting from the cyclodextrin column.
5. Calculate the individual enantiomer by the following equation:

$$(+)\text{-}\alpha\text{-HCH or }(-)\text{-}\alpha\text{-HCH concentration, ng/g} = (ACFG)/(BEDW)$$

where the symbols are defined as follows: A is the peak area of (+)-α-HCH or (−)-α-HCH in the sample; B is the peak area of (+)-α-HCH or (−)-α-HCH in the standard; C is half of the concentration of the racemic α-HCH standard (nanograms per milliliter); W is the initial sample weight (grams); D is the peak area of the internal standard (ε-HCH) in the sample at *m/z* 255; E is the concentration of the internal standard (ε-HCH) in the standard solution (nanograms per milliliter); F is the peak area of the internal standard (ε-HCH) in the standard solution at *m/z* 255; and G is the weight of internal standard (ε-HCH) added to the sample (nanograms).
The recovery (R) of the internal standard (ε-HCH) is:

$$R = [(EDU)/(FG)] \times [V/(V-10)] \times 100\%$$

where U is the final volume of the sample for GC–MS analysis (milliliters) and V is the volume of the sample extract before taking 10 mL for lipid determination (milliliters).

3.8. Quality Control

The following limits are set for a positive identification: (1) agreement of ER values at m/z 255 and 253 should be within ±5%, (2) the area ratios of the monitored ions for samples and standards should be within ±10%, and (3) the GC retention time should be within ±1% compared to a reference standard. α-HCH is manufactured as a racemic mixture of the two enantiomers. If no metabolism occurs, the ERs of α-HCH should be 1.00 (*see* **Note 16**). Replicate ($n = 3$) injections of analytical standards yields racemic ER values with a standard deviation of ±0.02, demonstrating that chiral phase GC–MS is capable of highly precise enantiomeric analysis (**Fig. 2A**). Average ERs are calculated from replicate injections of two or more separately extracted portions of the food sample, and standard deviations are calculated. All ERs meeting these requirements are considered reliable and are normalized by dividing by the average ER of the racemic standards for a given set of samples. This removes any bias that the column or instrument may introduce. To determine whether matrix effects alter the ERs, a control sample is spiked with racemic α-HCH standard. The ERs for the extracted spiked α-HCH should also be racemic if there is no matrix effect. If the sum of enantiomers determined on an achiral column is known, the concentrations of the (+) and (−) enantiomers can be calculated from the ERs.

For quantitative analysis, the samples are spiked with the surrogate internal standard of ε-HCH prior to extraction to assess the efficacy of the technique and to account for procedurally related analyte losses. Surrogate recoveries must fall between 60 and 120%.

With each set of six samples, a matrix spike and a matrix blank are also extracted. All quality control samples undergo the same cleanup procedure as the samples. The matrix spike contains a known amount of the target analyte at a concentration similar to that expected in the samples. Spike recoveries must range between 60 and 120%.

4. Notes

1. ε-HCH, a minor component in the technical HCH mixtures and not commonly observed in biological samples, was used as surrogate internal standard during development of this method. ε-HCH is consistently baseline resolved in the presence of native pesticides during GC. PCBs 30, 65, and 204; δ-HCH; and perdeuterated α-HCH have been used by other research groups as internal standards and are potentially suitable, provided that the chosen standard is also absent from commercial Aroclor mixtures or technical HCH mixtures and thus does not occur in environmental samples. However, because most PCBs are eluted in the first 3-mL fraction (which is discarded) of the SPE fractionation step and are only partially recovered in the HCH fraction, they will not be suitable as surrogate internal standards in this method.

2. Balances are required that are capable of weighing neat standards (milligrams), sample material (grams), and centrifuge bottles (>100 g). Two balances are used for this work, one high weight range for sample and extraction vessel weighing and the other for standard(s) preparation.

3. It is recommended that only 10- to 60-mesh or similar granular sodium sulfate be used because some analytes can be adsorbed onto the powdered material. Also, it is important

that this material be contaminant free, so it should be purchased as pesticide residue grade in glass containers or baked in a muffle furnace if purchased in bulk packages, for which exposure to plastic is an issue. Sodium sulfate must be treated as follows: Place it in a porcelain evaporating dish in a muffle furnace at 600°C for a minimum of 6 h; next, place it into a glass container while still hot (< 60°C) and seal with a Teflon®-lined cap to prevent the material from readsorbing contaminants from the atmosphere.

4. α-HCH enantiomers are present in samples at particularly low levels and in a complex mixture. Use of a highly selective detection technique, either GC/ECD or GC–MS/SIM, is mandatory. If injections of different concentrations of samples lead to variations in the ERs caused by interferents, MS will be a better detection method than ECD because the interferents have different m/z values that are not recorded by the MS/SIM. For this work, GC–MS/ECNI has been used. Numerous studies have shown the applicability of GC–MS/electron impact or ECNI to α-HCH enantiomer analysis as well *(7,9,17–22)*.

5. At present, permethylated β-cyclodextrin seems to be the phase of choice for the enantiomeric separation of α-HCH in biological samples. This type of column is commercially available from a number of manufacturers. The immobilized permethylated cyclodextrins of the Chirasil-Dex type have the advantages of solvent compatibility, resistance to temperature shock, and longevity. Usually, columns more than 30 m long are not recommended which may lead to extended retention times.

6. Prefacing the coated separation column with a section of uncoated but deactivated fused silica tubing can limit contamination of the separation column (nonvolatile residues tend to accumulate in the "guard column"). "Press-fit" connectors offer convenient low thermal mass fittings that, properly installed, can be highly dependable. Tubing ends should be freshly cut square, wiped with an acetone- or methanol-wetted tissue to remove any fragments, and examined under suitable magnification. When chromatographic peaks start to tail or are otherwise bad, about 30 cm can be cut from the beginning of the retention gap to restore good peak shapes. This can be done only two or three times before needing to change to a completely new retention gap.

7. Standards can be prepared from neat standards or by serial dilution of commercially prepared standards. All standards should be prepared in clean, solvent-rinsed volumetric glassware and stored in a freezer when not in use.

8. Soxhlet and ultrasonic extraction are the most common extraction techniques for the semivolatile compounds in solid samples. Sonication is a faster technique but requires constant operator attention. In both techniques, problems usually are caused by contaminated reagents (especially sodium sulfate) or poor laboratory practice when transferring sample extracts. The sample must be mixed with the sodium sulfate to achieve a sandy consistency prior to solvent addition.

9. It is critical that all extracts be reduced to as small a volume as possible before reconstitution in hexane since any residual acetone will affect separation in the next cleanup step. If the solvent (acetone/hexane) does not exchange to hexane after extraction, the remaining acetone in the extract will render the subsequent cleanup process irreproducible; also, the eluate from acid-modified silica becomes yellow, which may result in interferences in GC.

10. Because sulfuric acid may react with many double-bond, triple-bond, or aromatic compounds, this reaction can be used to convert many compounds found in pesticide and PCB extracts, from organic-soluble compounds to aqueous-soluble compounds, resulting in an extract that is much less contaminated. The H_2SO_4/silica removes the bulk of the lipids and other oxidizable components from the hexane extract. It is important to note that many pesticides will undergo this reaction, so such a cleanup can only be used for analysis of PCBs and some pesticides.

11. The use of prepacked SPE tubes containing silica packing can increase laboratory efficiency and reduce the amount of solvents used for these processes. They are used for fractionation of HCH from similar organics after acid-modified silica cleanup and for retaining coextractants that may interfere with GC analysis of HCH. Slower gravity elution of the samples will minimize premature breakthrough or channeling and will ensure maximum recoveries in the recovered fraction. It is very important to evaluate each lot of tubes to ensure minimal background from the device itself and to verify that the packing is at maximum activity level to maintain the expected retention capacity.

12. Splitless injection is usually used for HCH analysis. However, in environmental samples that have relatively low concentration of α-HCH enantiomers, large-volume injection has advantages over splitless injection. The benefit of large-volume injection is the obvious improvement in sensitivity. A second benefit to large-volume injection is a savings of time and effort in the preparation of small volume of samples (concentration step) for GC analysis; this also improves the accuracy of the procedure by eliminating losses associated with the concentration step. The major technical challenge associated with large-volume injection is to remove the solvent without losing the analytes of interest. PTV injectors can be used for the injection of large sample volumes, either by multiple injections or by solvent venting in the cold split mode. For a multiple-injection method, large volumes are injected in several smaller fractions (up to eight injections), with the solvent evaporating at low temperatures and exiting through the open split vent, between one injection and the next. Once a sufficient quantity of trace analytes is built up, the injector temperature is rapidly increased to evaporate the solutes onto the column. The solvent venting process, time, flow, and temperature all need to be optimized to minimize the loss of the analytes of interest.

13. Because α-HCH is an early-eluting organochlorine compound, the number of potential coelutions is limited after a good sample cleanup (lipid removal and group separation by SPE). This allows the application of relatively short GC columns as well as low separation temperatures. Because of the relatively low maximal operation temperatures of chiral GC columns (T_{max} = 200°C and T_{max} = 225°C for isothermal and temperature programming, respectively), comparably long hold times at the final temperature are necessary before the next injection.

14. In most enantiomer analyses, a definitive order of elution can only be established by injection of enantio-enriched standards. In the case of α-HCH, the elution order of the enatiomers on various CSPs is well known. Loss of a chiral compound during the sample cleanup will not influence the ER. The ER is not influenced by sample manipulations (dilution, injection, detection, or chemical and physical losses). Therefore, an individual enantio-enriched α-HCH standard is not necessary for the determination of ER or EF values.

15. The enantiomeric composition of chiral chemicals can be expressed as ratios (ERs) or fractions (EFs). ERs are more commonly used; EFs are more easily compared and used in mathematical equations (23). The ER is defined as ER = A_+/A_- [A_+ and A_- correspond the peak areas of the (+) and (–) enantiomers] when the optical rotations are known; otherwise, the ER is expressed as ER = A_1/A_2 (the peak area of the first eluting enantiomer over the second eluting enantiomer). The corresponding EF is defined as EF = $A_+/(A_+ + A_-)$ or EF = $A_1/(A_1 + A_2)$. The EF can also be calculated from the ER value by the equation EF = ER/(ER + 1).

16. The verification of the expected 1:1 ratio of a racemic mixture is mandatory in enantiomer analysis by chromatography. Indeed, a racemic composition represents an ideal equimolar mixture and is useful for testing the precision of integration facilities, which should always produce a correct ER of 1.

References

1. Kutz, F. W., Wood, P. H., and Bottimore, D. P. (1991) Organochlorine pesticides and polychlorinated-biphenyls in human adipose-tissue. *Rev. Environ. Contam. Toxicol.* **120**, 1–82.
2. Li, Y.-F., McMillan, A., and Scholtz, M. T. (1996) Global HCH usage with 1° × 1° longitude/latitude resolution. *Environ. Sci. Technol.* **30**, 3525–3533.
3. Simonich, S. L. and Hites, R. A. (1995) Global distribution of persistent organochlorine compounds. *Science* **269**, 1851–1854.
4. Willett, K. L., Ulrich, E. M., and Hites, R. A. (1998) Differential toxicity and environmental fates of hexachlorocyclohexane isomers. *Environ. Sci. Technol.* **32**, 2197–2207.
5. Hegeman, W. J. M. and Laane, R. W. P. M. (2002) Enantiomeric enrichment of chiral pesticides in the environment. *Rev. Environ. Contam. Toxicol.* **173**, 85–116.
6. Mössner, S., Spraker, T. R., Becker, P. R., and Ballschmitter, K. (1992) Ratios of enantiomers of α-HCH and determination of α-, β-, and γ-HCH isomers in brain and other tissues of neonatal northern fur seals (*Callorhinus ursinus*). *Chemosphere* **24**, 1171–1180.
7. Ulrich, E. M., Willett, K. L., Caperell-Grant, A., and Hites, R. A. (2001) Understanding enantioselective processes: a laboratory rat model for α-hexachlorocyclohexane accumulation. *Environ. Sci. Technol.* **35**, 1604–1609.
8. Müller, M. D., Schlabach, M., and Oehme, M. (1992) Fast and precise determination of α-hexachlorocyclohexane enantiomers in environmental samples using chiral high-resolution gas chromatography. *Environ. Sci. Technol.* **26**, 566–569.
9. Jantunen, L. M. and Bidleman, T. F. (1995) Differences in enantioselective degradation of α-hexachlorocyclohexane in the Bering-Chukchi Seas and the Arctic Ocean. *Organohalogen Compounds* **24**, 425–428.
10. Pfaffenberger, B., Hühnerfuss, H., Kallenborn, R., Köhler-Günther, A., König, W. A., and Krüner, G. (1992) Chromatographic-separation of the enantiomers of marine pollutants. 6. Comparison of the enantioselective degradation of α-hexachlorocyclohexane in marine biota and water. *Chemosphere* **25**, 719–725.
11. Tanabe, S., Kumaran, P., Iwata, H., Tatsukawa, R., and Miyazaki, N. (1996) Enantiomeric ratios of α-hexachlorocyclohexane in blubber of small cetaceans. *Mar. Pollut. Bull.* **32**, 27–31.
12. Vetter, W., Klobes, U., Hummert, K., and Luckas, B. (1997) Gas chromatographic separation of chiral organochlorines on modified cyclodextrin phases and results of marine biota samples. *J. High Resol. Chromatogr.* **20**, 85–93.
13. Vetter, W. and Schurig, V. (1997) Review: enantioselective determination of chiral organochlorine compounds in biota by gas chromatography on modified cyclodextrins. *J. Chromatogr. A* **774**, 143–175.
14. Kallenborn, R. and Hühnerfuss, H. (ed.). (2001) *Chiral Environmental Pollutants. Trace Analysis and Ecotoxicology.* Springer-Verlag, New York.
15. Buser, H.-R. and Müller, M. D. (1994) Isomer- and enantiomer-selective analyses of toxaphene components using chiral high-resolution gas chromatography and detection by mass spectrometry/mass spectrometry. *Environ. Sci. Technol.* **28**, 119–128.
16. Hoekstra, P. F., O'Hara, T. M., Karlsson, H., Solomon, K. R., and Muir, D. C. G. (2003) Enantiomer-specific biomagnification of α-hexachlorocyclohexane and selected chiral chlordane-related compounds within an Arctic marine food web. *Environ. Toxicol. Chem.* **22**, 2482–2491.
17. Falconer, R. L., Bidleman, T. F., and Szeto, S. Y. (1997) Chiral pesticides in soils of the Fraser Valley, British Columbia. *J. Agric. Food Chem.* **45**, 1946–1951.

18. Wiberg, K., Letcher, R. J., Sandau, C. D., Norstrom, R. J., Tysklind, M., and Bidleman, T. F. (2000) The enantioselective bioaccumulation of chiral chlordane and α-HCH contaminants in the polar bear food chain. *Environ. Sci. Technol.* **34,** 2668–2674.

19. Moisey, J., Fisk, A. T., Hobson, K. A., and Norstrom, R. J. (2001) Hexachlorocyclohexane (HCH) isomers and chiral signatures of α-HCH in the Arctic marine food web of the Northwater polynya. *Environ. Sci. Technol.* **35,** 1920–1927.

20. Meijer, S. N., Halsall, C. J., Harner, T., et al. (2001) Organochlorine pesticide residues in archived UK soil. *Environ. Sci. Technol.* **35,** 1989–1995.

21. Fisk, A. T., Holst, M., Hobson, K. A., Duffe, J., Moisey, J., and Norstrom, R. J. (2002) Persistent organochlorine contaminants and enantiomeric signatures of chiral pollutants in ringed seal (*Phoca hispida*) collected on the east and west side of the Northwater Polynya, Canadian Arctic. *Arch. Environ. Contam. Toxicol.* **42,** 118–126.

22. Wong, C. S., Lau, F., Clark, M., Mabury, S. A., and Muir, D. C. G. (2002) Rainbow trout (*Oncorhynchus mykiss*) can eliminate chiral organochlorine compounds enantioselectively. *Environ. Sci. Technol.* **36,** 1257–1262.

23. Harner, T., Wiberg, K., and Norstrom, R. J. (2000) Enantiomer fractions are preferred to enantiomer ratios for describing chiral signatures in environmental analysis. *Environ. Sci. Technol.* **34,** 218–220.

IV

PESTICIDE ANALYSIS IN WATER

27

Automated Headspace Solid-Phase Microextraction and Gas Chromatography–Mass Spectrometry for Screening and Determination of Multiclass Pesticides in Water

Taizou Tsutsumi, Mitsushi Sakamoto, Hiroyuki Kataoka, Janusz Pawliszyn

Summary

The method for the determination of multiclass pesticides in water samples by headspace solid-phase microextraction and gas chromatography–mass spectrometry (HS-SPME–GC–MS) is described. Of the 174 pesticides, 158 are extracted with a polyacrylate-coated fiber at 30 to 100°C and are classified into four groups according to the shape of their extraction temperature profiles. Each group of pesticides is continuously analyzed by HS-SPME–GC–MS using the Combi PAL autosampler. This method is successfully applied to the analyses of drinking and environmental water samples. The automated HS-SPME–GC–MS system is an extremely powerful, flexible, and low-cost approach for multiclass pesticide analyses in various water samples. The recoveries and reproducibilities are good, and the results are in good agreement with those obtained by the solid-phase extraction method. Although it has not yet been theoretically useful, the physical property diagram may be a valuable tool in inductively predicting adaptability of untested compounds in the HS-SPME system.

Key Words: Automated analysis; Combi PAL; extraction temperature profile; headspace solid-phase microextraction; mineral water; multiclass pesticides; river water.

1. Introduction

Pesticides are used on a large scale for agricultural purposes. The adverse effects of pesticides on both human health and the environment are a matter of public concern. Thus, both the actual state and the transition of pesticide residues in various matrices, including water, soil, and agricultural products, should be extensively monitored to determine that they are within specified limits. Therefore, screening, identification, and determination methods for these pesticides in the environment are necessary. Although performance enhancement and improvement of sensitivity and specificity of

From: *Methods in Biotechnology, Vol. 19, Pesticide Protocols*
Edited by: J. L. Martínez Vidal and A. Garrido Frenich © Humana Press Inc., Totowa, NJ

the analytical instrument have been attempted to satisfy this demand, until now most analytical instruments could not directly handle complex matrices such as environmental samples. Specifically, sample preparation steps such as extraction, concentration, fractionation, and isolation of analytes are very complicated, and over 80% of analysis time is generally spent on these steps.

Solid-phase extraction (SPE) has been widely adopted for preparing samples in the analysis of pesticides. SPE offers the following advantages over liquid–liquid procedures: less organic solvent usage, no forming or emulsion problems, and easier operation. Although conventional sample preparation methods have been used to analyze pesticide residues in various matrices, many of these methods require expensive instrumentation and an expert analyst and are very time costly. Therefore, efficient analytical systems that do not have these drawbacks are required.

Solid-phase microextraction (SPME), developed by Pawliszyn and coworkers *(1)*, has received an increasing amount of attention as a simple and convenient sample preparation technique that can be applied to numerous environmental matrices *(2–10)*. It saves preparation time, solvent purchase, and disposal costs. The SPME consists of two extraction modes. One is the direct immersion mode (DI-SPME) in which analytes are extracted from the liquid phase onto an SPME fiber, and the other is the headspace mode (HS-SPME), in which analytes are extracted from the gaseous phase (headspace) over liquid or solid samples onto the SPME fiber. DI-SPME and HS-SPME techniques can be used in combination with any gas chromatography (GC), GC–mass spectrometry (MS), high-performance liquid chromatography, and liquid chromatography–MS systems. In general, DI-SPME is more sensitive than HS-SPME for analytes present in a liquid sample, although HS-SPME gives lower background than DI-SPME *(2)*. SPME methods with DI and HS mode have been reported for the determination of various pesticides in water *(11–21)*. Some compounds need to convert to their volatile derivatives prior to SPME extraction *(12,14)*.

Because it was previously thought that HS-SPME could be used to differentiate volatile compounds from less-volatile compounds *(22)*, the use of HS-SPME extraction for semi- (and less-) volatile compounds has seldom been applied *(4)*. Furthermore, pesticide extraction is not thought to be more efficient when the temperature increases to above 60°C *(4)*; therefore, SPME extractions are usually conducted at room temperature. However, the sensitivity to an analyte and the fundamental applicability of a method toward the analyte are two independent parameters. We considered that it is necessary to recognize anew the applicability of SPME to the fields in which researchers take little interest before pursuing the sensitivity of the analytical system. We attempted to evaluate the applicability of HS-SPME determination of semi- and less-volatile pesticides in water samples and to expand the application range to the pesticides in actual matrices.

In this chapter, a screening method for multiclass pesticides categorized into subgroups based on the extraction temperature and a quantitative method for the determination of 45 pesticides using automated HS-SPME–GC–MS with a commercial autosampler (Combi PAL) *(23)* are described. The 174 pesticides detectable with GC were selected objectively and impartially based on their physical properties: vapor pressure and partition coefficient between octanol and water (*see* **Note 1**). The extrac-

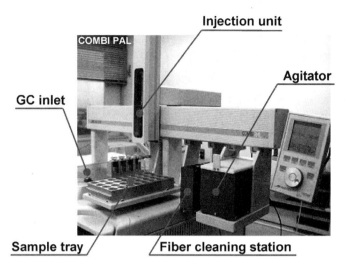

Fig. 1. Automated headspace solid-phase microextraction and gas chromatography–mass spectrometry system using the Combi PAL autosampler.

tion temperature profiles in HS-SPME were evaluated to classify the pesticides. The relationship between the features of the pesticide's extraction temperature profile and its physical properties provides us with a useful guide not only to systematic screening of pesticides not included in this study, but also to optimization of HS-SPME in actual matrices. We also developed a new multiple analytical system using the HS-SPME–GC–MS used in a great number of pesticide residue monitoring and compared it with the conventional SPE–GC–MS method. Optimum conditions for the automated HS-SPME–GC–MS method and typical problems encountered in the development and application of the method are discussed.

2. Materials

2.1. Equipment

1. GC ion trap MS system (e.g., Trace GC 2000-Polaris, Thermoquest Co., Austin, TX) equipped with programmable temperature vaporizing (PTV) injector fitted with a glass insert (1-mm id) (*see* **Note 2**).
2. Rtx-5MS fused-silica capillary column, 30 m × 0.25 mm id, 0.25 μm (e.g., Restek Co., Bellefonte, PA).
3. Combi PAL autosampler (**Fig. 1**) fully controlled with Cycle Composer software (e.g., CTC Analytics, Basel, Switzerland) (*see* **Note 3**).
4. SPME fibers: 85-μm polyacrylate (PA), 100-μm polydimethylsiloxane (PDMS), 65-μm PDMS–divinylbenzene (PDMS–DVB), 65-μm carbowax–DVB (CW–DVB), and 75-μm carboxen–PDMS (CAR–PDMS) (e.g., Supelco Co., Bellefonte, PA).
5. Sep-pack PS-2 plus cartridge (e.g., Waters Co., Milford, MA).
6. Crimp-top headspace vial (20 mL) equipped with a blue silicon/polytetrafluoroethylene (PTFE) septa and an open-center magnetic cap (e.g., La-Pha-Pack, Langerwehe, Germany).

2.2. Reagents

1. Pesticide standards (>95% pure). The pesticides are listed in **Table 1** (*see* **Note 4**).
2. Internal standards thiobencarb-d_{10} and chlornitrofen-d_4 (>96% pure).
3. Other materials: All solvents and anhydrous sodium sulfate are pesticide-analysis grade. Pure water is used after fresh purification with a Milli-Q Gradient 10A system (e.g., Millipore, Tokyo, Japan).

2.3. Solutions

1. Standard pesticide solutions: Standard stock solution (each 1 mg/mL) is prepared in acetone. Each pesticide is mixed at the concentration of 0.01–10 μg/mL in acetone. Several mixed standard solutions as vial packing are also purchased from Kanto Kagaku (Tokyo, Japan). These standard solutions are stored at 4°C.
2. Working mixed standard solutions: Appropriate amounts of the mixed standard solutions of **step 1** are spiked into the water–sample vial at a concentration of 0.01–10 ng/mL. These working standard solutions are stored at 4°C.
3. Internal standard solutions: Internal standard stock solutions (each 1 mg/mL) are prepared in acetone and are stored at 4°C.

2.4. Water Samples

Water samples are pure water, commercially available natural mineral waters (Volvic®, Evian®, and Contrex) and surface water collected from the river Katsuura located in Tokushima prefecture in Japan. The natural mineral water and the surface water are tested to ensure that they are free from the selected pesticides and stored at 4°C until use.

3. Methods

Two methods, HS-SPME and SPE, are proposed. HS-SPME is a new automated technique and is described in detail. SPE is widely adopted for preparing samples in the analysis of pesticides and is used for comparison with the HS-SPME method.

3.1. Automated HS-SPME–GC–MS System

Automation of the HS-SPME procedures is achieved with a Combi PAL autosampler. The computer controlled all the movements of the autosampler and the GC–MS. Run cycles of autosampler and GC-MS are shown in **Fig. 2** (*see* **Notes 5** and **6**).

3.1.1. Preparation of Samples

1. A 10-mL volume of a water sample and 4 g anhydrous sodium sulfate are placed in a 20-mL crimp-top headspace autosampler vial equipped with a PTFE-coated magnetic stir bar, and the solution is stirred with a magnetic stirrer at 235g.
2. For preparing the spiked water samples over the concentration range of 0.01–10 ng/mL, the appropriate amounts of the mixed standard solution of pesticides at 0.01–10 μg/mL in acetone are spiked into each vial (*see* **Note 7**).
3. The vials are sealed with both a blue silicon/PTFE septa and an open-center magnetic cap, and the solutions are stirred again.
4. In the calibration and quantitation studies, the mixed internal standard solution is also added to each sample.

3.1.2. Automated HS-SPME Procedure

1. Photographs of the SPME step using the Combi PAL autosampler are presented in **Fig. 3**. All movements of the SPME fiber in the processes of adsorption, desorption, and cleaning could be precisely controlled using this autosampler.
2. The SPME fibers are conditioned before use according to the supplier's instructions with the fiber cleaning station in the Combi PAL (*see* **Note 8**).
3. The sample vials are set on the sample tray in the Combi PAL, which is programmed to control the SPME extraction and subsequent desorption (*see* **Note 9**).
4. A sample vial is transported into the agitator in the Combi PAL, which keeps the vial oscillating at an optional fixed temperature during extraction.
5. The SPME needle pierces the septum of the sample vial, and the fiber is exposed in the headspace above the sample (*see* **Notes 10–12**).
6. A suitable SPME extraction temperature can be selected from four types of pesticides by the relationship between the characteristic features in extraction and their physical properties, such as vapor pressure and octanol/water partition coefficient. In the actual procedures for comprehensive monitoring, the same sample is divided into several vials for multigroup analysis. Each vial is sequentially examined by each group method in turn (*see* **Notes 13–21**).
7. After suitable extraction time (*see* **Note 21**), the fiber is withdrawn into the needle; the needle is removed from the septum and then inserted directly into the GC injection port.
8. The Combi PAL continuously inserts the fiber into the GC injection port in a short, fixed interval, and then the GC measurement starts simultaneously (**Fig. 2**).
9. The sample vial is then returned to the sample tray, and the fiber is thoroughly cleaned at 270°C under nitrogen for 10 min in the cleaning station (inert gas purging) in the Combi PAL, ready for the next extraction (*see* **Note 22**).

3.1.3. Gas Chromatography–Mass Spectrometry

1. GC–MS analysis of the pesticides is performed using a GC system with an ion trap mass spectrometer. It is equipped with a PTV injector fitted with a glass insert (1-mm id). It is used in the PTV splitless mode with a splitless time of 3 min and with the following temperature program: hold at 50°C for 0.5 min, 10°C/s to 250°C, hold for 3 min, 5°C/sec to 260°C, hold for 28 min (*see* **Note 23**).
2. The fused-silica capillary column follows a temperature program of 50°C held for 1 min, 25°C/min to 125°C, 10°C/min to 300°C, hold for 7 min. The carrier gas is helium at a constant flow (1 mL/min).
3. The GC–MS transfer line is held at 260°C and the ion source at 200°C. The full-scan mode (scan range 50–500 *m/z*) for all pesticides and the tandem mass spectrometry (MS/MS) mode for 11 pesticides (bitertanol, cyproconazole, fenvalerate, fludioxonil, isofenphos-oxon, metribuzin, myclobutanil, paclobutrazol, propiconazole, pyraclofos, pyridaphenthion) are used for detection and confirmation of the pesticides. One or two ions are selected from the spectrum of each pesticide to quantify the response. The ion energy used for electron impact is 70 eV (*see* **Note 23**).
4. The calibration curves are prepared as follows: The ratio of the peak area for the target pesticide to the peak area for the internal standard is plotted on the *y*-axis, and the initial concentration of the target pesticide is plotted on the *x*-axis. The concentrations of the pesticides in the samples are 0.01, 0.02, 0.05, 0.1, 0.2, 0.5, 1.0, 2.0, 5.0, and 10 ng/mL.

Table 1
CAS Registry Number, Physical Properties, Type of Extraction–Temperature Profile, and Optimum Extraction Temperature of the Studied Pesticides

Compound	CAS registry number	log P_v [a]	log P [b]	Type of extraction-temperature profile	Optimum extraction temperature (°C)
Detectable					
Acrinathrin	101007-06-1	-4.36	5.00	3	100
Alachlor	15972-60-8	0.32	3.09	2	80
Aldrin	309-00-2	1.20	6.50	1	60
Bendiocarb	22781-23-3	0.66	1.72	1	70
Benfluralin	1861-40-1	0.94	5.29	1	80
Benfuresate	68505-69-1	0.16	2.41	2	90
α-BHC	319-84-6	0.78	3.80	1	70
β-BHC	319-85-7	-1.32	3.78	1	70
γ-BHC	58-89-9	0.75	3.72	1	70
δ-BHC	319-86-8	0.67	4.14	1	70
Bifenox	42576-02-3	-0.49	4.50	2	80
Bifenthrin	82657-04-3	-1.62	6.00	3	100
Bitertanol	55179-31-2	-6.66	4.10	3	100
Bromobutide	74712-19-9	-1.08	3.47	2	80
Buprofezin	69327-76-0	0.10	4.30	2	90
Butachlor	23184-66-9	-0.41	4.50	2	90
Butamifos	36335-67-8	1.92	4.62	2	90
Butylate	2008-41-5	3.24	4.15	1	60
Cadusafos	95465-99-9	2.08	3.90	1	70
Carbaryl (NAC)	63-25-2	-1.39	1.59	2	80
Chlorfenapyr	122453-73-0	1.11	4.83	2	90
(*E*)-Chlorfenvinphos (α-CVP)	18708-86-6	0.00[c]	4.22	2	90
(*Z*)-Chlorfenvinphos (β-CVP)	18708-97-7	0.00[c]	3.85	2	90
Chlornitrofen (CNP)	1836-77-7	-1.15	4.97	2	90
Chlorobenzilate	510-15-6	-0.53	4.74	2	90
Chloroneb	2675-77-6	3.28	3.44	1	70
Chlorothalonil (TPN)	1897-45-6	-1.12	3.05	1	70
Chlorpropham	101-21-3	0.00	3.51	2	80
Chlorpyrifos	2921-88-2	0.43	4.70	2	80
Cinmethylin	87818-31-3	1.00	3.84	2	80
Clofentezine	74115-24-5	-3.89	4.10	2	70
Cyfluthrin	68359-37-5	-3.02	6.00	3	100
Cyhalofop-butyl	122008-85-9	-2.92	3.31	3	100
Cyhalothrin	68085-85-8	-3.00	6.80	3	100
Cypermethrin	52315-07-8	-3.64	6.60	3	100
Cyproconazole	94361-06-5	-1.46	2.91	3	100
Dichlorodiisopropyl ether (DCIP)	108-60-1	5.52	2.14	1	50
o,p'-DDD	53-19-0	[d]	[d]	2	80
p,p'-DDD	72-54-8	-0.74	6.02	2	80
p,p'-DDE	72-55-9	-0.10	6.51	2	80
o,p'-DDT	789-02-6	-0.74	6.79	2	80
p,p'-DDT	50-29-3	-1.67	6.91	2	80
Deltamethrin	52918-63-5	-4.91	4.60	3	100
Diazinon	333-41-5	1.08	3.30	1	80
Dichlofenthion (ECP)	97-17-6	1.87	5.14	1	80
Dichlofluanid	1085-98-9	-1.82	3.70	4	—
Dichrolvos (DDVP)	62-73-7	3.32	1.90	1	60
Dicofol	115-32-2	-1.28	4.30	2	70
Dieldrin	60-57-1	-0.11	5.40	2	80
Diethofencarb	87130-20-9	0.92	3.02	3	100

Table 1 *(Continued)*

Compound	CAS registry number	log P_v [a]	log P [b]	Type of extraction-temperature profile	Optimum extraction temperature (°C)
Difenoconazole	119446-68-3	-4.48	4.20	3	100
Dimethenamid	87674-68-8	1.56	2.15	2	90
(*E*)-Dimethylvinphos	71363-52-5	0.11[c]	3.12[c]	1	80
(*Z*)-Dimethylvinphos	67628-93-7	0.11[c]	3.12[c]	1	80
Dithiopyr	97886-45-8	-0.27	4.75	1	80
Edifenphos (EDDP)	17109-49-8	-1.49	3.83	1	80
Endrin	72-20-8	-0.40	5.20	2	80
α-Endosulfan	959-98-8	-0.40	3.83	1	80
β-Endosulfan	33213-65-9	-1.10	3.83	1	80
p-Nitrophenyl phenylphosphonothionate (EPN)	2104-64-5	-1.39	5.02	2	90
S-Ethyl *N,N*-dipropylthiocarbamate (EPTC)	759-94-4	3.51	3.20	1	60
Esprocarb	85785-20-2	1.00	4.60	2	80
Ethiofencarb	29973-13-5	-0.35	2.04	1	60
Ethion	563-12-2	-0.05	5.07	2	80
Ethofenprox	80844-07-1	-3.04	7.05	3	100
Ethoprophos	13194-48-4	1.67	3.59	1	70
Etobenzanid	79540-50-4	-1.68	4.30	3	100
Etoxazole	153233-91-1	-2.66	5.59	3	100
Etridiazole	2593-15-9	1.12	3.37	1	60
Etrimfos	38260-54-7	0.81	3.30	1	70
Fenarimol	60168-88-9	-1.19	3.69	3	100
Fenitrothion (MEP)	122-14-5	1.18	3.50	1	80
Fenobucarb (BPMC)	3766-81-2	0.20	2.79	1	70
Fenpropathrin	39515-41-8	-0.14	6.00	3	100
Fensulfothion	115-90-2	0.82	2.23	3	100
Fenthion	55-38-9	-0.13	4.84	2	80
Fenvalerate	51630-58-1	-1.72	5.01	3	100
Flucythrinate	70124-77-5	-2.92	6.20	3	100
Fludioxonil	131341-86-1	-3.41	4.12	2	80
Flusilazole	85509-19-9	-1.41	3.74	3	100
Flutolanil	66332-96-5	-2.19	3.70	3	100
Fluvalinate	69409-94-5	-7.05	4.26	3	100
Folpet	133-07-3	-1.68	3.11	4	—
Fthalide	27355-22-2	-2.52	3.20	4	—
Furametpyr	123572-88-3	-2.33	2.36	3	100
Halfenprox	111872-58-3	-3.11	4.10	3	100
Heptachlor	76-44-8	1.72	6.10	1	60
Heptachlor Epoxide	1024-57-3	0.41	4.98	2	80
Hexaconazole	79983-71-4	-1.74	3.90	3	100
Imazalil	35554-44-0	-0.80	3.82	3	100
Imibenconazole	86598-92-7	-4.07	4.94	3	100
Iprobenfos (IBP)	26087-47-8	-0.61	3.21	1	80
Isofenphos	25311-71-1	-0.66	4.04	2	90
Isofenphos P=O	31120-85-1	[d]	[d]	3	100
Isoprocarb	2631-40-5	0.45	2.30	1	70
Isoprothiolane	50512-35-1	1.27	3.30	3	100
Isoxathion	18854-01-8	-0.88	3.88	2	90
Kresoxim-methyl	143390-89-0	-2.64	3.40	2	80
Malathion	121-75-5	0.72	2.75	1	70
Mefenacet	73250-68-7	-3.19	3.23	3	100

Table 1 *(Continued)*
CAS Registry Number, Physical Properties, Type of Extraction–Temperature Profile,
and Optimum Extraction Temperature of the Studied Pesticides

Compound	CAS registry number	log P_v[a]	log P[b]	Type of extraction-temperature profile	Optimum extraction temperature (°C)
Mepanipyrim	110235-47-7	-1.63	3.28	2	90
Mepronil	55814-41-0	-1.25	3.66	3	100
Metalaxyl	57837-19-1	-0.13	1.65	2	90
Methabenzthiazuron	18691-97-9	-2.23	2.64	3	100
Methiocarb	2032-65-7	-1.82	3.08	1	80
Methyldymron	42609-73-4	[d]	3.01	1	80
Metolachlor	51218-45-2	0.62	2.90	2	90
Metribuzin	21087-64-9	-1.24	1.60	3	100
Molinate	2212-67-1	2.87	3.21	1	60
Myclobutanil	88671-89-0	-0.67	2.94	3	100
Napropamide	15299-99-7	-1.64	3.36	3	100
Paclobutrazol	76738-62-0	-3.00	3.20	2	90
Parathion	56-38-2	-0.05	3.83	2	80
Parathion-methyl	298-00-0	-0.70	3.00	1	80
Pencycuron	66063-05-6	-6.30	4.68	[e]	—
Pendimethalin	40487-42-1	0.60	5.18	2	80
Pentoxazone	110956-75-7	-1.95	[d]	1	80
Permethrin	52645-53-1	-1.15	6.10	3	100
Phenthoate	2597-03-7	0.72	3.69	1	70
Phosalone	2310-17-0	-1.22	4.01	2	90
Pirimicarb	23103-98-2	-0.01	1.70	3	100
Pirimiphos-methyl	29232-93-7	-0.17	5.00	2	80
Pretilachlor	51218-49-6	-0.88	4.08	2	90
Propiconazole	60207-90-1	-1.25	3.72	3	100
Propyzamide	23950-58-5	-1.24	3.43	2	90
Prothiofos	34643-46-4	-0.52	5.67	2	80
Pyraclofos	89784-60-1	-2.80	3.77	3	100
Pyributicarb	88678-67-5	-0.57	3.69	3	100
Pyridaben	96489-71-3	-0.60	6.37	3	100
Pyridaphenthion	119-12-0	-2.83	3.20	2	90
(*E*)-Pyrifenox	83227-22-9	0.23[c]	3.70[c]	3	100
(*Z*)-Pyrifenox	83227-23-0	0.23[c]	3.70[c]	3	100
Pyrimidifen	105779-78-0	-3.80	4.59	3	100
(*E*)-Pyriminobac-methyl	147411-69-6	-1.46	2.98	3	100
(*Z*)-Pyriminobac-methyl	147411-70-9	-1.57	2.70	3	100
Pyriproxyfen	95737-68-1	-0.05	5.55	3	100
Quinalphos	13593-03-8	-0.46	4.44	2	90
Quinomethionate	2439-01-2	-1.59	3.78	1	60
Silafluofen	105024-66-6	-2.60	8.20	3	100
Simazine (CAT)	122-34-9	-2.53	2.18	3	100
Simetryne	1014-70-6	-1.02	2.80	3	100
Tebuconazole	107534-96-3	-2.77	3.70	3	100
Tebufenpyrad	119168-77-3	-2.00	4.61	3	100
Tefluthrin	79538-32-2	0.90	6.50	1	80
Terbucarb (MBPMC)	1918-11-2	0.72	5.28	2	80
Terbufos	13071-79-9	1.54	2.77	1	70
Tetraconazole	112281-77-3	-0.74	3.56	3	100
Thenylchlor	96491-05-3	-1.55	3.53	3	100
Thiobencarb	28249-77-6	0.47	3.42	1	80
Thiometon	640-15-3	0.35	3.15	1	70

Table 1 *(Continued)*

Compound	CAS registry number	log P_v[a]	log P[b]	Type of extraction-temperature profile	Optimum extraction temperature (°C)
Tolclofos-methyl	57018-04-9	1.76	4.56	1	80
Tralomethrin	66841-25-6	-5.32	5.00	3	100
Triadimenol	55219-65-3	-3.22	3.08	2	90
Trichlamide	70193-21-4	1.00	d	3	100
Triclopyr-2-butoxyethyl	64700-56-7	d	d	1	70
Triflumizole	68694-11-1	-0.73	5.10	3	100
Trifluralin	1582-09-8	0.79	5.34	1	70
Uniconazole P	83657-17-4	0.72	3.28	3	100
Non-detectable					
Acephate	30560-19-1	-0.65	-0.89		
Acetamiprid	160430-64-8	-3.00	0.80		
Bensulide (SAP)	741-58-2	-0.97	4.20		
Cafenstrole	125306-83-4	-5.52	3.21		
Captafol	2425-06-1	-3.06	3.80		
Captan	133-06-2	-1.92	2.80		
Dimethoate	60-51-5	-0.60	0.70		
Fosthiazate	98886-44-3	-0.25	1.68		
Iprodione	36734-19-7	-3.30	3.00		
Lenacil	2164-08-1	-3.70	2.31		
Methamidophos	10265-92-6	0.36	-0.80		
Oxamyl	23135-22-0	-1.29	-0.44		
Probenazole	27605-76-1	0.19	1.40		
Propamocarb	24579-73-5	2.86	0.84		
Trichlorfon (DEP)	52-68-6	0.02	0.51		
Tricyclazole	41814-78-2	-1.57	1.40		

[a] Logarithmic value of vapor pressure (mPa).

[b] Logarithmic value of octanol-water partition coefficient.

[c] Data as mixture of (*E*)- and (*Z*)- isomers.

[d] No data.

[e] Not determined.

From **ref. 23**, with permission.

5. The limit of detection (LOD) is calculated as the concentration giving a signal-to-noise ratio of 3 (*S/N* = 3). Pure water samples that are spiked with target pesticides at concentration levels ranging from 0.01 to 10 ng/mL are analyzed to estimate the LOD (*see* **Notes 24–26**).

3.1.4. Analysis of Water Samples

Commercially available natural mineral waters (Volvic, Evian, and Contrex) and river surface water are tested to ensure that they are free from the selected pesticides. The recoveries for target pesticides spiked to water samples are determined as the ratio of peak area in each tested sample to that in pure water under the same conditions (*see* **Notes 27–29**).

3.1.5. Identification and Quantification

Detected analytes are quantified via each ion of pesticides (**Table 2**) in MS/MS mode. The concentration (nanograms per milliliter) of the pesticides in the samples is given as follows:

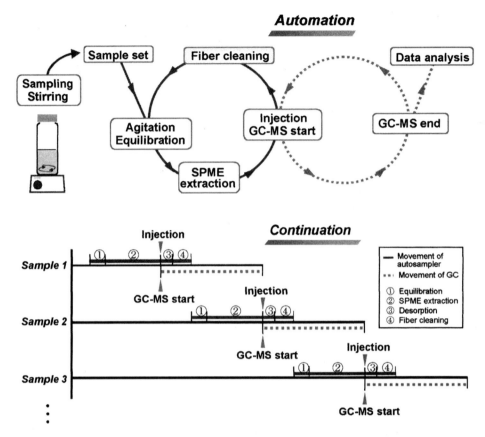

Fig. 2. Schematic diagram of the automated continuous analysis by headspace solid-phase microextraction and gas chromatography–mass spectrometry.

$$C = (y - b)/a$$

where y is the area count obtained from the sample; a is the slope of the regression line in the standard calibration curve; and b is the intercept of the regression line in the standard calibration curve.

3.2. Solid-Phase Extraction

1. A 500-mL volume of water is spiked with 0.5 mL of the mixed standard solution of pesticides at 0.5 µg/mL in acetone and mixed well.
2. A Sep-pack PS-2 plus cartridge is conditioned by sequentially rinsing with 5 mL each of dichloromethane, methanol, and pure water.
3. The 500-mL water sample is then pulled through the SPE cartridge at a flow rate of 10 mL/min.
4. The SPE cartridge is dried by introducing air into the cartridge for more than 30 min to dislodge the bound water.

Fig. 3. Photographs of each step of headspace solid-phase microextraction using the Combi PAL. (**A**) Transfer of vial from sample tray to agitator; (**B**) agitation for equilibration (heat block is actually covered); (**C**) insertion of the fiber into headspace of sample vial during agitation; (**D**) fiber cleaning in the fiber-cleaning station after desorption in the GC injection port.

353

Table 2
Quantitation Ion, Limits of Detection (LOD), Linearity, and Precision Data for 45 Pesticides by HS-SPME

Compound	Quantitation ion (*m/z*)	Extraction temperature (°C)	LOD (ng/mL)	Linearity Correlation coefficient	Concentration range (ng/mL)	RSD[d] (%)
Dichlorvos (DDVP)	185	60	0.05	0.999	0.1-2	6.1
Etridiazole	211 + 183	60	0.01	0.995	0.05-2	7.7
Chloroneb	193	60	0.01	[b]	—	—
Molinate	126	60	0.01	0.999	0.05-2	16.9
Fenobucarb (BPMC)	121	80	0.05	0.998	0.1-2	10.1
Benfluralin	292	80	0.01	0.999	0.05-2	4.9
Simazine (CAT)	201	100	0.1	0.995	0.2-5	11.9
Propyzamide	173	100	0.01	0.997	0.05-2	8.6
Diazinon	179	80	0.01	0.996	0.05-2	13.6
Chlorothalonil (TPN)	266	80	0.05	0.992	0.1-2	4.6
Iprobenfos (IBP)	204	80	0.01	0.995	0.05-2	9.2
Dichlofenthion	279	80	0.01	0.992	0.05-2	9.4
Bromobutide	232	100	0.01	0.986	0.05-2	2.5
Terbucarb (MBPMC)	205	80	0.01	0.995	0.05-2	3.6
Simetryne	213	100	0.1	0.992	0.2-5	15.4
Tolclofos-methyl	265	80	0.01	0.990	0.05-2	3.6
Carbaryl (NAC)	144	100	0.2	0.990	0.5-5	6.6
Metalaxyl	160	100	0.5	0.989	1-5	5.4
Dithiopyr	354	80	0.01	0.990	0.05-2	9.5
Fenitrothion (MEP)	260	80	0.05	0.991	0.1-2	14.4
Esprocarb	222	80	0.01	0.988	0.05-2	5.0
Malathion	173	100	10	[c]	—	—
Thiobencarb	100	80	0.01	0.994	0.05-2	5.3
Chlorpyrifos	314	80	0.01	0.991	0.05-2	7.8
Fthalide	243	60	5	[c]	—	—
Pendimethalin	252	80	0.01	0.994	0.05-2	6.1
Methyldymron	106	100	0.5	0.983	1-5	17.0
Isofenphos	213	100	0.05	0.990	0.1-2	18.2
Butamifos	286	100	0.05	0.995	0.1-2	15.2
Flutolanil	173	100	0.05	0.979	0.1-2	9.1
Napropamide	128	100	0.1	0.997	0.2-5	14.4
Isoprothiolane	204	100	0.05	0.990	0.1-2	13.2
Pretilachlor	262	100	0.05	0.982	0.1-2	14.9
Edifenphos (EDDP)	310	80	10	[c]	—	—
Buprofezin	175	100	0.05	0.988	0.1-2	9.9
Triclopyr-2-butoxyethyl	184	80	0.1	0.997	0.2-5	7.5
Isoxathion	177	80	0.1	0.990	0.2-5	18.1
Mepronil	119	100	0.2	0.991	0.5-5	6.3
Chlornitrofen (CNP)	317	80	0.01	0.984	0.1-1	15.5
Propiconazole	191 (259)[a]	100	0.1	0.984	0.2-1	9.1
Pyributicarb	165	100	0.05	0.996	0.1-2	15.4
Pyridaphenthion	156 (199)[a]	100	0.5	0.992	1-5	11.8
EPN	169	100	0.05	0.992	0.1-2	6.6
Mefenacet	192	100	5	[c]	—	—
Ethofenprox	163	100	0.05	0.987	0.1-2	11.2

[a] Product ion (precursor ion) selected in MS/MS mode.

[b] Nonlinear calibration curve was observed in the range of 0.05–2 ng/mL

[c] No calibration curve was determined.

[d] *n* = 3 determinations.

From **ref. 23**, with permission.

5. The SPE cartridge is then eluted with 3 mL dichloromethane.
6. The dichloromethane eluate is evaporated under nitrogen until its volume is condensed to less than 0.5 mL, and 0.5 mL of the solution containing thiobencarb-d_{10} and chlornitrofen-d_4 at 0.5 μg/mL is added as internal standard.
7. Adjust 1 μL of the solution to 1 mL with dichloromethane, then inject into the GC–MS (*see* **Note 30**).

4. Notes

1. Vapor pressure P_v and octanol/water partition coefficient P are taken from the literature *(24,25)*. Because the P_v and P vary widely, it is convenient to express them on a logarithmic scale. The log P_v values range from –7.0 to 5.5, and partition coefficient (log P) values range from –0.6 to 8.2.
2. A quadrupole ion trap mass spectrometer is adopted as a GC–MS detector because of its accurate identification of a target pesticide in both full-scan and MS/MS modes *(26)*.
3. The Combi PAL is selected as an SPME module, and mounted on a GC–MS system. It consists of a sample tray with valuable vial penetration depth for headspace or liquid extraction, heated agitator, fiber cleaning station (optional), and injection unit (**Fig. 1**).
4. Among the 174 pesticides tested, 158 pesticides were detectable at a concentration of 10 ng/mL in pure water. The other 16 pesticides could not be detected at 10 ng/mL under any condition used in this study.
5. Because the movements of the GC–MS and the autosampler are independent of each other, different procedures for two samples [GC procedure for *n*th sample and SPME procedure for (*n* + 1)th sample] could be performed simultaneously. Moreover, GC measurements for samples could be consecutively carried out by synchronizing the end of the GC–MS procedure for the former sample with the start of the GC injection in the SPME procedure for the latter sample (**Fig. 2**).
6. The entire SPME and GC–MS analysis of pesticides is accomplished automatically within 60 min.
7. The total concentration of acetone in each sample vial throughout the study was controlled within 0.05% (v/v).
8. The HS-SPME technique reduces background adsorption and matrix effects and consequently enhances the life expectancy of the SPME fiber because the fiber is not in contact with the sample *(27)*. The SPME fiber must be carefully handled because the coating is prone to strip from the needle during insertion and removal.
9. The sample vials can be set up to 32.
10. Five different commercially available SPME fibers (PA, PDMS, PDMS–DVB, CW–DVB, and CAR–PDMS) were tested for HS-SPME extraction of pesticides from water samples. Although the optimum fibers differed depending on the pesticide, all 158 pesticides except pencycuron could be sufficiently extracted with the PA fiber. The chromatogram of pencycuron extracted with the PA fiber is not suitable for quantitative treatment because of the small peak and the spread band.
11. The salt addition in HS-SPME increases the ionic strength of the solution, consequently decreasing the solubility of analyte, and the affinity of analyte to gaseous phase increases. The effect of the salt addition on the HS-SPME was tested at the concentration of 0–50% (w/v) of sodium sulfate with the PA fiber for 30 min at 80°C. The increase in the extraction yield for most pesticides declined when more than 40% (w/v) of sodium sulfate was

added. In addition, it was observed that the rates of increase for pesticides with log P values below 4 (log P < 4) are higher than those for pesticides with log P values above 4 (log P > 4). There was no change in the value of pH for each mixture solution, in which the 40% (w/v) of sodium sulfate is already added, before and after the working standard mixture was added. Therefore, no pH adjustment for the tested solution was carried out.

12. To optimize the extraction time and temperature, HS-SPME of pesticides was tested by triplicate extractions with PA fiber from 10-mL pure water samples spiked with the studied pesticides at 10 ng/mL containing 40% (w/v) sodium sulfate as the extraction time was increased in 10-min steps from 10 to 60 min and the extraction temperature was increased in 10° steps from 30 to 100°C. The optimum extraction temperature for each pesticide (*see* **Table 1**) was determined after due consideration of not only the extraction temperature profile, but also the extraction time profile.

13. The extraction temperature profiles are classified into four types according to their shapes (**Table 1**). Typical examples of types 1–3 profiles are shown in **Fig. 4A–C**, respectively. The log P_v and log P values of the pesticides classified into the same type aggregate into a cluster. The distribution of three types and boundary lines of the applicability are shown in **Fig. 5**.

14. Type 1: A peak shape is a characteristic feature of this type. The profile has one peak in which the apex lies at 60–80°C. The extraction yield increases by elevating the temperature from 30°C to the apex level at 60–80°C. As the temperature increases to a value higher than the apex temperature, the extraction yield rapidly decreases. Although we have not yet confirmed the mechanism behind their profiles experimentally, there seem to be two possibilities. One is a reduction in the distribution coefficients between the coating material and analytes at elevated temperature (*27*), and the other is hydrolysis of the pesticides (*28*). The log P_v and log P values of the pesticides in this type range from –2.0 to 5.5 and from 1.7 to 6.5, respectively. The physical property values of the pesticides in this type are distributed in circle A (**Fig. 5**) with one exception, dichlorodiisopropyl ether (DCIP).

15. Type 2: A plateau shape is a typical feature of this type. In contrast to type 1, the profile of this type does not have a well-defined peak. The extraction yield increases with elevating temperature until it reaches a maximum at 60–80°C. The extraction yield is steady or slightly decreases as temperature increases to higher than the maximum temperature. The log P_v and log P values of the pesticides in this type range from –3.9 to 1.9 and from 1.6 to 6.9, respectively. The physical property values of the pesticides in this type are distributed in circle B (**Fig. 5**). Circle B lies on the upper and left side of circle A.

16. Type 3: A steep slope is a characteristic feature of this type. An outstanding characteristic of this type is its uniqueness. There is no or a poor extraction yield of each pesticide that belonged to Type 3 at room temperature. Over the boundary temperature (about 60°C), the extraction yield of the pesticides continued to increase with temperature until 100°C. The log P_v and log P values of the pesticides in this type range from –7.1 to 1.3 and from 1.6 to 8.2, respectively. The physical property values of the pesticides in this type are distributed in circle C (**Fig. 5**) with a few exceptions, a larger area than both circles A and B. Circle C is upward and to the left of circles A and B. The discovery of type 3 contributed greatly to the enhancement of applicability from the viewpoint that the physical properties are spread over a wide range of values.

17. Type 4: This type has no defining feature. The profiles of three pesticides (dichlofluanid, folpet, and fthalide) are classified into this type as shown in **Table 1**. The phenomenon caused by hydrolysis is observed in folpet as follows: In the extraction time profile at 30–

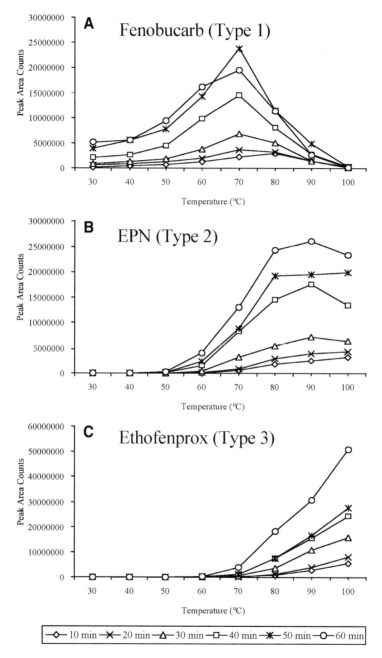

Fig. 4. Example extraction profiles given by analyzing pure water samples (10 mL) spiked at 1 ng/mL of each pesticide. Each point represents an average of triplicate extractions with a PA fiber. EPN, p-nitrophenyl phenyl phosphonothionate. From **ref. 23** with permission. Copyright 2004 Elsevier.

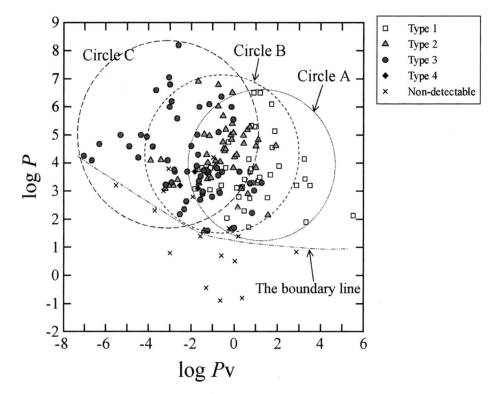

Fig. 5. Physical property diagram rearranged according to the classification of extraction temperature profiles of the selected pesticides. logP_v, logarithmic value of vapor pressure (mPa); log$P$, logarithmic value of octanol–water partition coefficient. From **ref. 23** with permission. Copyright 2004 Elsevier.

50°C, the extraction yields of folpet started decreasing at 40–60 min, whereas they continued to increase from 10 to 30 min. There were no pesticides observed in the extract over the course of the extraction (10–60 min) at temperatures higher than 50°C.

18. Although the regions in the circles overlap each other (**Fig. 5**), this diagram is useful for estimating the extraction feature in pesticides that have not been examined.

19. A line of demarcation separating detectable pesticides and nondetectable pesticides can be drawn in the diagram (**Fig. 5**). The observation that the demarcation line is curved upper and leftward suggests that the hydrophobic pesticides are apt to be extracted more easily than hydrophilic pesticides as the vapor pressure of pesticide becomes lower.

20. The decision of analytical condition for a pesticide in multigroup analysis does not necessarily coincide with the optimization for the pesticide but is more simple and flexible than the fractionation adjustment in such conventional sample preparations as liquid–liquid extraction and SPE. When a new pesticide should be added in a monitoring list, it is easy to select the grouping temperature for the pesticide because the preexamination for ob-

taining the extraction profile of the pesticide can be rapidly carried out by the autosampler device. Therefore, new pesticides can be easily classified into the suitable group without altering the conditions established for already-existing pesticides.

21. It is difficult to carry out the simultaneous determination of 45 pesticides under one set of conditions because their optimum conditions are different from each other (**Table 1**). Therefore, multigroup analysis is adopted for the simultaneous determination of 45 pesticides. The 45 pesticides are divided into three groups and extracted at 60, 80, or 100°C, as shown in **Table 2**. The pesticides are extracted with the PA fiber for 40 min in such a way that each sample could be analyzed within 1 h. The values for the quantitative evaluations are obtained under the conditions for the multigroup method, not under the optimum conditions for each pesticide.

22. One fiber can be used repeatedly at least more than 100 times.

23. To determine whether the automated HS-SPME–GC–MS system is quantitatively useful for detecting multiresidue pesticides in water, 45 of the 158 detectable pesticides were selected. The selected pesticides and the selected ions (*m/z*) for the determination of each pesticide with the mass chromatograms under full-scan and MS/MS modes are listed in **Table 2**.

24. The calibration curves are linear, ranging from 0.05 to 5 µg/mL. The correlation coefficient was calculated from the 15 data (triplicates in five points) on the calibration curve and ranged between 0.979 and 0.999, as shown in **Table 2**.

25. The LOD values for each pesticide are given in **Table 2**. The values of 0.01 ng/mL in LOD for the 17 pesticides are achieved only under approximate conditions for screening, not under optimized conditions for each pesticide.

26. The precision of the method is evaluated by calculating the relative standard deviation (RSD) in pure water. The RSD values for each pesticide in pure water are obtained by triplicate analysis of spiking target pesticides into pure water at the concentration of the lowest standard solution for each calibration curve. The RSD values ranged from 3.6 to 18% (**Table 2**). The HS-SPME system could provide superior productivity and reproducibility.

27. To confirm validity of the HS-SPME–GC–MS method, known amounts of pesticides are spiked at 0.5 ng/mL to water samples (pure water, three different natural mineral waters, and surface water), and their recoveries are calculated. Of the 45 pesticides tested, 7 (metalaxyl, malathion, fthalide, methyldymron, edifenphos, pyridaphenthion, and mefenacet) were excluded from subsequent examinations because the LOD values for these seven pesticides were more than 0.5 ng/mL. The mean recovery values and the RSD values for the 38 selected pesticides in water samples at the concentration of 0.5 ng/mL were obtained with five replicate analyses (**Table 3**).

28. Typical chromatograms for the spiked pesticides in pure water and in the surface water are shown in **Fig. 6**.

29. The recovery values and the RSD values of the HS-SPME technique are compared with those values of the SPE technique by analyzing pure water and surface water samples (five replicates) spiked with the tested pesticides at 0.5 ng/mL. The results are in good agreement with those obtained by SPE, as shown in **Table 3**.

30. The GC–MS conditions are the same as those of the HS-SPME–GC–MS method except for a splitless time of 1 min.

Table 3
Recoveries and Relative Standard Deviations (RSD) for 38 Pesticides in Water Samples by HS-SPME and SPE

| Compound | Extraction temperature (°C) | HS-SPME | | | | | | | | SPE | |
| | | Volvic | | Evian | | Contrex | | Katsuura River | | Katsuura River | |
		Recovery (%)	RSD (%)	Recovery (%)	RSD (%)	Recovery (%)	RSD (%)	Recovery (%)	RSD (%)	Recovery (%)	RSD (%)
Dichlorvos (DDVP)	60	125.9	15.6	87.8	4.7	110.5	8.7	73.2	4.9	117.8	7.1
Etridiazole	60	95.6	5.8	45.2	6.4	87.6	4.2	99.4	5.9	96.5	5.6
Chloroneb	60	94.1	4.1	54.3	6.8	94.2	3.4	125.5	6.9	89.5	1.6
Molinate	60	97.7	4.0	72.7	8.5	100.0	7.3	108.5	5.3	82.5	2.6
Fenobcarb (BPMC)	80	135.9	5.7	153.6	2.7	159.7	4.9	93.9	7.9	89.7	5.2
Benfluralin	80	100.7	3.9	109.3	3.3	96.8	7.3	94.3	6.7	80.4	9.3
Simazine (CAT)	100	83.9	10.4	108.4	9.3	103.7	11.4	117.5	2.3	100.4	6.0
Propyzamide	100	91.8	6.7	99.1	1.6	97.7	7.5	107.2	3.0	97.5	2.9
Diazinon	80	93.2	2.6	109.7	2.0	89.4	3.7	92.4	2.6	98.6	5.6
Chlorothalonil (TPN)	80	80.6	17.2	140.0	3.9	141.1	5.1	90.9	5.7	239.2	5.8
Iprobenfos (IBP)	80	84.3	3.0	92.0	3.4	100.2	6.7	80.0	9.9	91.9	10.9
Dichlofenthion	80	79.9	4.8	117.5	1.8	104.5	7.7	103.9	2.3	88.3	8.8
Bromobutide	100	80.7	6.8	93.5	2.5	114.0	8.3	103.3	1.4	97.0	6.5
Terbucarb (MBPMC)	80	83.2	3.4	112.2	2.8	102.1	4.3	97.1	3.2	95.1	4.0
Simetryne	100	80.6	8.4	101.8	5.9	113.0	10.8	130.9	3.7	88.0	5.5
Tolclofos-methyl	80	75.0	3.1	108.3	2.6	103.6	6.6	95.5	2.2	77.9	5.5
Carbaryl (NAC)	100	91.2	15.1	90.6	15.9	117.8	8.5	105.7	5.4	78.7	13.0
Dithiopyr	80	86.9	4.1	88.5	3.8	102.1	4.2	101.6	1.1	79.0	6.8
Fenitrothion (MEP)	80	107.3	5.9	124.9	4.8	66.1	12.5	88.1	11.7	115.2	4.6

Table 3 (*Continued*)
Recoveries and Relative Standard Deviations (RSD) for 38 Pesticides in Water Samples by HS-SPME and SPE

| Compound | Extraction temperature (°C) | HS-SPME | | | | | | | | SPE | |
| | | Volvic | | Evian | | Contrex | | Katsuura River | | Katsuura River | |
		Recovery (%)	RSD (%)	Recovery (%)	RSD (%)	Recovery (%)	RSD (%)	Recovery (%)	RSD (%)	Recovery (%)	RSD (%)
Esprocarb	80	83.7	2.8	88.6	3.3	94.7	7.9	97.2	6.7	90.6	5.2
Thiobencarb	80	96.9	3.9	90.4	5.5	93.2	4.9	99.0	3.2	93.8	7.5
Chlorpyrifos	80	92.2	5.1	75.4	3.8	78.4	4.1	96.1	3.9	87.4	7.6
Pendimethalin	80	76.4	2.5	106.1	6.5	80.9	6.7	86.4	7.0	81.0	1.6
Isofenphos	100	95.4	5.7	89.6	8.1	76.5	7.6	98.6	3.0	73.3	8.0
Butamifos	100	94.5	2.9	87.5	6.9	87.1	9.3	91.9	5.3	101.0	5.0
Flutolanil	100	114.5	5.8	112.7	15.4	101.5	3.2	118.1	9.2	102.8	7.8
Napropamide	100	112.7	5.4	111.6	4.2	100.2	4.7	102.8	3.9	117.0	4.1
Isoprothiolane	100	114.7	10.1	111.1	11.1	98.0	6.9	104.0	5.5	132.0	6.0
Pretilachlor	100	92.0	4.6	95.1	5.2	87.5	8.2	92.5	5.4	120.0	4.4
Buprofezin	100	89.5	5.4	94.1	6.5	93.1	8.9	99.8	6.6	105.6	3.8
Triclopyr-2-butoxyethyl	80	112.5	2.7	120.2	5.8	96.7	7.4	71.8	8.3	102.3	4.3
Isoxathion	80	86.4	3.9	88.2	6.9	70.5	6.5	77.0	4.4	106.2	7.6
Mepronil	100	119.0	4.9	120.6	12.7	107.2	7.8	128.4	8.3	106.8	6.7
Chlornitrofen (CNP)	80	77.4	5.4	101.2	4.3	86.3	8.2	93.4	6.4	85.1	6.0
Propiconazole	100	82.1	9.7	51.3	6.8	60.8	7.4	96.0	8.9	111.1	6.8
Pyributicarb	100	96.4	3.6	66.5	12.6	91.1	8.8	98.5	3.2	80.6	5.0
EPN	100	113.9	7.4	108.4	6.6	78.8	9.1	68.0	12.5	117.2	5.2
Ethofenprox	100	91.1	10.1	108.5	5.7	110.0	10.4	132.7	6.9	65.3	9.6

Spiking level of 0.5 ng/mL; n = 5 determinations.
From **ref. 23**, with permission.

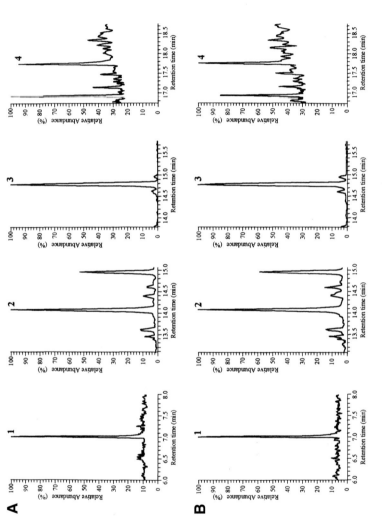

Fig. 6. HS-SPME–GC–MS mass chromatograms for the studied pesticides in (**A**) pure water spiked at 0.5 ng/mL of each pesticide and (**B**) surface water (Katsuura River) spiked at 0.5 μg/mL of each pesticide. GC–MS conditions: Rtx-5MS cclumn (30 m × 0.25 mm i., 0.25-μm film thickness); column temperature programmed at 50°C, hold for 1 min, 25°C/min to 125°C, 10°C/min :o 300°C, hold for 7 min; PTV splitless mode injection with a splitless time of 3 min and with the following temperature programmed at 50°C, hold for 0.5 min, 10°C/s to 250°C, hold for 3 min, 5°C/s to 260°C, hold for 28 min; helium carrier gas at a constant flow (1 mL/min); 260°C transfer line temperature; 200°C ion source temperature; 70-eV ionization voltage. Peaks: 1, dichlorvos (*m/z* 185); 2, simetryne (*m/z* 213); 3, thiobencarb (*m/z* 100); 4, mepronil (*m/z* 119).

References

1. Arthur, C. L. and Pawliszyn, J. (1990) Solid-phase microextraction with thermal desorption using fused silica optical fibers. *Anal. Chem.* **62,** 2145–2148.
2. Kataoka, H., Lord, H. L., and Pawliszyn, J. (2000) Applications of solid-phase microextraction in food analysis. *J. Chromatogr. A* **880,** 35–62.
3. Lord, H. L. and Pawliszyn, J. (2000) Evolution of solid-phase microextraction technology. *J. Chromatogr. A* **885,** 153–193.
4. Beltran, J., Lopez, F. J., and Hernandez, F. (2000) Solid-phase microextraction in pesticide residue analysis. *J. Chromatogr. A* **885,** 389–404.
5. Alpendurada, M. de F. (2000) Solid-phase microextraction: a promising techniques for sample preparation in environmental analysis. *J. Chromatogr. A* **889,** 3–14.
6. Pawliszyn, J. (2001) Solid-phase microextraction. *Adv. Exp. Med. Biol.* **488,** 73–87.
7. Kataoka, H. (2002) Automated sample preparation using in-tube solid-phase microextraction and its application—a review. *Anal. Bioanal. Chem.* **373,** 31–45.
8. Buldini, P. L., Ricci, L., and Sharma, J. L. (2002) Recent applications of sample preparation techniques in food analysis. *J. Chromatogr. A* **975,** 47–70.
9. Zambonin, C. G. (2003) Coupling solid-phase microextraction to liquid chromatography. A review. *Anal. Bioanal. Chem.* **375,** 73–80.
10. Krutz, L. J., Senseman, S. A., and Sciumbato, A. S. (2003) Solid-phase microextraction for herbicide determination in environmental samples. *J. Chromatogr. A* **999,** 103–121.
11. Lambropoulou, D. A. and Albanis, T. A. (2001) Optimization of headspace solid-phase microextraction conditions for the determination of organophosphorus insecticides in natural waters. *J. Chromatogr. A* **922,** 243–255.
12. Henriksen, T., Svensmark, B., Lindhardt, B., and Juhler, R. K. (2001) Analysis of acidic pesticides using *in situ* derivatization with alkylchloroformate and solid-phase microextraction (SPME) for GC–MS. *Chemosphere* **44,** 1531–1539.
13. Queiroz, M. E., Silva, S. M., Carvalho, D., and Lancas, F. M. (2001) Comparison between solid-phase extraction methods for the chromatographic determination of organophosphorus pesticides in water. *J. Environ. Sci. Health B* **36,** 517–527.
14. Gerecke, A. C., Tixier, C., Bartels, T., Schwarzenbach, R. P., and Muller, S. R. (2001) Determination of phenylurea herbicides in natural waters at concentrations below 1 ng/L using solid-phase extraction, derivatization, and solid-phase microextraction-gas chromatography-mass spectrometry. *J. Chromatogr. A* **930,** 9–19.
15. Ramesh, A. and Ravi, P. E. (2001) Applications of solid-phase microextraction (SPME) in the determination of residues of certain herbicides at trace levels in environmental samples. *J. Environ. Monit.* **3,** 505–508.
16. Boussahel, R., Bouland, S., Moussaoui, K. M., Baudu, M., and Montiel, A. (2002) Determination of chlorinated pesticides in water by SPME/GC. *Water Res.* **36,** 1909–1911.
17. Perez-Trujillo, J. P., Frias, S., Conde, J. E., and Rodriguez-Delgado, M. A. (2002) Comparison of different coatings in solid-phase microextraction for the determination of organochlorine pesticides in ground water. *J. Chromatogr A* **963,** 95–105.
18. Tomkins, B. A. and Ilgner, R. H. (2002) Determination of atrazine and four organophosphorus pesticides in ground water using solid phase microextraction (SPME) followed by gas chromatography with selected-ion monitoring. *J. Chromatogr. A* **972,** 183–194.
19. Souza, D. A. and Lancas, F. M. (2003) Solventless sample preparation for pesticides analysis in environmental water samples using solid-phase microextraction-high resolution gas chromatography/mass spectrometry (SPME–HRGC/MS). *J. Environ. Sci. Health B* **38,** 417–428.

20. Li, H. P., Li, G. C., and Jen, J. F. (2003) Determination of organochlorine pesticides in water using microwave assisted headspace solid-phase microextraction and gas chromatography. *J. Chromatogr. A* **1012,** 129–137.

21. Gonçalves, C. and Alpendurada, M. F. (2004) Solid-phase microextraction–gas chromatography–(tandem) mass spectrometry as a tool for pesticide residue analysis in water samples at high sensitivity and selectivity with confirmation capabilities. *J. Chromatogr. A* **1026,** 239–250.

22. Zhang, Z. and Pawliszyn, J. (1993) Headspace solid phase microextraction. *Anal. Chem.* **65,** 1843–1852.

23. Sakamoto, M. and Tsutsumi, T. (2004) Applicability of headspace solid-phase microextraction to the determination of multi-class pesticides in waters. *J. Chromatogr. A* **1028,** 63–74.

24. TomLin, C. D. S. (ed.). (2002) *The e-Pesticide Manual, A World Compendium*, 12th ed. British Crop Protection Council, Bracknell, U.K.

25. Research Center for Environmental Risk (provider). *Webkis-Plus Chemicals Database.* Available at: http://w-chemdb.nies.go.jp/index.html (2003). National Institute for Environmental Studies, Ibaraki, Japan.

26. Reyzer, M. L. and Brodbelt, J. S. (2001) Analysis of fire ant pesticides in water by solid-phase microextraction and gas chromatography/mass spectrometry or high-performance liquid chromatography/mass spectrometry. *Anal. Chim. Acta* **436,** 11–20.

27. Doong, R. A. and Liao, P. L. (2001) Determination of organochlorine pesticides and their metabolites in soil samples using headspace solid-phase microextraction. *J. Chromatogr. A* **918,** 177–188.

28. Zini, C. A., Lord, H., Christensen, E., de Assis, T. F., Caramão, E. B., and Pawliszyn, J. (2002) Automation of solid-phase microextraction–gas chromatography–mass spectrometry extraction of eucalyptus volatiles. *J. Chromatogr. Sci.* **40,** 140–146.

28

Analysis of Herbicides in Water by On-Line In-Tube Solid-Phase Microextraction Coupled With Liquid Chromatography–Mass Spectrometry

Hiroyuki Kataoka, Kurie Mitani, and Masahiko Takino

Summary

We have developed a method using in-tube solid-phase microextraction coupled with liquid chromatography–mass spectrometry (SPME–LC–MS) for the assay of herbicides in water samples. Chlorinated phenoxy acid herbicides have been automatically extracted into a DB-WAX capillary by repeated draw/eject cycles of sample solution and desorbed with acetonitrile. Triazine herbicides have also been extracted easily by this in-tube SPME method. Compared with direct sample injection into LC–MS, the in-tube SPME technique resulted in a greater than 25-fold concentration. The automated in-tube SPME–LC–MS system can continuously extract herbicides from aqueous samples, followed by LC–MS. It provides a simple, rapid, selective, sensitive, low-cost method for herbicide analyses and can be directly applied to various water samples without any pretreatment. Although the in-tube SPME technique has not yet been extensively applied, its application to various fields is expected because the analytical processes, from sample preparation to chromatography and data analysis, can be automated.

Key Words: Automated analysis; chlorinated phenoxy acid herbicides; in-tube solid-phase microextraction; liquid chromatography–mass spectrometry; river water; triazine herbicides.

1. Introduction

Throughout the world, pesticides and herbicides are used in agriculture on an increasingly large scale, with pollution from these agents spread throughout ecosystems by leaching and runoff from soil into groundwater and surface water. When released into the environment, these compounds can also disrupt normal endocrine function in a variety of aquatic and terrestrial organisms, making appropriate control of these compounds and prevention of related health hazards important (*1–3*). Thus, both the actual state and the transition of residues of pesticides and herbicides in various matrices,

From: *Methods in Biotechnology, Vol. 19, Pesticide Protocols*
Edited by: J. L. Martínez Vidal and A. Garrido Frenich © Humana Press Inc., Totowa, NJ

including water, soil, and agricultural products, have received increasing attention in the last few decades. These environmental samples have been analyzed by various techniques, including gas chromatography (GC), high-performance liquid chromatography (HPLC), and capillary electrophoresis using various detection systems, to obtain quantitative and qualitative information on the presence of these residues *(4–7)*. Most of these techniques, however, require extensive and time-consuming sample preparation methods, such as extraction, concentration, fractionation, and isolation, prior to analysis, leading to the need for a major simplification in sample preparation methods, including miniaturization in scale *(8–13)*.

Solid-phase microextraction (SPME) *(14)* is a simple and convenient sample preparation technique that uses a fused-silica fiber coated on the outside with an appropriate stationary phase. The analyte in a sample is directly extracted onto the fiber coating. SPME has been used routinely in combination with GC and GC–mass spectrometry (GC–MS) and successfully applied to a wide variety of compounds in gaseous, liquid, and solid samples, especially for the extraction of volatile and semivolatile organic compounds from environmental samples *(12,13,15–20)*. SPME has also been directly coupled with HPLC and LC–MS to analyze weakly volatile or thermally labile compounds not amenable to GC or GC–MS. In addition, a new SPME–HPLC system, known as in-tube SPME, which uses an open-tubular fused-silica capillary instead of the SPME fiber, has been developed *(16,21,22)*. In-tube SPME is suitable for automation, and extraction, desorption, and injection can be performed continuously using a standard autosampler. Automated sample-handling procedures not only shorten total analysis time, but also are more accurate and precise than manual techniques. With the in-tube SPME technique, organic compounds in aqueous samples are directly extracted from the sample into the internally coated stationary phase of a capillary and then desorbed by introducing a stream of mobile phase or static desorption solvent when the analytes are more strongly adsorbed to the capillary coating. Desorbed compounds are finally injected into the column for analysis. The in-tube SPME method has been used in the analysis of pesticides and herbicides from environmental matrices *(21–26)*.

In this chapter, we describe a method for the analysis of herbicides in environmental waters using automated on-line in-tube SPME coupled with LC–MS. LC–MS with electrospray ionization (LC–MS/ESI) was used to analyze six chlorinated phenoxy acid herbicides, whereas LC–tandem mass spectrometry (LC–MS/MS) was used to analyze five triazine herbicides. Optimum conditions for the in-tube SPME method and typical problems encountered in the development and application of this method are discussed.

2. Materials

2.1. Equipment

1. Liquid chromatograph (e.g., Agilent 1100 series, Agilent Technologies, Waldborn, Germany), consisting of a vacuum solvent degassing unit, a binary high-pressure gradient pump, an autosampler, a column thermostat, and a diode array detector connected on line to the MS system described here.
2. MS system single-quadrupole instrument for the analysis of chlorinated phenoxy acid herbicides (e.g., an Agilent 1100 series mass selective detector [MSD], equipped with

orthogonal spray/ESI (e.g., Agilent Technologies, Palo Alto, CA). LC–MS data are analyzed using LC/MSD ChemStation software (e.g., Agilent, Palo Alto, CA).

3. MS system triple-quadrupole mass spectrometer for the analysis of triazine herbicides (e.g., API 4000, Applied Biosystems, Foster City, CA) and equipped with a turbo ion spray interface. LC–MS/MS data are analyzed using software (e.g., Analyst Software 1.3.1, Applied Biosystems).

4. LC columns: An inertsil octadecylsilane 3 (ODS3) LC column 150 × 2.1 mm id, 5-μm LC column (e.g., GL Science, Tokyo, Japan) for the analysis of chlorinated phenoxy acid herbicides and a TSK gel Super-Octyl 100 × 2.0 mm id, 2-μm LC column (e.g., TOSOH, Tokyo, Japan) for the analysis of triazine herbicides.

5. GC capillaries (60 cm long) for in-tube SPME: DB-1 with 0.25-mm id, 1.0-μm film thickness (e.g., J&W Scientific, Folsom, CA); DB-5 with 0.25-mm id, 1.0-μm film thickness (e.g., J&W Scientific); DB-17 with 0.25-mm id, 1.0-μm film thickness (e.g., J&W Scientific); DB-50 with 0.25-mm id, 1.0-μm film thickness (e.g., J&W Scientific); DB-WAX, 0.25-mm id, 1.0-μm film thickness (e.g., J&W Scientific); Carbowax, 0.25-mm id, 1.0-μm film thickness (e.g., Supelco Co., Bellefonte, PA); Omegawax 250 with 0.25-mm id, 1.0-μm film thickness (e.g., Supelco Co.); and Supel Q-PLOT, 0.32-mm id, 12-μm film thickness (e.g., Supelco Co.) (*see* **Note 1**).

6. Screw-cap autosampler 2-mL vials equipped with silicon/polytetrafluoroethylene septa (e.g., Agilent Technologies, Palo Alto, CA).

2.2. Reagents and Solutions

1. Chlorinated phenoxy acid herbicides (>99% pure): 4-chloro-2-methylphenoxy acetic acid (MCPA); 2,4-dichlorophenoxy acetic acid (2,4-D); 2-(2,4-dichlorophenoxy) propionic acid (dichlorprop); 2,4,5-trichlorophenoxy acetic acid (2,4,5-T); 3-(2,4-dichlorophenoxy) butyric acid (2,4-DB); and 2-(2,4,5-trichlorophenoxy) propionic acid (2,4,5-TP) (e.g., Hayashi Pure Chemicals, Osaka, Japan). The chemical structures of these herbicides are shown in **Fig. 1A**.

2. Triazine herbicides (>99% pure): atrazine, simazine, ametryne, propazine, and prometryne (e.g., GL Science). These chemical structures are shown in **Fig. 1B**.

3. Standard herbicide solutions: Standard stock solution (each 1 mg/mL) is prepared in methanol, stored in the dark at 4°C, and diluted to the desired concentration with pure water prior to use.

4. Ammonium acetate, formic acid, dibutylamine acetate (DBA), and acetonitrile.

5. Other materials: solvents and pure water for mobile phase are HPLC grade. All other chemicals are of analytical reagent grade. Water is purified with a Milli-Q system (Millipore, Tokyo, Japan).

2.3. Water Samples

Water samples from a river, a lake, the pond of a golf course, and a paddy field were collected in glass bottles, filtered through 0.2-μm nylon syringe filters 13 mm in diameter (TOSOH, Tokyo, Japan), and stored at 4°C until tested.

3. Methods
3.1. Preparation of Samples

1. Water samples are adjusted to the appropriate pH to increase the extraction efficiency of the analyte to stationary phase in SPME (*see* **Note 2**).

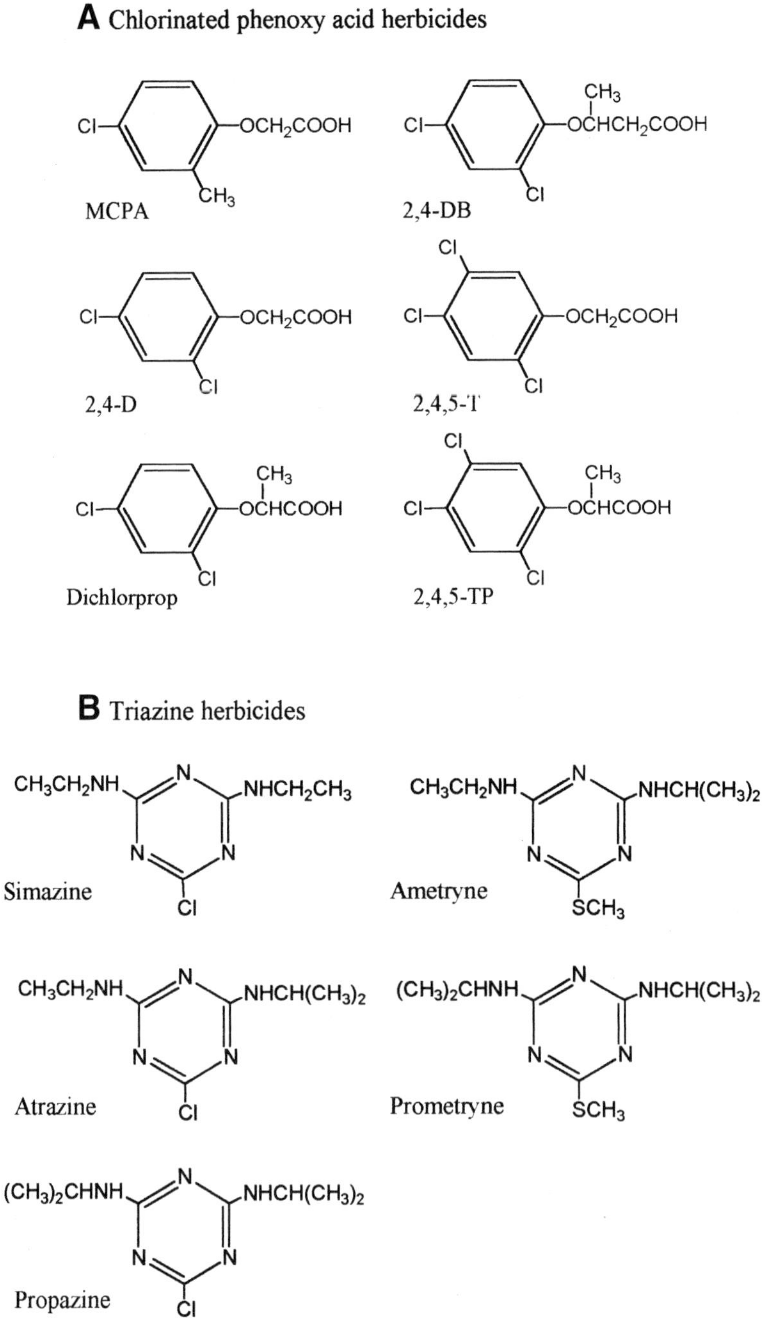

A Chlorinated phenoxy acid herbicides

MCPA

2,4-DB

2,4-D

2,4,5-T

Dichlorprop

2,4,5-TP

B Triazine herbicides

Simazine

Ametryne

Atrazine

Prometryne

Propazine

Fig. 1. Chemical structures of the chlorinated phenoxy acid and triazine herbicides used in this study.

2. For the analysis of chlorinated phenoxy acid herbicides, 0.9 mL of each water sample is transferred to a 2-mL screw-cap autosampler vial to which 0.1 mL of 20 mg/mL formic acid is added. The vial is vortexed and set onto the sample tray in the autosampler (*see* **Note 3**).

3. For the analysis of triazine herbicides, 1.0 mL of each water sample is transferred to a 2-mL screw-cap autosampler vial, and the vial is set on the sample tray in the autosampler.

3.2. Automated In-Tube SPME–LC–MS System

The in-tube SPME procedures are automated with an Agilent 1100 series autosampler. The computer controls all movements of the autosampler and LC–MS. A schematic diagram of the in-tube SPME–LC–MS system is shown in **Fig. 2**.

3.2.1. Automated In-Tube SPME Procedure

1. The GC capillary used as an in-tube SPME device is cut to the appropriate length with a ceramics cutter (*see* **Notes 4–7**).

2. The capillary is placed between the injection loop and the injection needle of the autosampler. A photograph of the connection of the GC capillary and the autosampler is presented in **Fig. 3** (*see* **Note 8**).

3. Vials (2 mL) are each filled with 1 mL sample to be extracted and set onto the autosampler, which is programmed to control the SPME extraction and desorption technique. In addition, 2-mL autosampler vials with septa and 1.5 mL each of methanol and water are set onto the autosampler (*see* **Note 9**).

4. The capillary column is washed and conditioned by two repeated draw/eject cycles (40 µL each) of methanol and water, and a 50-µL air plug is drawn prior to the extraction step (*see* **Note 10**).

5. The extraction of herbicides onto the capillary coating is performed by 20–25 repeated draw/eject cycles of 30–40 µL of sample at a flow rate of 100–200 µL/min with the six-port valve in the load position (**Fig. 2A**) (*see* **Notes 11–13**).

6. After washing the tip of the injection needle by one draw/eject cycle of 2 µL methanol, the extracted compounds are desorbed from the capillary coating with acetonitrile or mobile phase flow (*see* **Notes 14–16**).

7. The compounds are transported to the LC column by switching the six-port valve to the inject position (**Fig. 2B**) and detected by the diode array detector and MS system. During LC–MS analysis, the GC capillary is washed and conditioned with mobile phase for the next extraction (*see* **Notes 17** and **18**).

3.2.2. LC–MS for Analysis of Chlorinated Phenoxy Acid Herbicides

1. LC separation of chlorinated phenoxy acid herbicides is performed on a 150 × 2.1 mm column packed with 5 µm Inertsil ODS3 using linear gradient elution for 15 min with a mobile phase of acetonitrile–water containing 5 µmol/mL DBA (from 30 + 70 to 85 + 15 v/v). The flow rate is 0.2 mL/min (*see* **Note 19**).

2. All MS optimization steps are carried out in the ESI negative ion mode using the analytical column. Nitrogen as nebulizing and drying gas (350°C) is generated from compressed air using a nitrogen generator. The nebulizing gas pressure is set at 60 psi, and the drying gas is held at 12 L/min. The drying gas temperature is kept at 350°C, and the fragmentor voltage for in-source fragmentation is set at 100 V. The skimmers 1, 2, and entrance lens voltages in the ion source of the MSD are automatically optimized using a calibration standard and set at 23, 47, and 57 V, respectively. Mass spectra are acquired over the scan

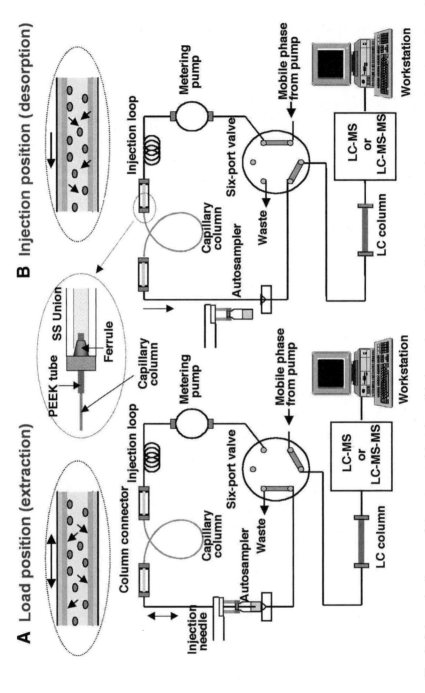

Fig. 2. Schematic diagram of the automated in-tube solid-phase microextraction–liquid chromatography–mass spectrometry system: (A) extraction step; (B) desorption step. SS, stainless steel.

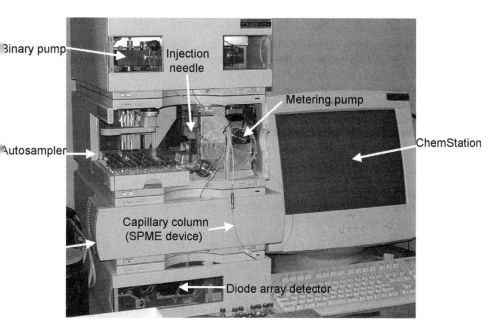

Fig. 3. Photograph of the connection between a capillary and the Agilent 1100 autosampler.

range *m/z* 100–500 unified atomic mass units (u) using a step size of 0.1 u and a scan speed of 0.5 scan/s (*see* **Notes 20** and **21**).

3. Quantitative analysis is carried out using the selected ion monitoring (SIM) mode of base ion peaks at *m/z* 199 (MCPA), 219 (2,4-D), 233 (dichlorprop), 255 (2,4,5-T), 161 (2,4-DB) and 267 (2,4,5-TP), with a dwell time of 250–500 ms. To verify the presence of target analytes in river water, the halogen isotopic ion of all target analytes is monitored (*see* **Note 22**).

4. The calibration curves are prepared in the range 0.05–50 ng/mL as follows: The ratio of the peak area for the target herbicide to the peak area count is plotted on the *y*-axis, and the initial concentration of the target herbicide is plotted on the *x*-axis. The concentrations of the chlorinated phenoxy acid herbicides in the samples are 0.05, 0.1, 0.2, 0.5, 1.0, 2.0, 5.0, 10, 20, and 50 ng/mL.

5. The limit of detection (LOD) is calculated as the concentration giving a signal-to-noise ratio of three ($S/N = 3$), and the limit of quantification (LOQ) is defined as 10 times the standard deviation with a spiked real sample, such as river water. Pure water and river water samples spiked with target herbicides at concentrations of 0.05 ng/mL and higher are analyzed to estimate the LOD and LOQ (*see* **Note 23**).

6. The concentration (nanograms per milliliter) of the pesticides in the samples is given as follows:

$$C = (y - b)/a$$

where *y* is the area count obtained from the sample, *a* is the slope of the regression line in the standard calibration curve, and *b* is the intercept of the regression line in the standard calibration curve.

3.2.3. LC–MS/MS for Triazine Herbicide Analysis

1. LC separation for triazine herbicides is performed on a 100×2.0 mm column packed with 2 µm TSK gel Super-Octyl. The HPLC column temperature is 40°C, and the mobile phase and flow rate are acetonitrile/2 mM ammonium acetate (50 + 50 v/v) and 0.2 mL/min, respectively (*see* **Note 24**).

2. The turbo ion spray interface is operated in the positive ion mode at 5500 V and 500°C. Nitrogen as a nebulizing and drying gas is generated from compressed air using a Kaken N_2 generator (System Instruments Co. Ltd., Tokyo, Japan). The ion sources gas 1 (GS1) and 2 (GS2) are set at 60 and 70 L/min, respectively. The curtain gas flow is set at 20 L/min, and the collision gas is set at 4.0 L/min (*see* **Note 25**).

3. Quantification is performed by multiple reaction monitoring (MRM) of the protonated precursor ion and the related product ion for triazine herbicides. Quadrupoles Q1 and Q3 are set on unit resolution. MRM in the positive ionization mode is performed using a dwell time of 150 ms per transition to detect ion pairs at m/z 228.04/186.05 (ametryne; 25-eV collision energy), m/z 215.90/173.95 (atrazine; 25-eV collision energy), m/z 201.93/131.95 (simazine; 27-eV collision energy), m/z 229.93/145.95 (propazine; 33-eV collision energy), and m/z 242.02/158.05 (prometryne; 33-eV collision energy). The analytical data are processed by Analyst software (version 1.3.1) (*see* **Note 26**).

4. The calibration curves are prepared at concentrations ranging from 0.05 to 10 ng/mL by plotting the ratio of the peak area for the target herbicide to the peak height or peak area count on the y-axis and the initial concentration of the target herbicide on the x-axis. The concentrations of the triazine herbicides in the samples are: 0.05, 0.1, 0.2, 0.5, 1.0, 2.0, 5.0, and 10 ng/mL.

5. The LOD is calculated as the concentration giving a signal-to-noise ratio of three ($S/N = 3$). Pure water and river water samples spiked with target herbicides at concentrations of 0.05 ng/mL and higher are analyzed to estimate the LOD (*see* **Note 27**).

6. The concentration (nanograms per milliliter) of the pesticides in the samples is given as follows:

$$C = (y - b)/a$$

where y is the area count obtained from the sample, a is the slope of the regression line in the standard calibration curve, and b is the intercept of the regression line in the standard calibration curve.

3.3. Analysis of Water Samples

Various environmental water samples are tested. The recoveries of target pesticides spiked into water samples are determined as the ratio of peak area in each tested sample to that in pure water under the same conditions (*see* **Note 28**).

4. Notes

1. For in-tube SPME, several commercially available capillary columns are usually used. Each column has different selectivity in its type of stationary phase, internal diameter, length, and film thickness, as well as during GC analysis. For example, in a low-polarity column with methyl silicon liquid phase, the hydrophobic compounds are selectively retained (extracted) in comparison with hydrophilic compounds, whereas in a high-polarity column with polyethylene glycol liquid phase, hydrophilic compounds are selectively extracted in comparison with the hydrophobic compounds.

2. The extractions of acidic and basic compounds are effective under acidic and alkaline conditions of the sample solution, respectively. The stability of each compound in the sample solution, however, must be verified beforehand. Chlorinated phenoxy acid and triazine herbicides are extracted at acidic pH and neutral pH (in water), respectively. Although salting out increases the extraction efficiency for fiber SPME, it causes blockage of the column by salt deposits with in-tube SPME. In addition, hydrophilic solvents such as methanol decrease the extraction efficiency by increasing the solubility of the compound in the sample. The extraction efficiency, however, is almost uninfluenced by methanol concentrations of 5% or less. The amount of compound extracted by the stationary phase depends on the concentration of the compound in the sample, and over 0.5 mL is necessary for the injection needle soak using a 2-mL autosampler vial.

3. To increase the extraction efficiency, 0.2% formic acid (pH 2.0) is added as a sample matrix modifier.

4. Because the capillary column is generally stable for the mobile phase usually used in HPLC, one capillary can be used repeatedly more than 500 times. To maintain the GC capillary after 50 samples, it should be rinsed with 100 µL acetonitrile and removed from the instrument, followed by purging of the remaining solvent with dry nitrogen gas. Although the capillary column with chemically bonded or cross-linked liquid phase is very stable to water and organic solvents, strong inorganic acids or alkali can easily cause deterioration. It is therefore necessary to monitor the durability of the coating to exposure to mobile phases and any desorption solvents required by the loss of response of known analytes. Thus, it is necessary to monitor the loss of response of known analytes. For the storage of the capillary, each end should be plugged using a stainless steel union to avoid oxidation of the stationary phase by air.

5. The internal diameter, length, film thickness, and so on of the column are related to sample load and the amount of compounds extracted. If these increase, the load and amount extracted will increase, and the extension of sample bandwidth will cause peak broadening and tailing. In addition, if the film thickness of the stationary phase is large, large amounts of compound can be extracted, but it may not be possible to desorb the compound quantitatively from the capillary column. In our experience, a 50- to 60-cm length of capillary column or a narrow-bore column in which the film thickness is comparatively small has been most suitable for maximizing the extracted amount, for minimizing peak tailing, and for preventing carryover with in-tube SPME.

6. We compared the extraction efficiencies of three capillaries (DB-WAX, DB-50, and DB-1) for chlorinated phenoxy acid herbicides from aqueous solution. As shown in **Fig. 4**, the relatively polar DB-WAX gives the best extraction efficiency compared with the less-polar DB-5 and DB-1 for all herbicides. The effect of film thickness on extraction is very important in SPME. When we evaluated three capillaries of different thickness (0.25, 0.5, and 1.0 µm), we found that the maximum extraction efficiency for each herbicide was obtained using the thickest (1.0 µm).

7. For in-tube SPME of triazine herbicides, a porous polymer-type capillary column, Supel-Q PLOT, gives superior extraction efficiency compared with the liquid phase-type capillary columns, Omegawax 250, DB-17, and DB-1 (**Fig. 5**).

8. In-tube SPME is carried out using the Agilent 1100 series LC system without modification of the autosampler itself. Capillary connections are facilitated using a 2.5-cm sleeve of 1/16-in. polyetheretherketone (PEEK) tubing at each end of the capillary. A PEEK tubing internal diameter of 330 µm is suitable for the capillary used. Normal 1/16-in. stainless steel nuts, ferrules, and connectors are then used to complete the connections.

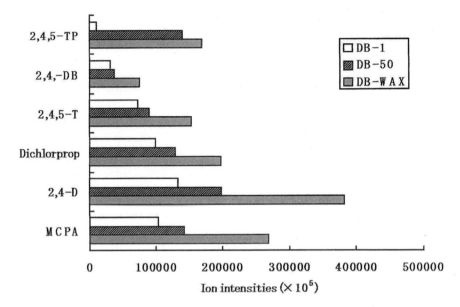

Fig. 4. Evaluation of three capillary columns for in-tube SPME of chlorinated phenoxy acid herbicides. In-tube SPME conditions included 100 ng/mL of each herbicide, a sample pH of 5.0, 15 draw/eject cycles, a draw/eject volume of 25 μL, a draw/eject rate of 0.2 mL/min, and mobile phase desorption (dynamic elution). From **ref. 26** with permission.

The injection loop is retained in the system to avoid fouling of the metering pump. The autosampler software is programmed to control the in-tube SPME extraction, desorption, and injection.

9. Up to 100 sample vials can be set up.

10. Air plugging before the extraction step is carried out to prevent not only sample mixing but also desorption of the analyte from the capillary coating by the mobile phase during the ejection step.

11. The draw/eject volume and the number of sample solutions are related to the extracted amount and are dependent on the capacity of the column. Complete equilibrium extraction is generally not obtained for any of the analytes, however, because the analytes are partially desorbed into the mobile phase during each ejection step. Although the amount extracted increases independent of the draw/eject volume, the bandwidth extends, and the peak becomes broad. For columns 60 cm long and 0.25 mm in internal diameter, the column capacity is 29.4 μL, and the capacity from the injection needle (10 μL) to the tip of the column is 39.4 μL. In our experiments, the optimum draw/eject volume is 30–40 μL for tested drugs, and the extraction efficiency did not increase even at high volumes. Although an increase in the number of draw/eject cycles can enhance extraction efficiency, peak broadening is often observed. In addition, the draw/eject speed corresponds to the agitation speed of fiber SPME, and the extraction efficiency increases with this speed. We find, however, that the optimal flow rate of draw/eject cycles is 50–100 μL/min. Below this level, extraction requires an inconveniently long time, and above this level, bubbles form inside the capillary and the extraction efficiency is reduced. Ideally,

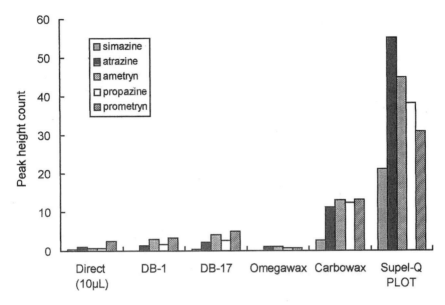

Fig. 5. Evaluation of five capillary columns for in-tube SPME of triazine herbicides. In-tube SPME conditions included 100 ng/mL of each herbicide, 20 draw/eject cycles, a draw/eject volume of 40 μL, a draw/eject rate of 0.1 mL/min, and mobile phase desorption (dynamic elution).

the compound has reached distribution equilibrium, and the extraction amount is maximized. To reduce analysis time, it is possible to stop the extraction prior to equilibrium, but only if sufficient sensitivity is obtained. The extraction time depends on the volume, speed, and cycle of the draw/eject procedure, and these conditions must be fixed to obtain quantitative reproducibility.

12. For chlorinated phenoxy acid herbicides, the optimum extraction conditions are 25 draw/eject cycles of 30 μL sample in 0.2% formic acid (pH 2.0) at a flow rate of 0.2 mL/min using a DB-WAX capillary (**Fig. 6**). Extraction equilibrium is not reached for all herbicides after 30 draw/eject cycles, and peak broadening for MPCA, 2,4-D, and dichlorprop is observed at 30 cycles. This peak broadening is likely because of broadening of the bandwidth of the analytes extracted into the capillary.

13. For triazine herbicides, the optimum in-tube SPME conditions are 20 draw/eject cycles of 40 μL of sample at a draw/eject rate of 100 μL/min using a Supel-Q PLOT capillary column.

14. There are two methods for desorption of the compound adsorbed to the capillary column: the dynamic method, which desorbs into the flow of the mobile phase, and the static method, which desorbs into the solvent aspirated from the outside. Static desorption is preferred when the analytes are more strongly adsorbed to the capillary coating. In either case, it is necessary to carry out quick and perfect desorption with a minimum volume of solvent. For a capillary column 60 cm long and 0.25 mm in internal diameter, desorption is usually carried out by aspirating 40 μL. For the static method, it is also necessary to consider the compound's solubility and miscibility with the mobile phase. For the dynamic method, it is possible to desorb to the mobile phase flow directly after switching the six-port valve. Although carryover may be observed after the analysis of highly con-

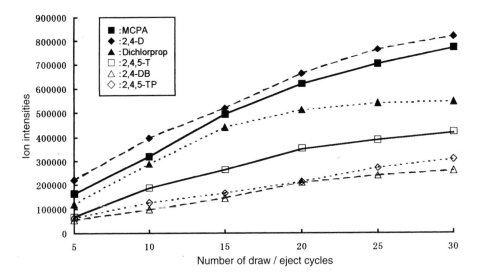

Fig. 6. Extraction time profiles of chlorinated phenoxy acid herbicides with the DB-WAX capillary column. Sample modifier is 0.2% formic acid. Other conditions are identical to those in **Fig. 4**. From **ref. 26** with permission.

centrated samples, it is possible to wash the injection needle and the capillary column by rinsing with methanol and the mobile phase several times prior to the next analysis via the draw/eject mode. Therefore, carryover with in-tube SPME is much lower, or eliminated, in comparison with fiber SPME. It is therefore possible that these conditioning, extraction, and desorption operations are automatically carried out by the overall injection program.

15. The effect of the desorption solvent on the desorption of chlorinated phenoxy acid herbicides from the capillary is examined using acetonitrile, methanol, and the mobile phase. The maximum intensities of all analytes are obtained using 10 µL acetonitrile by static desorption. Above 10 µL, peak broadening is observed, although the intensities do not increase.

16. The extracted triazine herbicides are easily desorbed from the capillary by dynamic desorption using a mobile phase flow, and no carryover is observed because the capillary column is washed and conditioned by draw/eject cycles of methanol and mobile phase prior to extraction.

17. The extraction yields estimated from the amounts of analytes extracted in the stationary phase by in-tube SPME under optimal conditions ranged from 23.9 to 30.7% for chlorinated phenoxy acid herbicides and 25.7 to 35.7% for triazine herbicides.

18. The entire automatic in-tube SPME extraction and desorption of chlorinated phenoxy acid herbicides took 10 min. The in-tube SPME–LC–MS method took 30–40 min, and automatic analysis of about 40 samples per day is possible during overnight operation.

19. The postcolumn addition of neutralization buffer is required to form ions in solution and to facilitate the charging of droplets. An equimolar amount of triethylamine is therefore added to the formic acid mobile phase at a flow rate of 0.1 mL/min by the tee union installed between the analytical column and the ion source. The best separation for all herbicides is obtained with 10 µmol/mL ammonium acetate, 0.1% formic acid, and 5 µmol/mL DBA as the mobile phase modifiers. However, the response factor is at least

twice as sensitive for all herbicides when 0.1% triethylamine rather than 0.1% formic acid is used as the postcolumn additive.

20. Although the mobile phase containing DBA as a volatile ion-pairing reagent is directly compatible with MS/ESI, the use of the ion-pairing reagent will cause analyte suppression and a high chemical background because of the formation of a complex with target analytes *(27)* if the MS/ESI condition is not optimized. The intensity of the analyte does not show a large variation when the drying gas flow rate is varied from 4 to 13 L/min. For the nebulizer gas pressure, when the volatile ion-pairing reagent is used, a higher value ensures the best sensitivity for the analytes because it can break the neutral ion pairs of the analytes *(28)*. These parameters are maintained at 12 L/min and 60 psi.

21. Fragmentor voltage (capillary exit voltage) is applied to the exit of the capillary and affects the transmission and fragmentation of sample ions. At higher fragmentor voltage, more fragmentation and better ion transmission will occur, and maximum structural information and sensitivity will be obtained. The intensities of the $[M-H]^-$ and $[M-RCOOH]^-$ ions for all herbicides are shown as a function of the fragmentor voltage. For all herbicides except 2,4-DB, which exhibited $[M-(CH_2)_3COOH]^-$ as the base peak, $[M-H]^-$ ions are observed as the base peak at less than 60 V and presented maximum sensitivities. At higher fragmentor voltage, the intensities of $[M-H]^-$ ions decreased, and the intensities of $[M-RCOOH]^-$ ions increased. Other ions observed in the mass spectra of all herbicides are isotopic ions, which derived from the chlorine element and are of maximum abundance for 2,3,5-T and 2,4,5-TP. The fragmentor voltage for all herbicides is set at 60 V, and **Table 1** shows the typical mass spectra of the six chlorinated phenoxy acid herbicides at 60 V.

22. The $[M-H]^-$ ions are selected for all chlorinated phenoxy acid herbicides, except that $[M-(CH_2)_3COOH]^-$ is selected for 2,4-DB. Typical total ion and SIM chromatograms of a standard mixture of six herbicides (each 1 µg/mL) are shown in **Fig. 7**.

23. As shown in **Table 1**, the linearity in the range 0.05–50 ng/mL is very good for all herbicides with correlation coefficients r^2 higher than 0.999. The LOD for each herbicide is in the range 0.005–0.03 ng/mL, whereas the LOQ of each herbicide is in the range 0.02–0.06 ng/mL. The intraday and interday precisions for the analysis of river water samples spiked with 0.1 ng/mL are 2.5–4.1% and 6.2–9.1%, respectively. The quantitative results for all herbicides spiked into river water, at 0.1 ng/mL and using external standards, are shown in **Table 1**. The accuracy of these quantitative results is 10–20%, and no significant interference peaks are observed.

24. Triazine herbicides are analyzed within 10 min on a TSK gel Super-Octyl column. Although ametryne and propazine are partially coeluted (**Fig. 8**), the tandem mass spectrometer allows the selective detection of substances with varying masses or fragments without chromatographic separation.

25. A solution of 10 ng/mL in 50% methanol containing 1% acetic acid infused at a flow rate of 5 µL/min produced a signal of appropriate abundance in the positive ion mode for the protonated precursor ion $[M+H]^+$ using the turbo ion spray. Parameters, including nebulizer gas stream, curtain gas, and ion spray voltage, are optimized by flow injection analysis with a mobile phase flow of 0.2 mL/min. Tuning is processed using an automatic tuning tool of the Analyst software to determine the declustering and focusing potentials, the fragmentation pattern, the collision energy, and the collision cell exit potential. The most abundant fragments are sufficiently separated by Q3 in unit resolution, so that could be used for quantification. The main fragments are formed by the secession of moieties of masses of m/z 41.99 (possible sum formulas C_3H_6).

Table 1
Sensitivity, Linearity and Precision of Chlorinated Phenoxy Acid Herbicides

| Herbicides | r^2 | Sensitivity (ng/mL) | | Quantitative[c] results (ng/mL) | Precision (RSD, %) | |
		LOD[a]	LOQ[b]		Repeatability[d]	Reproducibility[e]
MCPA	0.9998	0.01	0.04	0.12	3.5	8.3
2,4-D	0.9996	0.005	0.02	0.09	2.5	9.1
Dichlorprop	0.9996	0.01	0.04	0.11	3.3	7.9
2,4,5-T	0.9998	0.02	0.02	0.12	2.6	8.7
2,4-DB	0.9994	0.03	0.06	0.11	4.1	7.3
2,4,5-TP	0.9992	0.02	0.04	0.08	3.7	6.2

Source: From **ref. 26** with permission.

[a]Limit of detection (LOD) was defined as $S/N = 3$ for the river water spiked with 0.05 ng/mL.

[b]Limit of quantification (LOQ) was defined as 10 standard deviations for five replicates of the river sample spiked with 0.1 ng/mL.

[c]Calculated for river water spiked at 0.1 ng/mL.

[d]Repeatability was calculated by analyzing five water samples spiked with 0.1 ng/mL within 1 d.

[e]Reproducibility was calculated by analyzing three river water samples spiked with 0.1 ng/mL per day for 3 d.

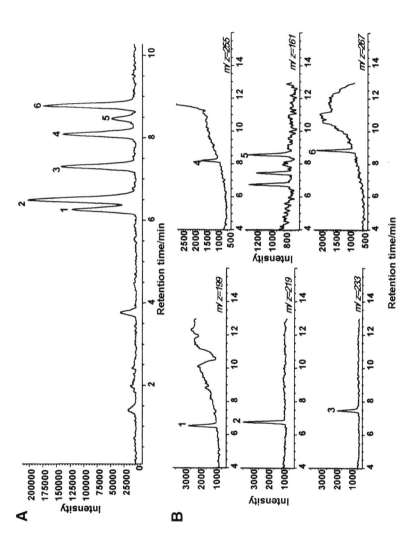

Fig. 7. Total ion and SIM chromatograms of the six chlorinated phenoxy acid herbicides: (**A**) Total ion chromatograms are obtained using dibutylamine acetate (DBA) as ion-pairing reagent. Eluents A (water) and B (acetonitrile) containing 5 mmol/L DBA. Linear gradient from 30% B/A to 85% B/A in 15 min, 15 min reequilibration. (**B**) SIM chromatograms are obtained by in-tube SPME–LC–MS of river water spiked with 0.05 ng/mL. Peaks: 1, MCPA; 2, 2,4-D; 3, dichlorprop; 4, 2,4,5-T; 5, 2,4-DB; 6, 2,4,5-TP. From **ref. 26** with permission.

379

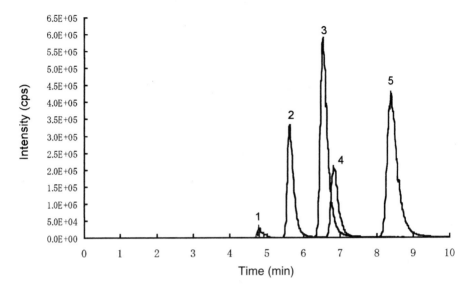

Fig. 8. Chromatogram of water sample spiked with 10 ng/mL standard triazine herbicides obtained in MRM in positive ionization mode by in-tube SPME–LC–MS/MS. Peaks: 1, simazine (m/z 201.93 → 131.95); 2, atrazine (m/z 215.90 → 173.95); 3, ametryne (m/z 228.04 → 186.05); 4, propazine (m/z 229.93 → 145.95); 5, prometryne (m/z 242.02 → 158.05).

26. To determine the optimal composition, different mobile phases consisting of acetonitrile–ammonium acetate were tested. A standard solution of 10 ng/mL of each triazine herbicide used for flow injection analysis was prepared in water. The best signal was achieved using acetonitrile–2 mM ammonium acetate (50:50 v/v). The development of the chromatographic system was focused on short retention times and coelution of triazine herbicides, paying attention to matrix effects as well as good peak shapes. A high proportion of organic solvent was used to coelute each substance. Thus, an increased flow rate of 0.2 mL/min produced a good peak shape and made a runtime of 10 min possible.

27. We observed good linearity for each herbicide in the concentration range 0.05–10 ng/mL, with correlation coefficients above 0.992. The LOD of each herbicide is in the range of 0.003–0.01 ng/mL, except for simazine (0.11 ng/mL). The in-tube SPME method shows 17–30 times higher sensitivity than the direct injection method.

28. The recoveries of herbicides spiked into water samples are each above 95%. MCPA and 2,4-D are detected in 7/30 river water samples tested. For each of these, good agreement is obtained between in-tube SPME–LC–MS and other methods, including SPE–GC–MS and SPE–HPLC. Triazine herbicides were not detected in the environmental water samples tested in this study.

References

1. Dearfield, K. L., McCarroll, N. E., Protzel, A., Stack, H. F., Jackson, M. A., and Waters, M. D. (1999) A survey of EPA/OPP and open literature on selected pesticide chemicals. II. Mutagenicity and carcinogenicity of selected chloroacetanilides and related compounds. *Mutat. Res.* **443**, 183–221.

2. Schantz, S. L. and Widholm, J. J. (2001) Cognitive effects of endocrine-disrupting chemicals in animals. *Environ. Health Perspect.* **109**, 1197–1206.
3. Waissmann, W. (2002) Health surveillance and endocrine disruptors. *Cad. Saude. Publ.* **18**, 511–517.
4. van der Hoff, G. R. and van Zoonen, P. (1999) Trace analysis of pesticides by gas chromatography. *J. Chromatogr. A* **843**, 301–322.
5. Karcher, A. and El Rassi, Z. (1999) Capillary electrophoresis and electrochromatography of pesticides and metabolites. *Electrophoresis* **20**, 3280–3296.
6. D'Ascenzo, G., Curini, R., Gentili, A., Bruno, F., Marchese, S., and Perret, D. (2000) Determination of herbicides in water using HPLC–MS techniques. *Adv. Chromatogr.* **40**, 567–598.
7. Hogendoorn, E. and van Zoonen, P. (2000) Recent and future developments of liquid chromatography in pesticide trace analysis. *J. Chromatogr. A* **892**, 435–453.
8. Sabik, H., Jeannot, R., and Rondeau, B. (2000) Multiresidue methods using solid-phase extraction techniques for monitoring priority pesticides, including triazines and degradation products, in ground and surface waters. *J. Chromatogr. A* **885**, 217–236.
9. Wells, M. J. and Yu, L. Z. (2000) Solid-phase extraction of acidic herbicides. *J. Chromatogr. A* **885**, 237–250.
10. Pico, Y., Font, G., Molto, J. C., and Manes, J. (2000) Solid-phase extraction of quaternary ammonium herbicides. *J. Chromatogr. A* **885**, 251–271.
11. Lee, N. A. and Kennedy, I. R. (2001) Environmental monitoring of pesticides by immunoanalytical techniques: validation, current status, and future perspectives. *J. AOAC Int.* **84**, 1393–1406.
12. Beltran, J., Lopez, F. J., and Hernandez, F. (2000) Solid-phase microextraction in pesticide residue analysis. *J. Chromatogr. A* **885**, 389–404.
13. Krutz, L. J., Senseman, S. A., and Sciumbato, A. S. (2003) Solid-phase microextraction for herbicide determination in environmental samples. *J. Chromatogr. A* **999**, 103–121.
14. Arthur, C. L. and Pawliszyn, J. (1990) Solid-phase microextraction with thermal desorption using fused silica optical fibers. *Anal. Chem.* **62**, 2145–2148.
15. Kataoka, H., Lord, H. L., and Pawliszyn, J. (2000) Applications of solid-phase microextraction in food analysis. *J. Chromatogr. A* **880**, 35–62.
16. Lord, H. and Pawliszyn, J. (2000) Evolution of solid-phase microextraction technology. *J. Chromatogr. A* **885**, 153–193.
17. Alpendurada, M. de F. (2000) Solid-phase microextraction: a promising technique for sample preparation in environmental analysis. *J. Chromatogr. A* **889**, 3–14.
18. Pawliszyn, J. (2001) Solid-phase microextraction. *Adv. Exp. Med. Biol.* **488**, 73–87.
19. Buldini, P. L., Ricci, L., and Sharma, J. L. (2002) Recent applications of sample preparation techniques in food analysis. *J. Chromatogr. A* **975**, 47–70.
20. Zambonin, C. G. (2003) Coupling solid-phase microextraction to liquid chromatography. A review. *Anal. Bioanal. Chem.* **375**, 73–80.
21. Eisert, R. and Pawliszyn, J. (1997) Automated in-tube solid-phase microextraction coupled to high-performance liquid chromatography. *Anal. Chem.* **69**, 3140–3147.
22. Kataoka, H. (2002) Automated sample preparation using in-tube solid-phase microextraction and its application—a review. *Anal. Bioanal. Chem.* **373**, 31–45.
23. Gou, Y., Tragas, C., Lord, H. L., and Pawliszyn, J. (2000) On-line coupling of in-tube solid-phase microextraction (SPME) to HPLC for analysis of carbamates in water samples: comparison of two commercially available autosamplers. *J. Microcol. Sep.* **12**, 125–134.
24. Gou, Y. and Pawliszyn, J. (2000) In-tube solid-phase microextraction coupled to capillary LC for carbamate analysis in water samples. *Anal. Chem.* **72**, 2774–2779.

25. Wu, J., Tragas, C., Lord, H., and Pawliszyn, J. (2002) Analysis of polar pesticides in water and wine samples by automated in-tube solid-phase microextraction coupled with high-performance liquid chromatography–mass spectrometry. *J. Chromatogr. A* **976,** 357–367.

26. Takino, M., Daishima, S., and Nakahara, T. (2001) Automated on-line in-tube solid-phase microextraction followed by liquid chromatography/electrospray ionization–mass spectrometry for the determination of chlorinated phenoxy acid herbicides in environmental waters. *Analyst* **126,** 602–608.

27. Takino, M., Daishima, S., and Yamaguchi, K. (2000) Determination of haloacetic acids in water by liquid chromatography–electrospray ionization–mass spectrometry using volatile ion-pairing reagents. *Analyst* **125,** 1097–1102.

28. Takino, M., Daishima, S., and Yamaguchi, K. (1999) Analysis of anatoxin-a in freshwaters by automated on-line derivatization-liquid chromatography-electrospray mass spectrometry. *J. Chromatogr. A* **861,** 191–197.

29

Coupled-Column Liquid Chromatography for the Determination of Pesticide Residues

Elbert Hogendoorn and Ellen Dijkman

Summary

This chapter demonstrates the versatility and feasibility of coupled-column reversed-phase liquid chromatography (LC/LC) for the rapid, selective, and sensitive determination of pesticides in environmental water samples. The work includes the setup and application of three different methods, covering the wide polarity range of pesticides as well as various LC/LC approaches and the important parameters and aspects to be considered in method development.

Key Words: Coupled column; liquid chromatography; pesticide residues.

1. Introduction
1.1. General Aspects

For a number of important features such as automation of sample pretreatment and improvement of sensitivity or selectivity, column switching in liquid chromatography (LC) has been widely adopted in the field of pesticide residue analysis. Column switching usually involves the use of two columns, which can be switched on- or off-line by a six-port high-pressure switching valve. Basically, the first column (C-1) serves for sample enrichment (injection) and cleanup, and the second column (C-2) provides the (final) separation of the compound(s).

Depending on the dimension of the first column C-1, one can discriminate between precolumn switching LC (LC/PC) or solid-phase extraction LC (SPE–LC) and coupled-column LC, LC/LC. The major difference between these two techniques is the type of the first column and the inherent important features dimension, particle size, and separation power.

In SPE–LC, the first column C-1 is a small precolumn, typically 2- to 4.6-mm id and less than 10 mm long and usually is packed with materials of a relatively large particle size (10–40 µm) to facilitate large volumes of aqueous samples. In LC/LC, the

From: *Methods in Biotechnology, Vol. 19, Pesticide Protocols*
Edited by: J. L. Martínez Vidal and A. Garrido Frenich © Humana Press Inc., Totowa, NJ

dimension of C-1 is that of an analytical column packed with highly efficient materials (particle size $\leq$ 5 µm).

Hence, the major difference between SPE–LC and LC/LC is the difference in separation power of C-1, which is low in SPE–LC (plate number $N < 50$) and high in LC/LC (plate number $N > 2000$). Reviews *(1–3)* extensively informed on the scope (applications), the type of configurations, and new developments (trends) of LC/LC in residue analysis of pesticides and related compounds.

In environmental trace analysis of polar organic contaminants, reversed-phase liquid chromatography (RPLC) is usually confronted with an excess of polar interferences (e.g., inorganic and organic ions and degradation products of organic matter like humic and fulvic acids). Especially for not very selective ultraviolet (UV) detection, determination of more polar analytes eluting in the first part of the chromatogram will be difficult or impossible.

This common situation is illustrated by the simulated chromatogram in **Fig. 1**, which presents the interferences usually encountered in pesticide residue analysis employing RPLC/UV. As indicated by S1, abundantly present early eluting sample constituents cause the major problem. Usually, this problem is (partly) solved by a laborious cleanup procedure.

S1 interferences can be partly removed via low-resolution cleanup, such as off-line SPE or on-line with LC/PC. However, for analytes of type A1 with a polarity in the range of the excess of interferences, such a cleanup will not provide sufficient selectivity with UV detection. For mass spectrometric (MS) or tandem mass spectrometric (MS/MS) detection, selectivity will be distinctly less influenced. However, the presence of an excess of matrix compounds can severely affect the ionization efficiency, which will usually result in signal suppression of the analytes, making reliable quantification difficult. Studies clearly demonstrated the improvement in MS detectability and quantification of analytes when using LC/LC for on-line cleanup of water samples *(4)* or plant extracts *(5,6)*.

The two remaining type of interferences, S2 and S3, are usually controlled by optimizing (changing) chromatographic conditions. For example, a separation between S2 and A3 might be achieved by tuning the LC selectivity with a change in eluotropic strength φ or type of modifier. Extension of the run time caused by a late-eluting S3 interference can be (partly) reduced by a fast, steep gradient, started after the elution of A3. Note that such a step also consumes additional analytical time for reconditioning the separation column prior to the next injection. Reversed-phase coupled-column LC (LC/LC) can effectively remove interferences of type S1 and S3.

1.2. Development of an Analytical Method Using LC/LC

The scheme of a reversed-phase LC/LC system is presented in **Fig. 2**. In comparison to the back-flush column-switching SPE–LC technique, schematically shown in **Fig. 3** (*see* **Note 1**), LC/LC is a forward-flush, column-switching, heart-cutting technique. As shown in **Fig. 2**, it consists of three different steps to perform on-line cleanup and analysis of aqueous samples or extracts.

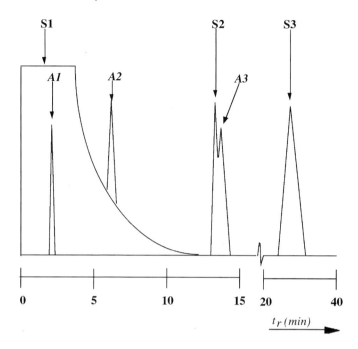

Fig. 1. Simulated chromatogram usually encountered in the RPLC/UV pesticide residue analysis of environmental samples. A1, A2, and A3, eluting target compounds; S1, S2, and S3, types of sample interferences (for further explanation, *see* text).

Step 1 involves the injection of an aliquot of uncleaned aqueous sample or extract onto the first column C-1, followed by a cleanup with a certain volume of M-1, the *cleanup volume*, which removes from C-1 a large part of S1 interferences with the mobile phase M-1.

In Step 2, just before the elution of the (first) analyte A1, takes place, C-1 is switched on-line with the second separation column C-2 during a time that the fraction containing the analyte(s) (the *transfer volume*) is completely transferred from C-1 to C-2.

Finally, in Step 3, the analytes A1, A2, and A3 are separated on C-2, and simultaneously the remaining part of S1 interferences as well as that of type S3 will be washed from C-1 with M-1 or a stronger eluent and reconditioned with a volume of M-1 prior to the next injection.

The procedure in LC/LC method development is straightforward and usually includes a number of consecutive steps as schematically presented in **Fig. 4**. First, information must be obtained regarding whether sufficient sensitivity will be available with the selected detection mode, which in most cases will be UV detection, MS detection, or fluorescence detection (FLD). Without sufficient sensitivity, the setup of an LC/LC method is not feasible. If information on detectability is not available, preliminary experiments will be necessary (e.g., in case of MS detection, by infusion experiments

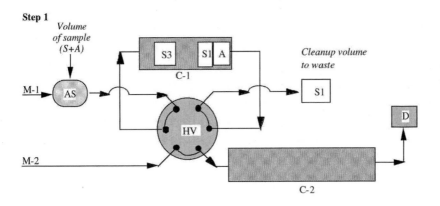

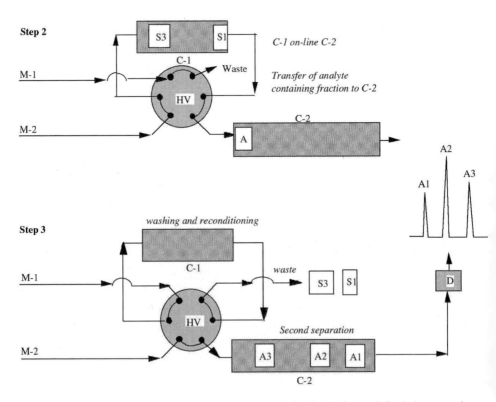

Fig. 2. Schematic presentation of the three steps applied in the forward-flush, heart-cutting, column-switching technique of coupled column LC (LC/LC). M-1 and M-2, first and second mobile phase, respectively; C-1 and C-2, first and second analytical separation column, respectively; HV, six-port high-pressure switching valve; AS, autosampler; S, S1, and S3, interferences (*see* **Fig. 1**); A, A1, A2, and A3, target compounds (*see* **Fig. 1**); D, detector; W, waste.

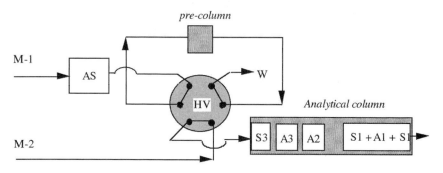

Fig. 3. Schematic presentation of the third step applied in the backflush, column-switching LC/SPE technique. M-1 and M-2, first and second mobile phase, respectively; C-1 and C-2, first and second analytical separation column, respectively; HV, six-port high-pressure switching valve; AS, autosampler; S1 and S3, interferences (*see* **Fig. 1**); A1, A2, and A3, target compounds (*see* **Fig. 1**); D, detector; W, waste.

with standard solutions in a suitable buffer). Unfortunately, only a few pesticides have a favorable native fluorescence property. Hence, the selection of FLD usually depends on the reactivity of the pesticide(s) with a suitable fluorogenic label, such as 9-fluoronylmethylchloroformate (FMOC-Cl) and *o*-phthalaldehyde. Regarding UV detection, the sensitivity depends on the value of the molar extinction coefficient ε (L.mol^{-1}.cm^{-1}) at a suitable wavelength λ (nanometers). The molar extinction coefficient (ε) of a species is defined by the equation $A = \varepsilon bc$, where A is the absorbance of the solution, b is the path length, and c is the concentration of the species. This information can easily be calculated from the data of the peak in the RPLC chromatogram (*see* **Note 2**). In our experience in the analysis of pesticides in water samples, UV detection becomes feasible with values for ε and λ above 2000 L.mol^{-1}.cm^{-1} and 200 nm, respectively.

The next important parameter to be considered is the retention on a C18 silica-bonded phase. Based on our experience, the k value of the compound on the C18 column with pure water as the mobile phase must be above 1 to aim at a successful LC/LC procedure.

As indicated in **Fig. 4**, the next steps involve the selection of columns and the establishment of suitable chromatographic conditions on the first and second columns. These experiments will be done without the use of the high-pressure switching valve (HV), and the outlet of the separation column (C-1 or C-2) will be directly connected to a detector. C-1 must provide sufficient separation to remove a major part of the abundantly present early eluting matrix constituents prior to elution of the analyte. Therefore, a column with high separation power is attractive. However, an unnecessary increase of separation power and dimension of the column must be avoided. First, the peak elution volume of the analyte should be kept as low as possible. Second, an increase of the column dimension will also increase the reconditioning time of C-1 between analyses. As a rule of thumb, the eluotropic strength of M-1 should be selected in such a way that the capacity factor of the analyte is in the range of $1 < k < 5$,

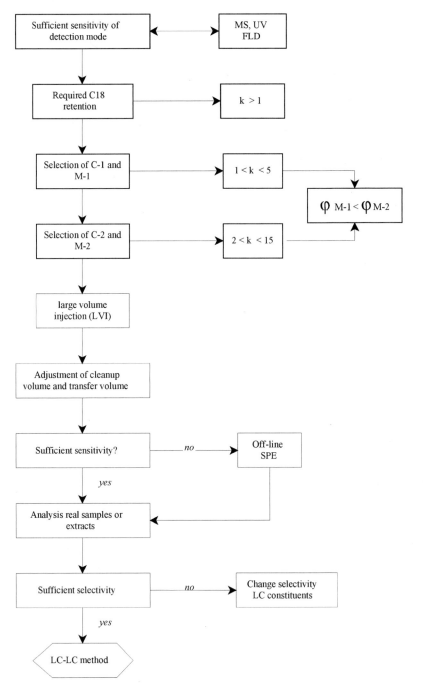

Fig. 4. Scheme of straightforward LC/LC method development in the trace analysis of pesticides.

rendering a cleanup volume of at least two times the dead volume of the column V_0 to remove a large part of the excess of early eluting interferences.

Especially in the analysis of compounds with little C18 retention, it is favorable to select a second column (C-2) with a higher separation power than C-1. This provides flexibility when optimizing the eluotropic strength of M-2 to reach an adequate compromise between the required separation efficiency on the second column and the desired peak compression (sensitivity of detection). The favorable feature of peak compression obtained by a step-gradient elution in column switching is a result of an increase in eluotropic strength φ (fraction of organic modifier) of the mobile phase on C-1. This favorable process has been clearly described and visualized *(7,8)*. It must be emphasized here that the φ of M-1 never must exceed the φ of M-2 ($\varphi_{M-1} < \varphi_{M-2}$) to avoid unnecessary band broadening of the compounds. For the selection of C-2, one should not overkill separation power. As a rule of thumb and to achieve fast analysis and good sensitivity, the capacity factor of the analyte(s) on C-2 should not exceed 15. Especially for UV detection, one selects the same modifier (methanol, acetonitrile, or mixtures) for both mobile phases (M-1 and M-2) and minimizes the size of the step gradient (viz. the difference in eluotropic strength of two eluents) to minimize baseline distortion in the chromatogram.

After the selection of adequate conditions, the attainable sensitivity is determined by large-volume injection (LVI) of standard solutions in high-performance liquid chromatography (HPLC) grade water or buffer (*see* **Note 3**). This is done by connecting C-1 to the UV detector, which simultaneously provides the values for cleanup volume and transfer volume, as illustrated in **Fig. 5**.

The maximum allowable sample injection volume, that is, the volume that can be injected without excessive band broadening during injection, largely depends on the C18 retention of the analyte. To maximize sample injection volume for the analysis of acidic pesticides, the initial pH of the mobile phase should be as low as possible for stationary phases based on modified silica's typically are around pH 2.5. Special attention must be paid to analytes with little C18 retention, such as ethylenethiourea (ETU; main metabolite of maneb and zineb), with a k value of 1.2 in pure water as the mobile phase. Such a property makes top-column focusing of the analyte impossible, and because of migration of ETU during injection, band broadening rapidly starts to increase with increasing injection volume. In general, the start of peak deformation of the analyte on injection on C-1 is selected as a criterion for the maximum allowable injection volume.

If sufficient sensitivity cannot be obtained with an LVI below 10 mL, an off-line concentration step will be necessary. An example of this approach involving the processing of extracts with LC/LC is shown in the second method (*see* below). After the selection of suitable conditions, the LC/LC method is applied to the analysis of real-life environmental samples or extracts. Of course, depending on the type of sample and the site of sampling, the concentration and nature of the ionic and/or highly polar interferences can vary strongly. During this part of the optimization one can, if required, tune selectivity by choosing another type of hydrophobic column packing material or one from a different manufacturer or, in most cases, firstly, by changing the

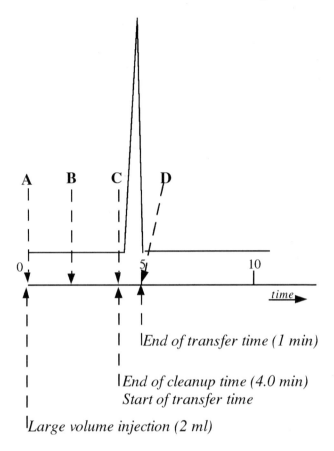

Fig. 5. Simulated chromatogram of an injection of a standard solution on C-1 connected to the detector for the selection of LC column-switching conditions. A–B, injection volume; A–C, cleanup volume; C–D, transfer volume.

mobile phase constituents such as type of modifier, buffer and pH. As shown in the determination of ETU in various types of ground water *(9)*, increasing the separation power of C-1 is most efficient for very polar compounds.

In the RPLC/UV analysis of acidic compounds employing a buffer at low pH (<4), no modifier gradient should be used because these will result in a large interfering peak (hump) caused by a release of humic and fulvic acids from the column during the gradient; it is demonstrated that the use of pH-based gradients offers higher selectivity *(10)*.

2. Materials
2.1. Reagents

1. Pesticides (purity > 99%) (e.g., Dr. S. Ehrenstorfer, Promochem, Wesel, Germany).
2. HPLC-grade acetone, acetonitrile, methanol, and pure water.
3. Phosphoric acid (89% pure), hydrochloric acid (37% pure), and trifluoroacetic acid (TFA; 99% pure).

4. Analytical-grade potassium hydroxide, potassium dihydrogen phosphate, and disodium tetraborate decahydrate.
5. FMOC-Cl.

2.2. Equipment

1. HPLC column-switching system (**Fig. 2**), basically consisting of two LC pumps (flow set at 1 mL/min); an autosampler (AS); a six-port HV; two analytical reversed-phase columns (C-1 and C-2); a detector (D; UV, MS, or fluorescence); and a recorder or integrator for quantitative measurements of the peak heights.
2. 3-mL SPE cartridges containing 500 mg 40 µm C18-bonded silica.
3. A 50×4.6 mm id and 100×4.6 mm id column packed with 3-µm C18 as C-1 and C-2, respectively, for Method 1.
4. A 50×4.6 mm id column packed with 3-µm C18 as C-1 and a 150×4.6 mm id column packed with 5-µm SPS-5PM-S5-100-ODS as C-2 for Method 2. (SPS, semipermeable surface; S5, sperical 5 µm particle; ODS, octadecyl silane.)
5. A 30×4.6 mm id column packed with 5-µm C18 as C-1 and a 250×4.6 mm id column packed with 5-µm Hypersil APS (NH2) as C-2 for Method 3.

2.3. Solutions and Mobile Phases

1. Stock standard pesticide solutions (~500 µg/mL) are made in acetonitrile or methanol. Dilutions are made in HPLC-grade water.
2. 0.05 M phosphate buffer, pH 5.7, prepared by dissolving 6.803 g potassium dihydrogen phosphate in water and adjusting pH with a 2 M potassium hydroxide solution.
3. 0.05 M disodium tetraborate, pH 9.0, solution in HPLC-grade water.
4. 0.025 M disodium tetraborate, pH 9.0, solution in HPLC-grade water.
5. FMOC-Cl solution of 2000 µg/mL in acetonitrile.
6. Mobile phase (M-1 and M-2) of acetonitrile–water (48:52 v/v) for Method 1 (isoproturon).
7. Mobile phase (M-1) of acetonitrile–0.03 M phosphate buffer, pH 2.4 (35:65 v/v), and a mobile phase (M-2) of acetonitrile-0.03 M phosphate buffer, pH 2.4 (40:60 v/v), for Method 2 (bentazone and bromoxynil).
8. Mobile phase (M-1 and M-2) of acetonitrile-0.05M phosphate buffer, pH 5.7 (35:65 v/v), for Method 3 (glyphosate and aminomethylphosphonic acid [AMPA]).

3. Methods

Three LC/LC methods for the rapid analyses of various pesticides in water samples involving on-line and off-line approaches are presented below. Information on the pesticides analyzed with the three selected methods is given in **Table 1**. As can be seen from this table, the three methods involve the trace analysis of (1) a rather neutral compound (isoproturon), (2) acidic compounds (bentazone and bromoxynil), and (3) amphoteric/ionic compounds (glyphosate and AMPA), thus covering a wide polarity range in the determination of pesticides.

3.1. Determination of Isoproturon

This application deals with the direct LC/LC/UV trace analysis of a phenyl ureum herbicide (isoproturon) in water. Phenyl urea herbicides form a group of compounds with favorable RPLC properties, such as sufficient C18 retention, and good elution performance without the use of a buffer. Sensitive detection is possible with both UV and MS, making a combination with RPLC the preferred technique of analysis.

Table 1
Information for Selected Methods and Pesticides

Method	Pesticide	Chemical family	pKa	Solubility in water (mg/L)
3.1.	Isoproturon	Urea	n.a.	65
3.2.	Bentazone	Benzothiadiazole	3.3	570
	Bromoxynil	Hydroxybenzonitrile	3.86	130
3.3.	Glyphosate	Phosphonic acid	0.78	12,000
			2.29	
			5.59	
	AMPA	Phosphonic acid	n.a.	>100,000

n.a., information not available.

The high selectivity of MS and moreover of tandem mass spectrometric detection (MS/MS) offers the possibility to perform quantification and identification simultaneously in one run. However, low-cost and robust UV detection can be a very attractive technique in monitoring studies involving a limited number of target pesticides to be analyzed in a large number of environmental samples with a low expectation to find positive samples.

3.1.1. Method Development

According to the scheme of **Fig. 4**, the following considerations and steps are made in LC/LC/UV method development:

1. Isoproturon has a good UV detectability (ε of about 20,000 L/mol.cm at about 244 nm).
2. With a k value of about 5 in a mobile phase of methanol–water (50:50 v/v), the C18 retention can be considered high. This means that there is top-column focusing of the analyte during the injection of a large volume of water sample.
3. Having sufficient C18 retention, a short cartridge column (50 × 4.6 mm id) packed with 3-μm C18 is selected as C-1. To enhance somewhat the separation power, a 100 × 4.6 mm id column with the same material is used for C-2. A cartridge housing system favors the flexibility of including a guard column before C-1.
4. Regarding the modifier of the mobile phase, acetonitrile is preferred above methanol because of slightly improved peak shape and a decrease in column pressure. For both M-1 and M-2, a mobile phase of acetonitrile–water (48:52 v/v) is selected, providing a k value of 4.1 for isoproturon.
5. A sample injection volume of 4.0 mL is selected to provide the required 0.1-μg/L limit of detection for isoproturon in water.
6. Selecting a 5.85-mL cleanup volume and a 0.45-mL transfer volume as accurately adjusted column switching conditions (*see* **Fig. 5**), it appeared that the LC/LC method is suitable for the analysis of isoproturon in surface and ground water samples.

3.1.2. Analytical Procedure (see **Note 4**)

1. Inject a 4.0-mL volume of water sample on C-1.
2. After cleanup with 5.85 mL M-1 (injection volume included), switch C-1 on-line with C-2 for 0.45 min for the transfer of the analyte containing fraction to C-2.

3. Do quantification of isoproturon by external calibration with standard solutions of the analytes in pure water, with UV detection at a wavelength of 244 nm.

3.2. Determination of Bentazone and Bromoxynil

This application involves multimodal LC/LC employing an analytical C18 column and an analytical column packed with efficient restricted-access material (RAM). This material combines two different selectivities in one column, that is, size exclusion of large molecular compounds and an efficient separation of the target compounds. The use of such a column in environmental analysis in LC/LC for the trace analysis of acidic pesticides is attractive to exclude (remove) large molecular weight compounds, such as dissolved organic compounds like humic and fulvic acids *(11–15)* (*see* **Note 5**). The major problem encountered in the trace analysis of acidic pesticides in water samples is visualized in **Fig. 6**. This figure clearly shows that the sampling at low pH of 100 mL of groundwater on a 100-mg SPE C18 cartridge results in a large increase of coextracted acidic interferences in comparison to the sampling at a neutral pH. In the RPLC analysis of acidic compounds involving a buffer at low pH (<4.0), the coextracted humic acid interferences always show up in the chromatogram as a big hump, obscuring the analytes at low levels and making quantification difficult.

3.2.1. Method Development

According to the scheme of **Fig. 4** and to the information given in **ref. *14***, the following considerations and steps are made in method development:

1. Bromoxynil and bentazone can be sensitively detected by UV detection (ε of about 22,000 L/mol.cm) at a wavelength of 217 nm.
2. The C18 retention of both compounds is considerably high (comparable to that of isoproturon) when using a buffer at low pH (2.4) as the aqueous constituent of the mobile phase.
3. The RPLC/UV characteristics of both bromoxynil and bentazone are attractive for investigation of the feasibility of the direct analysis of water samples with LVI. The injection of about 3–4 mL of sample should be sufficient to reach the 0.1-µg/L (ppb) level. However, this project involved the sampling before, during, and after agricultural applications of the herbicides. To encompass this period of several months, it was decided to apply SPE shortly after the arrival of samples and to analyze batches of extracts (*see* **Subheading 3.2.2.**).
4. As applied for isoproturon, the feasibility of the two cartridge columns (50 × 4.6 and 100 × 4.6 mm id) packed with 3-µm C18 for C-1 and C-2 was investigated. Rendering adequate *k* values of the acidic pesticides and providing some additional peak compression for the enhancement of sensitivity, mixtures of acetonitrile–phosphate buffer, pH 2.4, of 35:65 (v/v) and 40:60 (v/v) were selected for M-1 and M-2, respectively.
5. Using an injection volume of 200 µL of the SPE extract to reach the required limit of detection of 0.05 µg/L, the cleanup and transfer volumes are precisely determined. Unfortunately, as a result of an elevated background signal (hump) of the transferred interferences of these samples, the LC/LC method with the two C18 columns did not provide sufficient selectivity to reach the required sensitivity.
6. The feasibility of a RAM column is investigated. As a rule of thumb, separation power of both columns is minimized to speed analysis. However, available precolumns packed with RAM materials did not provide sufficient retention or resulted in too much band broadening of the analytes *(14)*. At the time of method development, only a 150 × 4.6 mm

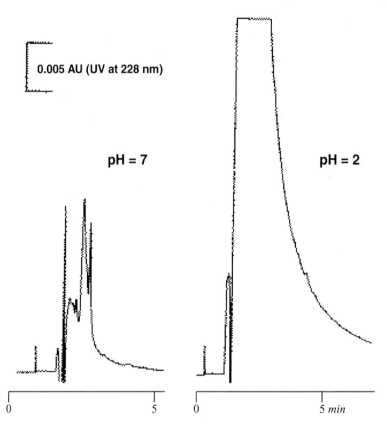

Fig. 6. Two RPLC/UV (228 nm) chromatograms of extracts of a groundwater sample obtained after loading of 25 mL sample adjusted at pH 7.0 and pH 2.0 on 100-mg C18 SPE cartridges. After evaporating the 1 mL of acetone used for desorption, the residue is redissolved in 1 mL 0.05% aqueous TFA. Injection of 100 µL of extract on a 3 µm C18 column (100 × 4.6 mm id) and a mobile phase of methanol–0.05% TFA in water (60:40 v/v) at 1 mL/min.

id RAM analytical column packed with 5-µm SPS-ODS material was available. To apply cleanup, the use of a RAM column as C-1 in LC/LC seems to be more adequate. However, the length of the available column makes this configuration unsuitable. Time for cleanup and transfer will be unnecessarily long, and the exposure of such an expensive column to the injection of uncleaned extracts is not attractive. Therefore, the analytical SPS column is used as C-2 in combination with the 50 × 4.6 mm id 3-µm C18 column as C-1. This C18/SPS column combination provided the required selectivity to analyze these types of water samples at the required sensitivity.

3.2.2. Analytical Procedure (see **Note 6**)

1. Precondition the SPE cartridges with 3 mL methanol, acetone, and methanol and 6 mL 0.1% phosphoric acid in water.
2. Bring a 200-mL water sample to pH 2.1 (±0.1) with 200 µL TFA and percolate through the preconditioned cartridge at a flow rate of approx 4 mL/min.

3. After sample loading, dry the cartridge by passing air for 20 min.
4. Transfer the cartridge above a calibrated tube and, by means of slight overpressure, pass 2 mL acetone through the cartridge and collect in the tube.
5. Evaporate, prior to the LC analysis, the acetone extract to dryness using a warm water bath and a gentle stream of nitrogen.
6. Redissolve the residue by first adding 400 µL acetonitrile, followed by 1600 µL 0.1% phosphoric acid in water (*see* **Note 7**).
7. Inject a 200-µL volume obtained after the SPE procedure on C-1.
8. Switch, after cleanup with 2.9 mL M-1 (injection volume included), C-1 on-line with C-2 for 0.90 min for the transfer of the analyte containing fraction to C-2.
9. Do the quantification of bentazone and bromoxynil by external calibration with standard solutions of the analytes in acetonitrile–0.1% phosphoric acid in water (20:80 v/v) with UV detection at a 217-nm wavelength.

3.3. Determination of Glyphosate and AMPA

This application is a successful multimodal LC/LC method developed for the efficient residue analysis of the highly polar phosphonic acid herbicides glufosinate and glyphosate and glyphosate's major metabolite AMPA in water, soil, and plant materials *(16–21)*. As reviewed *(22)*, the sample pretreatment procedures in conventional analytical methodologies still in use are very laborious for obtaining sufficient sensitivity and selectivity. In the early 1990s, an original methodology employing LVI and multimodel LC/LC was developed that eliminated almost completely the existing laborious sample pretreatment procedures (*see* **Note 8**).

3.3.1. Method Development

According to **Fig. 4** and published information *(16–22)*, the following steps in method development have been made:

1. Phosphonic herbicides like glyphosate and glufosinate are small molecules without a chromophore suitable for UV or FLD detection. For sensitive and selective detection, the technique of precolumn derivatization with FMOC-Cl is selected. The final solution of the formed FMOC–glyphosate and FMOC–AMPA is a 2.5-mL mixture of acetonitrile–borate buffer (20:80 v/v).
2. Because of the required ion exchange chromatography on C2, the suitable LC conditions on C-1 are determined by the required conditions on C-2. Changes in mobile phase composition as a result of column switching can severely affect the stability of the ion exchange retention of analytes. Therefore, the mobile phase consisting of acetonitrile–0.05M phosphate buffer, pH 5.7 (35:65 v/v) is used on both columns.
3. Regarding separation power, C-1 must perform an efficient preseparation between the very-fast-eluting derivatives and the less-polar excess of formed FMOC–OH reagent. In addition, an LVI should be performed to obtain sufficient sensitivity without the necessity of a tedious off-line concentration step. As indicated below, these conditions can be adequately obtained on a short 30 × 4.6 mm id column packed with 5-µm C18 material.
4. Because of the high percentage of acetonitrile (elution power), LVI of the solution obtained after derivatization will lead to excessive band broadening of the compounds during injection. This problem could be resolved by diluting the solution with aqueous buffer prior to injection. A fourfold dilution is sufficient to provide fully top-column focusing of the compounds during injection (up to 2 mL of diluted solution).

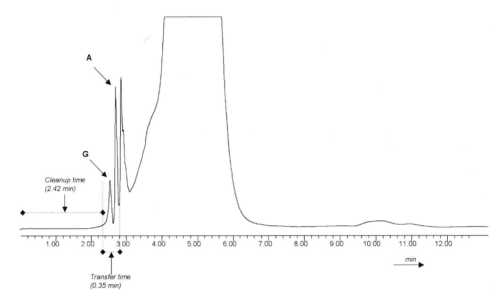

Fig. 7. Establishment of LC/LC column-switching conditions for the LC/LC/FLD analysis of glyphosate and AMPA. Injection of 2.0 mL of a standard solution obtained after derivatization and dilution. C-1, 30 × 4.6 mm id column packed with 5 μm C18 material, directly connected to FLD; mobile phase (M-1), acetonitrile–0.05 M phosphate buffer, pH 5.7 (35:65 v/v) at 1 mL/min.

5. Selecting a 2.0-mL sample injection volume to obtain sufficient sensitivity, the cleanup and transfer volumes are determined by performing LVI on C-1 connected to the FLD. The result of this experiment is displayed in **Fig. 7**. It nicely illustrates the top-column focusing of the analytes during the 2-mL injection of sample solution, followed by the elution as (almost) unretained analytes.

Under the selected column-switching conditions, the LC/LC/FLD method appeared to be very adequate and efficient for the fast processing of extracts of different types of soils and foodstuffs obtained after extraction, derivatization, and dilution.

3.3.2. Analytical Procedure (see **Note 9**)

1. Transfer a 1.5-mL volume of water sample into a glass tube together with 0.5 mL 0.05 M borate buffer (pH 9.0) and follow by 0.5 mL FMOC-Cl solution (2000 μg/mL).
2. Shake the tube and leave for 30 min at room temperature.
3. Add 7.5 mL of 0.025 M borate buffer to dilute this solution.
4. Inject 2.00 mL of the final solution on C-1.
5. After cleanup with 2.42 mL M-1 (injection volume included), switch C-1 on-line with C-2 for 0.35 min to transfer the fraction containing both glyphosate and AMPA derivates to C-2.
6. Do the quantification of the analytes by external calibration with standard solutions in water processed with the precolumn derivatization procedure with an FLD set at 263 (excitation) and 317 nm (emission).

4. Notes

1. As indicated in **Fig. 2**, LC/LC is (almost) always performed in the forward-flush mode. The main reason is to fully maintain the separation of analytes on C-1 and to avoid (partly) nullification of this separation by running it again on C-1 in a reverse direction. SPE–LC is preferably performed in the back-flush mode. The different connection of the SPE–LC mode is displayed in **Fig. 3**. SPE–LC aims at fully top-column focusing of the analytes on the precolumn, thus avoiding unwanted additional band broadening of the analytes by transport through the less-efficient large-size column-packing material. Obviously, the cleanup performance of SPE–LC regarding removal of S1 interferences will be distinctly less compared to LC/LC (*see also* **Fig. 3**).

2. Assuming a Gaussian peak shape and performing an isocratic elution, the molar extinction coefficient ε can be calculated from the peak in the chromatogram requiring the following information:

Description	Symbol	Unit
Flow of the mobile phase	F	mL/min
Half the peak width at 0.6 of the peak height	σ	min
Injection volume	V_0	mL
Concentration of the injected compound	C_0	mol/L
UV absorption of the compound	E	AU
Path length of the flow cell of the UV detector	b	cm

The dilution D of the compound in the flow cell is given by the equation

$$D = (F * \sigma * (2 * \Pi)^{1/2})/V_0 \tag{1}$$

The concentration of the compound (moles/liter) in the flow cell corresponding with the maximum of the Gaussian peak is given by the equation

$$C = C_0/D \tag{2}$$

The molar extinction coefficient ε of the compound at the given wavelength is given by the Lambert–Beer equation:

$$E = \varepsilon * b * C \tag{3}$$

Combining Eqs. (1)–(3) gives the value of ε by the equation

$$\varepsilon = (E * F * \sigma * (2 * \Pi)^{1/2})/(b * C_0 * V_0)$$

3. The high sample throughput of LVI-LC/LC originates from the fact that the sample injection is performed in the high-pressure mode with an autosampler equipped with a large-volume loop and a syringe pump. To perform efficient injection, a visible air gap is inserted at the front and end of the sample by the injector. This adequately replaces the three-times loop filling required for accurate injection, providing a fast and accurate LVI with minimal consumption of sample. For acidic analytes, before injection the pH value of the sample must always be brought about 2 units below the pKa value of the analyte(s).

4. The performance of the LC/LC procedure and the inherent important features of this technique are clearly demonstrated in **Fig. 8**. It must be emphasized that one would not expect such an improvement in selectivity of homomodal LC/LC using columns of similar selectivity. The multidimensional effect of LC/LC obtained by the efficient removal of the excess of matrix compounds allows the direct trace analysis of target analytes in aqueous

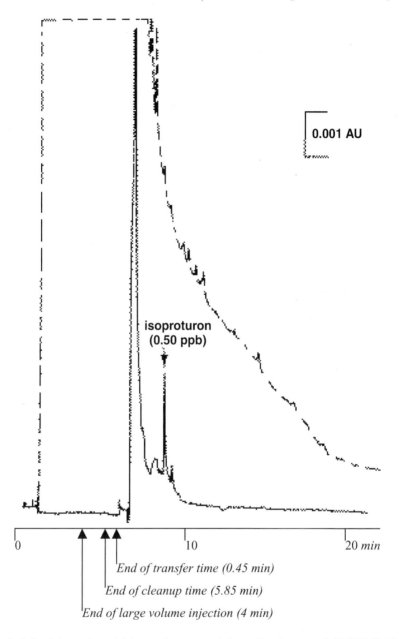

0.001 AU

isoproturon
(0.50 ppb)

0   10   20 *min*

End of transfer time (0.45 min)

End of cleanup time (5.85 min)

End of large volume injection (4 min)

Fig. 8. Selectivity and sensitivity performance of the on-line homomodal LVI-LC/LC (244 nm) of a surface water sample spiked with isoproturon at a level of 0.50 μg/L (ppb) and indicating the applied volumes for injection (4.0 mL), cleanup (5.85 mL), and transfer (0.45 mL). C-1, 50 × 4.6 mm id column packed with 3 μm C18; C-2, 100 × 4.6 mm ID column packed with 3 μm C18; M-1 and M-2, acetonitrile–water (48:52 v/v) at 1 mL/min; dashed line corresponds to the chromatogram obtained using the two columns connected in series without column switching.

samples in less than 10 min with nonselective UV detection. This application is an example of a single-residue method using a small transfer volume (typically < 1.0 mL), providing high selectivity. As shown in **Fig. 1**, the major problem concerns S1, the excess of early eluting interferences in the first part of the chromatogram. Obviously, in case of compounds with sufficient C18 retention (e.g., analytes of type A2 and A3), the separation power of C-1 eliminates the major part of the type S1 interferences. For such applications, an increase in transfer volume does not significantly affect the selectivity. This is clearly demonstrated for the multiresidue analysis of triazine herbicides *(23)*, carbamate pesticides *(24)*, and phenylurea herbicides *(25)* in environmental water samples, using the same LC/LC-UV approach and transfer volumes of 2.0, 5.9, and 10 mL, respectively.

5. RAM columns in LC/LC modes can greatly enhance the cleanup. Successful applications have been developed for the analysis of acidic pesticides in soils *(11,12)* and water *(13,14)*. Method development strategies for using various types of analytical RAM columns as regards the type of packing material and separation power (dimension) and the use in various types of LC/LC modes have been described *(15)*. The comprehensive study revealed that the combination of one analytical RAM column with a C18 column is most favorable in this field of analysis *(15)*.

6. The multidimensional effect clearly shown in **Fig. 9** makes it possible to determine the compounds to a level of 0.05 µg/L (ppb) in difficult types of surface water samples. As regards the total separation power of the two columns, the chromatographic analysis is significantly improved when LC/LC (with column switching) is used instead of only LC (without column switching).The combination of off-line SPE and the fast instrumental analysis of uncleaned extracts (less than 15 min) provides an efficient procedure with a high sample throughput.

7. After SPE and evaporation of the organic solvent, the obtained residue must be redissolved properly into an aqueous phase suitable for RPLC. This is always done by first adding the volume of the organic modifier (acetonitrile or methanol) followed by the volume of water.

8. This methodology is based on an unusual multimodal LC/LC approach first developed for the rapid determination of glufosinate in environmental waters *(16)*. In this new approach, the off-line tedious concentration step and the liquid–liquid extraction procedure after derivatization of the herbicides with FMOC-Cl followed by the instrumental analysis is replaced by a rapid on-line LVI-LC/LC/FLD procedure. In the LC/LC method, the powerful cleanup on C-1 is performed in a reversed way. Normally, the first step in LC/LC (*see* **Fig. 2**) is focused on the removal of an excess of the early eluting interferences (S1) and the transfer of a fraction containing the analyte(s) with sufficient C18 retention. The powerful cleanup performance of the new approach is obtained by advantageously using the amphoteric properties of the analytes and the unusual way of heart cutting. The ionized analytes will have very little C18 retention on C-1 and adequate retention on C-2, having a different (ion exchange) retention mechanism. The efficient cleanup is now obtained by transferring an almost-unretained small analyte-containing fraction, typically 200–250 µL, from C-1 to C-2. All interferences with more C18 retention (e.g., the excess of FMOC–OH) are retained on C-1 and sent to waste by the rinsing mobile phase during the separation of the analyte(s) on C-2.

9. The combination of a very simple sample pretreatment and the developed multimodal LC/LC/FLD method provides a very efficient methodology for the analysis of phosphonic acid herbicides in environmental waters with a high sample throughput. The performance of this approach is illustrated **Fig. 10**, showing the LVI-LC/LC/FLD analysis of a surface

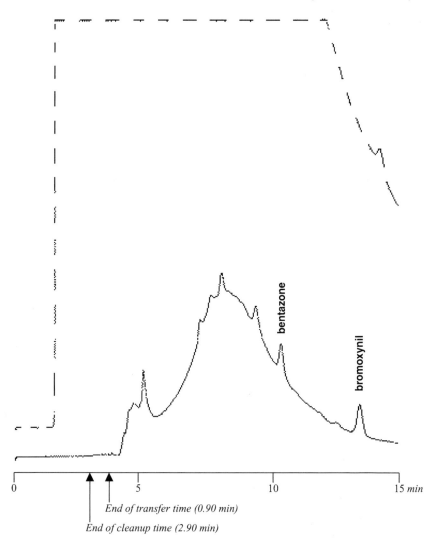

Fig. 9. Performance of multimodal LC/LC (217 nm) of 200 μL extract obtained after off-line SPE of a ditch water sample spiked with bentazone and bromoxynil at a level of 0.50 μg/L (ppb) with indication for the applied volumes for cleanup (2.90 mL) and transfer (0.90 mL). C-1, a 50 × 4.6 mm id column packed with 3 μm C18; C-2, a 150 × 4.6 mm id column packed with 5 μm SPS RAM material; M-1, acetonitrile–0.03 *M* phosphate buffer, pH 2.4 (35:65 v/v) at 1 mL/min; M-2, acetonitrile–0.03 *M* phosphate buffer, pH 2.4 (40:60 v/v) at 1 mL/min; dashed line corresponds to the chromatogram obtained using M-2 and the two columns connected in series without column switching.

water sample after derivatization and dilution and containing both AMPA (1.0 μg/L) and glyphosate (7.5 μg/L). Besides the analysis of environmental waters *(16,18,21)*, the meth-

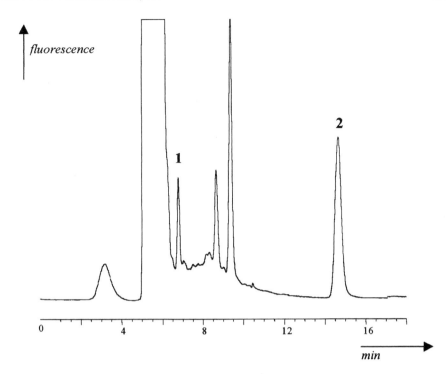

Fig. 10. Multimodal LC/LC/FLD (μex/μem, 263/317 nm) of a 2-mL injection of a surface water sample obtained after derivatization with FMOC and dilution and containing glyphosate (7.5 μg/L) and AMPA (1.0 μg/L). C-1, 30 × 4.6 mm id column packed with 5 μm C18; C-2, 250 × 4.6 mm id column packed with 5 μm Hypersil NH$_2$; M-1 and M-2, acetonitrile–0.05M phosphate buffer, pH 5.7 (35:65 v/v), both at 1 mL/min. For cleanup and transfer volume, *see* **Fig. 7**.

odology with a high sample throughput has been successfully applied to other matrices, such as soils *(17)* and plant materials *(19,20)*.

References

1. Hogendoorn, E. A. and van Zoonen, P. (2000) Recent and future developments of liquid chromatography in pesticide trace analysis. *J. Chromatogr. A* **892,** 435–453.
2. Hogendoorn, E. A. (2000) High-performance liquid chromatographic methods in pesticide residue analysis, in *Encyclopedia of Analytical Chemistry* (Meyers, R. A., ed.), Wiley, Chichester, UK, pp. 6264–6299.
3. Hogendoorn, E. A., van Zoonen, P., and Herández, F. (2003) The versatility of coupled-column LC. *LC–GC Europe* **16,** 44–51.
4. Dijkman, E., Mooibroek, D., Hoogerbrugge, R., et al. (2001) Study of matrix effects on the direct trace analysis of acidic pesticides in water using various liquid chromatographic modes coupled to tandem mass spectrometric detection *J. Chromatogr. A* **926,** 113–125.
5. Choi, B. K., Hercules, D. M., and Gusev, A. I. (2001) Effect of liquid chromatography separation of complex matrices on liquid chromatography–tandem mass spectrometry signal suppression. *J. Chromatogr. A* **907,** 337–342.

6. Pascoe, R., Foley, J. P., and Gusev, A. I. (2001) Reduction in matrix-related signal suppression effects in electrospray ionization mass spectrometry using on-line two-dimensional liquid chromatography. *Anal. Chem.* **73,** 6014–6023.

7. Verhij E.R. (1993) *Strategies for Compatibility Enhancement in Liquid Chromatography–Mass Spectrometry.* PhD thesis, University of Leiden, Febodruk-Enschede.

8. Gort, S. M., Hogendoorn, E. A., Dijkman, E., van Zoonen, P., and Hoogerbrugge, R. (1996) The optimization of step-gradient elution conditions in liquid chromatography: application to pesticide residue analysis using coupled-column reversed-phase LC. *Chromatographia* **42,** 17–24.

9. Hogendoorn, E. A., van Zoonen, P., Brinkman, U. A. T. (1991) Column-switching RPLC for the trace-level determination of ethylenethiourea in aqueous samples. *Chromatographia* **31,** 285–292.

10. Hogendoorn, E. A. and van Zoonen, P. (1995) Coupled-column reversed-phase liquid chromatography in environmental analysis. *J. Chromatogr. A* **703,** 149–166.

11. Parrilla, P., Kaim, P., Hogendoorn, E. A., and Baumann, R. A. (1999) An internal surface reversed phase column for the effective removal of humic acid interferences in the trace analysis of mecoprop in soils with coupled-column RPLC-UV. *Fresenius J. Anal. Chem.* **363,** 77–82.

12. Hogendoorn, E. A., Huls, R., Dijkman, E., and Hoogerbrugge, R. (2001) Microwave assisted solvent extraction and coupled-column reversed-phase liquid chromatography with UV detection. Use of an analytical restricted-access-medium column for the efficient multi-residue analysis of acidic pesticides in soils. *J. Chromatogr. A* **938,** 23–33.

13. Martínez Fernández, J., Martínez Vidal, J. L., Parrilla Vázquez, P., and Garrido Frenich, A. (2001) Application of restricted-access-media column in coupled-column RPLC with UV detection and electrospray mass spectrometry for determination of azole pesticides in urine. *Chromatographia* **53,** 503–509.

14. Hogendoorn, E. A., Westhuis, K., Dijkman, E., et al. (1999) Semi-permeable surface analytical reversed-phase column for the improved trace analysis of acidic pesticides in water with coupled-column reversed-phase liquid chromatography with UV detection. Determination of bromoxynil and bentazone in surface water. *J. Chromatogr. A* **898,** 45–54.

15. Hogendoorn, E. A., Dijkman, E., Baumann, R. A., Hidalgo, C., Sancho, J. V., and Hernández, F. (1999) Strategies in using analytical restricted access media columns for the removal of humic acid interferences in the trace analysis of acidic herbicides in water samples by coupled column liquid chromatography with UV detection. *Anal. Chem.* **71,** 1111–1118.

16. Sancho, J. V., López, F. J., Hernández, F., Hogendoorn, E. A., and van Zoonen, P. (1994) Rapid determination of glufosinate in environmental water samples using 9-fluoronylmethoxycarbonyl precolumn derivatization, large-volume injection and coupled-column liquid chromatography *J. Chromatogr. A* **678,** 59–67.

17. Sancho, J. V., Hidalgo, C., López, F. J, Hernández, F., Hogendoorn, E. A., and Dijkman, E. (1996) Rapid determination of glyphosate residues and its main metabolite AMPA in soil samples by liquid chromatography. *Int. J. Environ. Anal. Chem.* **62,** 53–63.

18. Sancho, J. V., Hernández, F., López, F. J, Hogendoorn, E. A., Dijkman, E., and van Zoonen, P. (1996) Rapid determination of glufosinate, glyphosate and aminomethylphosphonic acid in environmental waters using precolumn fluorogenic labeling and coupled-column liquid chromatography. *J. Chromatogr. A* **737,** 75–83.

19. Hogendoorn, E. A., Ossendrijver, F. M., Dijkman, E., and Baumann R. A. (1999) Rapid determination of glyphosate in cereal samples by means of pre-column derivatization with 9-fluorenylmethyl chloroformate and coupled-column liquid chromatography with fluorescence detection. *J. Chromatogr. A* **833**, 67–73.

20. Hernández, F., Hidalgo, C., and Sancho, J. V. (2000) Determination of glyphosate residues in plants by precolumn derivatization and coupled-column liquid chromatography with fluorescence detection. *J. AOAC Int.* **83**, 728–734.

21. Hidalgo, C., Rios, C., Hidalgo, M., Salvadó, V., Sancho, J. V., and Hernández, F. (2004) Improved coupled-column liquid chromatographic method for the determination of glyphosate and aminomethylphosphonic acid residues in environmental waters. *J. Chromatogr. A* **1035**, 153–157.

22. Stalikas, C. D. and Konidari, C. N. (2001) Analytical methods to determine phosphonic and amino acid group-containing pesticides. *J. Chromatogr. A* **907**, 1–19.

23. Hernández, F., Hidalgo, C., Sancho, J. V., and Lopez, F. J. (1998) Coupled-column liquid chromatography applied to the trace-level determination of triazine herbicides and some of their metabolites in water samples. *Anal. Chem.* **70**, 3322–3328.

24. Martínez Vidal, J. L., Parrilla Vázquez, P., and Martínez Fernández, J. (2000) Coupled column liquid chromatography for the rapid determination of *N*-methylcarbamates and some of their main metabolites in water. *Chromatographia* **51**, 187–192.

25. Van der Heeft, E., Dijkman, E., Baumann, R. A., and Hogendoorn, E. A. (2000) Comparison of various liquid chromatographic methods involving UV and atmospheric pressure chemical ionization mass spectrometric detection for the efficient trace analysis of phenylurea herbicides in various types of water. *J. Chromatogr. A* **870**, 39–50.

30

On-Line Admicelle-Based Solid-Phase Extraction–Liquid Chromatography–Ionization Trap Mass Spectrometry for the Analysis of Quaternary Ammonium Herbicides in Drinking Water

Dolores Pérez-Bendito, Soledad Rubio, and Francisco Merino

Summary

Ion pair admicelle-based solid-phase extraction (SPE) is proved here to be a valuable strategy to concentrate quaternary ammonium compounds (quats; chlormequat, mepiquat, paraquat, diquat, and difenzoquat) from drinking water. The approach is based on the adsolubilization of quats–surfactant ion pairs on sodium dodecyl sulfate (SDS)-coated alumina at pH 2.0. The on-line coupling of admicelle-based SPE to liquid chromatography (LC)–electrospray ionization trap mass spectrometry (MS/ESI) permits the identification and quantification of quats at the levels required by the international recommended levels (0.1 µg/L). The concentration factor achieved is about 250 using 50-mL sample volumes. The detection limits range from 10 to 30 ng/L. On-line regeneration of the sorbent in each run is easily achieved by disruption of SDS admicelles with methanol and posterior coating of the alumina with SDS.

Key Words: Admicelles; chlormequat; diquat; difenzoquat; drinking water; ion trap mass spectrometry; liquid chromatography; mepiquat; on-line LC/SPE; paraquat; pesticides; quaternary ammonium herbicides; quats; sodium dodecyl sulfate-coated alumina; solid-phase extraction.

1. Introduction

Hemimicelles and admicelles, largely studied by physical chemists (*1–10*) and applied in analytical processes to admicellar-enhanced chromatography (*11*) and to the preconcentration of heavy metals from aqueous samples (*12–17*), have been seldom used for the extraction/concentration of organic compounds (*18–21*). These sorbents are formed by the adsorption of ionic surfactants on metal oxides such as alumina, silica, titanium dioxide, and ferric oxyhydroxides. Hemimicelles consist of

From: *Methods in Biotechnology, Vol. 19, Pesticide Protocols*
Edited by: J. L. Martínez Vidal and A. Garrido Frenich © Humana Press Inc., Totowa, NJ

monolayers of surfactants adsorbing head down on the opposite-charged surface of the oxide. The formation of admicelles occurs after surface saturation by the adsorbed surfactant. Admicelles have the structure of bilayers, with the hydrophobic interaction of the nonpolar chain groups of the surfactants the driving force for their formation. So, the outer surfaces of the hemimicelle and admicelle are hydrophobic and ionic, respectively, providing a twofold mechanism for retention of organics. Benefits obtained with the use of these sorbents have been high extraction yields, easy elution of analytes, high breakthrough volumes, and high flow rate for sample loading *(22)*. An additional advantage is that there are many surfactants commercially available, so both the degree of hydrophobicity and the charge of the sorbent can be easily modified according to the nature of analytes.

The usefulness of these sorbents for the concentration of pesticides from environmental water samples is illustrated here for the analysis of quaternary ammonium herbicides (quats: chlormequat [CQ], mepiquat [MQ], paraquat [PQ], diquat [DQ], and difenzoquat [FQ]) in drinking water sources and finished drinking water. The method is based on (1) the preconcentration of quats on sodium dodecyl sulfate (SDS) admicelles on the basis of the formation of SDS–quats ion pairs, (2) the on-line separation of quats by liquid chromatography (LC), and (3) the quantification and identification of analytes using electrospray ionization (ESI) in the positive mode/ionization trap mass spectrometry (MS).

2. Materials

2.1. Reagents

1. High-performance liquid chromatography-grade acetonitrile.
2. Heptafluorobutyric acid (HFBA) (99% pure, reagent grade)
3. SDS (98% pure, reagent grade)
4. Reagent-grade concentrated hydrochloric acid.
5. Reagent-grade methanol.

2.2. Reagent Solutions

1. Admicelle formation solution: Dissolve 0.3200 g SDS in 490 mL bidistilled water, then adjust at pH 2.0 with hydrochloric acid and dilute to 500 mL in a volumetric flask.
2. Eluting solution: Dilute 325 μL HFBA to 25 mL with methanol in a volumetric flask.
3. Alumina surface-regenerating solutions: methanol and bidistilled water adjusted at pH 2.0 with hydrochloric acid.

2.3. Stock Standard Solutions

1. Stock standard 200-mg/L solutions of the following reagent-grade herbicides: DQ, MQ, FQ, PQ, and CQ. Prepare individual stock solutions in bidistilled water and keep them in polyethylene containers at 4°C under light protection conditions.
2. Internal standard solution: Prepare a stock 200-mg/L ethyl viologen dibromide solution in bidistilled water and store it in a polyethylene container at 4°C under light protection conditions.

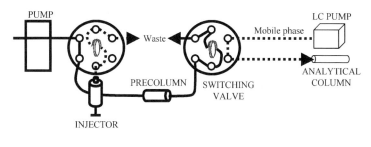

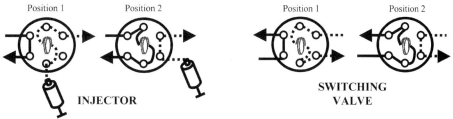

Fig. 1. On-line setup coupling admicelle-based SPE to LC.

2.4. Mobile Phase

1. Solvent A: Add 250 mL bidistilled water to 750 mL acetonitrile.
2. Solvent B: Add 3.25 mL HFBA to 800 mL bidistilled water and dilute to 1 L in a volumetric flask.

2.5. Sampling Equipment

1. Amber polyvinylchloride 0.1- to 1-L high-density or silanized glass bottles. If amber bottles are not available, protect samples from light.
2. Nylon 0.45-µm pore size, 47-mm diameter membrane filters.

2.6. On-Line Extraction System

The on-line setup coupling the admicelles-based solid-phase extraction (SPE) to LC can be easily performed in any laboratory and consists of the following elements: a speed-variable peristaltic pump, silicon rubber pumping tubes (1.14-mm id, 2.82-mm od), an injection valve with a 200-µL sample loop, a Teflon precolumn (35 mm long, 0.4-mm id) packed with 80 mg γ-alumina (155 m^2/g surface area; 8.5 point of zero charge [pcz]; 50- to 200-µm particle diameter range; 100-µm mean value; 58-Å mean pore size; 3.97 g/cm^3 density) and with cotton placed on either end to support the alumina and a six-port switching valve with a 200-µL sample loop. These elements are assembled according to the configuration shown in **Fig. 1** using polytetrafluoroethylene tubes (0.8-mm id).

2.7. LC–MS Equipment

An LC/ESI trap or quadrupole MS system (e.g., Agilent Technologies 1100 series, Palo Alto, CA, or equivalent) coupled on-line with the extraction system above described. The analytical column recommended is a 25-cm C8 column with 4.1-mm id and 5-μm particle diameter.

3. Methods

The methods described below outline (1) the way in which the sample must be collected and preserved, (2) the on-line extraction of quats from water samples by admicelles-based SPE, (3) the quantitation of analytes by LC–MS, and (4) their identification from tandem mass spectrometry (MS/MS) spectra.

3.1. Sample Collection and Preservation

1. Collect grab samples in either amber polyvinylchloride high-density or silanized glass bottles. The analytes are light sensitive, particularly diquat.
2. Filter samples through 0.45-μm filters to remove suspended solids.
3. Preserve samples by storage at 4°C and addition of hydrochloric acid to pH 2.0.
4. Analyze samples within 7 days of collection.

3.2. On-Line Admicelle-Based Extraction (see Note 1)

The general scheme for the on-line preconcentration of quats onto SDS admicelles, their elution, and regeneration of the sorbent is described in **Subheadings 3.2.1.–3.2.5.** This includes (1) the formation of SDS admicelles onto the alumina surface, (2) the sample preparation, (3) the percolation of the sample through the sorbent, (4) the elution of quats and their injection into the chromatographic system, and (5) the regeneration of the sorbent (*see* **Note 2**).

3.2.1. Formation of SDS Admicelles Onto the Alumina Surface (see **Note 3**)

The amount of SDS that must be passed through the mineral oxide to obtain maximum adsorbed surfactant (i.e., 176 mg SDS/g alumina) should be at least 200 mg SDS/g alumina to consider the losses of surfactant caused by the admicelles:aqueous surfactant monomers equilibrium.

1. Set the injector and switching valves in position 2 (**Fig. 1**)
2. Form admicelles by passing 25 mL admicelles formation solution (**Subheading 2.2.1.**) through the precolumn at a flow rate of 2.5 mL/min.

3.2.2. Sample Preparation

The sample volume to be analyzed for quats must be less than 75 mL, which is the breakthrough volume for the most polar quats (CQ and MQ) (*see* **Note 4**).

1. Add ethyl viologen (final concentration 1 μg/L) to 50 mL of the sample acidified with hydrochloric acid (pH 2.0) using the internal standard solution (**Subheading 2.3.2.**).
2. Add the appropriate amount of SDS to give a concentration of 460 mg/L (*see* **Note 5**).

3.2.3. Percolation of Samples Through Admicelles

1. Pass the 50-mL water samples containing the analytes, the internal standard, and SDS at pH 2.0 through the precolumn at a flow rate of 2.5 mL/min with the injector and switching valves set at position 2.

2. Meanwhile, load the 200-µL sample loop of the injector with the eluting solution (**Subheading 2.2.2.**) and equilibrate the analytical column with the mobile phase (15% of solvent A and 85% of solvent B, **Subheading 2.4.**).

3.2.4. Elution of Quats and Injection in the Chromatographic System

1. Switch the injection valve to position 2 and inject 200 µL of the eluting solution (**Subheading 2.2.2.**) (*see* **Note 6**)
2. Conduct the eluted analytes to the 200-µL sample loop of the six-port switching valve by distilled water using a flow rate of 2.5 mL/min for 15 s (*see* **Note 7**)
3. Set the switching valve at position 1 to inject the quats in the chromatographic system and simultaneously start the mobile phase gradient.
4. Return the switching valve to position 2 within 0.5–1 min to ensure the complete introduction of quats into the chromatographic system.

3.2.5. Regeneration of the Sorbent

1. Set the injection valve at position 1.
2. Regenerate the alumina surface into the precolumn by successively passing 1 mL of MeOH, which disrupts the admicelles and removes the surfactant, and 5 mL of 0.01 M HCl to recover the initial charge density on the alumina surface (*see* **Note 8**).
3. For initiation of a new run, proceed according to **Subheading 3.2.1.**

3.3. LS–MS Quantification

The steps described in **Subheadings 3.3.1.–3.3.3.** outline the procedure for the separation of quats by LC on a C8 reversed-phase column using HFBA as an ion pair reagent and their quantification by MS (*see* **Note 9**).

3.3.1. Operating Conditions for Analysis of Quats

1. Establish the following elution gradient program for the chromatographic separation of quats and methyl viologen: 0–20 min (linear gradient from 15 to 35% of solvent A, *see* **Subheading 2.4.1.**); 20–25 min (linear gradient to 100% of solvent A); 25–32 min (isocratic conditions with 100% of solvent A). The flow rate is 0.6 mL/min.
2. Program the diver valve to send the mobile phase to waste in the first 12 min of the initiation of the elution gradient program to remove the most polar matrix compounds, which can dirty the ionization source. After that, send the eluted components to the ESI source.
3. Set the parameters of the MS/ESI specified in **Table 1**.
4. Acquire full-scan spectra of quats and the internal standard in the range 105–255 m/z, with the target mass fixed to the following m/z values: 114 for MQ, 122 and 124 for CQ, 183 and 184 for DQ, 185 and 186 for PQ, 249 for FQ, and 212 and 213 for ethyl viologen.
5. Obtain smooth chromatograms by using the Gauss function (5-point width, 1 cycles).

3.3.2. Calibration

1. To closely match calibration standards to samples, process standards using the whole procedure, that is, with the same experimental conditions as selected for the analysis of unknown water samples (*see* **Note 10**). Working in this way, it is not necessary to know the recovery of the extraction process for each analyte.
2. Prepare 50-mL calibration solutions containing between 3 and 400 ng of each quat, 1 µg/L ethyl viologen, and 460 mg/L SDS; adjust with hydrochloric acid to pH 2.0 and make up to volume.

Table 1
ESI Parameters Set for the Analysis of Quats

	ESI(+) quats					
Parameter	CQ	MQ	DQ	PQ	IS	FQ
Capillary voltage (kV)	–1.0	–1.0	–3.5	–3.5	–3.0	–1.0
Capillary exit voltage (V)	50	50	30	30	40	40
Skimmer (V)	10	10	20	20	30	20
Trap drive	· 35	35	40	40	35	40
Source temperature (°C)			350			
Drying gas (L/min)			10			
Nebulizer gas (psi)			80			
Maximal accumulation time (ms)		100				

ISTD, internal standard.

3. Analyze a minimum of five calibration standards utilizing the LC–MS conditions specified in **Subheading 3.3.1.**
4. From the full-scan mass spectra of quats and methyl viologen (m/z scan range 105–255), obtain the extracted ion chromatograms at the m/z values specified in **Subheading 3.3.1.** for each herbicide and the internal standard (**Fig. 2**). Construct calibration curves from the quats/internal standard ratio of peak areas of the extracted ion chromatograms. The correlation coefficients r^2 obtained are higher than 0.994 for all the quats (*see* **Note 11**).

3.3.3. Method Performance

1. The practical detection limits are estimated from independent complete determinations of analyte concentrations in a typical matrix low-level material *(22)*. They are calculated by using a signal-to-noise ratio of 3 (the ratio between the quat peak area/internal standard peak area and the peak area of noise). The detection limits calculated from six determinations ranged between 20 and 30 ng/L for CQ, MQ, PQ, and DQ, and it was about 10 ng/L for FQ. These values were similar in distilled, tap, and mineral water.
2. Single-laboratory precision and accuracy results at several concentration levels in tap and mineral water using the whole procedure are presented in **Table 2**.
3. The absence of interferences from matrix components that could elute with quats causing ion suppression or space-charge effects on the ion trap was checked by comparison of the calibration curves obtained from standards and those obtained from tap and mineral water samples fortified with known amounts of quats. Because the analytical characteristics of both types of calibration curves were similar, ordinary calibration is recommended for analysis of quats in drinking water. Application of the method to other types of environmental water requires the assessment of the effect of matrix components (*see* **Note 12**).
4. Anionic and nonionic surfactant concentrations up to about 500 mg/L, which is the maximum level generally found in sewage effluents *(23)*, do not interfere. Sodium chloride concentrations below 0.04 M do not affect the adsorption of quats. Higher electrolyte concentrations, however, decrease their recoveries. Thus, recoveries decrease about 75% for CQ and MQ, 70% for DQ and PQ, and 55% for FQ in the presence of 0.2 M NaCl.

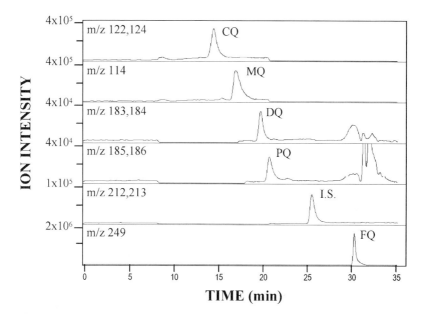

Fig. 2. LC–MS extracted ion chromatograms obtained from a fortified tap water sample containing 0.6 µg/L of each quat and has been treated using the whole procedure.

3.4. MS/ESI Identification of Quats

The steps that can be utilized for the identification of quats in water samples, on the basis of the comparison of their retention time and MS/MS spectra with those obtained from standard solutions, are described below (*see* **Note 13**).

1. Set in the ion trap the MS^2 fragmentation mode and fix the related instrumental parameters to the following values: 27% of ion parent cutoff mass; 40-ms collision-induced time; 4-*m/z* unit isolation width. The resonance excitation value used for each quat is shown in **Table 3**.
2. Isolate the precursor ion specified for each quat in **Table 3**.
3. For quats in sampling matching the retention time of standards identify them on the basis of the MS/MS spectra specified in **Table 3**.

4. Notes

1. The isotherm for the adsorption of SDS onto the positively charged alumina surface at pH 2.0 (**Fig. 3**) shows three regions available for the SPE of organic compounds: the hemimicellar, hemimicellar/admicellar, and admicellar regions. The last one is recommended for adsolubilization of ionic organic compounds because their retention is favored owing to the formation of analyte:surfactant ion pairs. The quats PQ and DQ, which have two quaternary nitrogens in their molecular structure, form with SDS ion pairs of stoichiometry 1:2, whereas CQ, MQ, and FQ yield 1:1 quat:SDS ion pairs.

Table 2
Mean Recovery (%) and (Relative Standard Deviation [%]) of Quats in Fortified ($N = 5$) Drinking Water Samples (Analyzed by On-Line Admicelle-Based SPE–LC–MS)

	Fortified tap water (mg/L)			Fortified mineral water (mg/L)		
	0.1	0.6	1.8	0.1	0.6	1.8
CQ	90 (11)	98 (10)	97 (6)	90 (12)	101 (9)	98 (6)
MQ	99 (13)	100 (8)	98 (5)	104 (11)	96 (9)	97 (7)
DQ	107 (8)	102 (8)	98 (6)	108 (9)	104 (7)	99 (5)
PQ	96 (10)	93 (9)	103 (4)	100 (7)	101 (8)	104 (6)
FQ	104 (9)	97 (7)	101 (3)	103 (10)	97 (8)	100 (4)

Table 3
Chemical Structure of Quats, Precursor Ions, and Excitations Resonance Values Selected for Fragmentation and Relative Abundance and Structure of the Product Ions Generated Under MS/MS/ESI(+) Analysis

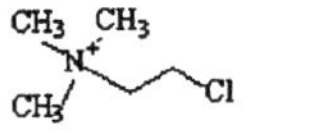

Chlormequat (CQ) — 122 — 1.05

CQ^+ 122 (100%)
$CH_2=N^+(CH_3)_2$ 58 (22%)
$[N(CH_3)_3]^{+\cdot}$ 59 (16%)
$[CH_2CH_2Cl]^+$ 63 (4%)
$[CH_2CHN(CH_3)_2]^+$ 86 (2%)
$CH_3-N^+(CH_3)-Cl$ 94 (5%)

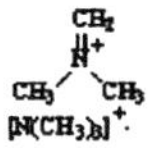

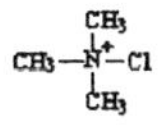

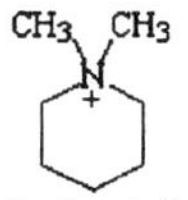

Mepiquat (MQ) — 114 — 0.95

MQ^+ 114 (100%)
$CH_2=N^+(CH_3)_2$ 58 (13%)
$[C_5H_9]^+$ 69 (4%)
98 (7%)
99 (11%)

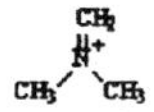

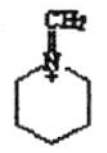

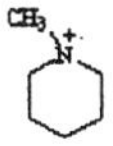

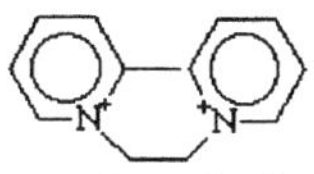

Diquat (DQ) — 184 — 0.90

DQ^+ 184 (100%)
$[DQ-C_2H_3]^+$ or $[DQ-CNH]^+$ 157 (80%)
$[DQ-C_2H_2]^+$ 158 (6%)
$[DQ-NH_2]^+$ or $[DQ-CH_4]^+$ 168 (8%)

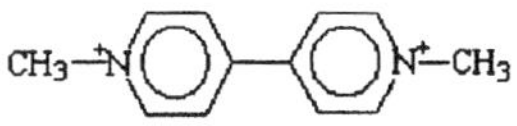

Paraquat (PQ) — 186 — 0.95

PQ^+ 186 (100%)
$[PQ^+-H^\cdot-CH_3CN]^+$ 144 (23%)
$[PQ^+-H^\cdot-HCN]^+$ 158 (100%)
$[PQ^+-H^\cdot-CH_3]^+$ 170 (83%)

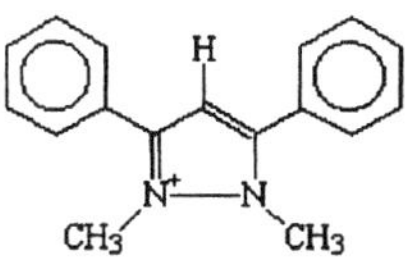

Difenzoquat (FQ) — 249 — 1.30

FQ^+ 249 (100%)
$[Ph_2C=N-CH_3]^+$ 118 (2%)
$[DF^+-C_6H_5CNCH_3]^{+\cdot}$ 131 (3%)
$[DF^+-C_6H_5CN]^{+\cdot}$ 146 (3%)
193 (8%)
$[DF^+-CH_3CN]^{+\cdot}$ 208 (6%)
233 (1%)
234 (1%)

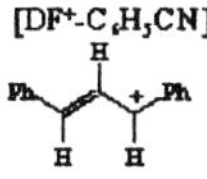

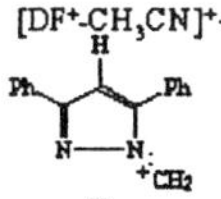

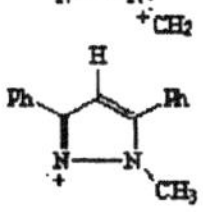

2. The SPE methods developed for quats have been revised by Picó et al. *(24)*. They are based on cation exchange resins, silica sorbents, and apolar phases. Ion pair formation using apolar SPE is the preferred method for isolating quats, and it has been proposed by the Environmental Protection Agency as the reference method for DQ and PQ determination *(25)*. However, some serious problems still remain (e.g., accuracy and precision are very poor, and the methods are often too time consuming and tedious) that make it difficult to optimize the methods for monitoring a large number of samples *(24)*. On the other hand, methods for the determination of the most recently commercialized quats (FQ, MQ, CQ) have been hardly developed.

3. According to the adsorption isotherm of SDS on alumina at pH 2.0 (**Fig. 3**), the amount of surfactant adsorbed in the admicellar region is constant, with a rough value of 176 mg SDS/g alumina. In this region, the SDS admicelles are in equilibrium with monomers of surfactant in the aqueous phase at concentrations in the range 300–1417 mg/L.

4. The breakthrough volume increases to 100 mL for DQ and PQ and 125 mL for FQ. When possible, it is better to handle a sample volume lower than the lowest breakthrough volume for more reproducible results. The preconcentration factor achieved for 50 mL of sample using the LC/SPE–MS approach described here is about 250.

5. The addition of surfactant to samples is essential to ensure that the concentration of admicelles in the mineral oxide remains constant during percolation of the sample because it can be inferred from data of the SDS adsorption isotherm (**Fig. 3**), when admicelles are used in on-line SPE applications, that the surfactant in the admicellar phase will continuously partition to the aqueous phase during percolation of the sample. As a result, the concentration of admicelles will progressively decrease during operation, and the partition constant for analytes and therefore their retention will change. The concentration of surfactant in the sample must be enough to replenish the losses of SDS caused by the admicelles:aqueous monomers equilibrium and below the critical micellar concentration of SDS at pH 2.0 (1417 mg/L) to avoid the partition of quats between aqueous micelles and admicelles.

6. The eluting solution must cause desorption of analytes from the admicelle and be compatible with the mobile phase (acetonitrile and the ion pair reagent HBFA). Organic solvents such as acetonitrile and methanol caused disruption of the admicelles and therefore elution of analytes. However, the eluted SDS–quats ion pairs are only partially substituted by the HFBA–quats ion pairs through the chromatographic process. As a result, the chromatogram shows peaks corresponding to both types of ion pairs. It is therefore necessary to add HFBA to the elution solvent at concentrations of about 0.1 *M* to produce complete destruction of the SDS–quats ion pairs and as a result quantitative recovery of analytes. The use of only aqueous HFBA does not achieve complete elution of quats. Sodium hydroxide, which also causes disruption of SDS admicelles by desorption of the surfactant (8.5 pcz of the alumina), is not a good eluent for quats because low recoveries are obtained for the most apolar analytes probably because of their high retention on the negatively charged surface of the alumina.

7. This time must be calculated for each on-line extraction system because it depends on the length of the tube connecting the precolumn and the switching valve.

8. The extraction precolumn works within a run for at least 20 cycles without deterioration. However, progressive loss of efficiency has been observed for runs between days. Therefore, it is advisable to substitute the alumina daily.

9. LC procedures for analyzing ammonium quaternary herbicides in various samples after SPE commonly use C18 reversed-phase columns and ion-pairing reagents in the mobile

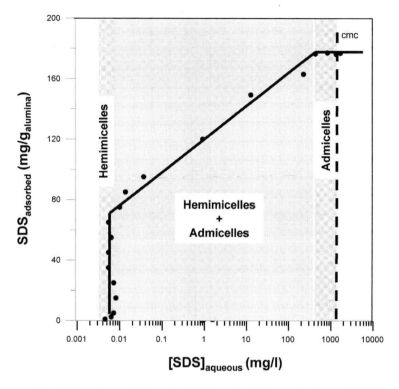

Fig. 3. Experimental adsorption isotherm for SDS on alumina at pH 2.0.

phase, such as heptanesulfonate *(26,27)*, octanesulfonate *(28)*, and orthophosphate *(29)*. Polymeric packings such as PS-DVD are recommended by the Environmental Protection Agency to avoid the free silanol group effect *(30–32)*. Coupling ion pair chromatography with MS is not a good approach because of the high concentration of nonvolatile conventional ion pair reagents in the mobile phase. Systems containing only volatile buffers such as HFBA *(33)* must be used. The utilization of ammonium acetate is not recommended because of the formation of quats–acetate adducts *(34,35)*.

10. With on-line systems, it is not advisable to carry out quantitative analysis by comparison with direct injections *(36)*. First, the volume of many injection loops is specified to an average accuracy of 20%, and calibration of a loop is a rather delicate and time-consuming operation. This does not have to be considered with off-line procedures because the same loop is used for both analysis of unknown extracts and construction of calibration curves. Second, slight but imperceptible band broadening may occur.

11. No gains in sensitivity are achieved by reducing the m/z scan range to 10 (termed selected ion monitoring [SIM] mode in the operation of the Agilent system), which is logical considering how an ion trap works *(37)*. Thus, mass analysis in an ion trap consists of two steps: (1) accumulation of ions in the trap and (2) production of a spectrum by resonantly ejecting the ions from the trap to detector. Reducing the acquisition range only shortens one part of the scan program (e.g., the time spent to production of the spectrum). As a

result, the number of microscans that can be acquired is increased, and the ion current observed for SIM would be greater. However, only slight gains in sensitivity are achieved (about two to three times). Although this modest gain in sensitivity may be beneficial in some cases, the additional information that would have been provided by a full scan is typically more significant. The inherent high sensitivity of the ion trap allows full-scan spectra to be obtained at lower levels than can be detected with SIM on conventional quadrupolar instruments. If an ion trap is not available and quantification is carried out with a quadrupole filter instrument, SIM greatly improves sensitivity vs full scan by continuously collecting current for a single mass.

12. *Space charge*, defined as the electrostatic forces that exist between ions that are held in closely to each other, can degrade severely the performance of the ion trap as a mass spectrometer. The effects of space charge are observed commonly as losses in mass resolution or mass shifts. Therefore, it is very important to avoid overfilling the trap with ions. However, the experts all agree that overfilling the trap is unlikely with commercial instruments because an automatic gain control test scans to determine the total ion current and adjusts accordingly the gate time for admitting ions into the trap. In real samples, coelution of ions from components of the matrix can decrease the sensitivity for the target ion because the trap is partially filled with other ions, but this possible interference can be detected by comparison of the results obtained from calibration using a fortified sample with known amounts of quats and using external standards.

13. Identification of quats in samples has been undertaken using thermospray *(38)*, fast atom bombardment *(39)*, and atmospheric pressure ionization interfaces *(40–42)*, typically in combination with separation by either LC *(38,40,41)* or capillary electrophoresis *(42,43)*. Reported product ions by the different MS/MS techniques used have been similar. Some detailed MS^n studies on ion trap instruments have allowed the examination of the fragmentation pathways and the identification of a number of transitions that are highly specific to each quat *(44)*. Assignation of structures to the product ions in **Table 3** is based on these previous studies.

Acknowledgment

We gratefully acknowledge financial support from the Spanish CICyT (project BQU 2002-01017).

References

1. Dobias, B. (ed.). (1993), *Coagulation and Flocculation. Surfactant Sciences*, Series 477. Marcel Dekker, New York.
2. Bohmer, M. R. and Koopal, L. K. (1992) Adsorption of ionic surfactants on variable-charge surfaces. 1. Charge effects and structure of the adsorbed layer. *Langmuir* **8,** 2649–2659.
3. Bohmer, M. R. and Koopal, L. K. (1992) Adsorption of ionic surfactants on variable-charge surfaces. 2. Molecular architecture and structure of the adsorbed layer. *Langmuir* **8,** 2660–2665.
4. Koopal, L. K., Lee, E. M., and Bohmer, M. R. (1995) Adsorption of cationic and anionic surfactants on charged metal oxide surfaces. *J. Colloid Interface Sci.* **170,** 85–97.
5. Goloub, T. P., Koopal, L. K., Bijsterbosch, B. H., and Sidorova, M. P. (1996) Adsorption of cationic surfactants on silica. Surface charge effects. *Langmuir* **12,** 3188–3194.
6. Valsaraj, K. T. (1989) Partitioning of hydrophobic nonpolar volatile organics between the aqueous and surfactant aggregate phase on alumina. *Sep. Sci. Technol.* **24,** 1191–2005.

7. Valsaraj, K. T. (1992) Adsorption of trace hydrophobic compounds from water on surfactant-coated alumina. *Sep. Sci. Technol.* **27**, 1633–1642.
8. Harwell, J. H., Hoskins, J., Schechter, R. S., and Wade, W. H. (1985) Pseudophase separation model for surfactant adsorption: isomerically pure surfactants. *Langmuir* **1**, 251–262.
9. Behrends, T. and Herrmann, R., (2000) Adsolubilization of anthracene on surfactant covered silica in dependence on pH: indications for different adsolubilization in admicelles and hemimicelles. *Colloids Surf. A* **162**, 15–23.
10. Bury, R., Favoriti, P., and Treiner, C. (1998) Specific interactions between salicylate derivatives and cetylpyridinium chloride both at the silica/water interface and in solution. A calorimetric investigation. *Colloids Surf. A* **139**, 99–107.
11. Pramauro, E. and Pelizzetti, E. (eds.). (1996) *Surfactants in Analytical Chemistry, Applications of Organized Amphiphilic Media.* Elsevier, New York.
12. Manzoori, J. L., Sorouraddin, M. H., and Schemirani, F. (1995) Chromium speciation by a surfactant-coated alumina microcolumn using electrothermal atomic absorption spectrometry. *Talanta* **42**, 1151–1155.
13. Hiraide, M. and Shibata, W. (1998) Collection of trace heavy metals on dithizone-impregnated admicelles for water analysis. *Anal. Sci.* **14**, 1085–1088.
14. Manzoori, J. L., Sorouraddin, M. H., and Schabani, A. M. H. (1998) Determination of mercury by cold vapor atomic absorption spectrometry after preconcentration with dithizone immobilized on surfactant-coated alumina. *J. Anal. At. Spectrom.* **13**, 305–308.
15. Manzoori, J. L., Sorouraddin, M. H., and Schabani, A. M. H. (1999) Atomic absorption determination of cobalt after preconcentration by 1-(2-pyridylazo)-2-naphtol immobilized on surfactant-coated alumina. *Microchem. J.* **63**, 295–301.
16. Hiraide, M., Iwasawa, J., and Kawaguchi, H. (1997) Collection of trace heavy metals complexed with ammonium pyrrolidinedithiocarbamate on surfactant-coated alumina sorbents. *Talanta* **44**, 231–237.
17. Hiraide, M. and Hori, J. (1999) Enrichment of metal–APDC complexes on admicelle-coated alumina for water analysis. *Anal. Sci.* **15**, 1055–1058.
18. Rubio, S. and Pérez-Bendito, D. (2003) Supramolecular assemblies for extracting organic compounds. *Trends in Anal. Chem.* **22**, 470–485.
19. Saitoh, T., Nakayama, Y., and Hiraide, M., (2002) Concentration of chlorophenols in water with sodium dodecylsulfate-γ-alumina admicelles for high-performance liquid chromatographic analysis. *J. Chromatogr. A* **972**, 205–309.
20. Merino, F., Rubio, S., and Pérez-Bendito, D. (2003) Solid-phase of amphiphiles based on mixed hemimicelle/admicelle formation: application to the concentration of benzalkonium surfactants in sewage and river water. Anal. Chem. **75**, 6799–6806.
21. Merino, F., Rubio, S., and Pérez-Bendito, D. (2004) Evaluation and optimization of an online admicelle-based extraction–liquid chromatography approach for the analysis of ionic organic compounds. *Anal. Chem.* **76**, 3878–3886.
22. Thompson, M. and Ellison, S. L. R. (2002) Harmonised guidelines for single laboratory validation of methods of analysis. *Pure Appl. Chem.* **74**, 835–855.
23. Matthew, J. S. and Jones, M. N. (2000) The biodefradation of surfactants in the environment. *Biochim. Biophys. Acta*, **1508**, 235–251.
24. Picó, Y., Font, G., Moltó, J. C., and Mañes, J. (2000) Solid-phase extraction of quaternary ammonium herbicides. *J. Chromatogr. A* **885**, 251–271.
25. Hodgeson, J. W., Bashe, W. J., and Eichelberger, J. W. (1997) Method 549.2 *Determination of Diquat and Paraquat in Drinking Water by Liquid–Solid Extraction and High Performance Liquid Chromatography With UV Detection.* Environmental Protection Agency, Cincinnati, OH, p. 101.

26. Corasaniti, M. T., Strongoli, M. C., and Nisticò, G. (1990) Determination of paraquat in rat brain using ion-pair solid-phase extraction and reversed-phase high-performance liquid chromatography with ultraviolet detection. *J. Chromatogr. B* **527**, 189–195.
27. Corasaniti, M. T. and Nisticò, G. (1993) Determination of paraquat in rat brain by high-performance liquid chromatography. *J. Chromatogr. A* **643**, 419–425.
28. Needham, L., Paschal, D., Rollen, Z. J., Liddle, J., and Bayse, D. (1979) Determination of paraquat in marihuana by reversed-phase paired-ion high performance liquid chromatography. J. Chromatogr. Sci. **17**, 87–90.
29. Ahmad, I. (1983) On-line trace enrichment of difenzoquat in water and its determination by HPLC. *J. Environ. Sci. Health,* **18**, 207–219.
30. Hodgeson, J. W. and Bashe, W. J. (1990) Method 549, in *Determination of Diquat and Paraquat in Drinking Water by Liquid–Solid Extraction and High Performance Liquid Chromatography With UV detection.* Environmental Protection Agency, Cincinnati, OH, p 101.
31. Hodgeson, J. W., Bashe, W. J., and Eichelberger, J. W. (1992) Method 549.1, in *Determination of Diquat and Paraquat in Drinking Water by Liquid–Solid Extraction and High Performance Liquid Chromatography With UV detection.* Environmental Protection Agency, Cincinnati, OH, p. 119.
32. 40 CFR Part 141. (1998) *National Primary Drinking Water Regulation: Analytical Methods for Certain Pesticides and Microbial Contaminants,* Proposed Rule. Environmental Protection Agency, Cincinnati, OH.
33. Marr, J. C. and King, J. B. (1997) A simple high performance liquid chromatography/ ionspray tandem mass spectrometry method for the direct determination of paraquat and diquat in water. *Rapid Commun. Mass Spectrom.* **11**, 479–483.
34. Catherine, S. E., Startin, J. R., Goodall, D. M., and Keely, B. J. (2001) Formation of gas-phase clusters monitored during electrospray mass spectrometry: a study of quaternary ammonium pesticides. *Rapid Commun. Mass Spectrom.* **15**, 1341–1345.
35. Taguchi, V. Y., Jenkins, S. W. D., Crozier, P. W., and Wang, D. T. (1998) Determination of diquat and paraquat in water by liquid chromatography (electrospray ionization) mass spectrometry. *J. Am. Soc. Mass Spectrom.* **9**, 830–839.
36. Hennion, M. C. (1999) Solid-phase extraction:method development, sorbents, and coupling with liquid chromatography. *J. Chromatogr. A* **856**, 3–54.
37. Yates, N. A., Booth, M. M.; Stepheson, J. L., and Yost, R. A. (1995) Practical Ion Trap Technology: GC/MS and GC/MS/MS, in *Practical Aspects of Ion Trap Mass Spectrometry* (Raymon, E. M. and Todd, J. F. J., eds.). CRC Press, New York, 1995, Chapter 4, pp. 121–186.
38. Barcelo, D., Durant, G., and Vreeken, R. J. (1993) Determination of quaternary amine pecticides by thermospray mass spectrometry. *J. Chromatogr. A* **647**, 271–277.
39. Todeur, Y., Sovocool, G. W., Mitchum, R. K., Niederhut, W. J., and Donnelly, J. R. (1987) Use of FAB MS/MS for analysis of quarternary amine pesticide standards. *Biomed. Environ. Mass Spectrom.* **14**, 733–736.
40. Evans, C. S., Startin, J. R., Goodall, D. M., and Keely, B. J. (2000). Optimisation of ion trap parameters for the quantification of chlormequat by liquid chromatography/mass spectrometry and the application in the analysis of pear extract. *Rapid Commun. Mass Spectrom.* **14**, 112–117.
41. Castro, R., Moyano, E., and Galceran, M. T. (1999) Ion-pair liquid chromatography–atmospheric pressure ionization mass spectrometry for the determination of quaternary ammonium herbicides. *J. Chromatogr. A* **830**, 145–154.

42. Moyano, E., Games, D. E., and Galceran, M. T. (1996) Determination of quaternary ammonium herbicides by capillary electrophoresis/mass spectrometry. *Rapid Commun. Mass Spectrom.* **10,** 1379–1385.

43. Wycherley, D., Rose, M. E., Giles, K., Hutton, T. M., and Rimmer, D. A. (1996) Capillary electrophoresis with detection by inverse UV spectroscopy and electrospray mass spectrometry for the examination of quaternary ammonium herbicides. *J. Chromatogr. A* **734,** 339–349.

44. Catherine, S. E., Startin, J. R., Goodall, D. M., and Keely, B. J. (2001) Tandem mass spectrometric analysis of quaternary ammonium pesticides. *Rapid Commun. Mass Spectrom.* **15,** 699–707.

31

Molecular Imprinted Solid-Phase Extraction for Cleanup of Chlorinated Phenoxyacids From Aqueous Samples

Claudio Baggiani and Cristina Giovannoli

Summary

Solid-phase extraction (SPE) represents a suitable way to clean up and preconcentrate samples containing traces of chlorinated phenoxyacids. High selectivity levels may be obtained using columns packed with materials based on well-defined molecular recognition mechanisms, such as immunoaffinity columns. SPE procedures involving molecular imprinted polymers (MISPE, molecular imprinted solid-phase extraction) have been proposed as a valid substitute for immunoaffinity procedures because of their simplicity of preparation and high stability to extreme chemical conditions. Here, we describe the preparation and the use of a polymer obtained by imprinting with the herbicide 2,4,5-trichlorophenoxyacetic acid as a molecular imprinted SPE sorbent. The polymer will be useful for extracting and concentrating selectively several strictly related phenoxyacids (2,4,5-trichlorophenoxyacetic acid; 2,4-dichlorophenoxyacetic acid; (R,S)-2-(2,4,5-trichlorophenoxy)-propionic acid; (R,S)-2-(2,4-dichlorophenoxy)-propionic acid; 2-methyl-4-chlorophenoxyacetic acid; and (R,S)-2-(2-methyl-4-chlorophenoxy)-propionic acid) from water samples before performing the analysis by capillary electrophoresis or high-performance liquid chromatography.

Key Words: Cleanup; 2,4-D; dichlorprop; fenoprop; halogenated phenoxyacids; herbicides; MCPA; mecoprop; molecular imprinting; molecular recognition; preconcentration; solid-phase extraction; 2,4,5-T.

1. Introduction

Legislation poses severe limits on the presence of chlorinated phenoxyacids in waters because of their high toxicity for mammals and aquatic organisms and long-term persistence in the environment. Thus, these pesticides are suitable subjects of trace analysis. Because of their insufficient sensitivity, the detection and quantification of these substances at trace levels cannot be performed directly using traditional analytical techniques, but it requires one or more preconcentration steps.

From: *Methods in Biotechnology, Vol. 19, Pesticide Protocols*
Edited by: J. L. Martínez Vidal and A. Garrido Frenich © Humana Press Inc., Totowa, NJ

Solid-phase extraction (SPE) represents a suitable way to clean up and preconcentrate samples containing traces of chlorinated phenoxyacids. This technique is more rapid, simple, economical, and environment friendly than traditional liquid–liquid extraction. The main drawback associated with SPE columns packed with ordinary stationary phases is the low selectivity of the retention mechanism. Also, with the desired analyte(s), many interfering substances of similar hydrophobicity/hydrophilicity are retained and concentrated because of the very limited selectivity of the partition equilibriums involved *(1–4)*.

Selectivity may be obtained using columns packed with materials based on well-defined molecular recognition mechanisms, such as immunoaffinity columns. These kinds of SPE columns are affected by problems that limit more general use. High production costs for specific antibodies from polyclonal or monoclonal antisera and the poor stability of the antibody-binding sites in the organic solvents make large-scale use of these sorbents impracticable *(5–7)*.

Of the many ways to make synthetic hosts able to imitate natural antibodies, one of the most interesting involves molecular imprinting. In this technique, functional monomers are assembled around a template (imprint) molecule by noncovalent interactions and then linked together using a cross-linking agent. Template removal leaves cavities within the molecularly imprinted polymer (MIP) that possess a shape and functional group complementarity to the imprint molecule, allowing its tight and selective uptake *(8–10)*. SPE procedures involving molecular imprinted polymers (MISPE, molecular imprinted solid-phase extraction) have been proposed as a valid substitute for immunoaffinity procedures because of their simplicity of preparation and high stability to extreme chemical conditions *(11–14)*.

Here, we describe the use of a polymer obtained by imprinting with the herbicide 2,4,5-trichlorophenoxyacetic acid (2,4,5-T) as a MISPE sorbent. This polymer is used to extract and concentrate several strictly related phenoxyacids from water samples before analysis by capillary electrophoresis or high-performance liquid chromatography (HPLC).

2. Materials

2.1. Halogenated Herbicides

1. Analytical-grade 2,4,5-T. Caution: Do not use technical 2,4,5-T because of the risk of contamination by chlorinated dioxins.
2. Analytical-grade 2,4-dichlorophenoxyacetic acid (2,4-D). Caution: Do not use technical 2,4-D because of the risk of contamination by chlorinated dioxins.
3. Analytical-grade (R,S)-2-(2,4,5-trichlorophenoxy)-propionic acid (fenoprop).
4. Analytical-grade (R,S)-2-(2,4-dichlorophenoxy)-propionic acid (dichlorprop).
5. Analytical-grade 2-methyl-4-chlorophenoxyacetic acid (MCPA).
6. Analytical-grade (R,S)-2-(2-methyl-4-chlorophenoxy)-propionic acid (mecoprop).
7. Mother solutions for **items 1–6**: 2 mg/mL in acetonitrile; store at −20°C.

2.2. Polymerization Reagents

1. Synthesis-grade 4-vinylpyridine (4-VP). Caution: Toxic for ingestion; skin sensitizer, risk of irreversible effects.

2. Synthesis-grade ethylene dimethacrylate (EDMA).
3. Synthesis-grade 2,2'-azobis-(butyronitrile) (AIBN). Caution: The solid decomposes violently when heated; store at −20°C)

2.3. Solvents and Chemicals

1. Analytical-grade methanol. Caution: Toxic for inhalation, ingestion, and contact. Risk of irreversible effects.
2. HPLC-gradient-grade acetonitrile. Caution: Highly flammable. Toxic for inhalation, ingestion, and contact.
3. Analytical-grade glacial acetic acid. Caution: Risk of severe burns.
4. Analytical-grade formic acid. Caution: Risk of severe burns.
5. Analytical-grade sodium phosphate monobasic dihydrate.
6. Analytical-grade sodium tetraborate decahydrate.
7. Ultrapure water.
8. α-Cyclodextrin.
9. β-Cyclodextrin.
10. γ-Cyclodextrin.

2.4. Laboratory Equipment

1. Ultrasound bath.
2. Mechanical mortar.
3. 90- and 25-μm sieves.
4. Soxhlet apparatus for continuous extraction.
5. Empty SPE cartridges (glass or polypropylene).
6. Vacuum manifold for SPE.
7. Capillary electrophoresis apparatus with ultraviolet (UV) detector.
8. HPLC apparatus with binary pumps and UV detector.

3. Methods

The methods described below outline (1) the preparation of the 2,4,5-T-imprinted polymer, (2) the preparation of the MISPE column, (3) the extraction of halogenated phenoxyacids from aqueous samples, and (4) the determination of extracted analytes by capillary electrophoresis and HPLC.

3.1. Preparation of the 2,4,5-T-Imprinted Polymer

This method describes the preparation of a 4-vinylpyridine-*co*-ethylene dimethacrylate polymer by radical thermopolymerization in the presence of a template molecule able to mimic the analytes. The use of a mimic is justified by the necessity to imprint the polymer with a molecule not present in the analytical samples to avoid interferences in the subsequent analysis because of the slow release of residual template.

The molecular recognition properties of the 2,4,5-T-imprinted polymer are caused by ionic and hydrophobic interactions between the molecules of template and the binding sites (*see* **Fig. 1**). To obtain strong hydrophobic interactions, it is necessary to use a polar porogenic solvent during the polymerization process. Solvents frequently used in molecular imprinting technology such as chloroform and acetonitrile are less efficient. In fact, polymers obtained with these solvents show poor binding capacity and reduced selectivity *(15)*.

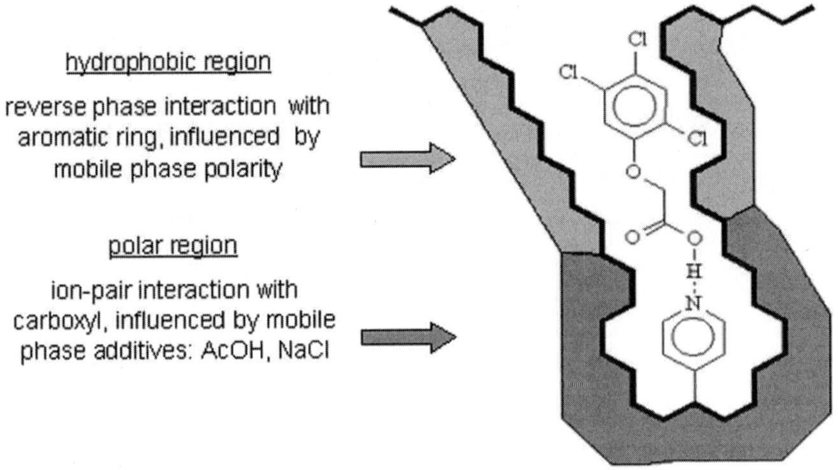

hydrophobic region

reverse phase interaction with
aromatic ring, influenced by
mobile phase polarity

polar region

ion-pair interaction with
carboxyl, influenced by mobile
phase additives: AcOH, NaCl

Fig. 1. Conceptual model of the molecular recognition properties of the 2,4,5-T-imprinted polymer are caused by ionic and hydrophobic interactions between the molecules of the template and the binding site.

1. In a 10-mL wall test tube, dissolve under sonication 0.600 g 2,4,5-T (2.35 mmol) into 3.70 mL methanol–water (3:1 v/v) (*see* **Note 1**). Then, add 0.243 mL (2.35 mmol, molar ratio template–functional monomer 1:1) 4-VP, 3.53 mL EDMA (15.7 mmol, molar ratio functional monomer–cross-linker 1:8) (*see* **Note 2**) and 0.050 g AIBN (*see* **Note 3**).
2. Purge the test tube with analytical-grade nitrogen for almost 5 min (*see* **Note 4**), seal with an airtight plug. and leave it to polymerize overnight at 65°C in a water bath (*see* **Notes 5 and 6**).
3. After complete polymerization of the mixture, open the plug (*see* **Note 7**) and grind the pale blue-white monolith obtained (*see* **Note 8**) in a mechanical mortar (*see* **Note 9**).
4. Separate the grinded powder through a double sieve (90-μm upper screen, 25-μm lower screen) to isolate a fraction suitable for SPE column packing.
5. Wet the fraction retained by the 25-μm screen with methanol and wash carefully with abundant deionized water to eliminate the submicrometric dust that adheres to the polymer particles (*see* **Note 10**).
6. Wash the obtained particle fraction with acetic acid by overnight extraction in a Soxhlet apparatus to eliminate the residual template (*see* **Note 11**). Dry the polymeric particles at 50°C overnight (*see* **Note 12**).

3.2. Preparation of the MISPE Column

This method describes the packing and the stabilization of SPE columns with the imprinted polymer. The amount of polymer used is suitable for the extraction/cleanup of aqueous samples not exceeding 100 mL. The stabilization of the columns is a necessary step to eliminate residual molecules of template that could interfere in the pesticide extraction and analysis (*see also* **Note 13**).

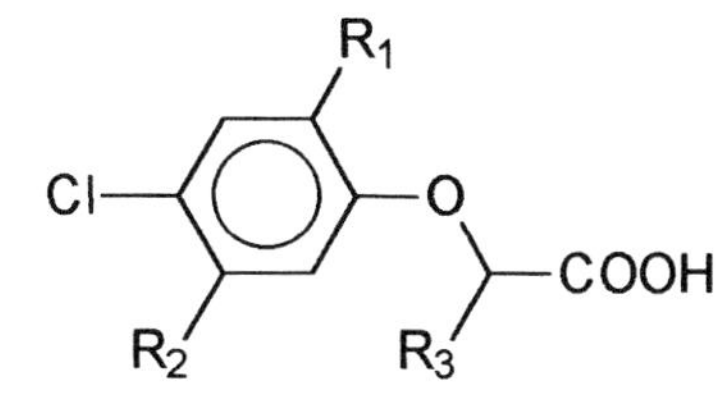

herbicide	R₁	R₂	R₃
2,4,5-T	Cl	Cl	H
2,4-D	Cl	H	H
MCPA	CH₃	H	H
mecoprop	CH₃	H	CH₃
dichlorprop	Cl	H	CH₃
fenoprop	Cl	Cl	CH₃

Fig. 2. Structure of halogenophenoxyacid herbicides recognized by the 2,4,5-T-imprinted polymer.

1. Mix 0.200 g dry polymer with 1 mL methanol in a test tube and sonicate until complete homogeneity of the suspension is reached (*see* **Note 14**).
2. Transfer the suspension rapidly into an empty 5-mL SPE column provided with outlet frit and stopcock to secure the packing (*see* **Note 15**).
3. Connect the column to a vacuum manifold and apply a gentle vacuum until the polymer is completely packed (*see* **Note 16**). Then, apply an inlet frit, gently pushing it into the column with a glass rod of proper dimension (*see* **Note 17**).
4. Wash the SPE column, adding sequentially 1 mL acetic acid, methanol, and water.
5. Check the presence of residual 2,4,5-T after washing with acetic acid (*see* **Note 13**; *see also* **Subheadings 3.4.** and **3.5.**).
6. Repeat the cycle if 2,4,5-T is present in the eluate.

3.3. Extraction of Halogenated Phenoxyacids From Aqueous Samples

This method describes the extraction of some halogenated phenoxyacids from aqueous samples. The polymer imprinted with 2,4,5-T shows selectivity toward a small group of halogenated phenoxyacids, strictly resembling the template in their molecular structures (*see* **Fig. 2**): 2,4-D, fenoprop, dichlorprop, MCPA, and mecoprop. Other halogenated phenoxyacids and related esters are not recognized (*see* **Table 1**). The same happens for acidic pesticides characterized by molecular structures very different from the template.

The extraction protocol is based on three distinct steps (*see* **Fig. 3**): (1) column loading, in which the analytes and other hydrophobic substances are retained and hydrophilic neutral substances and inorganic ions are eluted; (2) column washing, in which the analytes are retained and other hydrophobic substances are eluted; (3) column elution, in which the analytes are recovered.

1. Filtrate about 5 mL aqueous sample on 0.22-μm polyethylene or Teflon membranes mounted on cartridges (*see* **Note 18**). If the filtered sample is viscous, dilute 1:1 with deionized water.

Table 1
Recoveries on MISPE Column Using 10 mL Deionized Water Containing 2 mg/mL of Analytes

	Herbicides, recovery percentage for 20 mg in 10 mL deionized water							
	2,4,5-T	2,4-D	Fenoprop	Dichlorprop	MCPA	Mecoprop	Atrazine	Bentazone
Washing, first milliliter	n.d.	n.d.	n.d.	n.d.	n.d.	n.d.	98.5 ± 3	98 ± 2
Washing, second milliliter	n.d.	n.d.	n.d.	n.d.	n.d.	n.d.	n.d.	n.d.
Washing, third milliliter	n.d.	n.d.	n.d.	n.d.	6.5 ± 2.5	10.5 ± 3.5	n.d.	n.d.
Elution, first milliliter	97.5 ± 4	98.5 ± 3.5	96.5 ± 4	95 ± 7.5	87 ± 7	88.5 ± 8	n.d.	n.d.
Elution, second milliliter	6 ± 2	n.d.	n.d.	n.d.	n.d.	n.d.	n.d.	n.d.
Elution, third milliliter	n.d.	n.d.	n.d.	n.d.	n.d.	n.d.	n.d.	n.d.
Recovery percentage	103.5 ± 6	98.5 ± 3.5	96.5 ± 4	95.0 ± 7.5	87.0 ± 7	88.5 ± 8	0	0

Source: Data taken from ref. *16*. The recovery was calculated excluding the herbicide found in methanol after column washing. n.d., not detectable.

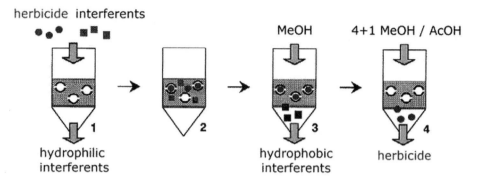

Fig. 3. Scheme of a typical molecularly imprinted solid-phase extraction.

2. Adjust the pH to 3.0 with 0.1 N aqueous hydrogen chloride (*see* **Note 19**). Samples containing organic solvents typically need to be diluted with water. The actual concentration of organic solvent that can be tolerated without affecting the column capacity must be less than 10% (v/v).
3. Rinse the column with 3 mL deionized water, pH 3.0, applying a flow rate of 3–5 mL/min.
4. Apply 1 mL sample (*see* **Note 20**) to the column with the vacuum turned off.
5. Aspirate the sample through the column at a flow rate of approx 1 mL/min or allow it to flow by gravity (*see* **Note 21**).
6. Wash the column three times with 1 mL methanol at a flow rate of approx 1 mL/min (*see* **Note 22**).
7. After the wash methanol has been passed through the column, dry the column under full vacuum for 2–5 min.
8. Turn off the vacuum and wipe the tips of the manifold needles to remove any residual wash solvent (*see* **Note 23**).
9. Insert collection tubes into the manifold rack.
10. Elute the column three times with 1 mL methanol–acetic acid or methanol–trifluoroacetic acid 4:1 (v/v) (*see* **Notes 22** and **24**) at a flow rate of approx 1 mL/min and collect the analytes (*see* **Note 25**).
11. Evaporate the eluate in a rotating evaporator or under a gentle stream of nitrogen.
12. Rinse the collection tubes three times under sonication with 0.200 mL of a solution suitable for the analytical detection (phosphate–borate buffer for capillary electrophoresis, acetonitrile–water 2:3 v/v containing 25 mM formic acid for liquid chromatography) (*see* **Note 26**).
13. Store the samples at −20°C.

3.4. Determination of Extracted Analytes by Capillary Electrophoresis

There is a large body of literature on the detection and quantification of halogenated phenoxyacids by capillary electrophoresis or liquid chromatography. The examples reported here (*see* **Figs. 4** and **5**) were chosen because of their simplicity. It should be noted that the capillary electrophoresis method is not suitable for a mass

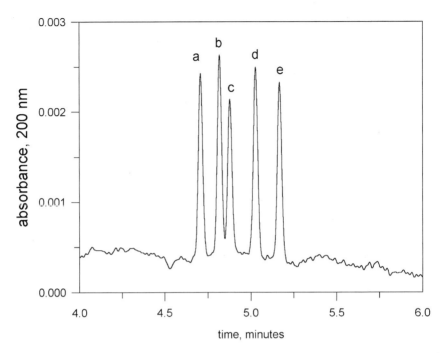

Fig. 4. Capillary electrophoresis of a 5-mL tap water sample spiked with a mixture containing 2 μg of a, dichlorprop; b, fenoprop; c, mecoprop; d, 2,4-D; and e, 2,4,5-T.

spectroscopy detector because of the presence of cyclodextrins in the running buffer; the liquid chromatography method can be easily adapted.

3.4.1. Capillary Electrophoresis

1. Prepare standard solutions of halogenated phenoxyacids in concentrations ranging from 0.25 to 25.0 mg/L by dilution of mother solutions with the running buffer (*vide infra*).
2. Filter 0.5 mL of each standard through a 0.22-μm polyethylene filter in suitable polypropylene autosampler vials (*see* **Note 27**).
3. Before each run (*see* **Note 28**), rinse the capillary (fused silica 50-μm id, 50 cm total length) with 0.1*M* sodium hydroxide for 90 s, with ultrapure water for 120 s, and with run buffer for 120 s.
   Running conditions: running buffer is phosphate–borate buffer (50 m*M* sodium phosphate monobasic, pH corrected to 7.5 with sodium tetraborate, containing 2.75 m*M* α-, β-, and γ-cyclodextrin); capillary thermostated at 25°C; hydrodynamic injection mode (0.07 MPa/s) for 3 s; cathodic run with applied voltage to 20 kV (current of about 60 μA); 200-nm UV detection at the cathodic end of the capillary.
4. For each standard and sample, perform a minimum of three runs. For each sample, integrate the peaks and calculate the mean area (*see* **Note 29**).
5. Build a calibration line by plotting the mean peak areas vs the concentrations of the standard solutions.

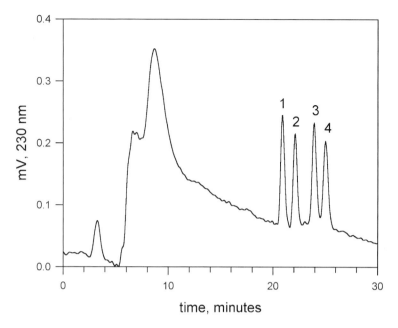

Fig. 5. Liquid chromatography of a tap water sample extracted on the MISPE column. 5 mL spiked with a mixture containing 2 μg of 2,4-D (1), MCPA (2), dichlorprop (3) and mecoprop (4).

6. The concentration of the phenoxyacids in the samples is given by the formula $d \times (V1/V2)$ $\times (A \times b_0)/b1$, where d is the dilution factor of the sample (**Subheading 3.3.**, **steps 1** and **2**); $V1$ is the volume of the sample introduced in the MISPE column (**Subheading 3.3.**, **step 4**); $V3$ the overall volume of the solution used to recover the eluate after the evaporation step (**Subheading 3.3.**, **step 12**); A is the mean peak area; and b_0 and b_1 are the intercept and the slope of the calibration line, respectively.
7. Use common statistical tests to verify method accuracy and reproducibility.

3.4.2. Liquid Chromatography

1. Prepare standard solutions of halogenated phenoxyacids in concentrations ranging from 0.20 to 20.0 mg/L by dilution of mother solutions with eluent A (*vide infra*).
2. Filter 0.5 mL of each standard through a 0.22-μm polyethylene filter in suitable polypropylene autosampler vials (*see* **Note 27**).
3. Before each run (*see* **Note 30**), equilibrate the column with eluent A following the absorbance at 230 nm until it falls below 0.001 UAFS and is stable for more than five column volumes.
   Elution conditions: column thermostated at 25°C; 5-μL loop injection; linear gradient from 60% acetonitrile–40% water (containing 25 mM formic acid, eluent A) to 80% acetonitrile–20% water (containing 25 mM formic acid, eluent B) in 30 min; mobile phase flow rate 1 mL/min; 230-nm UV detection.
4. For each standard and sample, perform a minimum of three runs. For each sample, integrate the peaks and calculate the mean area.

5. Build a calibration line by plotting the mean peak areas vs the concentrations of the standard solutions.
6. The concentration of the phenoxyacids in the samples is given by the formula $d \times (V1/V2) \times (A \times b_0)/b_1$, where d is the dilution factor of the sample (**Subheading 3.3., steps 1 and 2**), $V1$ is the volume of the sample introduced in the MISPE column (**Subheading 3.3., step 4**); $V3$ is the overall volume of the solution used to recover the eluate after the evaporation step (**Subheading 3.3., step 12**); A is the mean peak area; and b_0 and b_1 are the intercept and the slope of the calibration line, respectively.
7. Use common statistical tests to verify method accuracy and reproducibility.

4. Notes

1. 2,4,5-T cannot be appreciably dissolved in deionized water. To obtain a solution rapidly, dissolve 2,4,5-T in methanol and then slowly add the proper amount of water under continuous stirring or sonication. Direct dissolution of 2,4,5-T in the hydroalcoholic mixture is less efficient.
2. Commercial 4-VP and EDMA are stabilized with alkylphenols to suppress polymerization. They must be purified by distillation under reduced pressure immediately before use. Liquid–liquid extraction with diluted sodium hydroxide solutions is not convenient with the water-soluble 4-VP and less efficient with EDMA.
3. Other suitable radical initiators are 2,2′-azobis-(2,2,3-trimethylbutyronitrile) and 2,2′-azobis-(3-methoxy-2,2,3-trimethylbutyronitrile). When these initiators are used instead of AIBN, the polymerization temperature must be decreased to 50°C and 30°C, respectively.
4. Molecular oxygen is a powerful inhibitor of radical polymerization. Attention should be paid to maintain an inert atmosphere during the polymerization stage. Nitrogen can be conveniently substituted with argon or helium.
5. Imprinted polymers can be obtained by photoinduced radical polymerization using UV light sources (high- or medium-pressure mercury vapor lamps with 366-nm line of emission). The polymerization process is noticeably faster, and useful polymers are obtained within 1–6 h working at room temperature or in an ice bath. It should be taken into account that materials prepared by photopolymerization show enhanced selectivity but reduced sample capacity with respect to materials prepared by thermopolymerization.
6. Radical polymerization is an exothermic process. To avoid dangerous superheating of the polymerization mixture (with risk of an explosive runaway), do not try to prepare very large quantities of polymer (exceeding 250 mL of prepolymerization mixture) at once.
7. Very frequently, the polymeric monolith cannot be easily taken out of the test tube. In this case, smash the test tube with a steel spatula. The use of hammers or related devices should be avoided because of the risk of loss of polymer and contamination by glass fragments.
8. The polymer should be a white to very pale blue homogeneous monolith, sometimes with cracks. Excessive translucency, bubbles, or marked discontinuities in the polymeric texture indicate incomplete polymerization. Yellow-to-reddish appearance indicates 4-VP oxidation. Such polymers should be discarded.
9. The grinding/sieving step produces many very fine, submicrometric particles. This dust is potentially dangerous for the lungs. Undertake the operation under a hood or wear a suitable antidust mask. A porcelain mortar can also be used for manually grinding the polymer. However, this option is suitable only with a limited amount of material and takes a longer time.

10. Very fine, submicrometric powder could cause excessive high back pressures in the SPE column during and after the packing procedure. Multiple washings are fundamental to obtain dust-free polymeric particles.

11. Polymers for SPE applications require exhaustive Soxhlet extraction with the highly polar glacial acetic acid to obtain materials sufficiently free from residual template. Do not use methanol or acetonitrile as an extracting solvent because the amount of residual template in the polymer makes this material suitable for column chromatography but not for SPE.

12. Imprinted polymers are very stable under ordinary laboratory conditions and are resistant to acidic or basic conditions, organic solvents, high temperature (not exceeding 150°C), and pressure. They can be stored dry at room temperature for an indefinite amount of time without loss of binding properties. Avoid chemical oxidation.

13. Changing solvents from acetic acid to methanol and water causes cycles of polymer swelling/shrinking, with subsequent release of residual template molecules. We observed that 2,4,5-T-imprinted polymers are less prone to slow release of residual template than other imprinted polymers.

14. The suspension should be very homogeneous without colored particles or visible aggregates. After transferring the suspension in the column, polymer particles remaining in the test tube can be recovered by washing with some aliquots of methanol.

15. Glass SPE columns are expensive, but they can be accurately washed and recycled. Polyethylene SPE columns are very economical and commercially available in many formats but are more easily contaminated by several chemicals.

16. Never let the packed layer of polymer dry. Control the vacuum aspiration, slowly sucking the liquid layer of mobile phase. Close the stopcock immediately before the column runs dry. Extractions of analytes performed on dry columns are less reproducible and recoveries are not quantitative. If the column goes dry, it should be discarded, recovering the polymer eventually.

17. When pushing the inlet frit, do not press the polymer layer. A compressed layer offers more flow resistance to the liquid samples, and it makes column elution more difficult and, sometimes, very slow.

18. In most cases, particulates and other insoluble material can rapidly clog the column. The use of filtration membranes is efficient for elimination of these materials. It should be taken into the account that many biomacromolecules are too small to be retained by the membrane and potentially may interfere with the SPE and successive detection of the analytes. Avoid direct extraction of biological fluids, applying this protocol without a proper sample pretreatment.

19. The extraction column works efficiently only when aqueous samples are acidified. Aqueous hydrogen chloride may be substituted by diluted nitric or phosphoric acid. Avoid organic acids such as acetic, trifluoroacetic, or chloroacetic acid because they interfere with the retention mechanism. If the sample is buffered, use more concentrated acid.

20. Sample volumes between 0.5 and 10 mL do not require modifications to the method. Extraction of water samples with a volume up to 10 mL requires optimization of the washing and extraction volumes.

21. A slow rate of sample application is generally preferable because the kinetics of sorbent-analyte interactions is highly variable and strongly influenced by the composition of the sample.

22. Three small aliquots of solvent ensure improved analyte recovery rather than a single, larger volume aliquot.

23. Drying removes all traces of residual sample and washes solvent from the polymer. This produces a more concentrated final extract with a constant volume and composition and ensures better reproducibility and higher recoveries. Avoid overdrying the column because analyte recovery may decrease with excessive vacuum or drying times above 5 min. Note that dry columns look distinctly different from moist ones. Make a note of the column appearance before and after conditioning.

24. Trifluoroacetic acid is more efficient for eluting the analytes, but its methanolic solutions are unstable and should be prepared each day.

25. Allow the elution solvent to remain and soak the polymeric bed (without vacuum) for approx 1 min, which may slightly increase the analyte recovery.

26. Evaporation under a stream of nitrogen is simpler and can be performed simultaneously on several samples using a multiple nitrogen sparger. There is a risk of cross-contamination between samples. Repeated rinsing under continuous sonication is necessary to ensure quantitative recoveries.

27. The standard solutions of halogenated phenoxyacids should be prepared each day from stock solutions and fresh running buffer.

28. The protocol is reported for a Bio-Rad Capillary Electrophoresis model BioFocus 2000. Different instruments could require slightly different procedures.

29. Many commercial software programs calculate peak area corrected for the time corresponding to the peak maximum. This correction is necessary because in capillary electrophoresis the velocity of the analytes is not constant.

30. The protocol is reported for an Econosphere 25 cm × 4.6 mm id filled with 5-µm C18 reversed-phase packing (Alltech). Different C18 columns require slightly different elution conditions.

References

1. Thurman, E. M. and Mills, M. S. (1986) *Solid-Phase Extraction—Principles and Practice*. Wiley-Interscience, New York.
2. Pichon, V. (2000) Solid-phase extraction for multiresidue analysis of organic contaminants in water. *J. Chromatogr. A* **885,** 195–215.
3. Wells, M. J. M. and Yu, L. Z. (2000) Solid-phase extraction of acidic herbicides. *J. Chromatogr. A* **885,** 237–250.
4. Bruzzoniti, M. C., Sarzanini, C., and Mentasti, E. (2000) Preconcentration of contaminants in water analysis. *J. Chromatogr. A* **902,** 289–309.
5. Hage, D. S. (1998) Survey of recent advances in analytical applications of immunoaffinity chromatography. *J. Chromatogr. B* **715,** 3–28.
6. Van Emon, J. M., Gerlach, C. L., and Bowman, K. (1998) Bioseparation and bioanalytical techniques in environmental monitoring. *J. Chromatogr. B* **715,** 211–228.
7. Hennion, M. C. and Pichon, V. (2003) Immuno-based sample preparation for trace analysis. *J. Chromatogr. A* **1000,** 29–52.
8. Sellergren, B. (ed.). (2000) *Molecular Imprinted Polymers—Man Made Mimics of Antibodies and Their Applications in Analytical Chemistry*. Elsevier, Amsterdam.
9. Haupt, K. (2003) Imprinted polymers—tailor-made mimics of antibodies and receptors. *Chem. Commun.* **21,** 171–178.
10. Zimmerman, S. C. and Lemcoff, N. G. (2004) Synthetic hosts via molecular imprinting—are antibodies realistically possible? *Chem. Commun.* **7,** 5–14.
11. Andersson, L. I. (2000) Molecular imprinting for drug bioanalysis. A review on the application of imprinted polymers to solid-phase extraction and binding assay. *J. Chromatogr. B* **739,** 163–173.

12. Lanza, F. and Sellergren, B. (2001) Molecularly imprinted extraction materials for highly selective sample cleanup and analyte enrichment. *Adv. Chromatogr.* **41,** 138–173.
13. Martin-Esteban, A. (2001) Molecularly imprinted polymers: new molecular recognition materials for selective solid-phase extraction of organic compounds. Fresenius J. Anal. Chem. **370,** 795–802.
14. Ensing, K., Berggren, C., and Majors, R. E. (2001) Selective sorbents for solid-phase extraction based on molecularly imprinted polymers. *LC–GC* **10,** 942–954.
15. Baggiani, C., Giraudi, G., Giovannoli, C., Trotta, F., and Vanni, A. (2000) Chromatographic characterization of molecularly imprinted polymers binding the herbicide 2,4,5-trichlorophenoxyacetic acid. *J. Chromatogr. A* **883,** 119–126.
16. Baggiani, C., Giovannoli, C., Anfossi, A., and Tozzi, C. (2001) Molecularly imprinted solid-phase extraction sorbent for the clean-up of chlorinated phenoxyacids from aqueous samples. *J. Chromatogr. A* **938,** 35–44.

32

Automated Trace Analysis of Pesticides in Water

Euphemia Papadopoulou-Mourkidou, John Patsias, and Anna Koukourikou

Summary

An automated system appropriate for the analysis of a variety of chemical classes of pesticides and conversion products in water is presented. The system is based on the on-line solid-phase extraction (SPE) of target solutes followed by high-performance liquid chromatography (HPLC) tandem photodiode array/postcolumn derivatization/fluorescence detection (PDA/PCD/FLD). SPE is carried out on Hysphere-GP (10 × 2 mm) reversed-phase cartridges using the automated PROSPEKT system. Aliquots of 100 mL acidified (pH 3.0) samples are processed. The on-line HPLC analysis is carried out on a C18 analytical column eluted with a binary gradient of 10 mM phosphate buffer and an acetonitrile/water (90:10 v/v) mixture. Solutes are tentatively identified by tandem PDA/PCD/FLD. The PDA detector is used as a general-purpose detector; the FLD is useful for confirmation of N-methyl-carbamate and N-methyl-carbamoyloxime pesticides, which are postcolumn derivatized with o-phthalaldehyde and N,N-dimethyl-2-mercaptoethylamine-hydrochloride after hot alkaline hydrolysis. The entire analytical system is controlled and monitored by a single computer operated by the chromatographic software. Solute recoveries from water samples are better than 80%, except for some highly polar and highly hydrophobic compounds; for these compounds, the respective recoveries are in the range of 10–80%. The limits of identification, based on the PDA detector, are in the range 0.02–0.1 µg/L for the majority of target analytes; the limits of detection of the PCD/FLD system are in the range 0.005–0.02 µg/L for all amenable analytes.

Key Words: Automation; liquid chromatography; pesticides; postcolumn derivatization; solid-phase extraction; water analysis.

1. Introduction

Automation is of significant importance in modern pesticide trace analysis in water, and fully automated systems operated in user-unattended mode have been reported. Most of them are based on combined solid-phase extraction (SPE) of analytes from water samples with high-performance liquid chromatographic (HPLC) analysis involving on-line elution of the adsorbed analytes from the SPE cartridges by the flow of the mobile phase of the HPLC system (1–4). SPE on reversed-phase materials pro-

From: *Methods in Biotechnology, Vol. 19, Pesticide Protocols*
Edited by: J. L. Martínez Vidal and A. Garrido Frenich © Humana Press Inc., Totowa, NJ

vides, when compared to conventional liquid–liquid preconcentration techniques, significant advantages, such as less solvent consumption, limited operator involvement, and high potential for automation, especially when combined with on-line HPLC analysis. Automated systems based on on-line SPE–gas chromatographic (GC) analysis have also been reported *(5,6)*, but their application range is limited because parent compounds of many chemical groups of pesticides, especially those of the herbicides, and most major known conversion products of pesticides, are not GC amenable because of low volatility and increased thermolability. Even though GC and liquid chromatography (LC) are viewed as supplementary techniques in pesticide trace analysis, liquid chromatography is the first choice, especially regarding analysis of polar and thermolabile compounds *(7)*.

Analyte detection in LC analysis has been performed traditionally using an ultraviolet-visible (UVis) detector. Even though this detector is of medium sensitivity and selectivity, it is preferred because of the ease and low cost of operation, the good signal stability, the high application range, and in the case of use of a photodiode array detector, the on-line acquisition of UV spectra is possible, providing valuable information for the tentative identification of sought analytes. Depending on the physicochemical properties of the target analytes, other means of detection have also been exploited. Fluorescence detectors are quite selective and sensitive; however, their application range in pesticide trace analysis is very limited because a few pesticides exhibit inherent fluorescence; these are used mainly in combination with a derivatization step, producing fluorescent derivatives before (precolumn) or after (postcolumn) the liquid chromatographic analysis *(1,2)*. Mass spectrometric detectors have also been introduced for analyte detection in liquid chromatographic analysis, and many applications have been already reported *(8–14)*.

In this chapter, a multiresidue analytical method for pesticide determination in water at trace levels based on the on-line coupling of SPE and HPLC is described. Analyte detection is performed by photodiode array and fluorescence detectors operated in tandem mode. The photodiode array detector is used as a general-purpose detector; the fluorescence detector is considered only for the confirmation of *N*-methyl-carbamate and *N*-methyl-carbamoyloxime insecticides after hot alkaline hydrolysis and derivatization with *o*-phthalaldehyde and *N,N*-dimethyl-2-mercaptoethylamine-hydrochloride with the derivatization step performed on-line right after the elution of solutes from the diode array detector.

2. Materials

2.1. Reagents

1. Stock solutions of target pesticides: Prepare stock solutions of individual analytes at 1 mg/mL in methanol; add a few drops of 3 N hydrochloric acid in solutions of hydroxyatrazine, hydroxysimazine, and simazine to facilitate dissolution. Stock solutions are prepared annually and stored in Teflon-lined capped vials under deep freeze ($<-20°C$).
2. Mixed working calibration solutions: Prepare mixed calibration solutions at concentrations of 50, 25, 10, 5, 2.5, 1, 0.5, 0.25, 0.1, and 0.05 µg/mL in methanol; calibration solutions are prepared every 2 mo and stored under deep freeze (*see* **Note 1**).
3. Mobile phase solvent A: prepare a 10 mM KH_2PO_4/H_3PO_4 aqueous buffer of pH 3.2; prepare solution daily.

4. Mobile phase solvent B: Prepare an acetonitrile/water (90:10 v/v) mixture.
5. Derivatization reagent 1: Prepare a 0.2% NaOH aqueous solution, pH 13.0.
6. Derivatization reagent 2: Prepare a 0.01% *o*-phthalaldehyde and 0.2% *N,N*-dimethyl-2-mercaptoethylamine-hydrochloride solution dissolved in an aqueous buffer consisting of 0.3% boric acid and 0.09% NaOH at pH 9.1; prepare solution daily.
7. Prepare 3 *N* sulfuric acid and 3 *N* hydrochloric acid solutions.
8. Membrane filters for aqueous solutions with 0.2-μm pore diameter.
9. A reversed-phase C18 analytical column (150 × 4.6 mm, dp = 5 μm) (e.g., Nucleosil 100/5 C18 reversed-phase analytical column, Macherey-Nagel, Duren, Germany).
10. Disposable SPE cartridges suitable for the PROSPEKT automated solid-phase extraction system (e.g., Hysphere-GP, 20 × 2 mm, Spark Holland, Emmen, The Netherlands).

2.2. Instrumentation

1. PROSPEKT automated solid-phase extraction system (e.g., Spark Holland) or other equivalent.
2. HPLC system appropriate for at least linear binary gradient elution.
3. Autosampler suitable for loop flow injections in HPLC systems equipped with a six-port, two-position switching valve and a 20-μL or lower injection loop.
4. UV photodiode array detector (e.g., model 996, Waters, Milford, MA) with spectrum resolution of at least 1.2 nm.
5. Millenium software (Waters) for data treatment.
6. Postcolumn derivatization system (e.g., model Prometheus 300, Rigas Labs, Thessaloniki, Greece) equipped with two reaction coils, with the first coil heated and thermostated with an accuracy of ±0.1°C; two pulse-dampened reagent pumps with one-way valves in the reagent influent lines; and a 500-psi adjustable backpressure regulator. All parts of the system must be constructed by proper material to withstand strong alkaline conditions.
7. Fluorescence detector (e.g., model Spectroflow 980 fluorescence detector, Kratos, Ramsey) with variable and adjustable excitation and emission wavelengths.
8. Personal computer operated under a proper HPLC software supporting the installed analytical equipment.
9. The HPLC system must be connected with the PROSPEKT system in such a way that the adsorbed analytes on the SPE cartridge to be eluted by the mobile phase flow onto the analytical column in the back-flush mode. A flow diagram of the analytical system is presented in **Fig. 1**.

3. Methods
3.1. Protocol and Overview
3.1.1. Sample Preparation

Centrifuge water samples for the precipitation of suspended solids; centrifugation for 30 min at 3000*g* is proposed as the starting point (*see* **Note 2**). Transfer the supernatant to a clean glass container and adjust the pH to 3.0 by addition of 3 *N* sulfuric acid (*see* **Note 3**).

3.1.2. SPE of Samples

Program the PROSPEKT to perform sample SPE on disposable reversed-phase cartridges following four steps: (1) cartridge conditioning with a water-miscible organic solvent, followed by pure water; (2) sample loading; (3) cartridge washing; and (4) on-line analyte desorption.

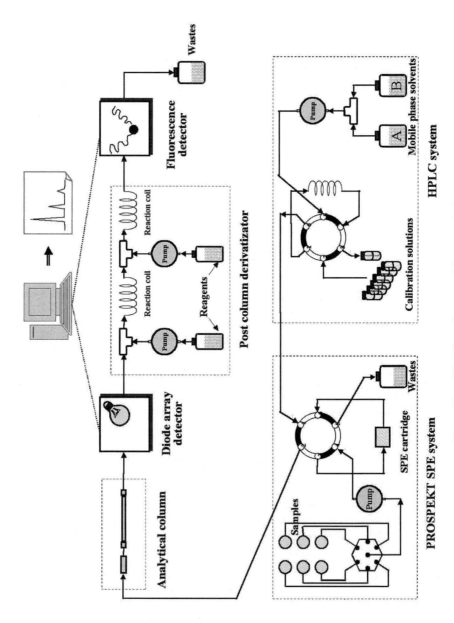

Fig. 1. Simplified flow diagram of the analytical system.

Table 1
Mobile Phase Composition and Flow Rate Settings

Time (min)	Solvent B (%)	Flow rate (mL/min)
00	5	1.00
03	20	1.00
35	100	1.00
40	100	1.40
45	5	1.25
50	5	1.00

The proposed method was validated by using 100-mL sample aliquots extracted on Hysphere-GP (10×2 mm) disposable reversed-phase SPE cartridges at a flow rate of 4 mL/min. SPE cartridges were previously conditioned by elution with 10 mL acetonitrile and 10 mL water at a flow rate of 4 mL/min. After sample loading, SPE cartridges were washed with 2 mL of water at a flow rate of 2 mL/min and placed on-line with the mobile phase flow of the HPLC system in such a way that the adsorbed analytes were eluted onto the analytical column in the back-flush mode (*see* **Note 4**).

3.1.3. HPLC Analysis

Prepare the HPLC system equipped with an alkaline-bonded silica reversed-phase analytical column eluted with a binary mobile phase consisting of aqueous buffer/acetonitrile mixture.

The proposed method was validated using a C18 reversed-phase analytical column (1.5×4.6 mm, dp = 5 µm) and a binary gradient mixture of a 10 mM phosphate buffer at pH 3.2 (solvent A) and an acetonitrile/water (90:10 v/v) mixture (solvent B) as a mobile phase; the gradient composition and the flow rate settings are presented in **Table 1** (*see* **Notes 5** and **6**).

3.1.4. Photodiode Array Detection

Operate the photodiode array detector in the range 190–400 nm and at a spectrum resolution of 1.2 nm. The data acquisition rate must be in compliance with the width of the eluted peaks; a value of 2 spectra/s is proposed (*see* **Note 7**). After data acquisition, chromatograms at various wavelengths (depending on the absorption spectra of the target analytes) are reconstructed, and the eluted peaks with a signal-to-noise ratio (S/N) ≥ 3 are located. Analyte detection is based both on retention time and UV absorption spectra match. Spectra from the top of each chromatographic peak, properly corrected with the background spectrum, are compared with target analyte reference spectra stored in custom-created libraries, and matches above a predetermined threshold fit and within a predetermined retention time window are reported (*see* **Note 8**). The identification limits of the target analytes in water are determined as the lower concentration levels at which the respective peaks could be detected at $S/N \geq 3$ and identified according to their UV spectra.

3.1.5. Postcolumn Derivatization of the N-Methyl-Carbamates

The postcolumn derivatization system is connected at the outlet of the photodiode array detector. Hot alkaline hydrolysis of the N-methyl-carbamates and N-methyl-carbamoyloximes producing methylamine is performed in a 300-μL reaction coil thermostated at 100°C by a 0.2% NaOH aqueous solution at pH 13.0 (reagent 1) introduced before the reaction coil at a flow rate of 0.3 mL/min. The produced methylamine is subsequently reacted in a second 500-μL coil operated at room temperature with a solution consisting of 0.01% o-phthalaldehyde and 0.2% N,N-dimethyl-2-mercaptoethylamine-hydrochloride dissolved in an aqueous buffer consisting of 0.3% boric acid and 0.09% NaOH at pH 9.1 (reagent 2) introduced before the reaction coil at a flow rate of 0.3 mL/min to produce a highly fluorescent indole derivative (*see* **Note 9**).

3.1.6. Fluorescence Detection

Fluorescence detection of the N-methyl-carbamates and N-methyl-carbamoyloximes after their postcolumn derivatization is performed using an excitation wavelength of 330 nm and an emission wavelength of 465 nm. The data acquisition rate, as for the photodiode array detector, must be in compliance with the width of the eluted peaks; a value of 2 measurements/s is proposed. Analyte detection is based strictly on the retention time of the eluted peaks (*see* **Note 10**). The detection limits of the target analytes in water are determined as the lower concentration levels at which the respective elution peaks could be detected at $S/N \geq 3$.

3.1.7. Analyte Quantification

Detected analytes are quantified via external standard calibration curves prepared by loop flow injections of the mixed working calibration solutions; 10-point external standard calibration curves corresponding to a 1- to 1000-ng injected amount (for 20-μL injections) are prepared for each analyte and for both detectors when appropriate (N-monomethyl-carbamates). The concentration of the pesticides in the samples is given by the formula:

Analyte concentration (μg/L) =

$\qquad$ Calculated analyte amount (absolute value, ng)/Sample volume (mL)

3.1.8. Automation

The analytical system is fully automated and can be operated user unattended. This is accomplished either by employing a proper software controlling all different parts of the analytical equipment or via the "external events" of the PROSPEKT system delivering simple run/stop commands to other devices of the analytical system ordering the run of preprogrammed triggered cycles.

3.1.9. Method Validation

After the selection of operating conditions, the performance of the analytical method is fully validated by calculating all performance parameters like chromatographic behavior (retention time, peak width, and asymmetry factor of the eluted peaks), recoveries, and detection limits for all target analytes (*see* **Notes 11** and **12**). Method variability and reproducibility are also calculated. Complete method validation is subsequently required only when a significant change concerning the instrumentation or the operating parameters that seriously affects the performance of the method is made. During normal continuous operation, the performance of the analytical method is recorded by analyzing daily three calibration solutions and three spiked samples made in environmental water at various concentrations and calculating the performance parameters mentioned above. Their values must be in the range of the predetermined windows; otherwise, the method must be fully revalidated.

3.2. Work Program for Routine Sample Analysis

1. Prepare mobile phase solvents; filter water and acetonitrile prior to solvent preparation; degas mobile phase solvents by bath sonication for at least 10 min in case of an on-line mobile phase degasser is not available. In the last case, during normal daily operation mobile phase solvents are further degassed by purging helium through the mobile phase reservoirs for 5 min for each hour of operation.
2. Turn system on; increase linearly the mobile phase flow to the operating value within 5 min or less; turn the first reaction coil heater on and wait for steady-state temperature to be achieved (*see* **Note 13**). Turn the flow of the derivatization reagents on and wait for the signal of the fluorescence detector to be stabilized; the analytical system is then ready for operation.
3. Analyze a calibration solution and a spiked field sample and make sure that method performance parameters are in the range of the desired levels.
4. Proceed with daily routine sample analysis sequence by analyzing a mixed working calibration solution and a spiked field sample every five unknown field samples; the concentrations of calibration solutions and those of spiked field samples must alternate between high and low levels of calibration and measurement, respectively.
5. To turn the analytical system off, proceed as follows: Turn the first reaction coil heater off and wait for its temperature to decrease below 50°C; turn the flow of the derivatization reagents off; decrease slowly the mobile phase flow rate to zero within 5 min or less; turn the system off.

4. Notes

1. Calibration solutions are made in mixtures of 10–30 target analytes, taking into account their retention times from the selected chromatographic column so that solutes included in each calibration mixed solution are adequately separated with resolution (R) values between adjacently eluting solutes higher than 1.
2. Precipitation of the suspended solids is required to avoid problems with clogging of the SPE cartridges during sample loading. Sample filtration must be avoided when highly hydrophobic pesticides are included in the target analyte list because they are amenable to adsorption onto membrane filter surfaces. When hydrophobic compounds are not included in the target analyte list of the indented user, filtration through a membrane filter with a pore size of 0.45 μm or lower is preferred.
3. Adjustment of sample pH at 3.0 or lower is required when acidic pesticides are included in target analyte list to suppress their ionization and therefore increase their affinity on reversed-phase SPE sorbents. Because sample acidification also increases the amount of the coextracted humic material contained in the field water samples, when acidic compounds are not included in the target analyte list, pH modification is not required.

4. The type of the sorbent material, the dimensions of the SPE cartridge, the elution mode, and the sample volume play a predominant role not only on the retention capacity of the target analytes, but also on the chromatographic performance of the overall analytical system *(15)*. Therefore, the initial selection and later modification of these parameters must be accompanied by a complete and consistent optimization and validation process. The type of the SPE sorbing material is selected according to the polarity and water solubility of the target solutes, and the sample volume is selected mainly according to the desired detection level, taking into account the sensitivity of the detector(s) used. By increasing sample volume, better detection limits are obtained when solutes are quantitatively retained. Polar solutes, however, are not quantitatively retained when a sample volume above the respective breakthrough volumes is adopted because of their early breakthrough from the SPE cartridge. In many cases, however, sample volumes above the breakthrough volumes of the target solutes result in increased method sensitivity for polar compounds regardless of their reduced extraction recoveries *(15,16)*. On the contrary, highly hydrophobic solutes exhibit higher extraction recoveries as the sample volume is increased *(15,17)*, probably because of the partial saturation of the secondary retention sites on the surfaces of the transfer lines prior to introduction onto the SPE cartridge. Sample volume also plays a predominant role in the shape of the eluted peaks. The width and asymmetry factor of the eluted peaks are increased with increasing sample volume. This is more significant in the case of cartridges packed with sorbent materials of high retention capacity (i.e., Hysphere-GP) and less significant or negligible for other types of sorbents with lower retention capacity *(15)*. Cartridge geometry (length and diameter) plays a predominant role in the shape of the eluted peaks. Large-dimension cartridges contain more sorbent material and provide higher solute retention; however, they result in poor chromatographic performance when coupled on-line with HPLC analysis. Long cartridges provide wide elution peaks from the analytical column; thus, cartridges longer than 10 mm are rarely employed. Cartridge diameter also must be below a specific percentage of the diameter of the analytical column for narrow elution peaks to be achieved because of better refocusing of the eluted analytes onto the analytical column. The effect is more intense in the case of forward-flush elution and in the case of cartridges packed with sorbents of high retention capacity when compared to the sorbing capacity of the packing material of the analytical column (as in the case of cartridges packed with Hysphere-GP material and C18 analytical column) *(15)*. In any case, back-flush elution provides narrower analyte elution peaks and must be preferred.

5. Acidification of the mobile phase is required when acidic pesticides are to be analyzed to suppress their ionization. Phosphate buffer has been preferred because of its low UV absorbance. Acidification of the mobile phase is also suppressing the ionization of the residual silanol groups on the bonded-silica surface and permits the usage of lower cost, not end-capped analytical columns. The retention times of ionizable pesticides of acidic (phenoxy-alkanoic acids, sulfonylureas, picolinic acids, imidazolinones, and other herbicides) and basic (i.e., triazine herbicides) behavior depend strongly on the mobile phase pH, and it is therefore strongly recommended that the pH of the mobile phase be adjusted precisely to obtain reproducible data. When acidic compounds are not included in the target analyte list, acidification of the mobile phase is not required; however, in this case end-capped analytical columns or proper mobile phase modifiers should be included to avoid secondary retention phenomena caused by the presence of ionized residual silanol groups.

6. A different type of reversed-phase analytical column with respect to the packing material or column dimensions and different mobile phase composition or gradient regime could

also be used according to the intended user's special needs and availability of new commercial chromatographic products. Every change in the operational parameters, however, should be first optimized and validated carefully, taking into account not only the chromatographic performance of the column, but also the compatibility with the selected sorbent type and geometry of the SPE cartridges (*see* **Note 4**).

7. Increasing the acquired spectrum range, spectrum resolution, and acquisition rate increases the detection quality; however, the amount of raw data and the required time for data processing are also increased. During initial selection or later modification of these parameters, the storage capacity and the processing rate of the available computer hardware must be taken into account.

8. UV absorption maxima of some analytes are affected by the analytical conditions used during data acquisition (mainly the mobile phase pH), and it is therefore strongly recommended to compare spectra acquired only under the same experimental conditions.

9. The flow rates of the derivatization reagents must be optimized with respect to the HPLC mobile phase flow rate for maximum sensitivity to be achieved.

10. Even though the derivatization reactions are very selective for the *N*-methyl-carbamates and *N*-methyl-carbamoyloximes, interferences from the matrix still appear in the respective chromatograms *(1,2)*, and data derived from the fluorescence detector cannot be used solely for analyte identification; data derived from the fluorescence detector are used in complementary support of the findings of the photodiode array detector.

11. The proposed method using the experimental conditions described in **Subheading 3.** and operated on a system consisting of the apparatus described in **Subheading 2.** and shown in **Fig. 2** has been validated for more than 170 pesticides and conversion products; included among the target solutes are representative members of most chemical families of pesticides, also including 23 *N*-methyl-carbamate and *N*-methyl-carbamoyloxime compounds amenable for postcolumn derivatization/fluorescence detection. The target analyte list, however, can be expanded furthermore depending on the special needs of the intended users. Analyte recoveries derived from the analysis of spiked water samples are presented in **Tables 2** and **3**. Recovery values are better than 80% for the majority of targeted analytes, except for the highly polar and the highly hydrophobic compounds. The highly polar analytes clopyralid, picloram, hydroxysimazine, and hydroxyatrazine exhibit reduced recoveries (11, 49, 19, and 75%, respectively, at a concentration level of 2 µg/L; *see* **Table 2**), because of their early breakthrough from the SPE cartridges. The recoveries of the highly hydrophobic compounds (organochlorines, pyrethroids, hydrophobic organophosphates, and other hydrophobic compounds) are below 80% because of sorption onto surfaces of transfer lines of the SPE system; recoveries increase as the sample volume increases *(15,17)*. Analyte identification limits calculated on the basis of data acquired by the photodiode array detector are presented in **Table 2**. Values range from 0.02 to 0.1 µg/L for the majority of the targeted analytes. Fluorescence detection provides better detection limits (0.005–0.02 µg/L) for amenable compounds; however, as mentioned in **Note 7**, these findings cannot solely support analyte identification in water, and the identification limits of the proposed analytical method are those provided by the photodiode array detector. A sample chromatogram derived from the analysis of a field water sample (River Loudias, Macedonia, Greece) is presented in **Fig. 2**. By means of the photodiode array detector, atrazine and molinate (peaks 2 and 3, respectively) at respective concentrations of 0.03 and 15.9 µg/L are determined. In the same sample, caffeine (peak 1) is also detected and quantified (0.67 µg/L). The "hill-type" peak in the first half of the chromatogram (**Fig. 2A**) is caused by elution of humic substances present in sur-

Table 2
Retention Times t_R, Quantification Wavelengths λ_{quan}, Recoveries % (Relative Standard Deviation [RSD] %), and Detection Limits of the Target Analytes

No.	Analyte	t_R (min)	λ_{quan} (nm)	Recovery % (RSD %, $n = 3$) 0.2 µg/L	2 µg/L	Detection limit (µg/L)
1	Clopyralid	4.4	221, 280	12 (10)	11 (4)	0.20
2	Aldicarb sulfoxide	7.7	215	99 (8)	97 (2)	0.10
3	Picloram	7.7	223	54 (11)	49 (3)	0.10
4	Aminocarb	8.2	215	101 (14)	97 (3)	0.20
5	Butoxycarboxim	8.7	215	99 (11)	98 (3)	0.10
6	Aldicarb sulfone	8.8	215	105 (12)	98 (4)	0.10
7	Hydroxysimazine	8.8	240	21 (9)	19 (6)	0.10
8	Oxamyl	8.9	220	103 (9)	98 (3)	0.10
9	Methomyl	9.1	233	102 (9)	100 (2)	0.10
10	Imazapyr	10.1	215, 240	108 (10)	100 (5)	0.10
11	Desisopropylatrazine	10.4	214	108 (4)	99 (1)	0.02
12	Hydroxyatrazine	10.6	241	78 (7)	75 (5)	0.05
13	Dicamba	11.1	215	108 (11)	100 (5)	0.05
14	Carbendazim	11.5	215	91 (10)	86 (7)	0.10
15	Metamitron	12.2	215, 308	106 (8)	101 (2)	0.02
16	Imidacloprid	12.5	270	99 (5)	97 (3)	0.05
17	Mevinphos 1	12.5	215	104 (5)	99 (3)	0.10
18	Diketometribuzin	12.6	259	90 (4)	86 (1)	0.02
19	3-OH carbofuran	12.9	215	94 (8)	91 (6)	0.10
20	Desethylatrazine	12.9	215	105 (4)	100 (1)	0.02
21	Chloridazon	13.0	228, 285	108 (6)	100 (2)	0.02
22	Metolachlor sulfonic acid	13.5	215	106 (7)	99 (2)	0.10
23	Desaminodiketometribuzin	13.5	260	102 (4)	100 (2)	0.02
24	Dioxacarb	13.7	215	101 (8)	95 (2)	0.10
25	Dimethoate	13.9	215	96 (7)	95 (5)	0.10
26	Mevinphos 2	14.5	216	99 (6)	96 (3)	0.10
27	Butocarboxim	15.1	215	104 (6)	99 (3)	0.10
28	Oxycarboxin	15.2	253	103 (6)	99 (3)	0.05
29	Desaminometribuzin	15.4	237	103 (6)	100 (3)	0.02
30	Aldicarb	15.6	215	96 (8)	94 (5)	0.10
31	Bentazon	15.7	240, 310	104 (7)	98 (2)	0.10
32	Metoxuron	16.3	215, 245	103 (6)	99 (2)	0.02
33	Imazamethabenz	16.4	215	102 (6)	100 (2)	0.05
34	Bromacil	16.4	215, 276	103 (8)	100 (4)	0.05
35	Pirimicarb	16.7	219	102 (3)	99 (1)	0.05
36	Terbacil	16.8	215, 280	107 (7)	101 (3)	0.05
37	Metolcarb	16.9	215	100 (10)	99 (1)	0.10
38	Paraoxon methyl	16.9	215, 273	102 (2)	99 (1)	0.05
39	Imazaquin	17.2	249	103 (7)	99 (3)	0.05
40	Metribuzin	17.2	215, 294	102 (4)	99 (2)	0.02
41	Simazine	17.3	222	104 (6)	101 (3)	0.02
42	Carbofuran	17.8	215	99 (2)	98 (2)	0.10
43	2,4-D	18.0	229, 284	104 (4)	100 (1)	0.10
44	Thophanate methyl	18.0	215, 265	88 (12)	83 (7)	0.05

(continued)

Table 2 *(Continued)*
Retention Times t_R, Quantification Wavelengths λ_{quan}, Recoveries % (Relative Standard Deviation [RSD] %), and Detection Limits of the Target Analytes

		t_R	λ_{quan}	Recovery % (RSD %, $n = 3$)		Detection limit
No.	Analyte	(min)	(nm)	0.2 µg/L	2 µg/L	(µg/L)
45	Bendiocarb	18.5	215, 276	100 (8)	99 (2)	0.10
46	Simetryne	18.5	222	102 (3)	99 (1)	0.02
47	Propoxur	18.7	215	101 (4)	100 (1)	0.10
48	MCPA	18.7	228, 279	101 (6)	98 (2)	0.10
49	Thiodicarb	18.9	235	105 (3)	101 (2)	0.10
50	Triasulfuron	19.2	223, 286	100 (7)	99 (1)	0.05
51	Carbaryl	19.7	215	100 (7)	99 (1)	0.02
52	Thiofanox	19.9	215	99 (11)	96 (8)	0.20
53	Terbumeton	19.9	219	103 (8)	100 (3)	0.05
54	Fluometuron	20.0	243	102 (5)	100 (1)	0.02
55	Ethiofencarb	20.2	215	89 (5)	86 (1)	0.10
56	Chlortoluron	20.2	215, 243	102 (4)	99 (1)	0.02
57	Carboxin	20.2	252, 293	107 (9)	101 (1)	0.05
58	2,4,5-T	20.3	215, 288	106 (5)	100 (3)	0.05
59	Atrazine	20.3	222	103 (3)	101 (1)	0.02
60	Dichlorprop	20.4	230, 285	109 (6)	101 (1)	0.10
61	Isoproturon	20.4	215	103 (5)	98 (3)	0.02
62	Monolinuron	20.6	245	102 (7)	100 (2)	0.02
63	Isoprocarb	20.9	215	107 (6)	98 (2)	0.10
64	Hydroxymetolachlor	20.9	215	103 (9)	100 (3)	0.10
65	2,4-Dichlorophenol	21.2	229, 286	102 (8)	100 (4)	0.05
66	Diuron	21.2	250	104 (3)	10.0 (1)	0.02
67	Paraoxon ethyl	21.4	215, 273	103 (5)	98 (2)	0.05
68	Metobromuron	21.5	248	102 (8)	101 (1)	0.02
69	Propachlor	21.8	215	106 (6)	101 (2)	0.10
70	Ametryne	22.3	222	106 (7)	100 (1)	0.02
71	Fenobucarb	22.4	215	106 (9)	100 (5)	0.10
72	Thiophanate ethyl	22.5	215, 267	089 (10)	85 (5)	0.05
73	Nuarimol	22.5	215	100 (6)	97 (3)	0.05
74	Methiocarb	22.6	215	101 (8)	97 (1)	0.10
75	Propazine	22.9	222	101 (6)	101 (2)	0.02
76	Promecarb	23.1	215	103 (7)	101 (4)	0.10
77	Propanil	23.1	215, 250	103 (5)	99 (1)	0.02
78	Dicloran	23.2	215, 360	101 (7)	98 (4)	0.05
79	2,4-DB	23.4	230, 285	98 (5)	97 (1)	0.10
80	Phosmet	23.5	215	102 (6)	100 (1)	0.05
81	Desmedipham	23.9	236	102 (6)	98 (2)	0.05
82	Linuron	24.0	215, 249	102 (7)	100 (1)	0.02
83	Terbuthylazine	24.1	223	102 (7)	101 (1)	0.02
84	2,4,5-Trichlorophenol	24.3	215, 292	100 (6)	96 (2)	0.02
85	Fenarimol	24.3	215	107 (8)	101 (2)	0.05
86	Naptalam	24.5	222	108 (9)	102 (3)	0.02
87	Ethofumesate	24.6	226, 279	106 (4)	101 (2)	0.10
88	Azinphos methyl	24.8	223, 285	104 (5)	100 (2)	0.10

(continued)

Table 2 *(Continued)*
Retention Times t_R, Quantification Wavelengths λ_{quan}, Recoveries % (Relative Standard Deviation [RSD] %), and Detection Limits of the Target Analytes

No.	Analyte	t_R (min)	λ_{quan} (nm)	Recovery % (RSD %, $n = 3$) 0.2 µg/L	2 µg/L	Detection limit (µg/L)
89	Propyzamide	24.9	215	106 (7)	101 (4)	0.05
90	Prometryne	24.9	222	106 (5)	99 (2)	0.05
91	Molinate	25.0	215	103 (6)	101 (2)	0.10
92	Diflubenzuron	25.1	215, 257	106 (9)	100 (1)	0.02
93	Napropamide	25.1	215, 292	105 (8)	100 (3)	0.05
94	Phenmedipham	25.1	236	93 (9)	89 (7)	0.05
95	Fenoxycarb	25.2	228, 279	105 (7)	100 (3)	0.05
96	Fenamiphos	25.2	215, 252	101 (6)	99 (2)	0.10
97	Bitertanol	25.2	215, 250	100 (4)	98 (2)	0.02
98	Parathion methyl	25.4	215	102 (3)	98 (1)	0.05
99	Alachlor	26.4	215	106 (9)	101 (2)	0.10
100	Acetochlor	26.6	215	107 (8)	98 (3)	0.10
101	Fenitrothion	26.7	215, 269	104 (3)	99 (1)	0.10
102	Metolachlor	26.7	215	103 (6)	101 (2)	0.10
103	Chlorothalonil	26.8	232	100 (5)	98 (3)	0.02
104	Diphenylamine	27.0	284	105 (7)	98 (6)	0.02
105	Azinphos ethyl	27.2	223, 286	103 (5)	99 (2)	0.10
106	Bupirimate	27.2	245	103 (4)	100 (1)	0.10
107	Penconazole	27.3	215	99 (7)	96 (3)	0.05
108	Diclofluanid	27.4	215	89 (10)	83 (7)	0.10
109	EPTC	27.6	215	104 (7)	100 (3)	0.10
110	Tolyfluanid	28.5	215	97 (8)	93 (5)	0.10
111	Iprodione	28.6	217, 249	105 (7)	100 (3)	0.10
112	Fenthion	28.7	215	101 (6)	98 (3)	0.10
113	Prochloraz	28.8	215	103 (6)	99 (2)	0.10
114	Parathion ethyl	29.0	215, 275	102 (3)	99 (2)	0.05
115	Pyrazophos	29.2	245	103 (4)	100 (1)	0.02
116	Hexaflumuron	29.4	215, 255	103 (10)	97 (6)	0.02
117	Benzoylprop ethyl	29.6	215	103 (2)	100 (1)	0.10
118	Pirimiphos methyl	30.2	247	101 (4)	98 (2)	0.02
119	Chlorpyriphos methyl	30.5	229, 289	100 (7)	97 (3)	0.05
120	Cycloate	30.5	215	105 (4)	100 (1)	0.10
121	Chlorobenzilate	30.8	215	102 (7)	99 (2)	0.10
122	Thiobencarb	30.9	220	93 (7)	89 (2)	0.05
123	Pretilachlor	31.0	215	102 (7)	99 (2)	0.10
124	Dialifos	31.5	222	96 (9)	95 (6)	0.05
125	Methoxychlor	31.9	228	107 (8)	101 (4)	0.10
126	Chloropropylate	32.3	225	108 (10)	102 (6)	0.10
127	Furathiocarb	32.5	215, 281	85 (8)	82 (2)	0.10
128	Diclofop methyl	32.6	215	92 (4)	89 (1)	0.10
129	Tetradifon	33.2	215	96 (8)	94 (5)	0.05
130	Temephos	33.3	215, 252	68 (11)	63 (6)	0.10
131	Fluazifop butyl	33.3	224, 268	87 (5)	85 (3)	0.10
132	Tralkoxydim	33.3	254	110 (9)	99 (3)	0.10

(continued)

Table 2 *(Continued)*
Retention Times t_R, Quantification Wavelengths λ_{quan}, Recoveries % (Relative Standard Deviation [RSD] %), and Detection Limits of the Target Analytes

No.	Analyte	t_R (min)	λ_{quan} (nm)	Recovery % (RSD %, $n = 3$) 0.2 µg/L	2 µg/L	Detection limit (µg/L)
133	Pendimethalin	33.5	240	92 (7)	89 (4)	0.05
134	Dicofol	33.7	230	76 (11)	71 (7)	0.05
135	Oxadiazon	33.8	215, 292	101 (8)	99 (1)	0.05
136	Benfluralin	33.9	215, 275	83 (9)	80 (5)	0.05
137	Trifluralin	34.1	215, 274	78 (8)	77 (5)	0.05
138	Pirimiphos ethyl	34.2	247	99 (7)	90 (2)	0.02
139	Carbophenothion ethyl	34.5	215, 263	64 (8)	61 (7)	0.05
140	Flucythrinate	35.5	215	39 (11)	34 (7)	0.10
141	λ-Cyhalothrin	35.7	215	39 (10)	37 (5)	0.10
142	Isopropalin	35.8	215, 242	76 (9)	70 (4)	0.05
143	*p,p′*-DDT	35.9	236	58 (10)	52 (6)	0.10
144	Fenvalerate	36.8	215	46 (9)	42 (4)	0.10
145	*p,p′*-DDE	37.1	248	52 (11)	47 (7)	0.10
146	Decamethrin	37.1	215	48 (8)	41 (3)	0.10
147	*o,p′*-DDT	37.5	235	49 (12)	45 (6)	0.10
148	Permethrin-*trans*	37.8	215	44 (7)	40 (5)	0.10
149	Permethrin-*cis*	38.1	215	39 (10)	36 (6)	0.10
150	*o,p′*-DDE	38.4	246	47 (10)	41 (7)	0.10
151	Bifentrin	38.6	215	46 (7)	40 (4)	0.10
152	Carbosulfan	39.0	215, 282	59 (8)	54 (4)	0.10

Data were recorded by the diode array detector. Validation data were calculated by analyzing spiked samples made in Axios River water samples.

face water samples. Its position on the chromatogram depends on the pH of the mobile phase *(17)*. By means of the fluorescence detector, 3-OH-carbofuran is determined at a concentration of 0.009 µg/L. However, the presence of this compound in the sample was not confirmed by the photodiode array detector as the concentration was lower than its respective identification limit, and its detection by the fluorescence detector alone is of limited value. Two other peaks (annotated by an asterisk) not corresponding to any compound from the target analyte list are also recorded. Another chromatogram derived from the analysis of a groundwater sample is presented in **Fig. 3**. Carbaryl is detected in the sample by both fluorescence and diode array detectors at concentrations of 0.11 and 0.12 µg/L, respectively. In the inset, the UV adsorption spectrum match provided by the photodiode array detector is shown. The detection of carbaryl by both detectors increases the possibility for its positive identification.

12. In unknown samples, poorly separated solutes or solutes with the same retention time are discriminated on the basis of their UV spectra. In this respect, the postcolumn hydrolysis, derivatization, and fluorescence detection of eluted solutes provide an additional piece of information taken into account for the tentative identification of solutes. Among the target analytes included in **Table 2**, there are certain coeluting solutes that are difficult to be discriminated on the basis of their respective UV spectra; such a case is the analysis of the

Table 3
Retention Times tR, Recoveries % (Relative Standard Deviation [RSD] %), and Detection Limits for N-Methyl-Carbamates and N-Methyl-Carbamoyloximes Spiked in Water Samples

No.	Analyte	t_R (min)	Recovery % (RSD %) (0.2 µg/L)	2 µg/L	Detection limit (µg/L)
1	Aldicarb sulfoxide	8.7	91 (11)	87 (8)	0.010
2	Aminocarb	9.1	93 (15)	90 (7)	0.020
3	Butoxycarboxin	9.4	95 (8)	93 (3)	0.010
4	Aldicarb sulfone	9.6	98 (4)	96 (2)	0.010
5	Oxamyl	9.7	97 (5)	95 (4)	0.005
6	Methomyl	9.9	101 (6)	99 (3)	0.005
7	3-OH carbofuran	13.6	93 (7)	90 (4)	0.005
8	Dioxacarb	14.3	91 (9)	87 (7)	0.005
9	Butocarboxin	15.9	99 (4)	98 (3)	0.005
10	Aldicarb	16.2	94 (8)	92 (5)	0.005
11	Metolcarb	17.5	100 (4)	98 (2)	0.005
12	Carbofuran	18.4	99 (3)	98 (2)	0.005
13	Bendiocarb	19.3	100 (2)	99 (1)	0.005
14	Propoxur	18.8	101 (4)	99 (2)	0.005
15	Thiodicarb	19.5	100 (3)	99 (1)	0.005
16	Carbaryl	20.3	99 (2)	98 (1)	0.005
17	Thiofanox	20.5	95 (9)	93 (7)	0.010
18	Ethiofencarb	20.7	87 (9)	83 (6)	0.010
19	Isoprocarb	21.4	99 (4)	98 (3)	0.005
20	Fenobucarb	23.0	100 (3)	99 (1)	0.005
21	Methiocarb	23.2	98 (5)	97 (2)	0.005
22	Promecarb	23.7	100 (4)	98 (2)	0.005
23	Furathiocarb	33.1	83 (3)	82 (1)	0.005

Data were recorded by the fluorescence detector.

choroacetanilide herbicides alachlor, acetoclor, and metolachlor, which are commonly found in field water samples; in such cases, samples should be also analyzed by either HPLC or GC with mass spectrometry detection.

13. Care must be taken not to heat the reaction coil before the establishment of the mobile phase flow.

References

1. Patsias, J. and Papadopoulou-Mourkidou, E. (1999) A fully automated system for analysis of pesticides in water: on-line extraction followed by liquid chromatography–tandem photodiode array/postcolumn derivatization/fluorescence detection. *J. AOAC Int.* **82,** 968–981.
2. Hiemstra, M. and de Kok, A. (1994) Determination of *N*-methylcarbamate pesticides in environmental water samples using automated on-line trace enrichment with exchangeable cartridges and high performance liquid chromatography. *J. Chromatogr.* **667,** 155–166.

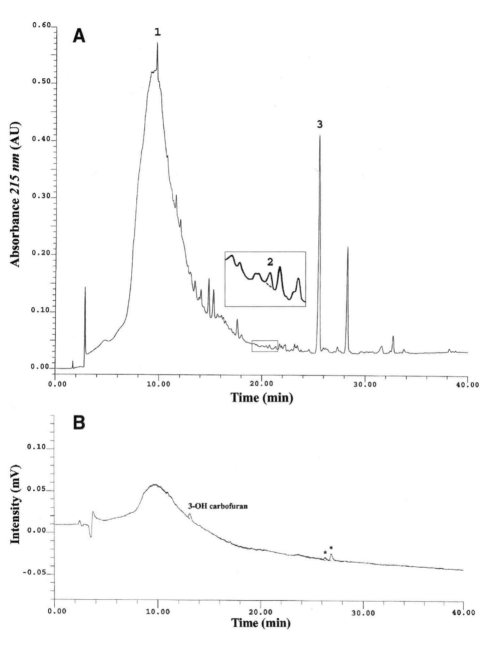

Fig. 2. Chromatogram from the analysis of a water sample taken from the river Loudias (Macedonia, Greece). (**A**) data recorded by the diode array detector at 215 nm; peaks are annotated as follows: 1, caffeine; 2, atrazine; 3, molinate. (**B**) data recorded by the fluorescence detector; *, matrix interferences (*see* **Note 11**).

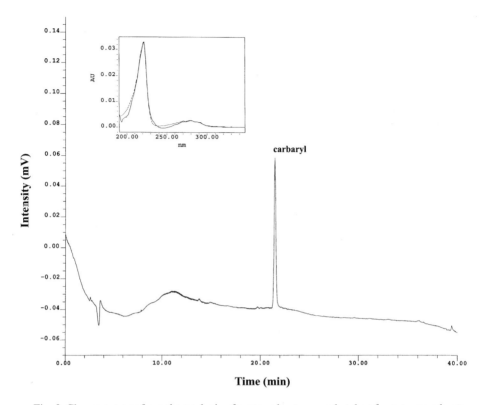

Fig. 3. Chromatogram from the analysis of a groundwater sample taken from a groundwater well from the region of Kavala (Macedonia, Greece). Data recorded by the fluorescence detector; the insert of the figure is the UV adsorption spectrum match provided by the diode array detector.

3. Slobodnik, J., Croenewegen, M. G. M., Brouwer, E. R., Lingeman, H., and Brinkman, U. A. T. (1993) Fully automated multi-residue method for trace level monitoring of polar pesticides by liquid chromatography. *J. Chromatogr.* **642,** 359–370.

4. Liska, I., Brouwer, E. R., Ostheimer, A. G. L., Lingeman, H., and Brinkman, U. A. T. (1992) Rapid screening of a large group of polar pesticides in river water by on-line trace enrichment and column liquid chromatography. *Int. J. Environ. Anal. Chem.* **47,** 267–291.

5. Pocurull, E., Aguilar, C., Borull, F., and Marce, R. M. (1998) On-line coupling of solid-phase extraction to gas chromatography with mass spectrometric detection to determine pesticides in water. *J. Chromatogr. A* **818,** 85–93.

6. Louter, A. J. H., van Beekvelt, C. A., Montanes, P. C., Slobodnik, J., Vreuls, J. J., and Brinkman, U. A. T. (1996) Analysis of organic microcontaminants in aqueous samples by fully automated on-line solid-phase extraction–gas chromatography–mass selective detection. *J. Chromatogr. A* **725,** 67–83.

7. Liska, I. and Slobodnik, J. (1996) Comparison of gas and liquid chromatography for analyzing polar pesticides in water samples. *J. Chromatogr. A* **733,** 235–258.

8. Santos, T. C. R., Rocha, J. C., and Barcelo, D. (2000) Determination of rice herbicides, their transformation products and clofibric acid using on-line solid-phase extraction followed by liquid chromatography with diode-array and atmospheric pressure chemical ionization mass spectrometric detection. *J. Chromatogr. A* **879,** 3–12.

9. Hu, J.-Y., Aizawa, T., and Magara, Y. (1999) Analysis of pesticides in water with liquid chromatography/atmospheric pressure chemical ionization mass spectrometry. *Water Res.* **33,** 417–425.

10. Crescenzi, C., Di Corcia, A., Guerriero, E., and Samperi, R. (1997) Development of a multiresidue method for analyzing pesticide traces in water based on solid-phase extraction and electrospray liquid chromatography mass spectrometry. *Environ. Sci. Technol.* **31,** 479–488.

11. Hogenboom, A. C., Jagt, I., Vreuls, R. J. J., and Brinkman, U. A. T. (1997) On-line trace level determination of polar organic microcontaminants in water using various pre-column–analytical column liquid chromatographic techniques with ultraviolet absorbance and mass spectrometric detection. *Analyst* **122,** 1371–1377.

12. Slobodnik, J., Hogenboom, A. C., Vreuls, J. J., et al. (1996) Trace-level determination of pesticide residues using on-line solid-phase extraction–column liquid chromatography with atmospheric pressure ionization mass spectrometric and tandem mass spectrometric detection. *J. Chromatogr. A* **741,** 59–74.

13. Chiron, S., Papilloud, S., Haerdi, W., and Barcelo, D. (1995) Automated on-line liquid–solid extraction followed by liquid chromatography–high-flow pneumatically assisted electrospray mass spectrometry for the determination of acidic herbicides in environmental waters. *Anal. Chem.* **67,** 1637–1643.

14. Slobodnik, J., van Baar, B. L. M., and Brinkman, U. A. T. (1995) Column liquid chromatography–mass spectrometry: selected techniques in environmental applications for polar pesticides and related compounds. *J. Chromatogr. A* **703,** 81–121.

15. Patsias, J. and Papadopoulou-Mourkidou, E. (2000) Development of an automated on-line solid-phase extraction high-performance liquid chromatographic method for the analysis of aniline, phenol, caffeine and various selected aniline and phenol compounds in aqueous matrices. *J. Chromatogr. A* **904,** 171–188.

16. Subra, P., Hennion, M. C., Rosset, R., and Frei, R. W. (1988) Recovery of organic compounds from large-volume aqueous samples using on-line liquid chromatographic preconcentration techniques. *J. Chromatogr.* **456,** 121–141.

17. Papadopoulou-Mourkidou, E. and Patsias, J. (1996) Development of a semi-automated HPLC-DAD system for screening pesticides at trace levels in aquatic systems of the Axios River basin. *J. Chromatogr. A* **726,** 99–113.

33

Gas Chromatography–High-Resolution Mass Spectrometry-Based Method for the Simultaneous Determination of Organotin Compounds in Water

Michael G. Ikonomou and Marc P. Fernandez

Summary

A gas chromatography–high-resolution mass spectrometry (GC–HRMS)-based method has been developed for the accurate and sensitive determination of nine organotin compounds in water: tetrabutyltin, tributyltin, dibutyltin, monobutyltin, triphenyltin, diphenyltin, monophenyltin, tricyclohexyltin, and dicyclohexyltin. Details on procedures for glassware treatment, sample conditioning, extraction, and derivatization are all provided. *In situ* derivatization with sodium tetraethylborate is employed to overcome difficulties associated with extracting ionic compounds into an organic solvent from an aqueous matrix. GC–HRMS conditions under selective ion monitoring mode and quality control criteria for the positive identification of analytes are presented. Linearity of response and minimal detectable limits are provided for each of the nine compounds, and the estimates of method limits of detection are 7–29 ng/L based on a 100-mL water sample. All GC–HRMS quantitative formulas for the calculation of internal standard recovery and recovery-corrected organotin concentration in unknown samples are provided. Method precision and accuracy are determined using spiked clean water samples as well as spiked seawater samples used in bioconcentration experiments.

Key Words: Antifouling paints; ethylation; GC–HRMS; organometallic compounds; seawater; speciation; tributyltin; ultratrace analysis.

1. Introduction

Organotin (OT) compounds, characterized by containing at least one carbon–tin bond and generally represented as $R_nSnX_{(4-n)}$ ($n = 1$–4; R = alkyl or aryl; X = H, OR', halogen, etc.), have been incorporated into many industrial and agricultural biocides and wood-preserving and antifouling agents. Because of the former widespread use of these compounds in such products, parent OT compounds and their breakdown products (i.e., di- and monosubstituted OT) have been found in both fresh and saltwater sources. The importance of accurate OT chemical speciation with a low part-per-bil-

From: *Methods in Biotechnology, Vol. 19, Pesticide Protocols*
Edited by: J. L. Martínez Vidal and A. Garrido Frenich © Humana Press Inc., Totowa, NJ

lion limit of detection (LOD) in environmental samples has been acknowledged by various scientists performing monitoring and ecotoxicological studies on these compounds. Most OT methods published determine environmental concentrations of tributyltin (TBT), dibutyltin (DBT), monobutyltin (MBT), triphenyltin (TPhT), diphenyltin (DPhT), monophenyltin (MPhT), tricyclohexyltin (TCyT), dicyclohexyltin (DCyT), tetrabutyltin (TeBT), fenbutatin oxide (FBTO), or various methyl-, ethyl-, or propyl-tin compounds (reviewed in **refs.** *1* and *2*).

Gas chromatography (GC) analysis is performed for OT analysis in most cases, utilizing one of many suitable detection methods, including atomic absorption spectrometry (AAS), atomic emission detection (AED), microwave-induced plasma–atomic emission spectroscopy (MIP–AES), flame photometric detection (FPD), pulsed flame photometric detection (PFPD), mass spectrometry (MS), and inductively coupled plasma–mass spectrometry (ICP–MS) (*1,3,4*; reviewed in **refs.** *5 7*). However, the main limitations and drawbacks of photometric emission/absorption-based detectors used in OT analysis include a lack of selectivity or sensitivity, limited dynamic range, or poor calibration stability. In addition, low-resolution MS suffers from interference in the low molecular mass region in which most diagnostic ions for OT appear *(2)*.

To date, high-resolution mass spectrometry (HRMS) has not been evaluated for OT analysis in environmental samples. This technique is very sensitive and extremely selective because of the narrow mass channels (10,000 resolution) used to monitor the target analytes. GC–HRMS-based analysis in conjunction with isotope dilution quantitation is one of the most robust methodologies used in ultra-trace environmental analyses. Also, because of the high specificity of the technique, minimum sample size is required for processing, and this is a substantial advantage in terms of minimizing potential matrix interferences and cost of analysis.

In this chapter, we describe a GC–HRMS-based methodology for the simultaneous determination in water of nine OT compounds: TeBT, TBT, DBT, MBT, TPhT, DPhT, MPhT, TCyT, and DCyT. Details such as (1) sample preparation, (2) instrumental analysis conditions, (3) quantitation and data treatment, and (4) overall method performance are presented.

2. Materials

1. Laboratory glassware: 100-mL graduated cylinders; 250-mL separatory funnels; 125-mL round-bottom flasks; hexane-rinsed 15-mL centrifuge tubes (polyethylene); precleaned glass microvials for GC analysis.
2. Magnetic sector high-resolution mass spectrometer (e.g., Micromass Autospec, EBE geometry).
3. Gas chromatograph (e.g., Hewlett-Packard HP 5890, Wilmington, DE) equipped with a split/splitless injector and a temperature-controlled GC–MS interface. The column used is a 30 m × 0.25 mm id (0.1-μm film) DB-5 capillary column (e.g., J&W, Folsom, CA).
4. OT standard mix: A mix of 1 ng cation/μL mono(*n*-butyl)tin trichloride (>99% pure), tri(*n*-butyl)tin monochloride (>99% pure), mono(*n*-phenyl)tin trichloride (>99% pure), di(*n*-phenyl)tin dichloride (>98% pure), tri(*n*-phenyl)tin monochloride (>99% pure), di(*n*-cyclohexyl)tin dichloride (>99% pure), and tri(*n*-cyclohexyl)tin monochloride (>99% pure); 2 ng cation/μL di(*n*-butyl)tin dichloride (>99% pure) (Quasimeme programme, Vrije Universiteit Amsterdam, Amsterdam, Netherlands); and 10 ng cation/μL tetra(*n*-

butyl)tin (>94% pure) (e.g., Alpha Aesar, Ward Hill, MA) in MeOH at −20°C protected from light for 4 months.

5. Method surrogate internal standard (ISTD): 1.00 ng cation/μL di(*n*-propyl)tin (DPrT) dichloride (e.g., Aldrich Chemical Company Inc., Milwaukee, WI) in MeOH at −20°C protected from light for 4 months.

6. Performance standard: 2.77 ng cation/μL tetra(*n*-pentyl)tin (TePeT; Quasimeme programme) in toluene at −20°C protected from light for 4 months.

7. Buffer: 1*M* NaOAc/acetic acid, pH approx 4.5 at room temperature.

8. Derivatization reagent: 1% (w/v) sodium tetraethylborate [NaB(Et)$_4$] (Alpha Aestar) in methanol made fresh as required. Note: this reagent is spontaneously flammable in air and therefore was prepared under nitrogen.

9. Miscellaneous reagents: cyclohexane (reagent grade or better); ultrahigh-purity helium and nitrogen; double-deionized water (on-site Milli-Q® water purification system); 2% nitric acid (HNO$_3$) bath; 5% dichlorodimethylsilane (DCDMS) in dichloromethane (DCM).

3. Methods

3.1. Sample Preparation

The preparation of liquid water samples for ultratrace GC-based analysis requires a number of steps: (1) glassware cleaning for ultratrace OT analysis; (2) sample spiking and ISTD addition; (3) sample conditioning, derivatization, and extraction; and (4) preparation of calibration standards.

3.1.1. Glassware Cleaning Protocol (see **Note 1**)

1. Rinse each piece of glassware with hot tap water several times.
2. Soak in 2% nitric acid (HNO$_3$), preferably overnight.
3. Rinse with hot tap water again.
4. Rinse in a laboratory dishwasher using distilled water without detergent.
5. Rinse twice with acetone and twice with DCM.
6. Treat with 5% DCDMS in DCM to deactivate the glassware surfaces.
7. Rinse twice with DCM to remove excess DCDMS.
8. Oven bake for 6 h at 325°C (*see* **Note 2**).
9. Rinse twice with acetone and twice with hexane before use.

3.1.2. Batch Composition, Method Validation, and Addition of Standards

1. Each sample processing batch should consist of 13 samples. The batch composition is a procedural blank (i.e., 100 mL double-deionized water or high-performance liquid chromatography-grade water); nine "real" 100-mL samples (*see* **Note 3**), of which one should be analyzed in duplicate; a spiked sample (i.e., 100 mL double-deionized water or high-performance liquid chromatography-grade water spiked with 50 μL OT standard mix); and a certified reference material (CRM) (*see* **ref. 8**) as well as a calibration standard (*see* **Subheading 3.1.4.**).

2. The overall analytical method is validated in terms of percentage recovery of the spiked compounds (i.e., accuracy), precision, and long-term stability using spiked samples analyzed in triplicate. Environmental water samples or matrix-free water samples need to be spiked with 50 μL of the OT standard mix (*see* **Note 4**) and be processed in triplicate through the entire method. Spiked samples should also be included in each batch in which real samples are to be processed.

3. For quality control purposes, a CRM for the compounds of interest should be included in the batch.
4. Add 100 μL of surrogate ISTD solution (*see* **Note 4**) to all samples, spikes, CRM, and blanks.

3.1.3. Sample Conditioning, Derivatization, and Extraction

1. Add 10 mL of NaOAc buffer to each sample so that the final pH of the sample is approx 4.5.
2. Shake samples and add 1 mL of the derivatization reagent (*see* **Note 5**).
3. Immediately shake samples for 1 min, add 50 mL hexane, and shake for 1 min again.
4. Allow samples to react at room temperature for 30 min and finally shake for 1 min.
5. Collect upper organic layer into a round-bottom flask and add another 0.5-mL aliquot of NaB(Et)$_4$ to the aqueous layer of each sample to ensure complete derivatization (*see* **Note 6**).
6. Repeat hexane extraction with an additional 50-mL aliquot.
7. Combine both organic extracts and reduce by rotary evaporation to approx 10 mL (*see* **Note 7**).
8. Transfer the reduced extract to a prerinsed centrifuge tube with hexane, add 2–3 drops of toluene, and evaporate under a gentle stream of nitrogen (temperature <30°C in both **steps 7** and **8**) to approx 0.3 mL (*see* **Note 7**).
9. Transfer the residue to a microvial suitable for GC analysis with toluene and add 10 μL of TePeT performance standard (PS) (*see* **Note 4**).

3.1.4. Preparation of Calibration Standard

1. Add 50 μL of the OT standard mixture (*see* **Note 4**) along with 100 μL ISTD (*see* **Note 4**) solution into a prerinsed 15-mL centrifuge tube.
2. Add 1 mL NaOAc buffer followed by 1 mL of 1% NaB(Et)$_4$ to this mixture and mix on a vortex stirrer.
3. This mixture is then extracted three times with 2 mL hexane.
4. The hexane extracts are combined and reduced to approx 300 μL, and 10 μL of PS (*see* **Note 4**) is then added.

3.2. GC–HRMS Analysis

To ensure high-quality data, rigorous quality assurance/quality control procedures need to be in place to ensure ultimate performance of the GC–HRMS system. These are (1) check if the instrument needs to be optimized, if necessary, daily to ensure that it is operating at 10,000 resolving power under both static and dynamic conditions; (2) check column resolution; (3) check ultimate sensitivity by analyzing the lowest concentration standard (*see* **Table 1**); (4) check linearity by analyzing a series of calibration standard solutions (*see* **Table 1**); (5) check sample carryover by injecting toluene after a calibration standard solution; and (6) check stability over the course of the analysis of real samples by analyzing calibration standard CS3 (*see* **Table 1**) every 8–10 h of operation. Although the information provided here may be adapted to other GC–MS techniques, both high and low resolution (*see* **Note 8**), HRMS provides benefits in terms of lower detection limits and higher analyte specificity. Both GC and HRMS conditions and parameters are provided along with details of the selected ion monitoring (SIM) method.

3.2.1. Operating Parameters for GC

GC conditions are optimized (*see* **Table 2**) to provide maximum separation and delivery of analyte species together with minimization of possible sample degradation throughout the run.

Table 1
Calibration Curves for the Nine Target OT Analytes

Compound	Concentration of calibration standard (CS) (µg cation/µL)							$R^2 [R_{(critical)}^2 = 0.3]$	MDL[a] (pg)
	CS1	CS2	CS3	CS4	CS5	CS6	CS7		
MBT	6	28	85	226	283	453	849	0.986	2.4
DBT	11	57	171	457	571	914	1714	0.998	1.8
TBT	5	26	78	208	260	416	780	0.996	2.4
TeBT	57	286	857	2286	2857	4571	8571	0.992	5.9
MPhT	6	30	89	238	297	475	891	0.998	0.13
DPhT	7	33	100	267	334	535	1003	0.991	0.53
TPhT	6	29	86	229	286	457	857	0.976	0.50
DCyT	6	30	91	242	303	485	909	0.985	1.4
TCyT	5	27	82	218	272	435	816	0.809	0.87

[a]MDL calculation based on a 1-µL injection.

Table 2
Operating Parameters for GC

Parameter	Value
Injection volume	1 µL sample plus 0.5 µL air
Carrier gas	Helium at constant pressure (~60 kPa)
Injector temperature	250°C
Column temperature program:	
Initial temperature	80°C
Ramp 1	5°C/min for 10 min
Ramp 2	10°C/min for 15 min
Final hold	280°C for 5 min

Table 3
Operating Parameters for HRMS

Parameter	Value
GC–MS interface temperature	250°C
Ionization source temperature	280°C
Electron impact ionization energy	33 eV
Detector voltage	350 V
HRMS resolution	10,000

3.2.2. Operating Parameters for HRMS (Magnetic Sector of EBE Geometry)

The mass spectrometer should be operated in the electron impact positive (EI$^+$) ionization mode. OT compounds under electron impact ionization conditions fragment readily, and the molecular ions are of very low intensity. The mass spectra are dominated by fragment ions that are characteristic for each analyte. In our previous publication (*see* **ref. 8**), we provided details on the profiles of the fragment ions of the analyte described in this procedure. The fragmentation is compound dependent and can vary depending on ion source temperature, ionization energy, and instrument geometry.

The temperature of the GC–MS interface, the ion source, and the ionization conditions are optimized to provide maximum sensitivity for the target analytes (**Table 3**). The aim is to establish conditions for controlled fragmentation (*see* **Note 9**) and to maximize the relative abundance of the diagnostic ions produced by each OT compound.

3.2.3. SIM Method

In the selection of SIM channels for each analyte, there is generally a trade-off between using the most intense ions and thus maximizing sensitivity, and choosing distinct ions (i.e., not present in any coeluting interference components) and thus minimizing interference. The ions selected for the SIM analysis of all nine OT analytes are listed in **Table 4**. Two distinct isotopic fragment ions containing tin (i.e., Sn^{118} and Sn^{120}) should be monitored for each analyte as a quality assurance/quality control measure to ensure positive target analyte identification (*see* **Subheading 3.3.1., step 3**).

Table 4
Summary of Positive Ions Monitored for Each OT Analyte by GC–HRMS

Analyte	194.94 m/z	196.94 m/z	233.05 m/z	235.05 m/z	231.04 m/z	233.04 m/z	261.08 m/z	263.08 m/z
TePeT[a]							$Sn^{118}(Pe)_2H$	$Sn^{120}(Pe)_2H$
DPrT[b]			$Sn^{118}(Pr)_2Et$	$Sn^{120}(Pr)_2(Et)$				
MBT			$Sn^{118}(Bu)(Et)_2$	$Sn^{120}(Bu)(Et)_2$				
DBT			$Sn^{118}(Bu)_2H$	$Sn^{120}(Bu)_2H$				
TBT			$Sn^{118}(Bu)_2H$	$Sn^{120}(Bu)_2H$				
TeBT			$Sn^{118}(Bu)_2H$	$Sn^{120}(Bu)_2H$				
MPhT	$Sn^{118}(Ph)$	$Sn^{120}(Ph)$						
DPhT	$Sn^{118}(Ph)$	$Sn^{120}(Ph)$						
TPhT	$Sn^{118}(Ph)$	$Sn^{120}(Ph)$						
DCyT					$Sn^{118}(Cy)(Et)H$	$Sn^{120}(Cy)(Et)H$		
TCyT					$Sn^{118}(Cy)(Et)H$	$Sn^{120}(Cy)(Et)H$		

Et, ethyl; Bu, butyl; Pe, pentyl; Cy, cyclohexyl; Ph, phenyl.
[a]Performance standard.
[b]Internal standard

459

Most ethylated OT analytes are resolved well by the chromatographic conditions recommended here (**Subheading 3.2.1.**). **Figure 1** shows the SIM (eight-channel) profiles of the OT standard mixture run by the GC–HRMS method. Because of similarities in boiling point and polarity to some degree, the phenyltins (F and H) coelute with cyclohexyltins (G and I) of the same order. However, because the phenyltins produce a predominant $[Sn(C_6H_5)]^+$ (monitored at 194.94 and 196.94 m/z) that occurs at a separate m/z than $[Sn(C_6H_{11})H_2]^+$ produced by the cyclohexyltins, these analytes can be quantified separately. Furthermore, the cyclohexyltins produce a predominant $[Sn(C_6H_{11})(C_2H_5)H]^+$ (monitored at 231.04 and 233.04 m/z) by EI$^+$ MS, which is free of interference from phenyltin fragment ions.

3.3. Data Treatment and Method Performance

Subheadings 3.3.1.–3.3.3. present relevant criteria for acceptance/rejection of GC–HRMS OT data, along with calibration procedures and formulae for calculating the recovery-corrected concentration of OT in water samples. **Subheading 3.3.4.** illustrates actual recoveries achieved for the ISTD and all nine targeted OT compounds spiked in clean water along with recoveries for the ISTD in actual seawater samples.

3.3.1. Quality Control Criteria for GC–HRMS Data

1. The HRMS must maintain a minimum resolution of 10,000 during the entire chromatographic run.
2. Two isotopes of the specific analyte must be detected at their exact m/z. Both of the isotopic peaks must be present, and their intensities must maximize within ±2 s of one another.
3. The ratio of the two isotopic peaks (i.e., Sn^{118}/Sn^{120}) must be within ±20% of their expected theoretical value (i.e., 0.73 ± 0.15). If this criterion is not met, it is very likely that the isotopic ion ratio is impacted by coeluting isobaric compounds. In such cases, the concentration of the target analyte should be flagged as not detected because of incorrect isotopic ratio.
4. The retention time of a specific analyte must be within 3 s to that obtained during analysis of the authentic compounds in the calibration standards.
5. Signal-to-noise ratio of each of the isotope m/z channels must be 3 or above.
6. To ensure that all samples are analyzed under optimum instrument performance, the samples are analyzed in a specifically designed batch. The batch composition follows this order:
   a. The CS3 standard is analyzed first to ensure that the relative response factors (RRFs) obtained are within ±20% of those established during analysis of the calibration curve standards (i.e., CS1 to CS7, **Table 1**).
   b. Following the CS3 analysis, 1 μL of toluene is injected to assess carryover, which should be less than 0.4% of the peak area of the preceding CS3 (*see* **Note 10**)
   c. 10 to 12 samples are analyzed.
   d. Subsequently, the CS3 solution is analyzed again, and the sequence is followed as indicated above. After 24 to 28 h of analysis, the instrument may be checked for resolution and sensitivity before another batch of samples is loaded for analysis.

3.3.2. Calibration and LODs

To ensure that analyte concentrations fall within the linear range of instrument response for each targeted OT, a multipoint calibration curve should be established at the beginning of the OT analysis. An example of the results of this type of multipoint

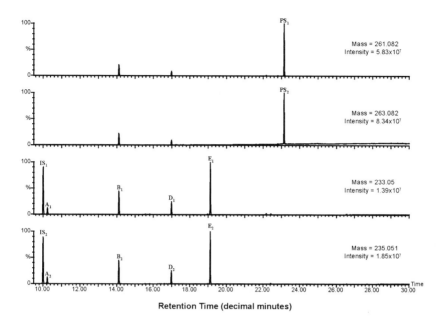

Retention Time (decimal minutes)

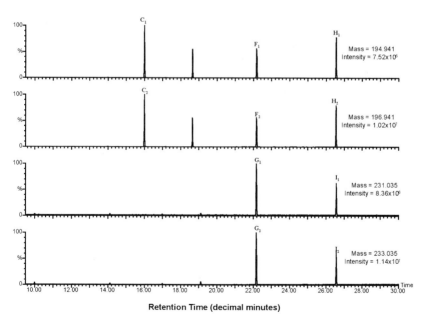

Retention Time (decimal minutes)

Fig. 1. SIM profiles of OT calibration standard (**Subheading 3.1.4.**) run by GC–HRMS. IS, internal standard TPrT at 10.00 min; PS, performance standard TePeT at 23.15 min; A, MBT at 10.24 min; B, DBT at 14.11 min; C, MPhT at 16.00 min; D, TBT at 17.00 min; E, TeBT at 19.12 min; F and G, DPhT and DCyT, respectively, at 22.18 min; H and I, TPhT and TCyT, respectively, at 26.55 min. Unidentified impurity at 18.65 min.

calibration is provided in **Table 1**. Linear calibration curves are established using seven concentrations of the OT standard mixture, each in triplicate with an additional blank (i.e., 0 μg/μL). The coefficient of determination R^2 value is given for the linear regression of each analyte together with the $(R_{critical})^2$ for significance at a 99% confidence level. This value is indicative of the goodness of fit of the relative response (response of analyte divided by response of ISTD) vs concentration data to the linear function $mX + b$.

In addition, minimum detection limits (MDLs) are provided for a signal-to-noise level of 3 based on the least concentrated calibration solution. MDLs range from 0.13 pg for MPhT to 5.9 pg for TeBT and are well within the range published for other sensitive and selective detection methods as evaluated by Aguerre et al. *(3)*. Method LODs are estimated at 7–29 ppt in water based on a 100-mL sample.

3.3.3. Quantitation

Quantitation of the nine OT analytes (TeBT, TBT, DBT, MBT, TPhT, DPhT, MPhT, TCyT, and DCyT) monitored in this work is based on the method of ISTD use, DPrT is used as the ISTD and a PS, TePeT, added following sample preparation, is used to determine the amount of ISTD lost throughout the sample preparation process. A single-point calibration is established using the calibration standard (prepared as described in **Subheading 3.1.4.**) run at the beginning of every instrument batch (*see* **Note 11**) resulting in RRF for all the nine analytes monitored relative to the ISTD, which in turn has an RRF relative to the PS. RRFs are calculated as indicated Equation (1b):

$$[A_1/A_2]/RRF_{(1\ to\ 2)} = [C_1/C_2] \tag{1a}$$

This rearranges to

$$RRF_{(1\ to\ 2)} = [C_1/C_2]/[A_1/A_2] \tag{1b}$$

where A is the summed response area of both SIM isotope channels, and C is the concentration of analyte in the solution analyzed. Subscript 1 refers to analyte and subscript 2 to ISTD (or PS if the analyte is ISTD).

Once RRFs are determined and linearity of response is established (**Subheading 3.3.2.**), Eq. (2) is used to determine the percentage recovery of the ISTD, and Eq. (3) is used to determine the recovery-corrected concentrations for each OT compound in the sample.

$$\text{Recovery }(\%) = [(27.7\ ng \times A_{IS})/(A_{PS} \times 100\ ng \times RRF_{IS\ to\ PS})] \times 100\% \tag{2}$$

$$*C = (100\ ng \times A_{unk}/A_{IS} \times RRF_{unk\ to\ IS})/SW \tag{3}$$

where $*C$ is the recovery-corrected concentration (nanograms per gram) of analyte in the sample, A is the peak area, RRF is the relative response factor, and SW is the sample weight (grams). Subscripts IS and unk refer to ISTD and unknown (i.e., sample), respectively.

All OT concentrations or amounts in this work should be based on weight of cationic species that is the convention for these compounds.

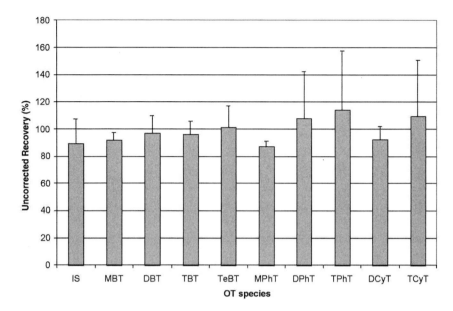

Fig. 2. Absolute recovery (not corrected for recovery of ISTD) of OT analytes from water by *in situ* derivatization with $NaB(Et)_4$ ($N = 4$, and error bars represent standard deviation) in DMQ water spiked with 50 µL of OT standard mixture.

3.3.4. OT Recovered in Water Sample

The average recovery for each of the nine target analytes is between 80 and 120%, and the total method standard deviation is less than 20% for the ISTD, first five target OTs, and DCyT (**Fig. 2**). However, TCyT, DPhT, and TPhT appear to suffer from high variability (*see* **Note 12**), which is also evident in the RRF variation (interbatch) reported in **Table 5**.

The method was also applied to the determination of OT compounds in TBT-spiked seawater samples using in a bioconcentration experiment, and the average ISTD recovery was determined as 94% (relative standard deviation 23%) for 14 samples and duplicates analyzed.

4. Notes

1. To eliminate the problem of high background levels because of analyte adsorption on glassware surfaces, commonly encountered in trace metal analysis, a rigorous cleaning procedure should be adopted for all routine glassware used in this work. Without acid washing and silanization (with DCDMS) of glassware, the OT analysis may suffer from background levels of MBT, DBT, TBT, TeBT, and MPhT. However, even when these procedures are adopted, the background levels of MBT and DBT may still be approx 4 ng and 0.5 ng, respectively. The levels of these two analytes in the procedural blanks may be quite consistent, in which case the samples may be easily background corrected for these two analytes.

Table 5
Average Relative Response Factor (RRFs) and Relative Retention Times
(RRTs) for the Target Compounds Based on Internal Standard (*N* = 6)

Analyte	RRT	RRF (RSD %)
DPrT (ISTD)	1.00	0.15^a (27)
MBT	1.02	0.36 (11)
DBT	1.41	0.48 (6)
MPhT	1.60	0.51 (37)
TBT	1.70	0.49 (14)
TeBT	1.91	0.18 (17)
DCyT/DPhT	2.22	1.23 (19)/0.61 (38)
TCyT/TPhT	2.66	0.90 (44)/0.63 (43)

RSD, relative standard deviation.
^{a}Relative to performance standard TePeT.

2. Because of potential loss of calibration of volumetric flasks on heating at high temperatures, this step, as well as prior silylation with DCDMS, should be avoided and replaced by simply air drying for this type of glassware.
3. Should preliminary analysis suggest that a sample contains OTs at levels outside the linear range of calibration, the initial sample amount may be adjusted so that the level of these OTs may be brought back into linear range for accurate quantitation.
4. Standard solutions should be brought out of cold storage and warmed to room temperature, followed by vortexing prior to dispensing.
5. To analyze OTs by GC, tri-, di-, and monosubstituted OT salts must first be made volatile by derivatization, which normally will involve introduction of alkyl or hydride group(s) to the Sn complex (*2*). Ashby et al. (*9*) developed a derivatization method employing sodium $NaB(Et)_4$ that converts ionic organotin compounds (i.e., $R_nSnX_{(4-n)}$) into nonpolar volatile derivatives (i.e., $R_nSn(Et)_{(4-n)}$) *in situ*, eliminating the problems associated with extracting ionic compounds into an organic solvent from an aqueous matrix. This technique of OT derivatization has replaced alkylation with Grignard reagents (*10–12*), which involves several steps, and is sensitive to even trace amounts of water present in the sample or sample extract (*13*).
6. In the presence of excessive nontarget consumption of $NaB(Et)_4$ as indicated by poor ISTD or spike recoveries or obvious presence of a large amount of dissolved, colloidal, or particulate matter, more of this reagent should be used as necessary.
7. Great care and patience must be exercised in performing this step as blowing the samples down too fast or bringing them down too low will result in poor recoveries.
8. For a comparison between magnetic sector HRMS and quadrupole MS for the nine OTs, *see* **ref. 8**.
9. Most EI^+ mass spectra that are published or included in MS libraries are generated at an electron energy of 70 eV (*14*). However, an electron energy of 33 eV should be used in this work. These softer ionization conditions are ideal to produce a favorable fragmentation pattern for all nine OT analytes.
10. An injection of toluene is made after the CS3 solution to verify that there is minimum carryover of analytes, that is, to check whether there are any residues in the system, including the injector, an etched syringe, GC column, GC–MS interface, and ion source. The intensities/response of all native compounds in the toluene should be less than 0.4% from those obtained from the analysis of the CS3 solution.

11. Percentage deviation from calibration is calculated as percentage difference between opening calibration and either middle calibration or final (or closing) calibration throughout the run. If this percentage deviation is greater than 20%, then RRFs should be calculated based on the nearest calibration standard in the run order.

12. This high variability could most likely be improved by the incorporation of a second ISTD that elutes closer to these analytes (retention time = 22–27 min).

Acknowledgments

We thank Tim He and Maike Fischer from DFO's Regional Dioxin Lab (RDL) for help with the analytical work and Fisheries and Oceans Canada and Environment Canada for financial support.

References

1. Bancon-Montigny, C., Lespes, G., and Potin-Gautier, M. (2000) Improved routine speciation of organotin compounds in environmental samples by pulsed flame photometric detection. *J. Chromatogr. A* **896**, 149–158.

2. Abalos, M., Bayona, J.-M., Compañó, R., Cgranados, M., Leal, C., and Prat, M.-D. (1997) Analytical procedures for the determination of organotin compounds in sediment and biota: a critical review. *J. Chromatogr. A* **788**, 1–49.

3. Aguerre, S., Lespes, G., Desauzeirs, V., and Potin-Gautier, M. (2001) Speciation of organotins in environmental samples by SPME–GC: comparison of four specific detectors: FPD, PFPD, MIP–AES and ICP–MS. *J. Anal. At. Spectrom.* **16**, 263–269.

4. Minganti, V., Capelli, R., and De Pellegrini, R. (1995) Evaluation of different derivatization methods for the multi-element detection of Hg, Pb and Sn compounds by gas chromatography–microwave induced plasma–atomic emission spectrometry in environmental samples. *Fresenius J. Anal. Chem.* **351**, 471–477.

5. De la Calle-Guntiñas, M. B., Scerbo, R., Chiavarini, S., Quevauviller, P., and Morabito, R. (1997) Comparison of derivatization methods for the determination of butyl- and phenyltin compounds in mussels by gas chromatography. *Appl. Organomet. Chem.* **11**, 693–702.

6. Tao, H., Rajendran, R. B., Quetel, C. R., Nakazato, T., Tominaga, M., and Miyazaki, A. (1999) Tin speciation in the femtogram range in open ocean seawater by gas chromatography/inductively coupled plasma mass spectrometry using a shield torch at normal plasma conditions. *Anal. Chem.* **71**, 4208–4215.

7. Vercauteren, J., Pérès, C., Devos, C., Sandra, P., Vanhaecke, F., and Moens, L. (2001) Stir bar sorptive extraction for the determination of ppq-level traces of organotin compounds in environmental samples with thermal desorption-capillary gas chromatography–ICP mass spectrometry. *Anal. Chem.* **73**, 1509–1514.

8. Ikonomou, M. G., Fernandez, M. P., He, T., and Cullon, D. (2002) Gas chromatography–high-resolution mass spectrometry based method for the simultaneous determination of nine organotin compounds in water, sediment and tissue. *J. Chromatogr. A* **975**, 319–333.

9. Ashby, J., Clark, S., and Craig, P. J. (1988) Methods for the production of volatile organometallic derivatives for application to the analysis of environmental samples. *J. Anal. At. Spectrom.* **3**, 735–736.

10. Arnold, C. G., Berg, M., Müller, S. R., Dommann, U., and Schwarzenbach, R. P. (1998) Determination of organotin compounds in water, sediments, and sewage sludge using perdeuterated internal standards, accelerated solvent extraction, and large-volume-injection GC/MS. *Anal. Chem.* **70**, 3094–3101.

11. Moens, L., Smaele, T. D., Dams, R., Van den Broeck, P., and Sandra, P. (1997) Sensitive, simultaneous determination of organomercury, -lead, and -tin compounds with headspace solid phase microextraction capillary gas chromatography combined with inductively coupled plasma mass spectrometry. *Anal. Chem.* **69,** 1604–1611.
12. Ceulemans, M., Witte, C., Lobinski, R., and Adams, F. C. (1994) Simplified sample preparation for GC speciation analysis of organotin in marine biomaterials. *Appl. Organomet. Chem.* **8,** 451–461.
13. Michel, P. and Averty, B. (1991) Tributyltin analysis in seawater by GC-FPD after direct aqueous-phase ethylation using sodium tetraethylborate. *Appl. Organomet. Chem.* **5,** 393–397.
14. Mosi, A. A. and Eigendorf, G. K. (1998) Current mass spectrometric methods in organic chemistry. *Curr. Organ. Chem.* **2,** 145–172.

34

Determination of Triazine Herbicides and Degradation Products in Water by Solid-Phase Extraction and Chromatographic Techniques Coupled With Mass Spectrometry

Hassan Sabik and Roger Jeannot

Summary

At least 29 triazines and 36 degradation products have been reported in the literature. For many reasons, no single method enables the simultaneous extraction and analysis of all of these compounds. Three methods are described for monitoring triazines and degradation products in filtered and raw water (0.5–1 L). Analytes are extracted from water samples by solid-phase extraction using cartridges packed with C18 bonded silica (Isolute Triazine) or graphitized carbon black (Carbopack B), and the eluate is analyzed by gas chromatography–mass spectrometry equipped with a large-volume injection system (LVI/GC–MS) and liquid chromatography with mass spectrometry and tandem mass spectrometry equipped with an atmospheric pressure chemical ionization source (LC/APCI–MS, LC–MS/MS/APCI). Tests show that most (>60%) of the recoveries are satisfactory. They are similar for all target pesticides for given types of water (filtered or raw water), sorbents, and chromatographic techniques. The coefficient of variation is less than 10% for most of the pesticides, regardless of the method used, and detection limits range from 0.2–3.7 ng/L and 2–5 ng/L for LVI/GC–MS and LC–MS/MS/APCI, respectively.

Key Words: C18; Carbopack B; degradation products; environmental analysis; gas and liquid chromatography; mass spectrometry; large-volume injection; solid-phase extraction; tandem mass spectrometry; triazines; water analysis.

1. Introduction

There is growing concern about both triazine herbicides and their degradation products because of their toxicity and their increased use and occurrence in natural waters *(1–3)*. Some triazine degradation products are as toxic as, or even more toxic than, their parent compounds *(2–4)*. At least 29 triazines and 36 degradation products have been reported in the literature *(4)*. Unfortunately, no single method enables the simultaneous extraction and analysis of all of these compounds *(5–22)*. In this chapter, we

From: *Methods in Biotechnology, Vol. 19, Pesticide Protocols*
Edited by: J. L. Martínez Vidal and A. Garrido Frenich © Humana Press Inc., Totowa, NJ

describe three methods that can be used to monitor triazines and degradation products in filtered and raw water (0.5–1 L). All begin with solid-phase extraction (SPE) using cartridges packed with either C18-bonded silica (Isolute Triazine) or graphitized carbon black (Carbopack B). The analytical methods used to determine the pesticides are gas chromatography–mass spectrometry equipped with a large-volume injection system (GC/LVI–MS), liquid chromatography with mass spectrometry and with tandem mass spectrometry equipped with an atmospheric pressure chemical ionization (LC–MS/APCI and LC–MS/MS/APCI, respectively) ion source.

2. Materials

1. Vacuum manifold.
2. 6.5 × 1.4 cm internal diameter (id) polypropylene cartridges packed with 1 g C18-bonded silica (Isolute Triazine) and a depth filter (IST Ltd, U.K.).
3. 6.5 × 1.4 cm id polypropylene cartridges packed with 500 mg graphitized carbon black (120–400 mesh) (Supelco).
4. Closed cartridges packed with 2.5 g anhydrous sodium sulfate (IST Ltd, U.K.).
5. Polytetrafluoroethylene filters (0.45 μm, 47-mm id) (Millipore).
6. Gas chromatograph equipped with a split/splitless programming temperature injector working in the LVI mode and coupled with a mass spectrometer (Varian; Thermo).
7. Electron ionization (EI) ion source.
8. Capillary columns (5% phenyl/95% methylsilicone; 30 m × 0.25 mm id, 0.25-μm coating thickness) (Supelco).
9. Deactivated capillary guard column (5 m × 0.53 mm id) (Supelco).
10. Liquid chromatograph coupled with a quadrupole mass spectrometer or an ion trap tandem mass spectrometer (Thermo).
11. APCI ion source.
12. LC system equipped with a quaternary pump and an autosampler.
13. LC columns filled with a C18 silica phase (25 cm × 4.6 mm id, Hypersil ODS 2 column packed with 5-μm particles) or other phases providing similar separation (Thermo).
14. Reagents: acetone, acetonitrile, dichloromethane (DCM), methanol (MeOH) (distilled-in-glass grade or residue pesticide quality) and Milli-Q water.
15. Pesticide and deuterated standards (>95% pure).

3. Methods

The three methods described here are (1) SPE using Carbopack B cartridges (protocol 1) followed by GC/LVI–MS, (2) SPE using Isolute Triazine cartridges (protocol 2) followed by GC/LVI–MS, and (3) SPE using Isolute Triazine cartridges (protocol 3) followed by LC–MS/APCI and LC–MS/MS/APCI.

3.1. Sample Collection and Preservation

The water samples are collected in 1-L amber glass containers (*see* **Note 1**) and stored at 4°C until extraction. Sequential extraction should be done within 48 h to avoid any transformation in the structure of the target analytes.

3.2. Solid-Phase Extraction

SPE is done to extract triazines and degradation products from filtered or unfiltered surface water (suspended particulate matter = 2–60 mg/L) using cartridges packed with various sorbents. Protocols 1 and 2 are used prior to GC/LVI–MS analysis, and protocol 3 is used prior to LC–MS/APCI and LC–MS/MS/APCI. Protocol 2 is a simultaneous filtration-and-extraction technique. The following procedure describes the steps used in the three protocols. **Table 1** gives specifications for each individual protocol.

1. Filter surface water sample (protocols 1 and 3) through a polytetrafluoroethylene filter (0.45 µm, 47-mm id) to eliminate suspended particulate matter (*see* **Note 2**).
2. Condition the cartridges by rinsing with solvents in the order indicated under "Conditioning" in **Table 1** (*see* **Notes 3** and **4**).
3. Load the sample by attaching a connector to the top of the cartridge. The other end of the connector should be fitted with a 1/8-in Teflon tube with a stainless steel weight at its end. Feed the weighted end of the Teflon tube into the sample container until it touches the bottom. Place the cartridge on the vacuum manifold.
4. Pass the sample through the cartridges: Turn on the vacuum pump to draw the sample through the cartridge. This takes approx 50 min (*see* flow rates in **Table 1**) and can be done with either a water or a vacuum pump.
5. Rinse the cartridge with water: Add 6 mL of Milli-Q water to the cartridges and pump for 15 min to remove any residual water (*see* **Note 5**). Discard washes.
6. Protocols 1 and 2 only: Place a closed cartridge filled with 2.5 g anhydrous sodium sulfate below the rinsed cartridge.
7. Protocol 1 only: Pass 1 mL methanol through the cartridge system (Carbopack–sodium sulfate) and discard this fraction.
8. Elute the cartridge system with selected eluent (**Table 1**) at a rate of 1 mL/min. Collect the eluate in a 15-mL conical test tube, then evaporate most of the solvent with a nitrogen stream at 25°C. Add 100 µL (20 µg) of the internal standard solution (atrazine-d₅; *see* **Note 6**) to the extract, then add enough of the appropriate solvent to reach the final volume (**Table 1**) for chromatographic analysis.

3.3. GC and LC Coupled With MS

3.3.1. GC/LVI–MS

Sample extracts from SPE protocols 1 and 2 are analyzed using a gas chromatograph coupled with an ion trap mass spectrometer (*see* **Note 7**) equipped with a split/splitless programming temperature injector working in the LVI mode (*see* **Note 8**). **Figure 1** shows a chromatogram obtained by injecting 40 µL of an extract of 1 L of water from a tributary of the Loire River (following SPE) into a GC/LVI–MS. The analysis steps are preparation of the calibration curve, chromatographic analysis, and identification and quantification of analytes.

1. Set the initial column temperature at 40°C for 5 min, increase it to 300°C at a rate of 5°C/min, and hold it steady at this temperature for 5 min.
2. Set the initial injector temperature at 40°C for 1.5 min (*see* **Note 9**), increase it to 300°C at a rate of 180°C/min, and hold it steady at this temperature for 62 min.
3. Very slowly (2 µL/s, *see* **Note 10**), inject 20–40 µL of the standard solutions (0.02–2 mg/L) and the extracts obtained from SPE protocols 1 and 2. Both the standard solutions and the extracts should include the internal standard at a concentration of 0.20 mg/L DCM. The flow rate of the carrier gas (helium) is set at 1 mL/min for a split flow of 50 mL/min. The split/splitless valve is operated in split mode from 0 to 1.5 min, splitless mode from 1.5 min to 3 min, and split mode after 3 min.

Table 1
Specifications for the Three SPE Protocols

		Protocol 1	Protocol 2	Protocol 3
Sorbent		GCB Carbopack B	C18 Isolute Triazine	C18 Isolute Triazine
Sample volume		1 L	0.5 L	0.5 L
Filtration		Yes	Not necessary	Yes
Conditioning	Solvent 1	3 × 6 mL DCM	3 × 6 mL DCM	6 mL acetone
	Solvent 2	6 mL MeOH	6 mL acetone	10 mL Milli-Q water
	Solvent 3	6 mL Milli-Q water	6 mL Milli-Q water	NA
Flow rate		20 mL/min	10 mL/min	10 mL/min
Eluent		15 mL DCM	15 mL DCM/acetone (80:20 v/v)	2 × 2 mL MeOH
Final volume		1 mL DCM	0.250 mL DCM	0.5 mL MeOH/water (50:50 v/v)

GCB, graphitized carbon black; NA, not applicable.

470

Abundance

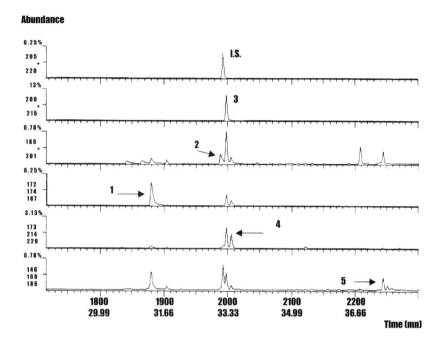

Fig. 1. A chromatogram obtained by injecting 40 µL into a GC/LVI–MS: Ion chromatograms of an extract of 1 L water from a tributary of the Loire River after SPE (Carbopack B, 120-400 mesh). Peaks: 1, desethyl-atrazine (17 ng/L); 2, simazine (<1 ng/L); 3, atrazine (96 ng/L); 4, terbuthylazine (<1 ng/L); IS, atrazine-d$_5$ (206 ng/L). Peak for other compound: 5, alachlor (10 ng/L).

4. Perform the mass spectrometric analysis with EI interface in full-scan mode from 47 U to 450 U.
5. Identify each peak by its mass spectrum and retention time (*see* **Note 11**) and extract the signal at specific m/z.
6. Quantify the analyte by measuring the peak areas (or the peak area/internal standard peak area ratio) of the sample extracts and comparing them to those of the standard solutions.
7. The m/z used for the quantification of the target compounds are given in **Table 2**.
8. Calculation: The concentration C (micrograms per liter) of the target analyte is calculated using the following equation:

$$C = C_e V_f / V_w$$

where C_e is the concentration in the final extract in micrograms per milliliter, V_f is the final volume of extraction solvent in milliliters, and V_w is the sample volume in liters.

3.3.2. LC–MS/APCI and LC–MS/MS/APCI

Sample extracts from SPE protocol 3 are analyzed using a liquid chromatograph coupled with a quadrupole mass spectrometer (*see* **Note 12**) or an ion trap mass spectrometer running in MS/MS mode (*see* **Note 13**) equipped with an injection valve (20–100-µL injection volume), a gradient pump, and an APCI ion source (*see* **Note 14**). **Figures 2** and **3** show LC/APCI plus full-scan MS/MS chromatograms of a stan-

Table 2
Selected Ions (*m/z*) Used for Triazine Quantification by GC/LVI–MS and LC–MS/APCI: Examples of Transition Ions for the Compounds Analyzed by LC–MS/MS/APCI Using Triple Quadrupole

Compound	Selected ions (*m/z*)	Precursor ion (*m/z*)	Daughter ions (*m/z*)	
	GC/LVI–MS	LC–MS/APCI	LC–MS/MS/APCI	
Ametryn	212, 227	228	228	186
Atraton	169, 196, 211	212	NA	NA
Atrazine	173, 200, 215	216	216	174
Cyanazine	225, 240	241	241	214
Desethyl-atrazine	172, 174, 187	188	188	146
Desmetryn[a]	NA	NA	214	172
Metribuzin	144, 182, 214	215	215	187
Desipropyl-atrazine	145, 158, 173	174	174	104
Hydroxyatrazine[a]	NA	NA	198	156
Hydroxysimazine[a]	NA	NA	184	114
Prometon	168, 210, 225	226	NA	NA
Prometryn	184, 226, 241	242	NA	NA
Propazine	172, 214, 229	230	230	188
Simazine	173, 186, 201	202	202	132
Terbuthylazine	NA	230	230	174

NA, not available.
[a]Not quantified.

dard solution and a detailed signal on each ion, respectively. The analysis steps are preparation of the calibration curve, chromatographic analysis, and identification and quantification of analytes.

3.3.2.1. LC–MS/APCI

1. Set the flow rate of the mobile phase (85% water/15% acetonitrile) at 1 mL/min.
2. Use an acetonitrile/water binary gradient of 15/85 to 60/40 in 50 min.
3. Inject 20–50 µL of the standard solutions (25, 50, 130, 260, 520, 1040, 2600 µg/L) and the extracts obtained from SPE protocol 3.
4. Set the APCI ion source in the positive ion mode (*see* **Note 15**).
5. Perform the MS analysis in full-scan mode from 130 to 450 U (*see* **Note 16**).
6. Identify each peak by its mass spectrum and retention time (*see* **Note 17**) and extract the signal at specific *m/z*.
7. Quantify the analyte by measuring the peak areas (or the sample area/internal standard area ratio) of the sample extracts and comparing them to those of standard solutions.
8. The *m/z* used for the quantification is given in **Table 2**.
9. Calculation: The concentration *C* (micrograms per liter) of target analyte is calculated using the following equation:

$$C = C_e V_f / V_w$$

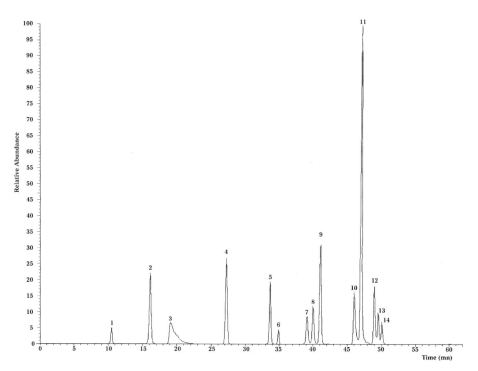

Fig. 2. LC/APCI plus full-scan MS/MS chromatogram of a standard solution (1 μg/mL of each compound in methanol) scanning from *m/z* 50 to approx 10 u above the precursor ions u. Peak assignments for triazines: 1, desipropyl-atrazine; 2, desethyl-atrazine; 4, simazine; 5, atrazine; 9, terbuthylazine. Peak assignments for other compounds: 3, carbendazim; 6, isoproturon; 7 and 8, triadimenol; 10, tebuconazole; 11, flusilazole; 12, penconazole; 13 and 14, propiconazole.

where C_e is the concentration in the final extract in micrograms per milliliter, V_f is the final volume of extraction solvent in milliliters, and V_w is the sample volume in liters.

3.3.2.2. LC–MS/MS/APCI

1. Set the flow rate of the mobile phase (85% water/15% acetonitrile) at 1 mL/min.
2. Use an acetonitrile/water binary gradient of 15/85 to 60/40 in 50 min.
3. Inject 20–50 μL of the standard solutions (1, 5, 10, 25, 50, 100 μg/L) and the extracts obtained from SPE protocol 3.
4. Set the APCI ion source in the positive ion mode (*see* **Note 15**).
5. Perform the MS analysis in MS/MS by (1) selecting the protonated ion of each compound for each time segment of the chromatographic separation; (2) setting, in the ion trap, the collision energies of the protonated ions with helium producing a mass spectra in MS/MS; and (3) analyzing the resulting daughter ions by scanning *m/z* between 50 and 10 u above the molecular mass of each compound (**Table 3**).
6. Identify each peak by its mass spectrum in MS/MS and retention time (*see* **Note 18**) and extract the signal of daughter ions at specific *m/z* (**Table 3**).

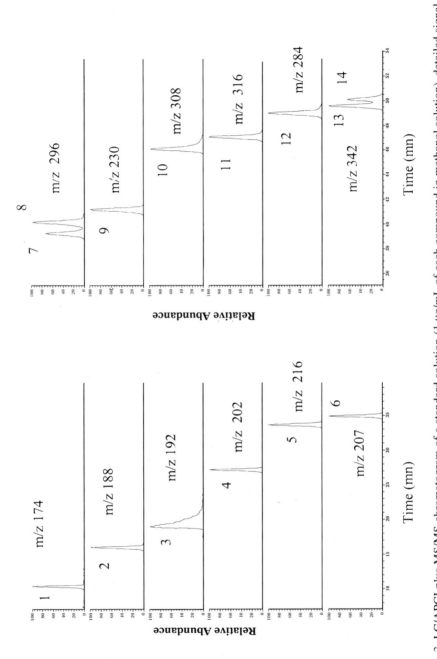

Fig. 3. LC/APCI plus MS/MS chromatogram of a standard solution (1 µg/mL of each compound in methanol solution), detailed signal on each ion. Peak assignments for triazines: 1, desethyl-atrazine; 2, desisopropyl-atrazine; 4, simazine; 5, atrazine; 9, terbuthylazine. Peak assignments for other compounds: 3, carbendazim; 6, isoproturon; 7 and 8, triadimenol; 10, tebuconazole; 11, flusilazole; 12, penconazole; 13 and 14, propiconazole.

Table 3
General Conditions Used for LC–MS/MS/APCI Analysis Using Ion Trap Mass Spectrometer

Segment	Time (min)	Protonated parent ion selected m/z	Collision voltage in ion trap detector (V)	Compounds	MS/MS full-scan interval	Retention time (min)
1	0.0–13.18	174	0.80	Desipropyl-atrazine	50–84	10.44
2	13.20–17.48	188	0.80	Desethyl-atrazine	50–98	16.01
3	17.50–31.10	202	0.90	Simazine	55–212	27.22
4	31.20–37.36	216	0.90	Atrazine	60–226	33.69
5	37.40–43.70	230	0.75	Propazine	60–240	39.23
		230	0.75	Terbuthylazine	60–240	40.06

Table 4
Detection Limits (DLs) and Mean Recovery Obtained for Target Pesticides in Surface Water Using the Proposed Methods

Pesticide	DL (ng/L)			Mean recovery (%) in surface water (n = 3)		
	Method 1	Method 2	Method 3	Method 1	Method 2	Method 3
Ametryn	1.6	0.6	NA	90 ± 11	105 ± 2	NA
Atraton	1.1	NA	NA	87 ± 3	NA	NA
Atrazine	1.5	0.2	2	86 ± 2	80 ± 3	92 ± 4
Cyanazine	3.7	0.8	NA	95 ± 7	125 ± 6	NA
Desethyl-atrazine	1.8	0.3	3	71 ± 4	55 ± 3	60 ± 3
Desipropyl-atrazine	2	0.3	5	78 ± 2	29 ± 2	35 ± 4
Metribuzin	1.5	NA	NA	78 ± 8	NA	NA
Prometon	0.6	NA	NA	91 ± 2	NA	NA
Prometryn	0.9	0.8	NA	58 ± 2	97 ± 2	NA
Propazine	0.7	0.2	2	84 ± 2	96 ± 6	96 ± 7
Simazine	2	0.2	3	95 ± 8	97 ± 2	98 ± 4

NA, not available.

Method 1: SPE using Carbopack B (120- to 400-mesh) cartridges followed by GC/LVI–MS. 1-L sample volume; 1-mL final volume.

Method 2: SPE using C18 Isolate Triazine cartridges followed by GC/LVI–MS. 0.5-L sample volume; 0.25-mL final volume.

Method 3: SPE using C18 Isolate Triazine cartridges followed by LC–MS/APCI and LC–MS/MS/ APCI. 0.5-L sample volume; 0.5-mL final volume.

7. Quantify the analyte by measuring the peak areas (or the sample area/internal standard area ratio) of the sample extracts and comparing them to those of standard solutions.

8. The m/z used for the quantification is given in **Table 3**.

9. Calculation: The concentration C (micrograms per liter) of target analyte is calculated using the following equation:

$$C = C_e V_f / V_w$$

where C_e is the concentration in the final extract in micrograms per milliliter, V_f is the final volume of extraction solvent in milliliters, and V_w is the sample volume in liters.

3.4. Quality Control

The compounds targeted in each protocol are given in **Table 4**.

1. Method blanks are analyzed after every five samples using a volume of 0.5 L of Milli-Q water, and blank samples are extracted using the same protocol as for surface water samples. No traces of target chemicals are detected in the blanks, and there is no interference except for propazine (0.9 ng/L), which is present in the internal standard solution (atrazine-d₅) at a ratio of 1%.

2. Recovery is measured by extracting and analyzing target pesticides previously spiked at 0.05- to 0.1- ppb levels from 0.5 to 1 L of surface water samples using the proposed methods (**Table 4**). Overall percentage recoveries are satisfactory for all target pesticides

(>60%) except desisopropyl-atrazine (more polar). The coefficient of variation is below 10% for most of the pesticides regardless of the method used.

3. Detection limits are determined for each analyte at a concentration providing a signal-to-noise ratio of 3 (**Table 4**). Detection limits ranged from 0.2 to 3.7 ng/L and 2 to 5 ng/L for GC/LVI–MS and LC–MS/MS/APCI, respectively.

4. Notes

1. The analytical samples are taken in amber glass bottles certified free of organic compound residues or in glass bottles wrapped in aluminum foil and previously washed with a detergent solution, rinsed with distilled water, and cleaned with petroleum ether and acetone. A blank should be used for verifying the efficiency of the bottle washing.
2. The sampling site must be homogeneous and representative. An all-Teflon pneumatic pump and Teflon tubing should be used when sampling surface water.
3. Carbopack B packing has positive charges on its surface that enable it to behave as both a reversed-phase and an anion exchanger sorbent. To activate the anion exchange sites, the cartridge must be conditioned with an acidic reagent. We do not recommend this step for routine analysis when using GC–MS because the acidic reagent remains in the eluate, contaminating the liner and the capillary precolumn or column.
4. The cartridge must not be allowed to go dry at any time during the conditioning and extraction steps.
5. Increase drying time (30 min) if there is any residual water in the eluate.
6. Various deuterated internal standards can be used (e.g., simazine-d_6 or terbuthylazine-d_5).
7. If a quadrupole mass spectrometer in selected ion monitoring mode is used instead of an ion trap spectrometer, similar performance can be obtained using three m/z ions (one identifier and two qualifiers).
8. Various LVI devices are available (e.g., on column and septum-equipped programmable injectors). LVI are devices that can be retrofitted to any GC and replace conventional split/splitless injectors.
9. The injection temperature should be 10–15°C above the vaporization temperature of the solvent used for injection.
10. The sample is injected very slowly with the injector temperature set a few degrees below the boiling point of the solvent. The solvent is vented at the same time as it is injected through the injector split valve.
11. Identification criteria in GC/LVI–MS: retention time (does not differ from the peak measured in the external standard by less than ±0.2%) plus mass spectrum [the most sensitive diagnostic ion is present, the two selected diagnostic ions do not deviate by more than $\pm(0.1 \times I_{st} + 10\%)$] in which I_{st} is the relative intensity of the diagnostic ions in the external standard.
12. An ion trap mass spectrometer can be used instead of a quadrupole mass spectrometer.
13. A triple-quadrupole mass spectrometer can be used instead of the ion trap mass spectrometer. This technique using time-scheduled selected reaction monitoring mode allows obtaining similar results to ion trap mass spectrometry in MS/MS mode.
14. An electrospray ionization ion source in positive mode could be used in place of the APCI. Generally, both are appropriate for nonpolar and low-polar compounds. Electrospray ionization is preferred for more polar compounds.
15. APCI conditions: 400°C vaporizer temperature; 225°C transfer capillary temperature; 5-µA corona discharge intensity; 0.6-MPa sheath gas pressure; 10-L/min auxiliary gas flow rate; 5-V octapole voltage. Positive ionization mode is used for these compounds produced in APCI protonated ions.

16. MS detection could be done in selected ion monitoring mode to improve the quantification limit. However, in this detection mode, the specificity is decreased.
17. Identification criteria in LC–MS/APCI: retention time, which does not differ from the peak measured in the external standard by less than ±0.2%; mass spectra, for which the protonated ion must be present in the sample and in the external standard.
18. Identification criteria in LC–MS/MS/APCI: retention time, which does not differ from the peak measured in the external standard by less than ±0.2%; mass spectra, for which the precursor ion and the most abundant daughter ion must be present.

References

1. Hapeman, C. J., Bilboulian, S., Anderson, B. G., and Torrents, A. (1998) Structural influences of low molecular weight dissolved organic carbon mimics on the photolytic fate of atrazine. *Environ. Toxicol. Chem.* **17,** 975–981.
2. Kolpin, D. W., Thurman, E. M., and Linhart, S. M. (1998) The environmental occurrence of herbicides—the importance of degradates in ground water. *Arch. Environ. Contam. Toxicol.* **35,** 385–390.
3. Thurman, E. M., Mills, M. S., Meyer, M. T., Zimmerman, L. R., Perry, C. A., and Goolsby, D. A. (1994) Formation and transport of deethylatrazine and deisopropylatrazine in surface water. *Environ. Sci. Technol.* **28,** 2267–2277.
4. Sabik, H., Jeannot, R., and Rondeau, B. (2000) Multiresidue methods using solid-phase extraction techniques for monitoring priority pesticides, including triazines and degradation products, in ground and surface waters. *J. Chromatogr. A* **885,** 217–236.
5. Ashton, F. and Klingman, G. (eds.). (1982) *Weed Science, Principles and Practices.* Wiley, London.
6. Cai, Z., Ramanujam, V. M. S., Gross, M. L., Monson, S. J., Cassada, D. A., and Spalding, R. F. (1994) Liquid–solid extraction and fast atom bombardment high-resolution mass spectrometry for the determination of hydroxyatrazine in water at low-ppt levels. *Anal. Chem.* **66,** 4202–4209.
7. Crescenzi, C., DiCorcia, A., Passariello, G., Samperi, R., and Carou, M. I. T. (1996) Evaluation of two new examples of graphitized carbon blacks for use in solid-phase extraction cartridges. *J. Chromatogr. A* **733,** 41–55.
8. Ferrer, I., Barceló, D., and Thurman, E. M. (1999) Doubled-disk solid-phase extraction—simultaneous cleanup and trace enrichment of herbicides and metabolites from environmental samples. *Anal. Chem.* **71,** 1009–1015.
9. Forcada, M., Beltran, J., Lopez, F. J., and Hernandez, H. (2000) Multiresidue procedures for determination of triazine and organophosphorus pesticides in water by use of large-volume PTV injection in gas chromatography. *Chromatographia* **51,** 362–368.
10. Geerdnik, R. B., Kooistra-Sijpersma, A., Tiesnitsch, J., Kienhuis, P. G. M., and Brinkman U. A. T. (1999) Determination of polar pesticides with atmospheric pressure chemical ionization mass spectrometry-mass spectrometry using methanol and/or acetonitrile for solid-phase desorption and gradient liquid chromatography. *J. Chromatogr. A* **863,** 147–155.
11. Jeannot, R., Sabik, H., Sauvard, E., and Genin, E. (2000) Application of liquid chromatography with mass spectrometry combined with photodiode array detection and tandem mass spectrometry for monitoring pesticides in surface waters. *J. Chromatogr. A* **879,** 51–71.
12. Jeannot, R., Sabik, H., Amalric, L., Sauvard, E., Proulx, S., and Rondeau B. (2001) Ultra-trace analysis of pesticides by solid-phase extraction of surface water with Carbopack B cartridges, combined with large-volume injection in gas chromatography. *Chromatographia* **54,** 236–240.

13. Koskinen, W. C. and Barber, B. L. (1997) A novel approach using solid phase extraction disks for extraction of pesticides from water. *J. Environ. Qual.* **26,** 558–560.

14. Kumazawa, T. and Suzuki, O. (2000) Separation methods for amino-possessing pesticides in biological samples. *J. Chromatogr. A* **747,** 241–254.

15. Lerch, R. N. and Donald, W. W. (1994) Analysis of hydroxylated atrazine degradation products in water using solid-phase extraction and high performance liquid chromatography. *J. Agric. Food Chem.* **42,** 922–927.

16. Liška, I. and Slobodník, J. (1996) Comparison of gas and liquid chromatography for analysing polar pesticides in water samples. *J. Chromatogr. A* **733,** 235–258.

17. Loos, R. and Niessner, R. (1999) Analysis of atrazine, terbutylazine and their *N*-dealkylated chloro and hydroxyl metabolites by solid-phase extraction and gas chromatography–mass spectrometry and capillary electrophresis–ultraviolet detection. *J. Chromatogr. A* **835,** 217–229.

18. Magnuson, M. L., Speth, T. F., and Kelty, C. A. (2000) Degradation of interfering triazine degradation products by gas chromatography-ion trap mass spectrometry. *J. Chromatogr. A* **868,** 115–119.

19. Molina, C., Durand, G., and Barceló, D. (1995) Trace determination of herbicides in estuarine waters by liquid chromatography-high-flow pneumatically assisted electrospray mass spectrometry. *J. Chromatogr. A* **712,** 113–122.

20. Sabik, H., Rondeau, B., Gagnon, P., Jeannot, R., and Dohrendorf, K. (2003) Simultaneous filtration and solid-phase extraction combined with large-volume injection in GC/MS for ultratrace analysis of polar pesticides in surface water. *Int. J. Environ. Anal. Chem.* **83,** 457–468.

21. Spliid, N. H. and Køppen, B. (1996) Determination of polar pesticides in ground water using liquid chromatography-mass spectrometry with atmospheric pressure chemical ionization. *J. Chromatogr. A* **736,** 105–114.

22. Thurman, E. M. and Mills, M. S. (eds.). (1998) *Solid-Phase Extraction: Principles and Practices.* Wiley Interscience, New York.

35

An Optical Immunosensor for Pesticide Determination in Natural Waters

Sara Rodríguez-Mozaz, Maria J. López de Alda, and Damia Barceló

Summary

The river analyzer (RIANA) is an optical immunosensor able to determine different types of organic pollutants in water samples. Interaction between fluorescently labeled antibody and analyte takes place on a solid–liquid interface and is detected by measuring the fluorescence. The RIANA integrates the optical part with a fluidics part in a compact unit, which is coupled to an autosampler, allowing the performance of automatic measurements within 15 min.

In this chapter, the use protocol for the sensor is described, and guidelines about experiment performance are provided. Up to three different analytes (in this case, the pesticides atrazine and isoproturon and the natural estrogen estrone) can be detected simultaneously. This multianalyte configuration allows reduction in time of analysis and in volume of sample and other reagents required. Different aspects, such as assay performance, cross-reactivity, matrix effects, or validation, are also considered.

Key Words: Environmental monitoring; estrogens; immunosensor; multianalyte; pesticides; water.

1. Introduction

Monitoring of pesticides and other relevant pollutants in natural waters is commonly required for environmental control. To accomplish this requirement, new analytical tools able to provide fast and reliable data are demanded. In this sense, biosensors seem to be appropriate. In general, biosensors for environmental applications do not compete with official analytical methods. Instead, they are mostly designed for the screening of samples by both regulatory authorities and industry. These samples can further be sent to laboratories, if necessary, for detailed chemical analysis, thus reducing the costs of monitoring programs (1). The river analyzer (RIANA) biosensor can be applied to the determination of different pesticides in natural waters (2–4).

RIANA is an optical immunosensor based on the use of antibodies as specific recognizing agents and an optical transducer able to convert the recognition of the analyte in a readable signal. Interaction between fluorescently labeled antibody and analyte

From: *Methods in Biotechnology, Vol. 19, Pesticide Protocols*
Edited by: J. L. Martínez Vidal and A. Garrido Frenich © Humana Press Inc., Totowa, NJ

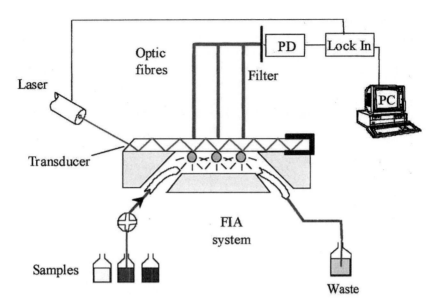

Fig. 1. Scheme of the immunosensor RIANA. The FIA system equipped with a six-port valve and a 1-mL syringe pump delivers the buffer solution, the sample, or the regeneration solutions to the flow cell. The transducer, a 1.5-mm thick surface polished sheet glass, is mounted on the flow cell and is sealed with an O-ring. The excitation light consists of a He-Ne laser. The collected fluorescent light is filtered and detected by photodiodes (PDs). PC, personal computer.

(which acts as an antigen) takes place on a solid–liquid interface and is detected by measuring the fluorescence. The device (scheme shown in **Fig. 1**) consists of a flow injection analysis system (FIA) with a six-way distribution valve equipped with a 1mL syringe pump connected through Teflon tubing to a flow cell. The transducer is mounted in this flow cell and sealed by an O-ring. Light from a collimated and modulated He-Ne laser source (633 nm, 7 mW) is directly coupled and guided through the length of the transducer by total internal reflection. An evanescent field is produced at each reflection spot and penetrates a few hundreds of nanometers into the external medium. This field causes the excitation of fluorescently labeled antibodies locally and specifically bound to the analyte derivatives. The fluorescence light is collected through optic fibers, filtered, and detected by photodiodes using lock-in detection. Automatic measurements are carried out with the system within 15 min, including the regeneration step. This system setup has been described previously (**5**).

On the other hand, RIANA also enables the simultaneous determination of several pollutants in the same sample. This feature can represent an interesting tool in environmental monitoring and screening because it allows a reduction in both time and volume of sample and other reagents required. This chapter describes the application of the RIANA immunosensor, in multianalyte configuration, to the simultaneous determination of two pesticides (atrazine and isoproturon) and a natural estrogen of growing environmental concern (estrone).

2. Materials
2.1. Chemicals

1. HPLC-grade methanol.
2. Pro analysis ethanol.
3. Ovalbumin (OVA) (e.g., Sigma; Deisenhofen, Germany).
4. Pesticide standards atrazine and isoproturon (e.g., Riedel-de-Haen, Seelze, Germany).
5. Estrone standard (e.g., Sigma-Aldrich; Deisenhofen, Germany).
6. Lyophilized antiestrone, antiisoproturon, and antiatrazine antibodies and the corresponding analyte derivatives used were kindly supplied by Dr. Ram Abuknesha (King's College, London, UK).

2.2. Working Solutions

1. Phosphate-buffered saline (PBS): 150 mM NaCl, 50 mM KH$_2$PO$_4$ adjusted to pH 7.4 with 2 M KOH in bidistilled water.
2. Regeneration solution: 0.5% sodium dodecyl sulfate (SDS) adjusted with an HCl solution (5 N) to pH 1.9.
3. 75:25 (v/v) Ethanol/bidistilled water rinsing solution.
4. Stock standard solutions: Stock standard solutions for each analyte are prepared in methanol at 1000 mg/L by dissolving 10 mg standard in 10 mL methanol. The solutions are kept at 20°C.

3. Methods
3.1. Preparation of the Transducer

1. As a solid-phase immunoassay, inmobilize the analyte derivative on a clean glass slide (transducer) following the protocol described by Barzen et al. *(6)*.
2. Briefly, clean glass interference layers (58 × 10 × 1.5 mm with 45° bevel on short side) by immersion in a freshly prepared hot mixture of concentrated piranha solution for 30 min and afterward rinse with water.
3. Achieve the silanization of the surface by treatment for 1 h with 30 µL of 3-glycidoxypropyltrimethoxysilanxysilane. Each aminodextran–hapten–conjugate (atrazine, estrone, and isoproturon) is then placed in the corresponding hole (~1-mm diameter) of a metal mask placed on top of the transducer surface. Conjugates from analyte derivative and polymer are thus coupled to the activated transducer surface in separated areas.

3.2. Preparation of Calibration

1. Prepare water standard solutions at various concentrations by appropriate dilution of the stock solutions with PBS. Final aqueous solutions should not contain more than 0.1% methanol.
2. Choose standard concentrations to match the estimated concentrations of the analytes in unknown samples.
3. Keep water standard solutions at 4°C up to 15 d.

3.3. Preparation of Antibodies Solution

1. Lyophilized anti-estrone, anti-isoproturon, and anti-atrazine antibodies (labeled with the fluorescent dye Cy5.5) are reconstituted with phosphate buffer solution before their use, giving a final mixture solution of the three antibodies (*see* **Note 1**). Concentrations obtained for anti-estrone, anti-isoproturon, and anti-atrazine are 3.5, 3.3, and 0.5 µg/mL, respectively.

2. The antibody solution must be strongly buffered since analyte–antibody binding is very sensitive to pH (*see* **Note 2**).

3. The antibody solution as well as the final incubated solution need to have a pH of approx 7.4.

4. Add the background protein OVA to reduce the nonspecific binding of proteins to the transducer surface. Therefore, prepare the antibody solution so that the OVA concentration is 2000 µg/mL (*see* **Note 2**).

5. Once the antibody solution is combined with the analyte solution, the final concentration of the OVA is about 200 µg/mL.

6. Final concentrations of the antibodies are 0.35, 0.33, and 0.05 µg/mL for antiestrone, antiisoproturon, and antiatrazine, respectively (*see* **Note 3**).

3.4. Preparation of the Samples

1. Water samples should show little turbidity; therefore, filter samples through 0.45-µm membrane filters (Millipore) to remove suspended solids that might cause dispersion of the emitted light and blockage of the FIA system.

2. As pointed out in **Subheading 3.3.**, **step 5**, the antibodies need a particular ionic strength and pH 7.4; therefore, adjust the water samples to the prescribed conditions before their analysis *(7)* (*see* **Note 4**).

3.5. Preparation of the System

3.5.1. Flow Cell and Rinsing

1. Place the transducer slice within the flow cell and subsequently seal with screws to prevent any leakage but allow the flow of the working and sample solutions through the active surface.

2. Connect the flow cell to the corresponding tubes of the fluidics system and rinse with PBS while checking for possible leaks and bubbles in the flow path (*see* **Note 5**).

3.5.2. Coupling the Laser

1. Align the laser with the transducer by manually adjusting the incidence angle of the laser beam on the polished 45° beveled end face of the glass slide so that discrete spots are formed on both sides of the transducer. This is caused by the total internal reflection of the laser beam within the transducer.

2. Align selected spots also with three of the six optical fibers, which separately conduce the light to the photodiodes.

3.6. Perform Binding Measurements

1. Incubate the analyte solution with the antibodies before each measurement. There are two types of analyte solutions: water samples containing unknown concentrations of analytes and standard solutions of known analyte concentrations used to calibrate the instrument. For both standard and unknown solutions, 100 mL of an antibody solution is added to 900 mL of the sample in a 1-mL Eppendorf tube and shaken with a vortex (*see* **Note 6**).

2. After incubation for at least 15 min to reach the binding equilibrium, the analyte–antibody solution is delivered into the flow cell by the flow injection system. During this time, antibodies bind to analyte molecules present in the sample.

3. After rinsing the flow cell with phosphate buffer solution, inject samples into the system and flow over the transducer surface for 5 min. Only unbound antibodies are able to bind to the analyte derivative covalently bound on the surface. Antibodies will only interact with the spots that contain analyte derivatives to which they are directed. During this process, the laser is switched off to avoid photobleaching of the dyes.

4. After a brief (30-s) rinse with PBS, the laser is turned on, and the evanescent field created excites locally the bound antibodies in the reflection spots.
5. Collect fluorescence by the optical fibers located under the sensor chip opposite the active spots *(8)*. The signal difference before and after the binding is measured to obtain an estimation of the number of molecules bound to the transducer layer.
6. After each measurement, rinsing with SDS solution achieves regeneration of the transducer, which allows the performance of a new run (*see* **Note 7**). Total analysis time for a single run is 15 min.
7. A common analytical sequence consists of nine blank measurements (analyte free), nine calibration standards (minimum of five) from 0 to 100 µg/L, and the samples to analyze. After incubation and mixing, samples are randomly placed in the autosampler and analyzed automatically (*see* **Note 8**). A entire calibration (standards, blanks, and real samples—all in triplicate) requires about 8 h.

3.7. Performance of the Data Analysis

1. The fluorescence signal decreases with increasing analyte concentration because of the reduced free antibody concentration that can bind to the surface. Prior to data analysis, sample fluorescent signals are normalized with the signal corresponding to a blank (maximum fluorescence). Average values (measurements made in triplicate) obtained for the calibration standards are fitted to a logistic function to plot a calibration curve:

$$Y = A + \frac{B - A}{1 + 10 \, [(\log IC_{50} - \log C)M]}$$

   where Y is the fluorescence signal difference (before and after the binding), C is analyte concentration, A and B are experimentally determined constants, IC_{50} is the signal corresponding to the inhibition concentration at 50% of absorbance, and M is the slope at the inflection point.
2. The lower and upper working ranges are defined as the concentrations at which the signal is quenched by 10 and 90%, respectively. This range depends on the sensitivity (slope) of the calibration curve.
3. The limit of detection (LOD) is defined as three times the standard deviation of the measurement without analyte (blank sample). The analytical parameters obtained for the target analytes are summarized in **Table 1**.
4. Normalized fluorescence values obtained with the unknown samples are interpolated in the calibration curve, and the concentration of the analyte in the sample is thus calculated.

3.8. Validation

Despite the practical advantages of biosensors, they must be verified by comparing the biosensor results with the results from chemical analysis. This verification process is known as *validation*. The RIANA immunosensor has to be checked by analyzing real samples or spiked real samples by both the multianalyte immunosensor and a conventional method, usually chromatographic (*see* **Fig. 2**).

4. Notes

1. Cross reactivity: The antibodies can react with cross-reactive substances in solution, leading to the determination of a wrong concentration of the target substance. The cross-reactivity of the antibody has to be determined before the immunosensor application to a

Table 1
Analytical Calibration Values Obtained for the Target Analytes
Using a Mixture of the Three Antibodies

Analytes	IC_{50} (µg/L)	LOD (µg/L)	RSD (%)	Working range (µg/L)
Estrone	0.79	0.08	6.98	0.17–10.7
Atrazine	1.84	0.16	1.07	0.35–1.47
Isoproturon	0.47	0.05	2.89	0.11–2.83

IC_{50}, inhibition concentration at 50% absorbance; LOD, limit of detection; RSD, relative standard deviation; working range, signals from 10 to 90% of the maximal signal at zero dose.

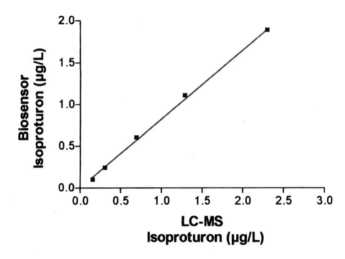

Fig. 2. Correlation between the immunosensor results and those obtained by SPE followed by LC–MS in the analysis of isoproturon on spiked river water. Regression equation was $y = 0.827 \times + 0.001$, $r^2 = 0.998$. From **ref. 12**, with permission from Elsevier.

certain compound. In multianalyte approaches, different analyte-specific antibodies are incubated together with the analytes. Therefore, the effectiveness of the assay is largely determined by the analyte specificity of the different antibodies. To evaluate the effect of using a mixture of different antibodies (instead of a single-antibody solution) on the assay performance, different assays were previously designed and statistical analysis performed *(9,10)*.

2. Stability: The antibody, as a biological reagent, is the weak component in the analytical process. The antibody solution decreases its activity after 12 h at room temperature; therefore, it is recommended to keep the solution at 4–8°C. Another aspect is the stability of the antibody in real samples. Antibodies lose their binding affinity in aggressive matrices with time. A 10% signal decrease (*see* **Fig. 3**) is observed after 17 measurements (one measurement cycle takes 15 min). Therefore, the time elapsed between the addition of the antibody to the wastewater and the measurement should never exceed 2 h *(7)*. On the other hand, the concentration of the background protein OVA can be increased from 0.2 to 2 mg/mL, depending on the matrix complexity, to protect the antibody against the

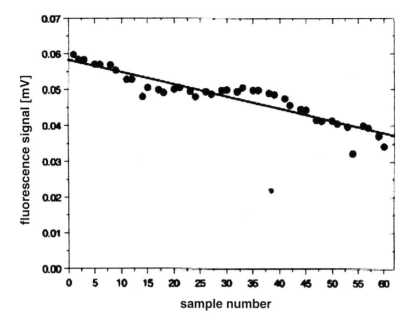

Fig. 3. Stability of the antibody in wastewater determined with the working immunosensor. From **ref. 7**, with permission from Elsevier.

matrix. In addition, the aqueous standard solutions have to be kept at 4–8°C and must be prepared regularly (*see* **Subheading 3.2.**) to avoid standard degradation to cause false quantification of the samples.

3. LOD: Reduction of the antibody amount per sample results in a decrease of the LOD and at the same time decreases the signal range of the calibration curve. A compromise has to be found by which the LOD is in a region of low concentration and the signal range is not too small; this means a system with high affinities, with the affinity of the antibody higher for the analyte than for the analyte derivative, and with a concentration of the antibody as low as possible.

4. Matrix effect: As in other analytical methods, matrix effects derived from their application to the analyses of complex mixtures should be taken into consideration *(11)*. Immunoassays are based on competitive interactions among antibodies, analytes, and analyte derivatives. These interactions can be affected by the pH or the ionic strength of natural water matrices. It has also been reported that the presence of other natural substances, such as dissolved organic carbon, that interact weakly with the antibody can induce an overestimation of the immunoassay response *(12)*. The influence of ground- and river water matrices on the immunosensor response has to be evaluated by spiking these matrices with standard solutions and observing the displacement of the calibration curves obtained (*see* **Fig. 4**). Interferences can then be removed by employing simple methods (diluting and buffering the sample, adjusting pH and conductivity, etc.) or by applying appropriate cleanup methods *(11)*. As explained, the background protein OVA is added to a real sample as an immolation protein. Thus, the probability of the antibody adhering to passive particles in the wastewater is reduced *(7)*.

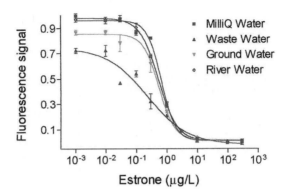

Fig. 4. Standard calibration curves obtained for estrone in bidistilled water, groundwater, river water, and wastewater. From **ref. _12_**, with permission from Elsevier.

5. Rinsing: The system should be washed every day before and after finalization of the measurements by rinsing thoroughly with PBS. Care should be taken to wash the flow cell repeatedly to avoid precipitation of any contaminants on the flow cell, which will increase light dispersion. Weekly washing procedures should be carried out using an ethanol rinsing solution to avoid algae formation in the fluidics. This solution must not pass through the flow cell because it can irreversibly damage the transducer surface.

6. Mixing: Good mixing of each solution is essential. Therefore, vortex mixing is recommended. However, vortex heating can easily denature the proteins. Therefore, it is recommended not to vortex longer than 5 s at a time.

7. Regeneration: In an ideal situation, immunosensors can be reused several times. Only when the antibodies are completely removed from the surface can a second measurement take place. Regeneration of the sensing layer takes place by equilibrium displacement of the immunoreaction or by using agents able to disrupt antibody–analyte association *(13)*. In this system, acidified SDS solution was used as a regeneration solution. The stability of the surface depends also on the regeneration of the transducer active surface. The transducers employed with the RIANA device exhibit a long operational time with the described optimized regeneration step. In the first 50 measurements with a new transducer, the signal decreases strongly, but in the subsequent measurements (up to 300) the signal stays quite stable. Nevertheless, blank samples are usually measured along the sequence to normalize the signals obtained.

8. Autosampler: An autosampler AS90/91 from PerkinElmer (Überlingen, Germany) is used to deliver the samples to the FIA system. Fluid handling and data acquisition are fully automated and computer controlled. A new autosampler from CTC Analytics, HTC PAS, has been used coupled to the RIANA. With this new system, there is no need to prepare and incubate samples and standards before analysis because the new autosampler is able to select, mix, and inject the samples following a programmed sequence. The new autosampler also permits better reproducibility with lower relative standard deviations. Lower LODs are thus also achieved. The RIANA prototype was constructed by PerkinElmer in the frame of the European Union project RIANA (ENV4-CT95-0066). At present, only three units are available and are at the laboratories of some of the partners that participated in the EU project. A new, more advanced prototype has been constructed by Central Research Laboratories (CRL; Middlesex, UK) within the AWACSS project

(EVK1-CT-2000-00045). This new multisensor system is designed for the monitoring of water pollution and will consist of a number of remote measurement stations linked by a communications network. In these stations, highly sensitive biosensors based on fluorescence detection will work unattended, measuring target pollutants levels in water and sending the results to a control station. The Internet-based control station will have direct access to each remote measurement unit. Part of the automatization of the new system involves reduced time and reactives.

Acknowledgments

This work has been supported by the Commission of the European Communities, AWACSS (EVK1-CT-2000-00045) and Ministerio de Ciencia y Tecnología (project PPQ 2000-3006-CE). Maria José López de Alda acknowledges a Ramon y Cajal contract from the Spanish Ministry of Science and Technology.

References

1. Parellada, J., Narvaez, A., Lopez, M. A., et al. (1998) Amperometric immunosensors and enzyme electrodes for environmental applications. *Anal. Chim. Acta* **362**, 47–57.
2. Mallat, E., Barzen, C., Abuknesha, R., Gauglitz, G., and Barcelo, D. (2001) Fast determination of paraquat residues in water by an optical immunosensor and validation using capillary electrophoresis–ultraviolet detection. *Anal. Chim. Acta* **427**, 165–171.
3. Mallat, E., Barzen, C., Abuknesha, R., Gauglitz, G., and Barcelo, D. (2001) Part per trillion level determination of isoproturon in certified and estuarine water samples with a direct optical immunosensor. *Anal. Chim. Acta* **426**, 209–216.
4. Mallat, E., Barzen, C., Klotz, A., Brecht, A., Gauglitz, G., and Barcelo, D. (1999) River analyzer for chlorotriazines with a direct optical immunosensor. *Environ. Sci. Technol.* **33**, 965–971.
5. Klotz, A., Brecht, A., Barzen, C., et al. (1998) Immunofluorescence sensor for water analysis. *Sens. Actuators B* **51**, 181–187.
6. Barzen, C., Brecht, A., and Gauglitz, G. (2002) Optical multiple-analyte immunosensor for water pollution control. *Biosens. Bioelectron.* **17**, 289–295.
7. Coille, I., Reder, S., Bucher, S., and Gauglitz, G. (2002) Comparison of two fluorescence immunoassay methods for the detection of endocrine disrupting chemicals in water. *Biomol. Eng.* **18**, 273–280.
8. Kroger, S., Piletsky, S., and Turner, A. P. F. (2002) Biosensors for marine pollution research, monitoring and control. *Mar. Pollut. Bull.* **45**, 24–34.
9. Vo-Dinh, T., Fetzer, J., and Campiglia, A. D. (1998) Monitoring and characterization of polyaromatic compounds in the environment. *Talanta* **47**, 943–969.
10. Hennion, M.-C. and Barcelo, D. (1998) Strengths and limitations of immunoassays for effective and efficient use for pesticide analysis in water samples: a review. *Anal. Chim. Acta* **362**, 3–34.
11. Marco, M.-P. and Barcelo, D. (1996) Environmental applications of analytical biosensors. *Meas. Sci. Technol.* **7**, 1547–1562.
12. Rodriguez-Mozaz, S., Reder, S., Lopez de Alda, M., Gauglitz, G., and Barceló, D. (2004) Simultaneous determination of estrone, isoproturon and atrazine in natural waters by an optical immunosensor river analyser (RIANA). *Biosens. Bioelectron.* **19**, 633–640.
13. Reder, S., Dieterle, F., Jansen, H., Alcock, S., and Gauglitz, G. (2003) Multi-analyte assay for triazines using cross-reactive antibodies and neural networks. *Biosens. Bioelectron.* **19**, 447–455.

Index